T0306132

FLUVIAL MEGAFANS ON EARTH AND MARS

Megafans are partial cones of river sediment that reach unexpectedly large dimensions, with the largest on Earth being 700 km long. Due to recent developments in space-based observations, global mapping efforts have shown that modern megafan features cover vast landscapes on most continents. This book provides a new inventory of nearly 300 megafans across 5 continents. The chapters focus on regional studies of megafans from all continents barring North America and Antarctica. The major morphological attributes of megafans and multi-megafan landscapes are discussed, and the principal controls on megafan development are examined. The book also compares megafans with alluvial fans, deltas, floodplains, and the recently recognised 'major avulsive fluvial system' (MAFS). The final part of the book discusses the application of megafan research to economic geology, aquifers, and planetary geology including layered deposits on Mars. This is an invaluable reference for researchers in geomorphology, sedimentology, and physical geography.

M. Justin Wilkinson is a senior research scientist in the Department of Geography at Texas State University, under contract in the Earth Science and Remote Sensing Unit at the NASA Johnson Space Center. Wilkinson's research focusses on 'unconventional' river landscapes, the end points of desert rivers, and extensive megafan depositional plains. The Astronaut Corps awarded him the 'Silver Snoopy' pin as Earth Observations trainer, and a book on Costa Rica illustrated with astronaut photos earned him NASA's Public Service Medal.

Yanni Gunnell is Professor of Physical Geography and Environmental Studies at the University of Lyon 2 Lumière, in France. He investigates environmental change and landscape evolution, and teaches geomorphology, Quaternary geology, geodynamics, geoarchaeology, land-cover change, and environmental history. He has served as editor for non-profit learned society journals, and as guest-editor of books and book chapters in several languages.

FLUVIAL MEGAFANS ON EARTH AND MARS

Edited by

M. JUSTIN WILKINSON

Texas State University, Jacobs Contract at NASA-Johnson Space Center

YANNI GUNNELL

University of Lyon 2 Lumière

Shaftesbury Road, Cambridge CB2 8EA, United Kingdom

One Liberty Plaza, 20th Floor, New York, NY 10006, USA

477 Williamstown Road, Port Melbourne, VIC 3207, Australia

314–321, 3rd Floor, Plot 3, Splendor Forum, Jasola District Centre, New Delhi – 110025, India

103 Penang Road, #05–06/07, Visioncrest Commercial, Singapore 238467

Cambridge University Press is part of Cambridge University Press & Assessment,
a department of the University of Cambridge.

We share the University's mission to contribute to society through the pursuit of
education, learning and research at the highest international levels of excellence.

www.cambridge.org
Information on this title: www.cambridge.org/9781108423373

DOI: 10.1017/9781108525923

First published 2023

A catalogue record for this publication is available from the British Library

Library of Congress Cataloging-in-Publication Data
Names: Wilkinson, M. Justin, 1946– editor. | Gunnell, Yanni, 1965– editor.
Title: Fluvial megafans on Earth and Mars : a significant mesoscale landform / edited by Justin Wilkinson, Yanni Gunnell.
Description: Cambridge, United Kingdom ; New York, NY : Cambridge University Press, 2023. | Includes bibliographical references and index.
Identifiers: LCCN 2022003888 (print) | LCCN 2022003889 (ebook) | ISBN 9781108423373 (hardback) | ISBN 9781108525923 (epub)
Subjects: LCSH: Fluvial geomorphology. | BISAC: SCIENCE / Earth Sciences / Geography
Classification: LCC GB561 .F545 2022 (print) | LCC GB561 (ebook) | DDC 551.3/55–dc23/eng20220322
LC record available at https://lccn.loc.gov/2022003888
LC ebook record available at https://lccn.loc.gov/2022003889

ISBN 978-1-108-42337-3 Hardback

Contents

Contributors

R. BRUCE AINSWORTH
Australian School of Petroleum, University of Adelaide, Adelaide, Australia

MARIO L. ASSINE
Instituto de Geociências e Ciências Exatas, Universidade Estadual Paulista, Rio Claro, Brazil

KEVIN BURKE
Department of Earth & Atmospheric Sciences, University of Houston, Houston Texas, USA (deceased)

EDGARDO CAFARO
College of Engineering and Hydrological Sciences, National University of the Litoral, Santa Fé, Argentina

TIMOTHY COHEN
GeoQuest Research Center, School of Earth and Environmental Sciences, University of Wollongong, Wollongong, Australia

NATHAN CURRIT
Department of Geography, Texas State University, San Marcos, Texas, USA

FRANK ECKARDT
Environmental & Geographical Science, University of Cape Town, Rondebosch, Rep. South Africa

ALESSANDRO FONTANA
Department of Geosciences, University of Padua, Padua, Italy

KUMAR GAURAV
Department of Earth and Environmental Sciences, Indian Institute of Science Education and Research (IISER), Bhopal, India

YANNI GUNNELL
Department of Geography, Université Lumière Lyon 2, Lyon, France

SURYA GUPTA
Department of Earth Sciences, Indian Institute of Technology Kanpur, Kanpur, India

STEPHEN T. HASIOTIS
Department of Geology, University of Kansas, Lawrence, USA

KAREN A. KAPTEINIS
Ochre Imprints Pty. Ltd., Abbotsford, Australia Environmental Geosciences, Department of Ecology, Environment and Evolution, La Trobe University, Bundoora, Australia

MIKHAIL A. KRESLAVSKY
Department of Earth and Planetary Sciences, University of California–Santa Cruz, Santa Cruz, California, USA

TESSA I. LANE
College of Science and Engineering, Flinders University, Adelaide, Australia

EDGARDO M. LATRUBESSE
Graduate Program in Environmental Sciences, Universidade Federal de Goiás, Goiânia, Goiás, Brazil

FALK LINDENMAIER
Federal Institute for Geosciences and Natural Resources – BGR, Hannover, Germany

CHRISTOPH LOHE
Federal Institute for Geosciences and Natural Resources – BGR, Hannover, Germany

JAN-HENDRIK MAY
School of Geography, University of Melbourne, Melbourne, Australia
Institute of Earth and Environmental Sciences, University of Freiburg, Germany

ROY MCG. MILLER
Consulting Geologist, Windhoek, Namibia

KANCHAN MISHRA
Department of Earth Sciences, Indian Institute of Technology Kanpur, Kanpur, India

ANDREA MOSCARIELLO
Department of Earth Sciences, University of Geneva, Geneva, Switzerland

PAOLO MOZZI
Department of Geosciences, University of Padua, Padua, Italy

RACHEL A. NANSON
Australian School of Petroleum, University of Adelaide, Adelaide, Australia

EDWARD PARK
Earth Observatory of Singapore and Asian School of the Environment, Nanyang Technological University, Singapore, Singapore

FRANK PREUSSER
Institute of Earth and Environmental Sciences, University of Freiburg, Freiburg, Germany

MARTIN QUINGER
Federal Institute for Geosciences and Natural Resources – BGR, Hannover, Germany

CARLOS RAMONELL
College of Engineering and Hydrological Sciences, National University of the Litoral, Santa Fé, Argentina

MARK SALVATORE
Department of Astronomy and Planetary Science, Northern Arizona University, Flagstaff, Arizona, USA

MAURÍCIO G. M. SANTOS
Department of Energy Engineering, Federal University of ABC, Santo André, São Paulo, Brazil

RAJIV SINHA
Department of Earth Sciences, Indian Institute of Technology Kanpur, Kanpur, India

M. ANWAR SOUNNY-SLITINE
Department of Geography, University of Florida, Gainesville, USA

JOSÉ C. STEVAUX
Departamento de Ciências Exatas, Universidade Federal de Mato Grosso do Sul, Campo Grande, Mato Grosso do Sul, Brazil

SAMPAT K. TANDON
Department of Earth Sciences, Indian Institute of Technology Kanpur, Kanpur, India

BOYAN K. VAKARELOV
Australian School of Petroleum, University of Adelaide, Adelaide, Australia

HEINZ VEIT
Geographisches Institut, University of Bern, Bern, Switzerland

DARIO VENTRA
Department of Earth and Environmental Science, University of Geneva, Geneva, Switzerland
Faculty of Geosciences, Utrecht University, Princetonlaan 8a, Utrecht, 3584 CB, The Netherlands

RICARDO VILALTA
Department of Computer Science, University of Houston, Houston, Texas, USA

JOHN A. WEBB
Environmental Geosciences, Department of Ecology, Environment and Evolution, La Trobe University, Bundoora, Australia

SUSAN Q. WHITE
Environmental Geosciences, Department of Ecology, Environment and Evolution, La Trobe University, Bundoora, Australia

M. JUSTIN WILKINSON
Department of Geography, Texas State University, San Marcos, Texas, USA
Jacobs JETS contract, Earth Science and Remote Sensing Unit, NASA–Johnson Space Center, Houston, Texas, USA

Foreword

Our first response when looking at the Earth from space, whether in person or in pictures, is wonder. Then we begin to ask questions. What are we seeing? What can we learn from these images? What is different about looking at thousands of square kilometres of the Earth's surface in a single view? Writing in 1994, historian Peter Whitfield observed that:

> The Earth was modeled [by people] who had never seen the earth. The force of this paradox should never be overlooked: it is as though a sculptor of the stature of Michelangelo should have been blind. ... We are justi- fied in regarding space photography and satellite imagery as marking a revolution in our perception of the planet.

Data in this book is based significantly on various kinds of imagery taken from space. It is gratifying to know that hundreds of astronaut images taken from low Earth orbit with handheld cameras have provided some raw material for this book. It is even more gratifying that new scientific ideas have been sparked by space imagery – in this case the idea that very large fans are a common feature of the planet's surface.

Africa's Okavango delta in the cover photo is a fea- ture well known to astronauts because its dark fingers of woodland stand out starkly against the orange sands of the Kalahari desert. Indeed, it may be the most photo- genic megafan that astronauts see. But it took views from space to reveal the surprising fact that this is only the most obvious of ten more neighbouring megafans.

Space-based imagery enabled global mapping by providing views large enough to encompass these large features, and has been a key in generating the world map of megafans. The megafan map reminds me of the pattern of lights we see from space at night: megafans mostly appear in the great dark regions of continents where there are few lights – so different from the brilliance of clustered cities where millions live. The accident of so many megafans existing in remote places may help explain why we have been slow to recognize megafans as common features on our planet.

Besides what space photography tells us, the level of detail provided by the Shuttle Radar Topography Mission (SRTM), also associated with human space flight, has assisted several authors in the present volume, illuminating subtle aspects of the smooth and flat country made by megafans. Several other kinds of space-based data have also been used to tease out attributes of some megafans.

I congratulate all the authors whose conclusions are presented in this first fully fledged book on the topic. In particular, most authors have combined space imagery with hard-won data logged from pits and boreholes on megafan surfaces.

And a next step is already in progress – the synthe- sised space and field data will be applied underground, to those parts of the planet that no one sees, for which no grand space views are available. The improved knowledge of the surface as represented in this volume, I am told, will assist geologists in reconstructing land- scapes long since buried from sight in the rock strata.

Spinoffs from chapters in this book have already generated new published research. I trust that the work will have many more such beneficial research results.

John E. Blaha
Former NASA astronaut
Space Shuttle Commander

Reference

Whitfield, P. (1994). *The Image of the World: 20 Centuries of World Maps*. London, UK: The British Library, 144 pp.

Alluvial fans have long been part of the geological and geomorphological literature. Originally, they were considered to be restricted to steep mountain fronts, especially in arid and semi-arid areas, notably in Death Valley, U.S.A., and other arid regions. They were observed to be relatively small, fan-shaped deposits consisting of coarse-grained, fluvially deposited gravel and sand, interbedded with poorly-sorted debris-flow deposits. The recognition of much larger fan-shaped deposits in more mesic climates, notably the Kosi Fan in India, gave rise to the notion of 'humid fans', which were considered to be entirely fluvial in nature, as opposed to the much smaller 'arid fans'. Humid fans have much gentler slopes than arid fans; slopes were considered to be a consequence of prevailing wetter environments.

Other large, low-gradient fans were subsequently recognised, notably on the Okavango, Niger and Nile rivers. These African fans, and the Kosi Fan too, had been referred to as 'inland deltas' by colonial cartographers. 'Inland deltas' occur in both humid and semi-arid areas, dispelling the notion that climate can be used for classification of fan-like features. The more appropriate term 'megafan' was introduced in the late 1980s to describe the Kosi Fan. This term has no genetic connotations and has become widely used to describe large, low-gradient fluvial fans.

Apart from the old colonial references to inland deltas in India and Africa, megafans went largely unrecognised, even in the early years of Earth observations from space, as pointed out in this volume (Chapter 1). Although satellite imagery was widely used by research workers from the earliest space missions, it was restricted to local areas of specific interest; satellite imagery was simply too costly for broad, exploratory landform studies, let alone at a global scale. In addition, significant specialist knowledge was required to process the imagery, which further restricted its use.

Two particular events radically transformed our ability to scrutinise the morphology of the planet using observations from space. The first was the release of Google Earth, which made relatively high-resolution, mostly de-clouded and 'zoomable' satellite imagery available to anyone, anywhere. All that was required was a personal computer and an internet connection. No special knowledge was needed to use it. The second was the release by NASA of the Shuttle Radar Topographic Mission (SRTM) data, which made high-quality topographic information available for most of the planet. This has proved particularly useful in outlying areas, making it possible to undertake terrain modelling even for the remotest locations, something previously impossible to do, even for governments. The discovery and study of megafans has increased exponentially and they are now known to have an almost global distribution, as illustrated in the excellent maps in Chapter 2. Strangely, they appear to be mostly absent from North America and Europe – raising very interesting research questions which no doubt will be addressed in the future. Some land surfaces on these continents are very young, having only recently emerged from beneath Pleistocene ice sheets, and megafan formation may still be in nascent stages, or the megafans may be buried beneath a mantle of glacial deposits.

Megafans lie at the interface between the disciplines of geomorphology and sedimentology. The former focuses more on landforms, whereas the latter focuses on what lies beneath the land surface and the ways in which this relates to depositional processes. It is rarely possible to deduce the nature of the landforms that existed during sediment deposition from the character of sedimentary rocks. Such rocks that were deposited on very large subaerial fans would simply be classified as fluvial without reference to the larger-scale depositional setting. As chapters in this volume demonstrate, megafan settings represent a common continental fluvial depositional environment in the present day. However, the calibre of the sedimentary material *per se* offers no clue as to the wider depositional setting. For example, in the western regions of Amazonia, megafans are active in wet environments, and are dominated by sand, silt and mud; whereas at the other end of the climatic spectrum fluvial and aeolian sediments of the Niger Inland Delta in the West Africa interfinger in the distal linear dune fields of the Sahara Desert. Intermediate depositional environments are found in the Chaco in South America, as described in this volume.

Many megafans show a proximal-to-distal decrease in the grain size of channel deposits, with similar decreases in channel-margin sediments. But these are by no means universal characteristics of megafan deposits. For example, the channel deposits of the

Okavango megafan show no change in grain size over its 150 km down-fan distance. Distal channel-margin deposits consist partly of organic material and mud derived from the burning of peat, but mainly of pedogenic calcrete and silcrete. Remarkably, such chemical sediments are more abundant than clastic sediments in the Okavango as a whole.

As demonstrated in this volume, many very large fluvial fans are now available for study, and it will become possible to establish more comprehensively the relationships between sediment calibre, planform geometry, fan gradients, fan hydrology, and climate. In addition, an important aspect of the study of megafans is which processes created (and possibly sustain) the accommodation space for the fan and how these relate to depositional processes and fan morphology. The study of megafans will provide models for structural geologists and sedimentologists that will be of value in exploration for economic minerals such as sandstone-hosted uranium, gas and oil, as well as for groundwater, especially in semi-arid to arid

environments — as represented here, for example, in the chapters relating to the economic applications of megafan research.

This volume contains a magnificent smorgasbord of megafan reviews, with examples from many parts of the Earth and with a pinch of Mars thrown in. It is about thirty years since Rachocki and Church published their edited book *Alluvial Fans – A Field Approach*. The present volume opens a completely new perspective on fan-like depositional features, perspectives that have been heavily influenced by observations from space, and it will form an excellent introduction to such systems for geomorphologists, sedimentologists and geologists. I expect that this volume will act as the spur to a significant expansion in research surrounding the topic of megafans.

Terence S. McCarthy
Emeritus Professor of Mineral Geochemistry,
University of the Witwatersrand, Johannesburg,
Republic of South Africa

Part I
Introduction

1

Megafans as Major Continental Landforms

M. JUSTIN WILKINSON and YANNI GUNNELL

Abstract

Discovery of the significance of fluvial megafans came about in the mid to late twentieth century. We suggest reasons why appreciation of their existence came late in the history of Earth science, even after the advent of space-based observation of planetary landscapes. The reasons are partly cultural: megafans are uncommon in the historic cradles of modern geology (Europe, North America). Reasons are also partly theoretical: rivers have been conceptualised chiefly as sediment bypass systems terminating in deltas, rather than as aggradational systems in their own right. Reasons are also perceptual: just as the megaflood origin of channeled scablands was held in disbelief, the inordinate size of megafans has stood in the way of accepting (i) the sheer magnitude of their unit-size and also (ii) their existence as active systems in modern landscapes, rather than just as stratigraphic features in the rock record. Post-1990, scientific activity around megafans accelerated and involved global mapping, classification, and regional investigations into patterns and processes. An overview of this take-off period is provided as a partial introduction to the remaining 17 chapters of this book, which are briefly outlined.

1.1 Outline, Purpose, and Scope

Megafans are fluvial sedimentary landforms of very low gradient and fan-shaped planform, with radial lengths of several tens, and up to hundreds of kilometres – i.e., significantly larger than the well-known smaller, mountain-front alluvial fans, but still much smaller than giant submarine fans such as the Indus and Bengal. However, megafans have long been poorly recognised in the literature on subaerial geomorphology and sedimentology. These landforms now require scientific attention, not only for reasons of fundamental Earth science understanding, but because of their importance as landscape elements occupied by major population centres around the world.

Our driving concern in this volume is to derive generalisations from the relatively limited total global population of ~270 identified megafans (as documented by remote means (Wilkinson and Currit, Ch. 2) compared with the thousands of examples of small alluvial fans detectable from the air, and many more in the near-surface Neogene geological record. The geomorphic and tectonic settings for megafans on each continent raise different kinds of research problems and approaches to solving them. Thus, in South America, the active Andean orogen has given rise to a large number of currently active megafans. By contrast, the Indo-Gangetic basin displays not only some active megafans but also others that are significantly incised and appear as successions of terraces. Cratonic Africa and Australia show patterns of isolated fans. Five chapters deal with groups of megafans (Africa, Asia, Central Europe, South America), the rest with more detailed studies of one or two megafans. The studies include megafans situated in basins of all major tectonic styles. Antarctica displays no subaerial megafans, and North America is excluded because so few modern examples exist and because of space limitations in this volume.

There seems little doubt about the importance of megafans as landforms on which very large human populations subsist – what Geddes (1960:253) termed the 'alluvial plains of profound significance [...] that have tended to be almost completely ignored in geomorphology, instead of providing a central theme for physical study...'. Fortunately, Geddes's (1960:258–259) comment of six decades ago ('even a general study of the world's plains is lacking (...) in spite of their immense environmental significance') has begun to be reversed. While focusing on northern India, he also suggested that perspectives from the Gangetic Plains might well apply to the study of similar alluvial settings in South America, North America, and parts of Europe, as is attempted in this volume. The vast agricultural potential of such plains, the relative ease of constructing irrigation and transport systems on their remarkably flat surfaces, and their associated vulnerability to extensive flooding – as witnessed in 2008 (and almost annually since 2013) by the Kosi River megafan in northern India (hereafter megafans are named simply after their formative rivers) – all argue for further study, scrutiny, and public awareness. Vast expanses of megafan terrain in African and South America (particularly the Chaco) are coveted for setting up irrigation projects, plantations, or tourism ventures. Water grabs in Mali (Niger megafan), and recent warfare over land in oil- and groundwater-rich South Sudan have also been prominent in current affairs (Pearce 2013). Until recently desolate, untamed and untenured, a number of megafans are among the last frontiers of this planet.

1.2 Expanding Perspectives on Megafans

1.2.1 Earlier Perspectives: Limited Recognition Pre-1990

Before space-based observation of Earth's landscapes was available, a few megafans had been described. Prime examples are the Kosi in India, which remains probably the most cited example (e.g., Geddes 1960; Parkash et al. 1983; Wells and Dorr 1987a, b; Gohain and Parkash 1990; Mohindra et al. 1992; Richards et al. 1993; Singh et al. 1993; Sinha and Friend 1994; Shukla et al. 2001; Goodbred 2003); the Okavango (for numerous early references, see historical review in McCarthy 2013); and megafans in the Andean

foreland (Cordini 1947, and early references in Iriondo 1993; Horton and DeCelles 2001; Latrubesse et al. 2012) and in the Pantanal region of SW Brazil (Klammer 1982; Tricart 1982; Souza et al. 2002; Assine 2005). In francophone literature, the 'inland delta' of the Niger River, in the Sahel region of Mali, was studied by Urvoy (1942) and later by Gallais (1967) prior to awareness of the fluvial megafan idiom *sensu hic*. This early recognition of deltas and large fans in continental interiors by isolated pioneers displays parallels in the history of science with other very large landforms such as megaflood scars in the Pacific Northwest of the United States (Bretz 1923) – in this case erosional rather than predominantly depositional landforms. At first critiqued by incredulous detractors (see review about the Channeled Scablands of the NW USA in Baker 1978), 'scablands' have now not only been validated, but also detected outside their type area by means of satellites and underwater sonar technology – from the very doorstep of modern geology's European homebase (the English Channel: Gupta et al. 2017) to remote regions such as Siberia and other areas in the solar system such as Mars (Burr et al. 2009).

Studies of individual megafans, however, often overlooked wider suites of neighbouring megafans, and thus the broader subregional-scale megafan setting. For example, the Okavango megafan in the Kalahari Basin of southern Africa, visually prominent in aerial imagery, has claimed perhaps even the bulk of attention for the past century. However, it is now known to be only one of a group of at least ten megafans in the region (Wilkinson et al., Ch. 4). Thus, only four multi-fan landscapes benefited from published studies prior to 1990, namely the Indo-Gangetic plains of northern India (Geddes 1960), the Chaco plains of Argentina and Paraguay (Iriondo 1984, 1987), the Pantanal (Tricart 1982; Tricart et al. 1984) in southwestern Brazil, and the Hungarian Plains (Borsy 1990). These studies were necessarily idiosyncratic to their local basins, with Geddes (1960:262) noting that some major Himalayan rivers such as the Ghaghara failed to display 'great alluvial fans or cones' compared with the continuous set of active megafans generated by major rivers in central South America. Experience from megafan landscapes of one continent was only tenuously transferred to other continents, partly because of language barriers and slow diffusion of the studies, and partly through the assumption that such landscapes were unusual or simply not representative of planetary landforms.

The widely held view of megafans as rare landforms was supported by the small number of known examples and by the scant scientific attention directed to these features, leading to cursory treatment in reference works – mostly as large end members of the spectrum of piedmont alluvial fans. Schumm's explicit opinion – in his influential 1977 book The Fluvial System – that large 'wet' fans [i.e., megafans] must have been widespread during pre-vegetation times probably reinforced the view that few such fans should be expected in modern landscapes. Experienced field geologists have noted that the very low slopes and occasionally immense size have made large fans difficult to recognise in modern landscapes (N. Cameron and R. Miller, pers. comm. to MJW). Lack of recognition was likely reinforced by the age of many megafans, as drainage patterns and fluvial morphology are progressively overprinted in remotely sensed imagery by eolian features, incision and terracing, and vegetation patterns. This has been especially the case within a broad geological mindset that had assumed that vast alluvial landscapes are specifically connected to coastlines in the form of deltas. This view still dogs research into fluvial landscapes and sediments on Mars (Wilkinson et al., Ch. 16).

A cultural component probably also played a part. Megafans are presently almost non-existent or inconsequential in the landscapes of Europe and North America, where Earth science matured as a modern discipline during the twentieth century. This coincidence has probably conditioned the pervasive view that incisional fluvial regimes, so dominant in these continents, are the norm on all continents. Thus, Schumm's (1977) classic three-zone model of the drainage system is based on the topographic sequence mountain–valley-confined floodplain–coastal delta, with the Mississippi drainage clearly in mind. This model reinforces the concept of rivers as sediment-bypass systems rather than as potentially aggradational systems in their own right, and implicitly excludes the vast megafan-dominated landscapes that are now attracting growing attention.

1.2.2 Accelerated Scientific Activity since ~ 1990: Global Mapping, Approaches, and Definitions

As mentioned above, a few examples of megafans were known before the 1990s. Blair and McPherson (1994) had specifically excluded large fluvial fans from the alluvial-fan designation. In their view, megafans belonged in the class of typical valley-confined floodplains, to be distinguished from short-radius, higher-gradient piedmont alluvial fans. Blair and McPherson (1994) reasserted the original definition of alluvial fans, namely as coarse-grained features with relatively steep slopes, of the type classically associated with the small desert alluvial fan less than 15–20 km in length, and distinguishable from larger river systems also in terms of sedimentary processes and products. Their view is interesting because they gave little validity to fanlike morphology, which is otherwise the overwhelmingly dominant approach.

Following widely held views, Miall (1996) took instead an inclusive stance of grouping large fans within a more broadly defined alluvial-fan class. However, large known megafans at that time, such as the Kosi megafan of northern India, the Chaco megafans, and the Pantanal, were not included by Miall (1996) (See Wilkinson, Ch. 17, Section 17.3). In a more detailed analysis, Stanistreet and McCarthy (1993) also took a more inclusive view, classifying all sizes of fan-like fluvial landforms as alluvial fans, with categories based on process and included small alluvial fans, braided fans, and the largest, so-called losimean (i.e., low sinuosity and meandering) Okavango type (150 km long).

Simultaneously, the increasing availability of satellite remote sensing products started to open up new potential for the identification of megafans worldwide. For example, starting in 1988 at the Johnson Space Center, astronaut-handheld imagery of continental surfaces revealed what may have been the first global perspective on megafans. It rapidly provided evidence of more than 150 examples, with some components of these inventories presented at conferences or in grey literature (e.g., Wilkinson 2001, 2005, 2006; Wilkinson et al. 2002, 2006, 2010; Sounny-Slitine and Latrubesse 2014).

The task of identifying from remote sensing products the global population of all medium and large fans (i.e., > 30 km long) was complemented by Weissmann et al. (2010, 2011), Hartley et al. (2010a, b) and Davidson et al. (2013), resulting in an overarching classification of fan-like fluvial deposits based on 415 examples. These authors applied the innovative term distributive fluvial systems (DFS) 'to encompass fluvial and alluvial distributive landforms at all scales' (Weissmann et al. 2015:189), in the attempt to circumvent the semantic issues associated with the definition of alluvial fans. Their purpose was first to identify modern DFS and describe what they deemed the important aspects of their morphology and structural setting; and then, given that they saw DFS to be the areally dominant landforms in present-day continental basins, to propose these as the basis for an 'alternative interpretation for much of the fluvial rock record' (Weissmann et al. 2010, 2011:329). These authors drew a major distinction between 'tributary' drainage patterns (Weissmann et al. 2011:329; also 'tributive' in Weissmann et al. 2015:214), which they saw as typical of regional degradational landscapes even though such landscapes include some of the largest and most active river floodplains in the world; and 'distributive' drainage patterns, which are typical of many landscapes of regional extent that are dominated by fan-like fluvial deposits of all dimensions.

Global data surveys supported the notion of a genetic continuum for these landforms, earlier demonstrated by Saito (2003), Saito and Oguchi (2005), and Hashimoto et al. (2008). The continuum questioned the 'natural depositional slope gap' that Blair and McPherson (1994) had argued must exist between alluvial fans and floodplains, and they reclassified megafans as 'rivers [i.e., floodplains] or river deltas' (Blair and McPherson 1994:457). Confusion was thus compounded because the slope gap does not exist between debris-flow-dominated 'torrential' fans and fluvial megafans, even though a *process* gap between these features does exist—given that smaller debris cones and alluvial fans are shaped by supercritical flow (and some even by non-Newtonian flow), whereas megafans are dominated by fluvial processes under a critical or subcritical flow regime.

The potential climatic conditions for the development of megafans has been another controversial topic. Earlier claims by Leier et al. (2005) are often quoted to support a climatic explanation for the distribution of

megafans. Results from different parts of the world demonstrate that megafans can be generated under a broad spectrum of climatic conditions, which range from periglacial to arid, semiarid and temperate climates. Some writers still invoke aridity or pronounced seasonality as an explanation for the existence of megafans (e.g., Fielding et al. 2012; Rossetti et al. 2014; Plink-Björklund 2015). This might also include equatorial regions covered today by dense tropical rainforest such as the Amazon.

The chapters in this volume thus address the following four dimensions in the study of megafans: (i) two-dimensional space, and thus the characterisation of present-day sky-view morphologies and other visually detectable patterns; (ii) process, by exploring the spectrum from less well understood local autogenic controls to wider allogenic controls such as tectonic setting, catchment geology, and climate; (iii) time, providing constraints on the age of deposits and landform assemblages; and (iv) stratigraphy, spanning the subsurface from shallow depths to depths of hundreds of metres. Due to disparities in documentation and purpose, it would be impossible for each chapter to address all these dimensions, but the list gives a sense of the approaches used thus far in the study of these large sedimentary bodies.

The present renewed attention to modern fluvial landscapes and their dominant fluvial styles, with potential for preservation in the geological record, has led to more detailed comparisons with large, but nevertheless confined floodplains (Fielding et al. 2012), and with related types of landform such as major avulsive fluvial systems, or MAFS (Latrubesse 2015), and large accretionary fluvial systems, or LAFS (R. Nanson, pers. comm. to MJW) (see Wilkinson, Ch. 17, Section 17.6.1).

Based on the universality of larger fans in modern sedimentary basins, Weissmann et al. (2010: 41) emphasised the extensive areal scale and distribution of 'DFS deposits [that] are probably more common than previously recognised in continental strata, and may form the bulk of the continental fluvial record'. This statement highlighted a critical distinction between rivers in long-term degradational settings (on which most facies and architectural models for fluvial deposits are based), and rivers in aggrading settings, the latter being heavily represented by DFS in modern landscapes. Weissmann et al. (2011) argued that DFS deposits have a particularly high preservation potential in the rock record. Hartley et al. (2010b) agreed that the many models based on converging river patterns at the channel scale 'provide a very valuable body of literature' (Weissmann et al. 2015:214), but citing scale considerations they noted that 'what we believe is missing in the literature on fluvial systems is an understanding of the larger-than-channel belt and basinal context in which fluvial systems are developed'.

The claim for a potentially dominant representation of DFS in the geological record was considered controversial or even rejected (Sambrook Smith et al. 2010; Fielding et al. 2012; Ashworth and Lewin 2012; Latrubesse 2015). The ensuing conversation refocused attention on the dimensions of very wide floodplains and their distinctiveness compared with DFS, especially megafans – a discussion ultimately aimed at the larger question of fluvial sedimentation styles and their preservation potential in the subsurface. Miall (2014), in particular, considered the debate on tributary vs. radial drainage patterns to be important because of its bearing on 'the mappability and predictability of fluvial systems in the subsurface' (p. 281). Such patterns are investigated in some detail in chapters in this book. Echoing the critique of Fielding et al. (2012), however, Miall (2014:281) stated that 'the most important counter argument to the importance of DFS [in dominating depositional patterns in active continental sedimentary basins] is the abundant documentation of the deposits of large rivers in the rock record'. This important and complex consideration, that of ultimate burial and preservation of fluvial sediment bodies (see especially Miall 2014, his chapters 2 and 6; and Miall et al. 2021), is a topic beyond the scope of this volume. Citing the Amazon, Paraná, and Magdalena rivers, Latrubesse (2015) has given evidence that very large axial rivers all display larger areas of active sedimentation than the largest megafans in central South America – illustrating the capacity of large rivers, even in erosional settings, to give rise to very significant zones of deposition. Latrubesse (2015) argued that some sub-environments of foreland tectonic depressions are inimical to preservation of DFS because they promote the erosional destruction of sediment bodies such as megafans due to the effects of tectonics-driven erosion.

Contrary to claims by Weissmann et al. (2011), Miall (2014) argued that bedforms and macroforms of facies models cannot serve as a basis for differentiating between degradational vs. aggradational (i.e., DFS-type) geomorphic systems because the processes that apply to these features operate in all rivers – whether valley-confined floodplains or unconfined megafan rivers. Miall (2014:280) reasoned that the difference in the setting was 'irrelevant' because bedforms and macroforms develop over time periods and scales small enough to operate in rivers of similar discharge range.

Over the last several years, a growing number of studies have nonetheless documented DFS successions in stratigraphic records from various ages and on all continents (Sáez et al. 2007; Latrubesse et al. 2010; Trendell et al. 2013; Gulliford et al. 2014; Klausen et al. 2014; Owen et al. 2015; Astini et al. 2018), and one chapter of this volume addresses these issues (Ventra and Moscariello, Ch. 14). We note that it is extremely difficult to differentiate in the geologic record between megafans and other large avulsive fluvial systems that are not DFS (Latrubesse et al. 2010; Valente and Latrubesse 2012, 2015). The existence of DFS in the rock record is not, however, the main focus of this volume, which is directed primarily at modern and submodern fans at the large end of the fan continuum. Nevertheless, because of the importance of the topic, four chapters are devoted partly or mainly to the deeper stratigraphy of surface megafan deposits (Ch. 8, Ch. 9, Ch. 11, and Ch. 15).

This volume is also an attempt to present the variety of research aims and ensuing methodologies that have been employed in the study of megafans. For example, significantly different results are derived from morphological mapping as opposed to geological mapping. In the former case, distal convergent drainage patterns have been excluded from the computation of area, either explicitly (Hartley et al. 2010a) or implicitly (Horton and DeCelles 1997; Barnes and Heins 2009); whereas geological mapping includes the entire unconfined zone occupied by fluvial landforms and sediments of the feeder river (e.g., Assine et al. 2014; Latrubesse et al. 2012, and chapters in this volume).

Scientific study of megafans has involved a variety of entry points. The most prominent has been their morphological similarity to alluvial fans, perhaps because the planform view had become so familiar in the voluminous literature on alluvial fans, with overviews presented by many authorities (e.g., Lecce 1990; Stanistreet and McCarthy 1993; McCarthy and Cadle 1995; Cooke et al. 2006). Stanistreet and McCarthy (1993) classified fans primarily by planform with a ternary subdivision by process, namely the elementary alpine debris cone (small range-front 'alluvial fan'), the braided fluvial fan, and a low-sinuosity/ meandering (Okavango) type, a subdivision broadly followed by Miall (1996). The literature nonetheless reveals other approaches. Applying a more strictly sedimentological approach, as noted earlier, Blair and McPherson (1994) simply grouped megafans as a type of landform constructed by fluvial aggradation, in contrast to the processes dominant on piedmont alluvial fans. By retaining the morphological and facies approaches to different degrees, classifications with many nuances and even contradictions have arisen.

Despite the attention paid to features of fan-like planform, the recognition of the full dimensions of many megafans was not immediately obvious. With the long tradition of geomorphic and geological research directed at small alluvial fans, and the relatively small Kosi and Okavango as examples of the few well-known megafans (both ~150 km long), simple dimensional attributes were often thought to be smaller than they are now known to be. For example, Horton and DeCelles (2001) gave significantly smaller dimensions for what they termed megafans in the northern Chaco Plains (which included the largest-known on the planet), compared with dimensions ascertained by Iriondo (1993), Weissmann et al. (2011) or Latrubesse et al. (Ch. 5). Under present climatic conditions, most Chaco Plains fan-forming rivers cease to flow hundreds of kilometres upstream of the megafan toe at the trunk Paraná River (Cafaro et al. 2010; Latrubesse et al. 2012). This led Horton and DeCelles (2001) to consider that river end points mark the distal margins of the megafans. Consequently, the areas they obtained for the Río Grande, Parapetí, and Pilcomayo megafans were much smaller than those now considered to be representative: ~12,600 km², ~5,800 km², and ~22,600 km², respectively, compared with 58,140 km², 59,656 km², and 216,210 km² measured for the full extent of the cones (Latrubesse, Ch. 5).

As commentary by Latrubesse (2015) reveals, overemphasis on planform as a unifying criterion has also diverted attention from the different sets of processes active on fans of different sizes.

1.3 Chapter Outlines

In the continuation of Part I, Introduction, Wilkinson and Currit (Ch. 2) provide a new map showing the distribution of 272 megafans (defined as fans with lengths greater than 80 km) worldwide, a total that more than doubles the number of features of similar dimension in a previously published distribution (Hartley et al. 2010a). The extreme variability by continent is apparent (one in North America, 87 in Africa), and the different tectonic styles are briefly mentioned. Building the map provided the raw material for the broad discussion in Chapter 17, Megafans in World Landscapes, which also concludes with an overview of possible future research directions.

Part II, Regional Studies, deals with the continents. Chapter 3 begins with mapping the megafans of the African continent and placing them in tectonic context; eighty-seven megafans are shown to be connected directly to the swells of Africa's unique basin-and-swell geomorphology. In Chapter 4, Wilkinson et al. identify ten megafans in the northern Kalahari Basin where until now only one was thought to exist, namely the well-known Okavango 'inland delta'. They show that six of these megafans sit astride basin divides such that the discharges of the six feeder rivers have flowed at times into two different basins. Three chapters are devoted to South America, where megafans are most widely developed and cover the largest contiguous area on the planet. Latrubesse et al. (Ch. 5) give a regional study of the Chaco megafans that stretch from central Bolivia to central Argentina through Paraguay, in which the discharge of the different fan-forming rivers is analysed and the several contributing allogenic controls are examined. In a similar study of somewhat smaller megafans of the Pantanal in southwest Brazil, Santos et al. (Ch. 6) map the intricately nested pattern of megafans and examine the relationship between catchment basin geology and megafan size. Avulsions are a key process on megafans, but their occurrence is sufficiently infrequent that little is known of their periodicity. On that topic, May et al.

(Ch. 7) document a detailed chronology of recent avulsions on the Rio Grande megafan of central Bolivia and discuss its possible connections with the Amazon.

In Europe, Fontana and Mozzi (Ch. 8) describe in detail the evolution of two major groups of fans, namely the largest five on the southern piedmont of the Alps in Italy, and those that have developed on the Pannonian Basin. The tributaries of the Danube River, feeding in from the Carpathian Mountains, reveal the effects of glaciation in the case of the Po Basin fans and the lack of glaciation effects in the Pannonian Basin. Gunnell (Ch. 9) gives a full history of the large Loire megafan in central France, from evolution of the shallow receiving basin, to the deposition of its major units, to the subsequent regional incision by major and minor rivers. Furthermore, the Loire River has acted as a 'divide megafan', flowing at different times westwards to the Atlantic and northwards through the Paris Basin towards the English Channel.

In southern Asia, Sinha et al. (Ch. 10) explore the major geomorphic difference between the incised western Gangetic Plains and the aggradational megafan country of the eastern Gangetic Plains. Sinha et al. (Ch. 11) update many aspects of the geomorphology of this well-known fan and map the detail of the modern course. In Australia, Lane et al. (Ch. 12) give a brief overview of the distribution of megafans on that continent, then illustrate the behaviour of a megafan on the coast of the Gulf of Carpentaria that enters the shallow marine realm. They map the greater number of avulsions near the present-day coastline and suggest the term megafan-delta for such features. Kapteinis et al. (Ch. 13) use radiometric satellite imaging to identify three megafans for the first time in Australia's state of Victoria. The flatness of the landscape has led to an apparent coalescence of the two larger megafans in the distal reaches, a comparatively unusual geomorphic occurrence.

In Part III Applications in Other Sciences, Ventra and Moscariello (Ch. 14) and Miller et al. (Ch. 15) report on subsurface fluvial sediments and stratigraphy, the former in a wide-ranging review of continental basins, the latter on the Cubango megafan in northern Namibia. The recent drilling of this megafan came about as a direct result of the identification of the megafan from a mapping study reported in Part I,

Chapter 2. Wilkinson et al. (Ch. 16) for the first time apply patterns seen in megafan landscapes to the kilometre-thick layered rocks in the Sinus Meridiani part of planet Mars. This new approach is based on a growing understanding of aggradational landscapes encapsulated in the 'megafan analogue'.

In Part IV, Megafans in World Landscapes, Wilkinson (Ch. 17) attempts a summary of the major attributes of continental megafans, especially of the drainage networks and large aggradational landscapes, which are so different from those of the more familiar 'dendritic' drainage patterns and valley-dominated morphologies of erosional landscapes. In the final chapter (Wilkinson and Gunnell, Ch. 18) broader conclusions are drawn from what proves to be a rich haul of future research topics, such as the still blurred divide between autogenic and allogenic controls over megafan evolution.

References

Ashworth, P. J. and Lewin, J. (2012). How do big rivers come to be different? *Earth-Science Reviews*, 114, 84–107.

Assine, M. L. (2005). River avulsions on the Taquari megafan, Pantanal wetland, Brazil. *Geomorphology*, 70, 357–371.

Assine, M. L., Corradini, F. A., Pupim, F. N., and McGlue, M. M. (2014). Channel arrangements and depositional styles in the São Lourenço fluvial megafan, Brazilian Pantanal wetland. *Sedimentary Geology*, 301, 172–184.

Astini, R. A., Martini, M. A., Oviedo, N. d. V., and Álvarez, A. (2018). El paleocañon de Tuc Tuca (Cordillera Oriental, Noroeste argentino); reconocimiento de una "zona de traspaso sedimentario" Cenozoic entre el interior cordillerano y un megaabanico en la región subandina. *Revista de la Asociación Geológica Argentina*, 75, 482–506.

Baker, V. R. (1978). The Spokane flood controversy. In V. R. Baker, and D. Nummedal, eds., *The Channeled Scabland, A Guide to the Geomorphology of the Columbia Basin*. NASA, Washington, D.C., 3–16.

Barnes, J. B. and Heins, W. A. (2009). Plio-Quaternary sediment budget between thrust belt erosion and foreland deposition in the central Andes, southern Bolivia. *Basin Research*, 21, 91–109.

Blair, T. C. and McPherson, J. G. (2009). Process and forms of alluvial fans. In A. J. Parsons, and A.D. Abrahams, eds, *Geomorphology of Desert Environments*. Springer, Berlin, 2nd edn., 354–402.

Borsy, Z. (1990). Evolution of the alluvial fans of the Alföld. In A. H. Rachocki, and M. Church, eds.,

Alluvial Fans: A Field Approach. Wiley, Chichester, 229–248.

Bretz, J. H. (1923). The channeled scabland of the Columbia plateau. *Journal of Geology*, 31, 617–649.

Burr, D. M., Baker, V. R., and Carling, P. A., eds. (2009). *Megaflooding on Earth and Mars*. Cambridge University Press, Cambridge, 330 pp.

Cafaro, E., Latrubesse, E., Ramonell, C., and Montagnini. M. D. (2010). Channel pattern arrangement along Quaternary fans and mega-fans of the Chaco plain, central South America. In M. Garcia, E. Latrubesse, and G. Perillo, eds., *River Coastal and Estuarine Morphodynamics, Vols. 1 and 2*, CRC Press, Netherlands, 349–354.

Cooke, R. U., Warren, A., and Goudie, A. S. (2006). *Desert Geomorphology*, University College London Press, London, 2nd edn, 526 pp.

Cordini, R. (1947). *Los Ríos Pilcomayo en la Región del Patiño*. Anales I, Dirección de Minas y Geología (Buenos Aires), 82 pp.

Fielding, C. R., Ashworth, P. J., Best, J. L., Prokocki, E. W., and Sambrook Smith, G. H. (2012). Tributary, distributary and other fluvial patterns: What really represents the norm in the continental rock record? *Sedimentary Geology*, 261–262, 15–32.

Gallais, J. (1967). Le delta intérieur du Niger et ses bordures, étude morphologique. Paris, Éd. du CNRS, 155 pp. (with 1:200 000 scale geomorphological maps).

Geddes, A. (1960). The alluvial morphology of the Indo-Gangetic plains. *Transactions of the Institute of British Geographers*, 28, 253–276.

Gohain, K., Parkash, B. (1990). Morphology of the Kosi megafan. In A. H. Rachocki, and M. Church, eds., *Alluvial Fans: A Field Approach*. Wiley, Chichester, 151–178.

Goodbred, S. L. (2003). Response of the Ganges dispersal system to climate change: a source-to-sink view since the last interstade. *Sedimentary Geology*, 162, 83–104.

Gulliford, A. R., Flint, S. S., and Hodgson, D. M. (2014). Testing applicability of models of distributive fluvial systems or trunk rivers in ephemeral systems: reconstructing 3D fluvial architecture in the Beaufort Group, South Africa. *Journal of Sedimentary Research*, 84, 1147–1169.

Gupta, S., Collier, J. S., Garcia-Moreno, D., et al. (2017). Two-stage opening of the Dover Strait and the origin of island Britain. *Nature Communications*, DOI: 10.1038/ncomms15101.

Hartley, A. J., Weissmann, G. S., Nichols, G. J., and Warwick, G. L. (2010a). Large distributive fluvial systems: characteristics, distribution, and controls on development. *Journal of Sedimentary Research*, 80, 167–183.

Hartley, A. J., Weissmann, G. S., Nichols, G. J., and Scuderi, L. A. (2010b). Fluvial form in modern continental sedimentary basins: distributive fluvial systems: reply. *Geology*, 38, e231.

Hashimoto, A., Oguchi, T., Hayakawa, Y., et al. (2008). GIS analysis of depositional slope change at alluvial-fan

toes in Japan and the American Southwest. *Geomorphology*, 100, 120–130.

Horton, B. K. and DeCelles, P. G. (1997). The modern foreland basins system adjacent to the Central Andes. *Geology*, 25, 895–898.

Horton, B. K. and DeCelles, P. G. (2001). Modern and ancient fluvial megafans in the foreland basin system of the central Andes, southern Bolivia: implications for drainage network evolution in fold-thrust belts. *Basin Research*, 13, 43–63.

Iriondo, M. H. (1984). The Quaternary of northeastern Argentina. *Quaternary of South America*, 2, 51–78.

Iriondo, M. H. (1987). Geomorfolgía y Cuaternario de la Provincia Santa Fé (Argentina). *D'Orbignyana* (Corrientes, Argentina), 4, 54 pp.

Iriondo, M. (1993). Geomorphology and late Quaternary of the Chaco (South America). *Geomorphology*, 7, 289–303.

Klammer, G. (1982). Die Palaeowüste des Pantanal von Mato Grosso und die pleistozäne Klimageschichte des brasilianischen Randtropen. *Zeitschrift für Geomorphologie*, 26, 393–416.

Klausen, T. G., Ryseth, A. E., Helland-Hansen, W., Gawthorpe, R., and Laursen, I. (2014). Spatial and temporal changes in geometries of fluvial channel bodies from the Triassic Snadd Formation of offshore Norway. *Journal of Sedimentary Research*, 84, 567–585.

Latrubesse, E. (2002). Evidence of Quaternary paleohydrological changes in middle Amazonia: the Aripuanã/Roosevelt and Jiparana fans. *Zeitschrift für Geomorphologie*, 129, 61–72.

Latrubesse, E. M. (2015). Large rivers, megafans and other Quaternary avulsive fluvial systems: A potential 'who's who' in the geological record. *Earth-Science Reviews*, 146, 1–30.

Latrubesse, E., Cozzuol, M., Rigsby, C., et al. (2010). The Late Miocene paleogeography of the Amazon basin and the evolution of the Amazon River. *Earth-Science Reviews*, 99, 99–124.

Latrubesse, E., Stevaux, J. C., Cremon, S., et al. (2012). Late Quaternary megafans, fans and fluvio–aeolian interactions in the Bolivian Chaco, Tropical South America. *Palaeogeography, Palaeoclimatology, Palaeoecology*, 356–357, 75–88

Lecce, S. A. (1990). The Alluvial fan problem. In A. H. Rachoki and M. Church, eds., *Alluvial Fans: A Field Approach*. Wiley, Chichester, 151–178.

Leier, A. L., DeCelles, P. G., and Pelletier, J. D. (2005). Mountains, monsoons, and megafans. *Geology*, 33, 289–292.

McCarthy, T. S. (2013). The Okavango Delta and Its Place in the Geomorphological Evolution of Southern Africa. *South African Journal of Geology*, 116, 1–54.

McCarthy, T. S. and Cadle, A. B. (1995). Alluvial fans and their natural distinction from rivers based on morphology, hydraulic processes, sedimentary processes, and facies assemblages – Discussion. *Journal of Sedimentary Research*, A65, 581–583.

Miall, A. D. (1996). *The Geology of Fluvial Deposits*. Springer, New York, 582 pp.

Miall, A. D. (2014). *Fluvial Depositional Systems*. Springer, New York, 316 pp.

Miall, A. D., Holbrook, J. M., and Bhattacharya, J. P. (2021). The stratigraphic machine. *Journal of Sedimentary Research*, 91, 595–610.

Mohindra, R., Parkash, B., and Prasad, J. (1992). Historical geomorphology and pedology of the Gandak megafan, middle Gangetic plains, India. *Earth Surface Processes and Landforms*, 17, 643–662.

Owen, A., Nichols, G. J., Hartley A. J., Weissmann, G. S., and Scuderi L. A. (2015). Quantification of a distributive fluvial system: the Salt Wash DFS of the Morrison Formation, SW U.S.A. *Journal of Sedimentary Research*, 85, 544–561.

Parkash, B., Awasthi, A. K., and Gohain, K. (1983). Lithofacies of the Markanda terminal fan, Kurukshetra district, Haryana, India. *International Association of Sedimentology, Special Publication*, 6, 337–344.

Pearce, F. (2013). *The Land Grabbers: The New Fight Over Who Owns the Earth*. Beacon Press, Boston, 336 pp.

Plink-Björklund, P. (2015). Morphodynamics of rivers strongly affected by monsoon precipitation: Review of depositional style and forcing factors. *Sedimentary Geology*, 323, 110–147.

Richards, K., Chandra, S., and Friend, P. (1993). Avulsive channel systems: characteristics and examples. In J. L. Best and C. S. Bristow, eds., *Braided Rivers*. Geological Society of London, Special Publication, 75, 195–203.

Rosetti, D. F., Zani, H., and Cremon, E. H. (2014). Fossil megafans evidenced by remote sensing in the Amazonian wetlands. *Zeitschrift für Geomorphologie*, 58, 145–161.

Sáez, A., Anadón, P., Herrero, M. J., and Moscariello, A. (2007). Variable style of transition between Palaeogene fluvial fan and lacustrine systems, southern Pyrenean foreland, NE Spain. *Sedimentology*, 54, 367–390.

Saito, K. (2003). Model of Alluvial Fan Development Based on Channel Pattern and Gravel Size. Report of Research Project, Grant-in-Aid for Scientific Research, 138 (in Japanese).

Saito, K. and Oguchi, T. (2005). Slope of alluvial fans in humid regions of Japan, Taiwan and the Philippines. *Geomorphology*, 70, 147–162.

Sambrook Smith, G. S., Best, J. L., Ashworth, P. J., et al. (2010). Fluvial form in modern continental sedimentary basins: distributive fluvial systems: comment. *Geology*, 38, e230.

Schumm, S. A. (1977). *The Fluvial System*. Wiley, New York, 338 pp.

Shukla, U. K., Singh, I. B., Sharma, M., and Sharma, S. (2001). A model of alluvial megafan sedimentation: Ganga Megafan. *Sedimentary Geology*, 144, 243–262.

Singh, H., Parkash, B., and Gohain, K. (1993). Facies analysis of the Kosi megafan deposits. *Sedimentary Geology*, 85, 87–113.

Sinha, R. and Friend, P. F. (1994). River systems and their sediment flux, Indo-Gangetic plains, Northern Bihar, India. *Sedimentology*, 41, 825–845.

Sounny-Slitine, M. A. and Latrubesse, E. M. (2014). Defining fluvial megafans through geomorphic mapping and metrics. American Geophysical Union Fall Meeting, 2014, Abstract EP51D-3551. https://ui.adsabs.harvard.edu/abs/2014AGUFMEP51D3551S/abstract

Souza, O. C., Araujo, M. R., and Mertes, L. A. K. (2002). Form and process along the Taquari River alluvial fan, Pantanal, Brazil. *Zeitschrift für Geomorphologie*, 129, 73–107.

Stanistreet, I. G. and McCarthy, T. S. (1993). The Okavango fan and the classification of subaerial fan systems. *Sedimentary Geology*, 85, 115–133.

Trendell, A. M., Atchley, S. C., and Nordt, L. C. (2013). Facies analysis of a probable large-fluvial-fan depositional system: the Upper Triassic Chinle Formation at Petrified Forest National Park, Arizona, U.S.A. *Journal of Sedimentary Research*, 83, 873–895.

Tricart, J. (1982). El Pantanal: un ejemplo del impacto geomorfologico sobre el ambiente. *Investigaciones Geográficas*, 29, 81–97.

Tricart, J., Frécaut, R., and Pagney, P. (1984). Le Pantanal (Brésil) : étude écogéographique. *Travaux et Documents de Géographie Tropicale, Bordeaux*, 52, 92 pp.

Urvoy, Y. (1942). *Les bassins du Niger : étude de géographie physique et de paléogéographie*. Larose, Paris, 134 pp.

Valente, C. and Latrubesse, E. (2012). Fluvial archive of peculiar avulsive fluvial patterns in the largest intracratonic basin of tropical South America: the Bananal basin, central Brazil. *Palaeogeography, Palaeoclimatology, Palaeoecology*, 356, 62–74.

Weissmann, G. S., Hartley, A. J., Nichols, et al. (2010). Fluvial form in modern continental sedimentary basins: distributive fluvial systems. *Geology*, 38, 39–42.

Weissmann, G. S., Hartley, A. J., Nichols, G. J., et al. (2011). Alluvial facies distributions in continental sedimentary basins—distributive fluvial systems. In S. K. Davidson, S. Leleu, and C.P. North, eds., *From River to Rock Record: The Preservation of Fluvial Sediments and their Subsequent Interpretation*. SEPM Special Publication, 97, 327–355.

Weissmann, G. S., Hartley, A. J., Scuderi, L. A., et al. (2015). Fluvial geomorphic elements in modern sedimentary basins and their potential preservation in the rock record: A review. *Geomorphology*, 250, 187–219.

Wells, N. A. and Dorr, J. A. (1987a). Shifting of the Kosi River, northern India. *Geology*, 15, 204–207.

Wells, N. A. and Dorr, J. A. (1987b). A reconnaissance of sedimentation on the Kosi alluvial fan of India, In F. G. Ethridge, R. M. Flores, and M. D. Harvey eds., *Recent Developments in Fluvial Sedimentology*. Society of Economic Palaeontologists and Mineralogists, Special Publication, 39, 51–61.

Wilkinson, M. J. (2001). Where large fans form: interim report of a global survey. 7th International Conference on Fluvial Sedimentology, Program and Abstracts, University of Nebraska-Lincoln, Lincoln, Nebraska (USA), 6–10 August 2001, p. 282. (University of Nebraska-Lincoln, Institute of Agriculture and Natural Resources, Conservation and Survey Division, Open-file Report 60).

Wilkinson, M. J. (2005). Large fluvial fans and exploration for hydrocarbons. NASA Tech Briefs 29, 64 [NASA Tech Briefs Online, No. MSC-23424 www.nasatech.com/Briefs/ps.html 30 March 2004].

Wilkinson, M. J. (2006). 'Method for Identifying Sedimentary Bodies from Images and Its Application to Mineral Exploration' – US Patent Office, #6,851,606, issue date 01/10/2006.

Wilkinson, M. J., Cameron, N. R., and Burke, K. (2002). Global geomorphic survey of large modern subaerial fans. *Houston Geological Society Bulletin*, 44, 11–13.

Wilkinson, M. J., Marshall, L. G., and Lundberg, J. G. (2006). River behavior on megafans and potential influences on diversification and distribution of aquatic organisms. *Journal of South American Earth Sciences*, 21, 151–172.

Wilkinson, M. J., Marshall, L. G., Lundberg, J. G., and Kreslavsky, M. H. (2010). Megafan environments in northern South America and their impact on Amazon Neogene aquatic ecosystems. In C. Hoorn, F. P. Wesselingh, eds., *Amazonia, Landscape and Species Evolution: A Look into the Past*. Blackwell, London, 162–184.

2

A Global Megafans Map

M. JUSTIN WILKINSON and NATHAN CURRIT

Abstract

Using a variety of remotely-sensed data, a world-wide survey of river-generated megafans is presented. Thus far, 272 partial cones reaching minimum lengths and widths of 80 km and 40 km, respectively, have been identified. They all indicate large areas of fluvially-laid sediment distant from present or past shorelines. This more than doubles prior counts of fans of these dimensions, partly as a result of using a different set of criteria. All are visible either as pristine or degraded features, and it is likely that more will be found as older, more eroded individual occurrences are identified. The greatest numbers are found in Asia ($n = 87$), from Iraq to the clusters of megafans on the south flank of the Himalaya, and also in central Asia. Africa displays a similar number ($n = 87$), almost all related to relatively low-relief topographic swells. In South America ($n = 60$), most megafans are found clustered along the east flanks of the Andes Mountains and include the longest known example (704 km). Few occur in Europe ($n = 9$) or North America ($n = 1$), in both cases as a result of incisional fluvial regimes operating almost continent wide. Australia hosts 28 megafans. River-generated fans with smaller dimensions are numerous on all continents, however.

2.1 Introduction

The global maps presented here are part of an ongoing mapping project to identify all active and relict megafans with radii > 80 km, using remote detection as a prime methodology. The data on megafans in this chapter are based on fan-like features that lie distant from shorelines in order to avoid confusion with deltas. Exceptions are megafans that enter smaller waterbodies such as various Asian lakes, e.g., the megafans entering Lake Balkhash in Kazakhstan (Fig. 2.1, C16, C17, SOD Table 2.1) and those that have been adequately researched to indicate that at least the proximal sectors have been emplaced solely by fluvial processes (examples are the Carpentaria megafans of Queensland, Australia – see Lane et al., Ch. 12). The maps (Fig. 2.2) include examples from several prior studies. In a global survey such as this it is likely that a few megafans mentioned in the scientific literature are unknown to us. The maps are probably incomplete because experience shows that some (older) megafans are more difficult to recognise.

The criteria listed in Table 2.1 were chosen for the purpose of identifying megafans worldwide as lithological entities (rather than as purely morphological features: e.g., Hartley et al. 2010a, b), an approach also used in chapters on Africa's megafans (Wilkinson et al., Ch. 3) and those of the northern Kalahari Basin of southern Africa (Wilkinson et al., Ch. 4). The criteria are derived mainly from the approximately twenty-five well documented megafans, which include studies of African megafans (Cubango, Kunene, and especially the Okavango – Africa S5–6, S10, Fig. 2.2a; SOD Table 2.1; Wilkinson et al., Ch. 4). Criteria are also derived from experience gained in the world survey in those cases for which surface and subsurface data has confirmed the fluvial origin of the sediments.

Other criteria are the existence of a river as a sediment source, and a smooth, low-gradient (usually $\ll 0.5°$)) surface of partial-cone morphology

Figure 2.1 **Ili and Karatal megafans, Southeastern Kazakhstan**. The Ili and Karatal rivers empty into the 600-km-long Lake Balkhash. Oblique northeast-looking astronaut image. Image STS039–085-OOE courtesy of the Earth Science and Remote Sensing Unit, NASA Johnson Space Center. A colour version of this figure is available in the SOD for Chapter 2.

occupying a lowland adjacent and sloping away from the source highland. Partial-cone morphology was considered a universal criterion for 'primary' megafans (for 'derived' megafans fed from primary megafans, coniform morphology was relaxed as a criterion; see Wilkinson, Ch. 17). The topographic apex of the cone was required to lie within a few kilometres of the point where the river exits the source catchment.

Some characteristics were considered as 'indicator attributes' rather than criteria – Table 2.1 section C – as being useful in guiding the researcher to some of the least recognisable megafans. Thus, radial drainage was treated as an attribute rather than a criterion because parallel and contributory drainage on medial and distal slopes is a pattern common on diamond-shaped fans (in their morphological approach Hartley et al. 2010a, excluded areas displaying these latter non-radial drainage patterns).

Concave-up longitudinal profiles are common, but our survey revealed a range from concave to convexo-concave to convex. Consequently, here too we included long-profile concavity as an attribute rather than a criterion. Topographic roughness, combining measures of smoothness and declivity (see Wilkinson et al. 2010), is a distinct identifier of depositional landscapes (e.g., Fig. 3.2 a–c, Wilkinson et al., Ch. 3; 17.1a, Wilkinson, Ch.17; Wilkinson et al. 2010). The vast megafans of the Andean foreland show that a characteristic signature of extreme smoothness is strongly related to megafan surfaces (Wilkinson et al. 2010).

Fluvial channel morphologies are often widespread on megafan surfaces. However, these can become so degraded and overprinted on relict surfaces as to be unrecognisable, so that they too are excluded as criteria. The large, relict Cubango megafan in northern Namibia (Africa S6, Fig 2.2a; SOD Table 2.1), fully 310 km long, was not recognised because it lacked overt radial channels as a prime visual cue (Wilkinson et al., Ch. 4). Many other features interpreted here as megafans display the gross morphology of a cone even though their surfaces are entirely, or almost entirely, covered with dunes, as in Saudi Arabia, the Tarim Basin (China) and central Australia.

In combination the criteria in Table 2.1 add confidence to the identifications presented in this study. Other than the abovementioned exception, the six

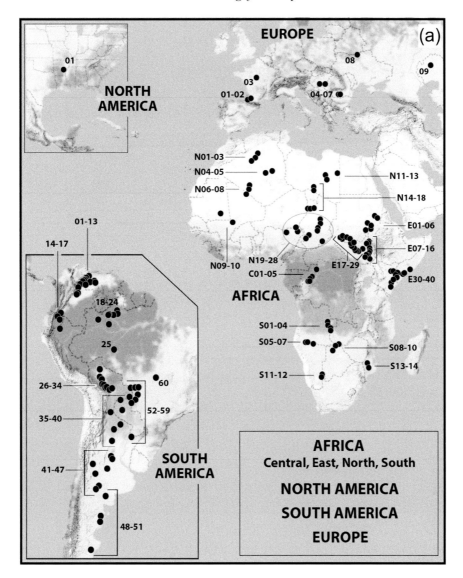

Figure 2.2 **Global distribution of megafans (> 80 km)**. Dots indicate the centre of each megafan. (a) West, (b) East. A colour version of this figure is available in the SOD for Chapter 2.

criteria used by Hartley et al. (2010a) in a global study of 415 fan-like features (with a lower bund of radii > 30 km) broadly overlap our criteria (see Discussion, Wilkinson, Ch. 17).

2.2 Methods

Our approach, based on identifying bodies of probable fluvial sediment, has the benefit of being testable: a megafan sedimentary body supplied by the fan-forming river can be expected to show a unique mineralogical signature in the fluvial component – e.g., by airborne radiometric methods (see Gunnell, Ch. 9 and Kapteinis et al., Ch. 13).

Analysis of modern and abandoned stream courses was originally based on remotely-sensed images from the image archive of astronaut-handheld images at the Johnson Space Center (Earth Sciences and Remote Sensing Unit, 2022). More recently Google Earth imagery and true- and false-colour infra-red Landsat imagery of the 1990 and 2000 series were major data sources. Field campaigns in northern Namibia, northern Argentina, and central France provided ground truth. Based on SRTM (Shuttle Radar Topography Mission) digital topographic data (available at ftp:// e0srp01u.ecs.nasa.gov/), contour maps with small vertical intervals (5 m) and topographic cross-sections, variously aligned, were constructed using digital

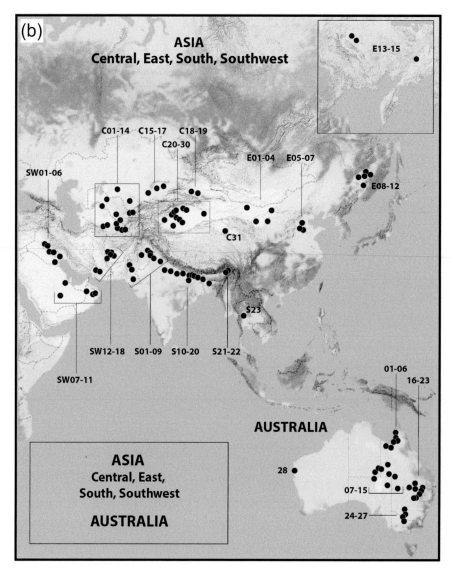

Figure 2.2 (*cont.*)

elevation models (DEMs) generated by the Microdem freeware package with a suite of visual displays, map images, GIS databases, etc. (www.usna.edu/Users/oceano/pguth/website/microdem.htm). Fan dimensions were measured using SRTM DEM 1 arcsec data interpolating 10-m contours and imagery showing stream orientation. On asymmetric fans we rendered the longest radius. ONC maps (1:100,000 scale – Office of Naval Cartography) gave further detailed information on dormant and discontinuous stream beds and some smaller river names.

Topographic roughness is a measure of slope variability along baselines of different lengths (0.6 km, 2.2 km, and 8.7 km in this study) (Wilkinson et al.

2010). The roughness algorithm was first developed to describe roughness characteristics of Mars's surface by Kreslavsky and Head (1999, 2002), and was applied to Earth based on SRTM data using a similar array of baselines (e.g., Wilkinson et al., Ch. 3, Fig. 3.5a, b). Lighter shades denote the higher roughness of mountainous landscapes; darkest tones indicate smooth plains landscapes of very low slope, especially megafans; shades of green indicate kilometre-scale roughness dominance, such as dune fields.

Hillshading, a topographic visualisation commonly available in DEM software, provided details and landscape syntheses that aided the mapping of megafan margins in particular. Published reports in a few cases

Table 2.1 *Criteria for megafan recognition by remote sensing means*

A Criterion: Setting – water and sediment source and supply

A1	Topographic margin: juxtaposed upland and lowland
A2	River: upland river flowing into neighbouring lowland

B Criterion: Morphological characteristic

B1	Plain-like surface morphology: smooth surface (low roughness signatures)
B2	_____: low declivity (< 1 degree)
B3	_____: sloping away from contiguous upland
B4	_____: > ~80 km longest radius; > ~40 km widest point
B5	_____: continuity of slope
B6	Partial cone: at least in proximal sector
B7	_____: apex (or remnant) near river exit-point from upland

C Common traits: Morphological and sedimentological characteristics

C1	Channels on plain: palaeochannel morphologies indicative of fluvial activity on original surface (common)
C2	_____: palaeochannel network covers entire low-slope surface (very common)
C3	_____: palaeochannel density significantly higher than in upland feeder basin
C4	_____: palaeochannel orientations radial proximally (medially often parallel to regional slope)
C5	_____: feeder river dimensions on megafan surface diminish medially/distally (common)
C6	Sediments: near-horizontal attitude of layers[a]
C7	_____: deposits coextensive with plain
C8	Water tables: higher proximally (i.e., nearer upland margin); or higher distally (near regional base level)
C9	_____: spring line located in medial/distal sector

Common traits: Subregional drainage and nesting patterns

C10	Drainage patterns: megafan axis usually orthogonal to upland front; as tributaries to basin trunk river
C11	Multi-megafan nesting patterns: orthogonal drainage gives lateral nesting along upland front; occasionally transverse to upland front
C12	Asymmetric basin fill: megafans on dominant sediment-supply flank of trunk river
C13	Basin trunk drainage line (where it exists) shows planform convex away from tributary megafans
C14	Modern river courses often located in inter-megafan depressions (where later incision is focused)

Note. [a] Attitude of layered rocks ascertainable from oblique stereoscopic imagery, especially of stream banks and valley walls.

provided some morphological data. A related geomorphic indication of megafan margins proved to be the easily recognised distinction between exposed rock outcrop and the smooth, lowland aggradational surfaces in remotely sensed images.

Radii were measured from the cone apex to the toe at the furthest point along the fan slope for those features whose radii are > 80 km and whose widths are > 40 km. Fans were classified as either active or relict. Active fans show clear evidence of modern activity, usually because of a strong vegetation response in visible wavelength and infrared imagery. Criteria for assigning inactive/relict status were primarily that the fan-forming river should be incised > 10 m into the fan surface (thereby broadly precluding flooding of the surface); or that the surface be covered

by pervasive dunefields; or, lastly, that an active fluvial lobe should cover less than ~10% of the area of the feature under scrutiny.

2.3 Results and Discussion

Two hundred and seventy-two megafans were identified (Fig. 2.2a, b), both active and relict (SOD Table 2.1) worldwide. Megafans are listed by continent with coordinates given for an apex and toe. The political entity within which the apex is located is also given. The megafan identifier assigned by Hartley et al. (2010a) is given where it exists, whether or not it met our length criteria.

The confidence of assigning a feature as a megafan declines as the age of the feature increases. Active fans that show clear evidence of modern activity, usually because of a strong vegetation response in visible wavelength and infrared imagery, are straightforward to identify. Where such evidence is absent, relict fans ('r' in SOD Table 2.1) in particular become more difficult to identify.

Criteria for assigning relict status were primarily that the fan-forming river should be incised > 10 metres into the fan surface (thereby broadly precluding flooding of the surface); or that the surface be covered by dunefields over > 80% of the fan surface; or that an active fluvial lobe should cover less than ~ 10% of the area of the feature under scrutiny. Between one in three and one in five megafans are relict as assigned by our criteria (Asia 18%, South America 25%, Africa 34%) by continent, excluding continents with low numbers of megafans.

A first-order classification by tectonic basin shows what has become the familiar pattern of a clustered distribution in compressional basins adjacent to the Andean, Himalayan and other highlands, a group that in all accounts for 30% (*n* = 82, e.g., South America 01–17, 26–40, 44–48; South Asia S01–22), and thus close to the 37% of fans (in the larger population > 30 km long) identified in forelands by Hartley et al. (2010). A smaller percentage is scattered in major strike-slip-related basins in Central and East Asia (16%: C01–31; E01–15, Fig. 2.2b; SOD Table 2.1). The combination of orogenic sediment supply and juxtaposed foreland basin provide the familiar nesting pattern Type 1, Fig. 17.7 (Wilkinson, Ch. 17). Such megafans include the widely quoted Kosi megafan (Asia S17, Fig. 2.2b; SOD Table 2.1) (e.g., Sinha et al., Ch. 11) and the longest megafan on the planet laid down by the Pilcomayo River in northern Argentina and western Paraguay (South America: 36, Fig. 2.2a; SOD Table 2.1), with a radius of more than 700 km (Iriondo 1987, and Latrubesse, Ch. 5).

The Hartley et al. (2010) and Weissmann et al. (2011) studies give a detailed breakdown of the tectonic settings in which fan-like forms > 30 km long are found. Their distributions differ significantly from ours in the case of Africa's 'basin-and-swell' tectonic environment (Burke and Gunnell 2008; Burke and Wilkinson 2016) where they counted 25 related to swells compared with 80 in this study. Our unexpectedly high count (29%, *n* = 80; seven of Africa's 87 megafans are not

swell-related) approaches that for foreland basins (30%, *n* = 82). By comparison with the strong clustering that characterises foreland settings, Africa's megafans are generally dispersed with only three clusters numbering more than four megafans (Wilkinson et al., Ch. 3). Many are erosionally degraded or dune covered, which may explain why they were not recognised until recently. Other broadly cratonic environments include interior Australia (8%, *n* = 21) and central Amazonia, broadly defined (3%, *n* = 9).

For good reason the megafans of the Himalayan piedmont are known as classics, especially the Kosi megafan (Asia S17, Fig. 2.2b; SOD Table 2.1), because of their pristine appearance, because they are active, but also because they are relatively small and clustered, allowing them, it seems, to be more easily recognised. Most of Asia's megafans are related to major mountain ranges, and are almost exclusively oriented and nested transverse to structural grain (nesting Types 1–3, Fig. 17.7, Wilkinson, Ch. 17). In southwest Asia, the great Mesopotamian megafans of Iraq (Asia SW01–04) are unusual in being aligned parallel with the foreland depression (nesting Type 4, Fig. 17.7, Ch. 17). In a few cases, megafans extend far onto the cratonic lowlands, such as those surrounding the Caspian Sea (Asia C06–09; Europe 09, Fig. 2.2a; SOD Table 2.1), with nesting patterns of Type 5, as seen in most of Africa and Australia (Fig. 17.7, Ch. 17).

Europe and North America are almost devoid of megafans that reach the minimum dimensions we used. North America's single megafan reaches 130 km in length (North America 01, Fig. 2.2a; SOD Table 2.1). Hartley et al. (2010) identified eleven fans arrayed along the north flanks of the Alaska Range, and eight in the Central Valley of California that reached their minimum criterion length of 30 km. The lack of megafans is attributable to the narrow dimensions of the basins available for fan development in the Rocky Mountains and Coastal Ranges of North America. The probable megafans of Neogene Ogallala sediments arrayed along the east flanks of the Rocky Mountains (Willett et al. 2018) stretch from Montana to Texas, but are now so erosionally degraded that length or width measurements cannot be ascertained; consequently these have not been included in our dataset.

More broadly, the explanation for the lack of megafans at a continental scale in North America relates to the fact that after 'central North America emerged from the Western Interior Seaway more than 70 million years ago, it entered a prolonged period of uplift and erosion that continues today…' (Willett et al. 2018), erosional regimes being the antithesis of region-wide fluvial aggradation. The lack of large fan-like features contrasts strongly with the very large number of small and iconic alluvial fans (*sensu stricto*), so well-known in the arid and semi-arid Basin and Range landscapes of the western USA. Similarly, Europe's megafans are also few in number, again partly because of the modern prevalence of vast region-wide erosional regimes in which fluvial aggradation is confined to relatively narrow valleys and coastlines. The megafans that do occur are specifically associated with orogenic belts such as the Alps, Pyrenees, and Carpathian Mountains (Fontana and Mozzi, Ch. 8). The Loire megafan of central France, also now incised, is an example in Europe of an African-type swell-flank megafan. We note that our length criteria are not used by all researchers: the large fans of the southern Alpine piedmont (< 80 km long) described by Fontana and Mozzi in Italy (Ch. 8), although falling just below the threshold length used in this study, are included in this volume as a counterpoint.

In geomorphic terms Wilkinson (Ch.17, Section 17.5) has suggested that three major 'geomorphic domains' characterise the world's sub-continental-scale lowlands. The relative lack of megafans in North America and Europe make these lowlands the type landscapes for an 'erosional domain'. At the opposite end of the erosion–aggradation continuum lies central South America, which provides the prime example of major lowlands that are dominated by an 'aggradational domain' – in the form of sets of nested megafans along the eastern flanks of the Andean mountain chain. Between these end-members, the vast alluvial floodplains of the central Amazon basin confined between erosional uplands (Latrubesse 2015) can be seen as an intermediate type. These three broad domains could act as baselines from which to evaluate subcontinental palaeogeographies of buried landscapes (Wilkinson, Ch. 17, Section 17.6.4).

Because there are relatively few megafans worldwide, the small size of the population substantially narrows the scope for statistical analysis. Accordingly we compared our count of megafans (radii > 80 km) with that of the larger population in the global survey

of Hartley et al. (2010a, b) and Weissmann et al. (2011, 2015) (in their categorisation they defined alluvial fans as < 20–30 km long, termed in this chapter 'alluvial fans (*sensu stricto*)', fluvial fans as 30–100 km long, and megafans as > 100 km long). We calculate that the number of fans > 80 km long in their dataset numbers close to 121 individuals (24%), whereas fans > 80 km long identified in this study more than doubles that total ($n = 272$). Part of the explanation for this higher total lies in the fact that we included a greater number of relict fans (but still congruent with our criteria) (9%; $n = 25$) ('r', SOD Table 2.1) that were included less frequently in their survey. Part of the explanation also probably lies in the identification of megafans that do not appear in visual or infrared wavelengths, but which readily appear as low-gradient features in DEMs and topographic roughness maps. These harder-to-detect megafans were extensively encountered in the Africa study (Wilkinson et al., Ch. 3).

Another explanation for our larger megafan total relates to differing definitions of fan length. Our results show that 52 fans recognised by Hartley et al. (2010a, b) and Weissmann et al. (2011) are somewhat longer by our criteria than the dimensions they assigned. Such reallocation increases the number of what we term megafans in their dataset from 121 to 173 ($121 + 52$) – a total that amounts to fully 42% (173/415) of the features they identified, rather than the nominal 24% mentioned above.

A yet broader perspective arises if those fans not listed by Hartley et al. (2010) but identified in our study (n = 151) are added to their total ($n = 415$). In this case, the global total of megafans (> 80 km long) amounts to 566 ($415 + 151$). The proportion of fluvial megafans then rises from 24% in the original count to 48% (total of 272 in this study as a proportion of all large fans, $n = 566$). This proportion appears unusually large in what is thought of as a heavily right-skewed distribution (e.g., Hartley et al. 2010, fig. 2 therein), and suggests that yet smaller fans are likely to be identified in future. It also means that megafans, in terms of combined area, could exceed the total of the very numerous small alluvial fans. Or these proportions may mean that alluvial fans (*sensu stricto*) are sufficiently different from megafans in terms of process (see Wilkinson, Ch. 17) that in fact they may be as part of the same population (on the grounds of process). With notable exceptions, the criteria for identifying features as megafans differ from study to study and are often left unstated, or are only loosely applied. It seems imperative, therefore, to specify the criteria being used, not only for purposes of better comparison, but also because it is possible that strict application of criteria will have the major benefit of revealing groups or subgroups (and their processes) that may not have been recognised previously. A benefit of applying this kind of definitional rigour has been the identification of the MAFS feature (major avulsive fluvial system: Latrubesse 2015, Wilkinson, Ch. 17) as different from megafans – even though several had been classed earlier as megafans on the basis of planform, radial drainage, very low slope, and dimension. But MAFS fail to display the generally accepted partial-cone morphology usually regarded as a key criterion for megafan identification. It appears that some processes unlike those that determine megafan morphology are likely operating on MAFS. Similarly, the 'derived' megafan type also fails to conform to the partial-cone criterion (Wilkinson, Ch. 17). Processes responsible for this phenomenon surely call for further research.

Although the global megafan total is likely to change as further research takes place, the present study offers what are regarded as robust proportions by tectonic setting. These proportions already suggest that continental landscape components (of megafan/multi-megafan scales) need to be reassessed. For example, Wilkinson et al. (Ch. 3) have calculated that the total area of African megafans is larger than the total subaerial extent of deltas of 17 of the world's largest rivers. When the total area of megafans from all continents is then known, it appears that they will probably constitute a landform type of greater significance in world subaerial landscapes than major coastal deltas. Megafan areas undoubtedly would have been even higher at times in the Pleistocene when sea levels were lower than they are during interglacials, such that many modern deltas can be expected to have acted as megafans (at least where low shelf slope geometry does not force river incision). We note also that rainforests, glacial outwash zones (Fontana and Mozzi, Ch. 8), and zones impacted by glacial megafloods in North America and central Asia – unaddressed in this volume – may also yield a number of new megafans.

References

Burke, K. and Gunnell, Y. (2008). *The African Erosion Surface: A Continental-scale Synthesis of Geomorphology, Tectonics, and environmental Change over the Past 180 Million Years*. Geological Society of America Memoir, 201, 66 pp.

Burke, K. and Wilkinson, M. J. (2016). Landscape evolution in Africa during the Cenozoic and Quaternary—the legacy and limitations of Lester C. King. *Canadian Journal of Earth Sciences*, 53, 1089–1102.

Earth Sciences and Remote Sensing Unit, NASA Johnson Space Center. https://eol.jsc.nasa.gov/ (accessed February 2022)

Hartley, A. J., Weissmann, G. S., Nichols, G. J., and Warwick, G. L. (2010a). Large distributive fluvial systems: characteristics, distribution, and controls on development. *Journal of Sedimentary Research*, 80, 167–183.

Hartley, A. J., Weissmann, G. S., Nichols, G. J., and Scuderi, L. A. (2010b). Fluvial form in modern continental sedimentary basins: distributive fluvial systems: reply. *Geology*, 38, e231.

Iriondo, M. H. (1987). Geomorfolgía y Cuaternario de la Provincia Santa Fé (Argentina). *D'Orbignyana* (Corrientes, Argentina), 4, 54 pp.

Kreslavsky, M. A. and Head, J. W. (1999). Kilometer-scale slopes on Mars and their correlation with geologic units: initial results from Mars Orbiter Laser Altimeter (MOLA) data. *Journal of Geophysical Research*, 104, 21,911–21,924.

Kreslavsky, M. A. and Head, J. W. (2002). Kilometer-scale roughness of Mars' surface: results from MOLA data analysis. *Journal of Geophysical Research*, 105, 26,695–26,712.

Latrubesse, E. M. (2015). Large rivers, megafans and other Quaternary avulsive fluvial systems: a potential "who's who" in the geological record. *Earth-Science Reviews*, 146, 1–30.

Weissmann, G. S., Hartley, A. J., Nichols, G. J., et al. (2011). Alluvial facies distributions in continental sedimentary basins—distributive fluvial systems. In S. K. Davidson, S. Leleu, and C. P. North, eds., *From River to Rock Record: The Preservation of Fluvial Sediments and their Subsequent Interpretation*. SEPM Special Publication, 97, 327–355.

Weissmann, G. S., Hartley, A. J., Scuderi, L. A., et al. (2015). Fluvial geomorphic elements in modern sedimentary basins and their potential preservation in the rock record: a review. *Geomorphology*, 250, 187–219.

Wilkinson, M. J., Marshall, L. G., Lundberg, J. G., and Kreslavsky, M. H. (2010). Megafan environments in northern South America and their impact on Amazon Neogene aquatic ecosystems. In C. Hoorn and F. P. Wesselingh, eds., *Amazonia, Landscape and Species Evolution: A Look into the Past*. Blackwell, London, 162–184.

Willett, S. D., McCoy, S. W., and Beeson, H. W. (2018). Transience of North American High Plains landscape and its impact on surface water. *Nature*, 561, 528–532.

Part II
Regional Studies

3

Megafans of Africa

M. JUSTIN WILKINSON, KEVIN BURKE, NATHAN CURRIT, and MIKHAIL A. KRESLAVSKY

Abstract

Mapping of Africa's megafans according to a set of criteria (radii $> 80\,\text{km}$, widths $> 40\,\text{km}$; high topographic smoothness; result: $n = 87$), suggests a direct relationship between fluvial megafans and the thirty relatively young tectonic swells of the continent. Although ten are barren of megafans, fully 85% display this relationship and are thus named 'swell-flank type'. Another control was also identified: almost two thirds of this group was related to swell flanks margined by a rift-related depression. Clustering is significant in this 'the swell-and-rift' subtype: 23 in South Sudan (Muglad-Melut troughs), nine in Kenya (Anza Rift), and four in southern Chad (Salamat, Dosseo, and Bongor rifts). The remainder ('swell-flank only' subtype) were found to be scattered widely. Only 3%, however, were exclusively related to rifts (e.g., the Okavango megafan of Botswana). Africa's megafans total at least $1.2\,\text{Mkm}^2$, average megafan unit area being $13{,}200\,\text{km}^2$. Flanks of the largest swells (e.g., Congo Basin flanks of the East African swell) are devoid of megafans, perhaps because of enhanced recent uplift. Coasts are similarly devoid of megafans, possibly for the same reason. Cratonic blocks where swell growth is less prominent are also devoid. Africa's largest rivers are associated with few megafans.

3.1 Purpose and Scope

Only recently has the significance of megafans been recognised, with > 272 mapped worldwide (Wilkinson and Currit, Ch. 2) of which 87 are found in Africa, on a par with the same number for South America. In a first continent-wide map, Africa's largest megafans ($> 80\,\text{km}$ long), with their fan-shaped geomorphic features composed dominantly of fluvial sediment, are shown to be widely distributed across the continent. A prime purpose of this study is to identify surfaces that can be mapped with some confidence as fluvial sediment bodies. These surfaces are almost all interpreted from remotely sensed data following several criteria (Table 3.1; see also discussion in Ch. 4). A second aim of this study is an analysis of the main tectonic setting of the megafans, settings which appear to explain the wide differences in distribution patterns. Fan areas were measured and active/relict status of each megafan were assigned. Only three megafans had been commonly known before ~ 1990 – the 'inland deltas' of the Okavango River in Botswana, the Niger River in Mali, and the White Nile's vast Sudd wetland of South Sudan.

It is now accepted that megafans have little relation in terms of process to the well-known, relatively short alluvial fan feature ($< 20\,\text{km}$ long) which falls within Group 8 in Miall's (1996, 2014) hierarchy of fluvial architectural forms (see SOD 3.1). In contrast, Wilkinson et al. (2006, 2010, Ch. 17) have suggested that megafans fit most appropriately at the midscale in Miall's hierarchy, that is, in Group 9 – a level equivalent to that of large deltas, major floodplains, and bajadas. Landscapes comprised of a set of nested megafans fit well at the Group 10 level (see Table 17.1 in Wilkinson, Ch. 17). Different sets of fluvial processes and architectures are associated with different levels in the hierarchy: alluvial fans are composed dominantly of channelised, sheetflood, hyper-concentrated flow and

Table 3.1 *Criteria for megafan recognition by remote sensing means*

A	**Criterion: Setting—water and sediment source and supply**	
	A1	Topographic margin: juxtaposed upland and lowland
	A2	River: upland river flowing into neighbouring lowland
B	**Criterion: Morphological characteristic**	
	B1	Plain-like surface morphology: smooth surface (low roughness signatures)
	B2	_____ : low declivity (< 1 degree)
	B3	_____ : sloping away from contiguous upland
	B4	_____ : > ~80 km longest radius; > ~40 km widest point
	B5	_____ : continuity of slope
	B6	Partial cone: at least in proximal sector
	B7	_____ : apex (or remnant) near river exit-point from upland
C	**Common traits: Morphological and sedimentological characteristics**	
	C1	Channels on plain: palaeochannel morphologies indicative of fluvial activity on original surface (common)
	C2	_____ : palaeochannel network covers entire low-slope surface (very common)
	C3	_____ : palaeochannel density significantly higher than in upland feeder basin
	C4	_____ : palaeochannel orientations radial proximally (medially often parallel to regional slope)
	C5	_____ : feeder river dimensions on megafan surface diminish medially/distally (common)
	C6	Sediments: near-horizontal attitude of layers [a]
	C7	_____ : deposits coextensive with plain
	C8	Water tables: higher proximally (i.e., nearer upland margin); or higher distally (near regional base level)
	C9	_____ : spring line located in medial/distal sector
		Common traits: Subregional drainage and nesting patterns
	C10	Drainage patterns: megafan axes usually orthogonal to upland front; as tributaries to basin trunk river
	C11	Multi-megafan nesting patterns: orthogonal drainage gives lateral nesting along upland front; occasionally transverse to upland front
	C12	Asymmetric basin fill: megafans on dominant sediment-supply flank of trunk river
	C13	Basin trunk drainage line (where it exists) shows planform convex away from tributary megafans
	C14	Modern river courses often located in inter-megafan depressions (where later incision is focused)

Note. [a] Attitude of layered rocks is ascertainable from oblique stereoscopic imagery, especially of stream banks and valley walls

mass wasting deposits, whereas megafans are dominated by alluvial ridges and overbank deposits, distinctions discussed in Wilkinson, Ch. 17.

3.2 Africa's Basin-and-Swell Topography

It is now recognised that non-orogenic land surfaces that reach altitudes greater than ~1 km above sea level, e.g., in shield regions, are generally dynamically maintained. Those in Africa are no exception (e.g., Burke 1996: fig. 25 therein; Burke and Gunnell 2008; Burke and Wilkinson 2016: figs. 2 and 3 therein). This recognition has revived interest in the basin-and-swell ideas that Krenkel (1922) applied to African landscapes a century ago. Du Toit (1933) applied Krenkel's ideas to southern Africa, and more recently Burke and Gunnell

(2008) synthesised them with modern tectonic theory. Their synthesis (SOD 3.2a) sees African landscapes reduced between ~100 and ~30 Ma to a continent-wide, mainly erosional platform, the African Surface, almost everywhere fairly near sea level (Burke and Gunnell 2008; Burke and Wilkinson 2016).

Applying insights from research on dynamics of the upper mantle (especially Burke and Wilson 1972; Burke and Dewey 1974; England and Houseman 1984; Burke 1996), Burke and Gunnell (2008) and Burke and Wilkinson (2016) argued that modern African landscapes have arisen as the African Surface has become gently deformed by the local uplift of swells and subsidence of the intervening basins. The resulting population of swells has affected the entire Africa Plate. The deformation is argued to have resulted in large part from the major event at ~32 Ma (Burke and Gunnell 2008; Burke and Wilkinson 2016), in which the Afar plume in Ethiopia emplaced hundreds of km³ of lava onto the surface in a geological instant (~2 Myr; Hofmann et al. 1997). Thirty swells are mapped in Africa, 29 of which are named (a thirtieth is the small 'unnamed swell' located between the Tibesti and Hoggar massifs; see Burke and Gunnell 2008: Frontispiece) (Fig. 3.1).

A major effect of the plume eruption appears to have been the 'pinning' of the Afro-Arabia Plate from below, resulting in Africa becoming a stationary continent (the 'Africa stationary' idea of McKenzie and Weiss 1975). As a consequence of plate motion ceasing, cellular circulation in the shallow mantle was initiated (SOD 3.2b). These circulations today provide the dynamic support for the small and large swells on the African continent (Figs. 3.1a, 3.2, 3.3, 3.4). In this view, individual swells of the continent reflect a pattern of shallow mantle convection on the 100–1,000 km horizontal length scale. The swells and their intervening basins today determine much of Africa's drainage

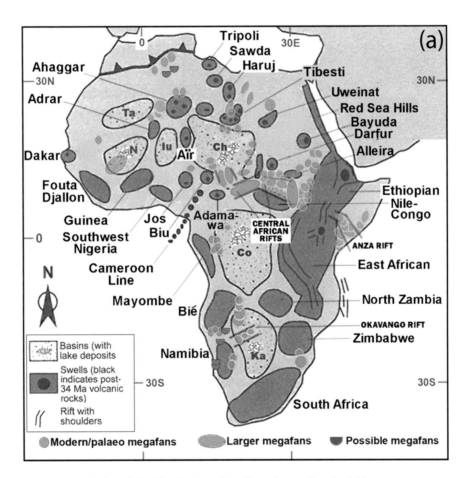

Figure 3.1 **Diagrammatic rendering of the distribution of swells and megafans in Africa.**
(a) Africa's thirty swells shown as dark-toned zones: twenty-nine named from Burke and Gunnell (2008). Inter-swell basins are stippled: major basins Chad (Ch), Congo (Co), and Kalahari (Ka); other basins are the Taoudeni (Ta), Niger (N) and Iullemeden (Iu). Eighty-seven megafans shown as light-toned ellipses, numbered in Fig. 3.2a–c. Half circles indicate six possible megafans. Megafans lie overwhelmingly on swell flanks (this study), especially where these are located adjacent to rift zones (rifts suggested by straight-line bounding faults in some cases). Two of the largest swells, East African and Ethiopian, are associated with the major rifts of the East African Rift system and the Red Sea. Some smaller swells have two peaks; nine volcanoes are subsumed as the Cameroon Line swell. Adapted from Burke and Gunnell (2008).
(b) Central Africa rift basins that have contributed to megafan location and development. Black: Cretaceous–Cenozoic continental rift basins. Grey: Cretaceous–Cenozoic marine and continental rift basins. Continental basins showing Muglad–Melut, White and Blue Nile and Atbara Rifts to the east, and the Salamat–Bongor–Chad Rifts to the west. The Anza Rift occupies the east side of Kenya. Dashed lines: international boundaries. Dotted lines: East African Rift. From Genik (1992).

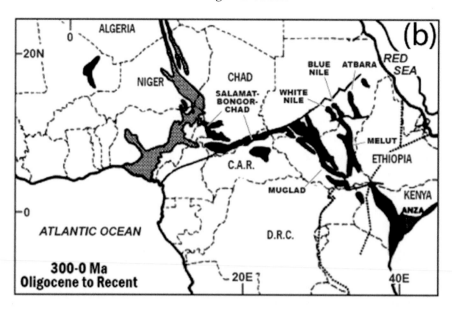

Figure 3.1 *(cont.)*

pattern. Only Africa's Mediterranean coast from Tunisia to Morocco differs in being dominated by Alpine collision in a Plate Boundary Zone (PBZ).

Support for Krenkel's (1922) model of presently active and independently rising swells has come from recent interpretations of hundreds of longitudinal river profiles in Africa (Paul et al. 2014). This interpretation also suggests that active swell-related uplift has been in progress since the ~32 Ma pinning event. Post ~32 Ma activity has reactivated ancient structures on many swells, and some activity has initiated new rifting.

King's (1942, 1967) better known theory of African landscapes revolved around the idea that very ancient erosion surfaces, dating to times before the break-up of Pangea and continuing to form until the present in stair-step fashion, are preserved most often as planar interior plateaus, and have continued until the present to dominate Africa's geomorphic evolution. A critique of King's theory can be found in Burke and Wilkinson (2016) (SOD 3.2c).

The modern expression of the swells is generally elliptical, with lengths from as little as 150 km in the case of the Dakar swell to more than 1,600 km in the case of the largest swells, the East African and South African (Fig. 3.1a). The swells have developed in various ways. The East African rifts are relatively narrow features that lie within the confines of the far wider Ethiopian and East-African swells, where older structures have been reoccupied in new or reactivated

rifts. The Shire and Rukwa rifts modify the surface of the North Zambian swell. The Red Sea and Gulf of Aden swells, crested by rifts of the East African system by ~28 Ma, have developed in the past 5 Ma into the young Gulf of Aden and Red Sea. However, most swells are not cut by rifts. Rifts now tend to be associated with inter-swell depressions. The Hoggar, Tibesti (Fig. 3.1a) and Darfur swells comprise double edifices. Some edifices are capped by volcanic rocks, but many display no evidence of volcanoes. Nine close-spaced volcanic edifices are grouped as the Cameroon Line swell. Because swell edifices are so widely scattered across Africa, the drainage of the continent, especially in headwater zones, is almost everywhere controlled by swell structures.

Interior basins have been localised as down-flexed, topographically low areas among the rising swells (Fig. 3.1a, and Burke 1976). They are today being loaded by sediments from the surrounding swells, especially in Africa's major Chad, Congo, and Kalahari basins, each of which extends over an area of ~1 Mkm^2. The Taoudeni, Niger, Iullemeden, and Sudd basins are smaller (Fig. 3.1a). The shallow but active Okavango and Makgadikgadi rifts overlie an extension of the mid-Zambezi rift (Podgorski et al. 2013; McCarthy 2013) that transects the Kalahari Basin.

Accelerated accumulation of sediment indicates swell initiation at ~32 Ma, as documented by the age of an offshore unconformity that underlies these younger

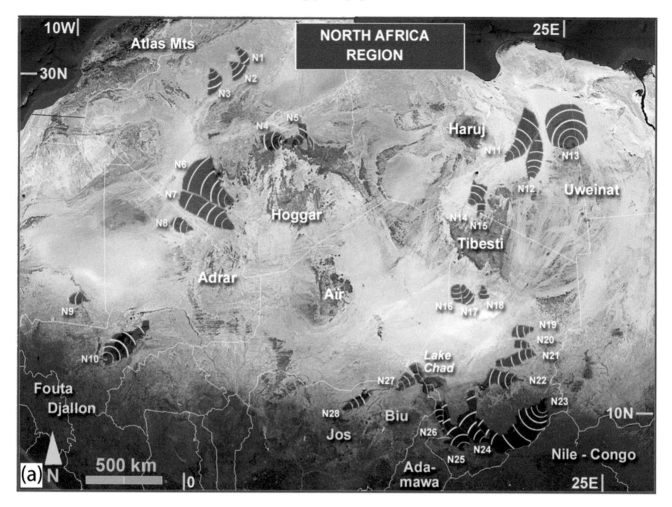

Figure 3.2 **Megafan distribution in Africa (numbered by region)**. Megafans are shown with stylised contours (convex in the downfan direction) with respect to swells and basins (all large-font capitalised names).
(a) North Africa Region: megafans concentrated around swells, with a relative absence in northwest Africa. The nested group in Southern Chad (N23–27) relates to several surrounding swells and the existence of major rift troughs. Names of swells are capitalised (with the exception of the Atlas Mts, a Plate Boundary Zone).
(b) East Africa Region: the largest concentrations of megafans in Africa are associated with both the Ethiopian and East African swells and adjacent rifts of the Muglad and immediately adjacent rift basins.
(c) Southern Africa Region: relatively few megafans are mostly concentrated in the northwest Kalahari Basin. Three 'rift-only' megafans (S8–10) located within the Okavango Rift appear to be unique in Africa (see text).
Congo Region inset: one megafan is sourced from the Nile-Congo and Adamawa swells (C1), the other four are fed directly from the Mayombe swell.

units. Interior basins have retained only small volumes of sedimentary rock deposited during the past 32 Myr (Burke 1976; Kadima et al. 2011; McCarthy 2013). Some basins make and break connection to the oceans with the rise and fall of base levels. This is an interpretation of the fact that the thickness of megafan sedimentary sections, accumulated over the past ~30 Ma, has been surprisingly small (Burke and Gunnell 2008). The explanation, we suggest, is that much of the sediment carried into steer's head basins (SOD 3.2d) has been evacuated via low spillways to the Mediterranean,

Atlantic and Indian ocean basins. Megafan stratigraphies are likely to contain many unconformities because the spillways are active episodically, for example in response to climatic variations.

3.3 Criteria and Methods

3.3.1 Criteria for Megafan Definition

The key criteria employed in identifying Africa's megafans (Table 3.1) follow those used for the world

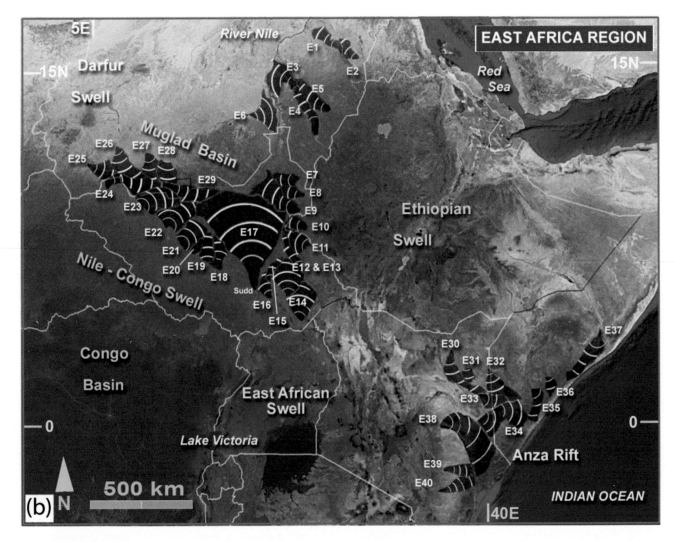

Figure 3.2 *(cont.)*

map (Wilkinson and Currit, Ch. 2) and the northern Kalahari megafan study area (Wilkinson et al., Ch. 4). The criteria were chosen for the purpose of identifying megafans as lithological entities (rather than for morphological reasons – e.g., Hartley et al. 2010). Our approach also has the benefit of being testable: for example, megafan bodies supplied by a single major river can be expected to show a unique mineralogical signature in the fluvial component – e.g., by airborne radiometric methods (see Gunnell, Ch. 9 and Kapteinis et al., Ch. 13). These criteria are derived mainly from the approximately twenty-five well documented megafans, which include studies of African megafans (Kunene, Cubango, and especially the Okavango – S5, S7, S10, Fig. 3.2c). Criteria are also derived from experience gained in the world survey reported by

Wilkinson and Currit (Ch. 2), for which surface and subsurface data has confirmed the fluvial character of the sediments.

The most important criteria are the existence of a river as a sediment source, and a smooth, very low-gradient (usually $\ll 0.5°$)) surface of partial-cone morphology, occupying a lowland adjacent to the source upland, and sloping away from the upland. Partial-cone morphology was considered to be a universal criterion, at least for the dominant 'primary' megafan type (compared with the small number of the 'derived' megafan type, see Wilkinson, Ch. 17). The apex of the cone was required to lie at or near the point where the river exits the source upland. Only cone-shaped surfaces $> \sim 80$ km long were included (except in the case of

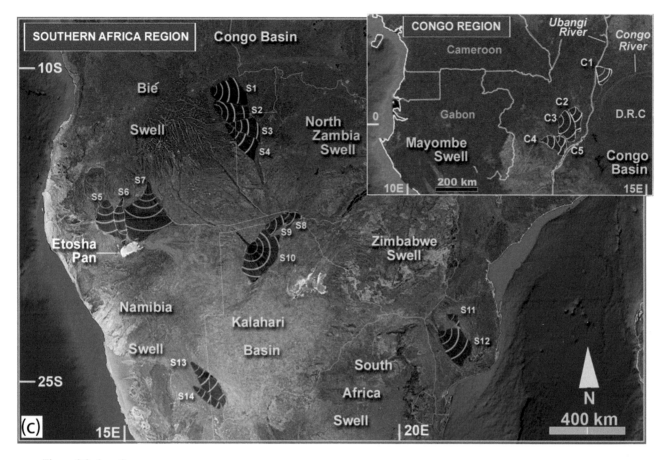

Figure 3.2 *(cont.)*

derived megafans: see SOD 3.3.1). Cone radii were measured from the cone apex to the toe at the furthest point of the continuous slope.

Radial drainage was applied only as an attribute rather than a criterion because parallel and contributory drainage on lower slopes is a pattern common on diamond-shaped fans. In their morphological approach to identifying megafans, Hartley et al. (2010) excluded areas of non-radial drainage patterns; they limited the measured fan dimension at the point where radial patterns give way to contributory patterns, even where long-profile continuity of surface is maintained. In megafan clusters, the nesting pattern within the cluster often involves evolution of elongated diamond-shaped planforms, which can display both radial and contributory drainage patterns. Furthermore, although radial drainage is very common (especially in proximal sectors), relict megafans do not always display prior drainage patterns, and alluvial ridges that characterise some megafans (e.g., the Rio Grande in Bolivia, Ch. 7)

display locally convergent drainage patterns between the ridges.

Concave-up longitudinal profiles are common, but our survey revealed a range, from concave to convexo-concave to convex. Consequently, we included long-profile concavity as an indicator attribute rather than a key criterion. However, continuity of the depositional surface is regarded as a criterion of fluvial sedimentation.

Fan perimeters can be variably interpreted because of truncation by fan-margin and axial rivers or burial beneath dune fields. Areas mapped are therefore conservative (judged to be up to 10–15% lower than the likely original extent), as including only sectors that are demonstrably part of the low-slope zone and thus arguably an indication of fluvial sedimentation.

Topographic roughness, combining measures of smoothness and declivity (see Wilkinson et al. 2010), is a distinct identifier of depositional landscapes (Fig. 3.5) (Wilkinson et al. 2010). The vast

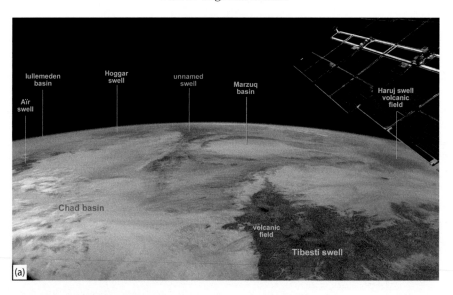

Figure 3.3 **Africa's megafans: regional distribution.**
(a) Astronaut view of the central Sahara, looking WNW from the Tibesti swell (southern Chad). Swells in Africa tend to be dominated by dark rocks, often recent volcanics, in comparison to light-toned intervening basin fills. From Burke and Gunnell (2008). Image ISS008-E-16999 courtesy of the Earth Science and Remote Sensing Unit, NASA Johnson Space Center (http://eol .jsc.nasa.gov). A colour version of this image is available in the SOD for Chapter 3.
(b) Upper Zambezi Basin megafans (western Zambia, easternmost Angola). Erosional upland of the Bié Swell (left) *vs.* aggradational megafan landscapes (centre). Megafans appear prominent in the infrared (purples, reds, blacks). Associated contour map (10 m contour interval) shows megafan landscapes significantly flatter than surrounding landscapes, and conicality of each megafan. Planform shape is controlled partly by proximity to one another of river exit points and partly by competition for space between fans.
(c) Swells and consequent drainage. ETOPO image (southern Angola, northern Namibia) shows intermediate-sized swells and associated simplified 'consequent' drainage patterns of major local rivers, orientation orthogonal to swell margins. Ellipses show zones of megafan accumulation in basins between the swells.
(d) Megafans of the Muglad and neighbouring depressions (South Sudan). This largest group of megafans in Africa is sourced from the Darfur, Nile–Congo, East Africa, Ethiopian, and Alleira swells. The White Nile has built one of Africa's largest megafans, the Sudd (E17: numbers as in Fig. 3.2b). Contour pattern (10 m interval) shows coniform morphology of the megafan cones (numbered as in Fig. 3.2b). Two unnumbered fans are smaller than the arbitrary lower limit (80 km radius) used in our study. Two axial megafans occupy the centre of the Muglad basin (E24, E29). Small arrows show entry points of fan-forming rivers at the apex of each megafan.
(e) Primary and derived megafans. Mountain-fed (fan-building) rivers, Q_{riv1} and Q_{riv2}, build 'primary' megafans at the point where they exit the erosional domain; 'derived' megafans form where basins are wide enough to accommodate another tier of megafans downstream of the primary megafans; derived megafans are built by fan-margin rivers, the discharge of which (Q_{margin}) combines that of mountain-fed and foothills-fed drainages (Q_{inter}).

megafan surfaces of the Andean foreland and Indo-Gangetic plains (Wilkinson et al. 2010) show that a characteristic roughness signature is strongly related to megafan surfaces. Fluvial channel morphologies are often widespread on megafan surfaces. However, these can become so degraded and overprinted on relict surfaces as to be unrecognisable, so that they too are excluded as criteria. The large, relict Cubango megafan in northern Namibia (S7, Fig. 3.2c), 310 km-long, was not recognised because it lacked overt radial channels as a prime visual cue (Wilkinson et al. 2008).

These criteria, in combination, add confidence to the megafan identifications in this study. Despite differences

noted in some criteria, the six criteria used by Hartley et al. (2010) in a global study of 415 fan-like features (radii > 30 km) broadly overlap our multiscale criteria (see Discussion in Ch. 17).

3.3.2 Methods and Hypotheses

Modern surfaces of Africa were examined for the presence of megafans at a resolution high enough to reveal the detail of sediment lobes at scales a few kilometres in length.

For detecting megafan surfaces and mapping their margins and dimensions, several georeferenced sources of data were used. GMTED2010 digital topographic

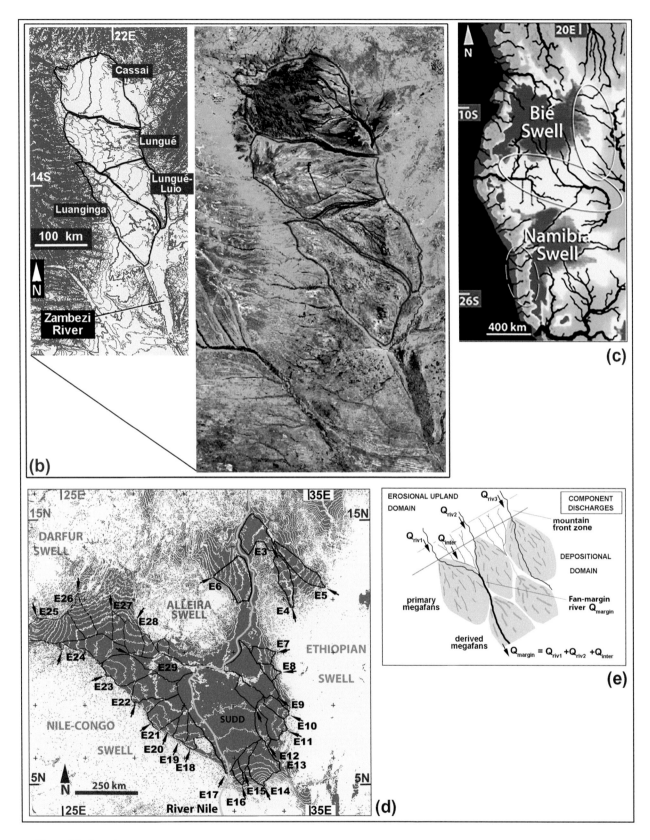

Figure 3.3 (*cont.*)

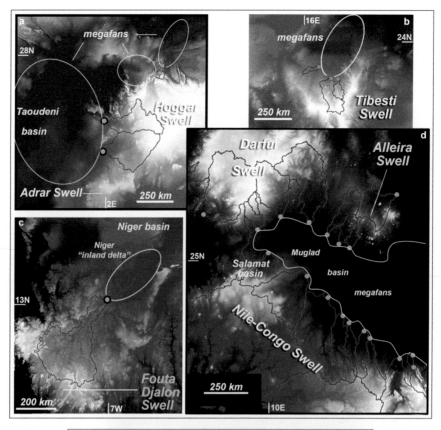

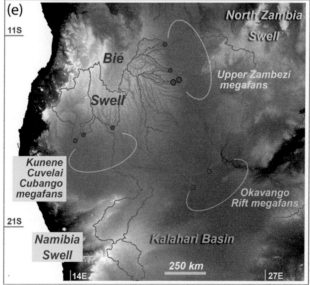

Figure 3.4 **Examples of eight swells whose consequent drainage basins feed megafans**. Africa's rising swell flanks are dominated by the catchments of consequent streams. Swell summits are brightest tones. Catchments of streams that feed megafans, outlined in red, extend up to or near summits of swells. Megafan zones are indicated by ellipses, megafan apexes indicated by blue dots. Examples of eight swells whose consequent drainage basins feed megafans: (a) Hoggar swell, western Sahara. (b) Tibesti swell, central Sahara. (c) Fouta Djalon swell, West Africa. (d) The Nile–Congo, Alleira and Darfur swells, South Sudan, feed the largest group of contiguous megafans on the continent. (e) The Bié and Namibia swells feed the largest concentrations in southern Africa. A colour version of image (e) is available in the SOD for Chapter 3.

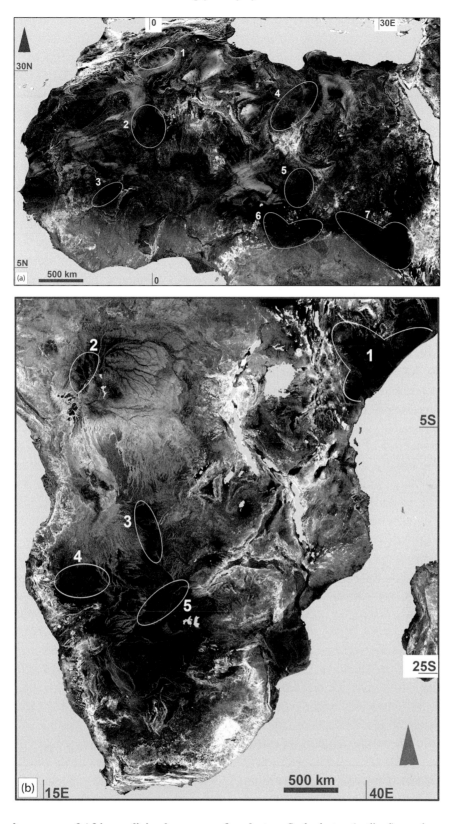

Figure 3.5 **Roughness map of Africa outlining larger megafan clusters**. Such clusters (outlined) are always associated with smooth, low-slope surfaces (dark tones). Rough, steep landscapes in light or very light tones characterise all swells as erosional zones (compare with Figs. 3.1, 3.2, 3.4); darkest areas are aggradational zones and smooth pedimented landscapes. A colour version of this figure is available in the SOD for Chapter 3. (a) Larger megafan concentrations in the northern half of Africa. 1: Atlas Mts; 2: Southern Algeria; 3: Niger 'inland delta', Mali; 4: Eastern Libya; 5: Eastern Chad; 6: Southern Chad; 7: South Sudan. (b) Larger megafan concentrations in the southern half of Africa. 1: Eastern Kenya–southern Somalia; 2: Western Democratic Rep. Congo; 3: Western Zambia; 4: Northern Namibia–southern Angola; 5: Northwest Botswana–northeastern Namibia.

data (https://earthexplorer.usgs.gov) at a 30 arc-second resolution were used to generate maps with 10 metre interpolated contour intervals, from digital elevation models constructed on a MICRODEM freeware platform (https://www.usna.edu/Users/oceano/pguth/web site/microdem/microdem.htm). Topographic profiles, variously aligned, were constructed from the DEMs.

Analysis of stream courses was based on remotely sensed images from the image archive of astronaut-handheld images at the Johnson Space Center (Earth Science and Remote Sensing Unit, 2019). Google Earth imagery and true- and false-colour Landsat 4-5 Thematic Mapper images with a spatial resolution of 30 m in the VNIR wavelengths (https://earthexplorer .usgs.gov) were major data sources. ONC maps (1:100,000 scale, DMAAC, various dates) gave further detailed information on dormant and discontinuous stream beds and the names of some smaller rivers. In complex multi-megafan landscapes such as exist in the southern slopes of the Atlas Mts and eastern flanks of the Chad basin, a few features assigned as megafans were interpreted from remnant surfaces.

In cases where the megafan surface is partly incised, the extant radial arrangement of drainages, combined with other criteria, are usually sufficient to classify a surface as a megafan. Delta surfaces were excluded as being processually at least partly generated by marine controls. Field campaigns in northern Namibia, northern Argentina, and central France provided ground truth for some of the remotely sensed interpretations.

As a first approximation for ascribing a tectonic setting to each megafan, we classified fan-forming rivers according to three tectonic environments, namely the plate boundary zone (PBZ) of northwest Africa, rift depressions, and the vast basin-swell zone that dominates most of the continental surface. Because of the density of swells on the continent, and the swell-related disruption of pre-swell drainage orientations, it is apparent that almost all Africa's rivers are now sourced on uplands classed as swells by Burke and Gunnell (2008) (Fig. 3.1a), with many larger rivers (but not all) occupying the intervening basin depressions. Where swells are dispersed or non-existent as in craton centres, other factors have influenced drainage history that generally do not impinge on issues dealt with here. The rift–depression subset includes overt control by fault scarps (as in the

case of the Okavango Rift megafans), but also rift-related depressions of far greater width dimension.

The 'primary', as opposed to 'derived', status of each megafan (see SOD 3.3.1) was also assigned to describe whether the megafan apex is situated at the source-upland margin (definition of primary status), or whether the megafan lies further downstream. The associated status of fan-forming drainage type, as either consequent or axial, was assigned in order to test the hypothesis that African megafans are primarily related to swell-slope topography (consequent drainage is headwater drainage whose orientation closely follows the azimuth of the swell flank – SOD 3.3.2a; axial drainage follows the axis of an inter-swell depression, gathering tributaries draining surrounding uplands).

Megafans were also grouped by their active or inactive/relict status. Active status was assigned if the formative river could be seen to be flooding some part of the fan (due to their large areal dimensions none are known whose entire surfaces are flooded; wetland vegetation is a good indicator of such flooded lobes), and/or if the river is at the same altitude as the surrounding megafan surface within a few metres (usually < 5 m). Relict status was assigned in those cases in which the formative river fails to be active on any part of the mapped megafan surface, or where the original surface has been largely erased by incision. This status most obviously includes fans into which the formative river has incised more than 10–15 m. Invasion by dunes is included as another criterion of relict status.

Planform shape was assigned only as triangular or diamond/elongated. This distinction was made despite the general assumption that the triangular planform is a norm.

We constructed a roughness map of the study area based on SRTM data. Topographic roughness is a measure of slope variability along baselines of different length (0.6 km, 2.2 km, and 8.7 km) (Wilkinson et al. 2010). The roughness algorithm was first developed to describe roughness characteristics of Mars's surface by Kreslavsky and Head (1999, 2002), and was used for Africa (Figs. 3.4, 3.5) based on a similar array of baselines (SOD 3.3.2b). Lighter shades denote the higher roughness of hilly landscapes such as the Bié Plateau; darkest tones indicate smooth plains landscapes of very low slope; shades of green indicate kilometre-scale roughness dominance, such as dune fields.

Hillshading, a topographic visualisation commonly available in DEM software, provided details that aided the mapping of megafan margins. Published reports in a few cases provided some morphological data. A related geomorphic indication of megafan margins frequently proved to be the easily recognised, often sharp dividing line between exposed rock outcrop and the smooth, aggradational surfaces in remotely sensed images.

3.4 Results

3.4.1 Megafan Distribution and Morphology by Tectonic Setting

The Africa survey yielded a total of 87 megafans (apex and toe positions of individual megafans can be downloaded via the Subsidiary Online Data, SOD, source at www.cambridge.com/megafans). Distribution is highly varied, from an absence of megafans in large areas to concentrations of a dozen or more clustered megafans in other parts. Most megafans occur in small groups of two to five. A first-order division reveals a large, swell-related Group A that all lie distant from plate margins in the swell-and-basin topography of Africa described above (94% of all Africa's megafans) and a small PBZ-related Group B of three megafans in the relatively minor plate-boundary zone on the lower slopes of Algeria's Saharan Atlas Mountains (Table 3.2). Two other megafans (Group C), one located at a regional scarp in southern Mauritania (N9) and one located

within the Messinian–Pliocene Sahabi palaeovalley (N12) described by Griffin (2006), show no overt relationship to local tectonic structures (Table 3.3). Six 'possible megafans' that failed to meet our criteria (Table 3.1) were not included in Tables 3.2 or 3.3. Many indications suggestive of megafan remnants were encountered, especially in Niger's Ténéré inter-swell basin. These await future investigation.

Megafans are categorised first by tectonic setting (swell flank or inter-swell depression) in Table 3.2, and then as single-source or multi-source in Table 3.3; a further subdivision indicates connection or lack of connection to rift zones. These categorisations yield slightly different percentages in each table, but the broad patterns are clear.

3.4.1.1 Swell Source and Geomorphic/ Tectonic Setting

(i) **Swell-Flank and Axial Megafans**. The swell-related Group A can be classified primarily by swell-flank location and inter-swell (axial) location: these are the dominant *swell-flank type* A.a (85%) and a smaller group, the *axial type* A.b (13%, Table 3.2). The swell flank is defined as a wide zone between the convex erosional topography of swell summits and the inter-swell basin floors. Axial-type A.b megafans are sourced by rivers emanating from two or three swells, and are thus usually located distant from swell flanks. The designation is difficult to assign in a few cases, but the frequency is far lower (13%) than that of the swell-flank type (85%).

Table 3.2 *Megafan tectonic setting. Percentages are derived from Table 3.3*

Megafan group		Megafan type		Megafan subtype	
Category	%	Type name	%	Subtype name	%
Group A Swell-related[a] ($n = 82$)	94	A.a. Swell-flank ($n = 70$)	85	A.a.1. Swell and rift ($n = 42$) A.a.2. Swell-flank-only ($n = 27$)	60 39
		A.b. Axial (inter-swell basins fed from more than one swell) ($n \approx 10$)[c]	13		
Group B PBZ-related[b]	3				
Group C Other	2				

Notes: [a] Excludes 'possible' megafans ($n = 6$); [b] Plate Boundary Zone; [c] designation unclear in two cases; [d] rift depression style is almost always steer's head (97% of cases); only 3% of cases involve the rift-only setting (Okavango triplet S8–S10)

Table 3.3 *Number of megafans grouped by upland type (plate boundary zone and swell). Megafan subgroups are those with single- vs. multi-swell source*

Source upland	Swell name	No. of mega-fans	Megafan code no.[a]	Megafans related to consequent streams No.	Axial megafans Code no.	Primary (*vs.* derived) megafans No.	Active (a) and relict (r) status
Group A	**Single-swell source**						
NW Africa	Tibesti	5	N14–18	3	-	4	2a, 3r
	Hoggar	4	N4-7	5	-	5	3a, 1r
	Adrar	1	N8	1	-	1	1r
	Uweinat	1	N13	1	-	1	1r
	Fouta Djalon	1	N10	1	-	1	1a
	Jos	1	N28	1	-	1	1a
	Darfur (West-facing)	5	N19–22, N23	5	-	5	2a, 3r
	Darfur (East-facing)	3	E25–27	3	-	3	3a
East Africa	Ethiopian[b]	16	E1-5, E7–11, E30–32, E35–37	13	E3	13	14a, 2r (+2r?)
	East-African	9	E14–17, E33–34, E38–40	7	E34	8	8a, 1r
	Nile–Congo	7	E18–24	7	-	7	7a
	Alleira	2	E6, E28	2	-	2	2a
Central Africa	Mayombe	4	C2–5	4	-	4	1a, 3r
Southern Africa	Bié	8	S1–7, S9	8	-	8	4a, 4r
	Namibia	2	S13, S14	2	-	2	2a
	Zimbabwe	1	S11	1	-	1	1a
	Single-swell subtotals	70		64		66	51a, 17r (+2r?)
	Multi-swell source						
NW Africa	Tibesti + Haruj	1	N11	-		1	1r?
	Jos + Biu	1	N27	-	N27	1	1a
	Nile–Congo + Adamawa	3	N25–26, C1	-		1	2a, 1r
East Africa	Nile–Congo + Darfur	1	N24	-	E24	1	1a
	Nile–Congo + Darfur + Alleira	1	E29	-	E29	1	1a
	East-African + Ethiopia	2	E12–13	-	E12, E13[c]	-	2a
Southern Africa	Bié + Namibia	1	S10	-		1	1a
	Bié + North Zambia	1	S8	-		?	1a
	South African + Zimbabwe	1	S12	-	S12	1	1r
	Multi-swell subtotals	12		0		7	
Group B	Atlas PBZ	3	N1–3	3		3	3r
Group C	Scarp between Taoudeni and Senegal basins[d]	1	N9	-	?	1	1r
	Palaeolake Chad[e]	1	N12	-	?	-	1r
	TOTALS	87		67	8	77	60a, 26r[f]

Notes: [a] see Fig. 3.2; [b] The Lugh Madera High (horst), source for megafans E35 and E36, is grouped with the Ethiopian swell; [c] Derived megafan; [d] Mauritania; [e] Griffin (2006) – see text; [f] one instance unclear

(ii) **Swell source patterns**. The large swell-related Group A can be subdivided also according to the number of swells from which each megafan is sourced, the large majority (85%) being supplied with sediment from a single swell ($n = 70$, Table 3.3) of the 82 swell-related group (Group A, Table 3.2); the rest are supplied from more than one swell. Table 3.3 also shows the broad association (78%) of the single-swell type with consequent rivers that are directly associated with swell flanks (SOD 3.3.2a) (67 of the 82 in the swell-related Group A, Table 3.2). Fig. 3.4 illustrates the simple consequent drainage pattern associated with swells, namely river basins (red outlines, Fig. 3.4) that rise from the summits of swells and are oriented orthogonal to the swell slope. In contrast, no megafans with multi-swell sources are associated with consequent rivers, but rather are associated with interswell, axial geomorphic settings.

Fully 94% of primary megafans are associated with single swells (66 of the 70 single-swell megafans, Table 3.3) versus those associated with the multi-swell type (58%, or 7 primary megafans of the 12 associated with multiple sources, Table 3.3). Similar proportions of both types are considered active (51 of 70 single-swell, 9 of 12 multi-swell, Table 3.3) rather than relict. Restated, there is a general overlap between megafan bodies deposited by swell-flank rivers and the primary and derived megafan types; these contrast with the small axial type A.b that is strictly related to rivers that flow along the axes of depressions, distant from the swell-flank sources of their formative rivers (e.g., N27, E29).

Megafans are unevenly distributed on the flanks of the twenty host swells: fifteen swells account for 90% of the swell-related Group A megafans, and six swells account for most of this group (65%, Table 3.3): the very extensive Ethiopian and East-African swells host sixteen and nine megafans, respectively, and the mid-sized Bié, Darfur, Nile–Congo and Tibesti swells host ten, eight, seven, and six megafans each. These six also contribute to some 'multi-source' megafans. Twelve swells give rise to three or fewer megafans. This distinctly skewed distribution is discussed in subsequent sections.

3.4.1.2 Rift-Related Megafans

One major reason for the asymmetric distribution lies in the tectonic settings of the two main subtypes,

namely the more complex, hybrid swell-and-rift subtype A.a.1 (60%), and the simpler swell-flank-only subtype A.a.2 (39%) (Table 3.2).

The swell-and-rift subtype A.a.1 encompasses megafans whose location is (i) geographically closely associated with swells as a sediment source, but (ii) which also lie within rift depressions, and (iii) are notably clustered contiguously – e.g., megafans of the Muglad–Melut group, Fig. 3.2b.

A small subset of three megafans of the Okavango Rift (Fig. 3.2c) is also rift-related. But it is unique because these megafans are (i) apexed at topographically exposed fault scarps (termed here the rift-only outlier in Table 3.2), (ii) situated geographically distant from their source swells, and (iii) remarkably, each megafan is supplied from a different combination of source swells (S8–10, Tables 3.2 and 3.3). Axial megafans also lie distant from their source swells in almost all cases but are not associated with exposed rifted fault margins.

Most of the contiguous subtype A.a.1 occurrences lie along component troughs of the West African and Central African Rift Subsystems (*W.A.S., C.A.S.*: Genik 1992) (southern Niger and southern Chad), the Mesozoic Muglad and Melut complex (South Sudan and southeastern Sudan), and the Anza Rift (Kenya) (Bosworth and Morley 1994). The clusters of subtype A.a.1 are: (i) the NNW–SSE-trending Muglad–Melut complex associated with the largest group of 23 megafans; (ii) the Anza Rift complex with 11 megafans; and (iii) the NNW- and WSW-trending family of rifts in southern Chad with 5 very large megafans (Figs. 3.1c and 3.2b). Three small troughs, the Atbara and Blue Nile rift systems, host a scatter of six megafans in northeast Africa (E1–2, E3–6 respectively). Subtype A.a.1 is strongly associated (97%, see Table 3.2) with the geomorphically complex depressions that typify fracture zones (i)–(iii), in which fault margins are not exposed subaerially but are buried by surficial deposits (usually of megafan origin). The unique Okavango Rift, part of the East Africa Rift (EAR), hosts three megafans (Fig. 3.2c).

3.4.2 Morphology: Area, Planform, Roughness

Individual megafan dimensions vary from 2,700 km^2 (S11) and ~80 km long (e.g., S11, E36) at the lower end, to $> 83,000\,km^2$ and $\sim 500\,km$ long—the largest being the Salamat megafan (N23, 505 km long, $87,000\,km^2$ – measured along the midline of this feature that occupies the right-angled Salamat–Bongor–Chad trough system the distance is $> 720\,km$) and the Nile/Sudd megafan (E17, 470 km long, $83,000\,km^2$) (SOD Table 3.1). Population averages are $15,000\,km^2$ and 185 km in length.

Factors affecting planform are apparently complex and involve competition for space in the large, nested clusters. Two controls, however, can be directly observed, namely the distal truncation of megafans by structural walls and/or trunk rivers, and the effects of fan orientation with respect to rift trough structures. Distal truncation leads almost everywhere to the triangular planform—examples are those in the Upper Zambezi basin (S1–4) and the Okavango Rift (S8–10)—which account for 34% of Africa's megafans. The diamond/trapezoid planform is significantly more numerous (58%), and some of these are oriented parallel with rift depression axes (N23–25), but others result from the specifics of interfan competition. More than twice as many megafans in Africa are active (69%) compared to relict (30%) (Table 3.3), the latter being either incised or partly dune-covered.

The darkest tones on the roughness map of Africa indicate smooth surfaces with low slopes over wide areas. Where these zones reach subregional extent, they coincide almost everywhere with the aggradational surfaces of megafans (Fig. 3.5). Other zones of flat, smooth topography are lake floors (Chad, Makgadikgadi, Etosha) and floodplains of large rivers, but these are small by comparison.

3.5 Discussion

3.5.1 Megafan Identification: Number and Size

Remotely sensed data of various kinds, using a set of criteria rigorous enough to provide a secure basis for identifying megafan sediment bodies (Table 3.1), has revealed eighty-seven megafans in Africa. These reach the startling proportions of at least $1.2\,M\,km^2$, equivalent to the areal extent of South Africa. This is a conservative estimate because six possible megafans and more than two dozen other suggestive features were encountered that do not meet all criteria. The average size of megafans is $13,200\,km^2$, varying from small ($2,700\,km^2$, near 80 km long: S11), to the two largest, being the Salamat megafan (N23), which reaches 503 km in length – or $> 720\,km$ as noted above—making this the longest megafan on the planet) and $87,000\,km^2$ in area; the other is the Nile/Sudd megafan (E17), which reaches 470 km and $83,000\,km^2$ in area (SOD Table 3.1).

A prior global study gave a total of thirty-one fan forms $> 80\,km$ long (i.e., megafans in our usage) in Africa (Hartley et al. 2010). We note that these were heavily weighted towards active megafans (twenty-eight), whose wetlands are visually apparent in remotely sensed imagery. Not unexpectedly, our survey revealed the progressive lack of clarity in the imagery of the megafan feature with increasing age of the feature, the loss of active wetlands being a first stage, a trend that ends with relict status in which the surface of the megafan is never fed by discharge of the fan-forming river. Relicts account for a significant portion of the identified megafan population (30%, Table 3.3).

Another reason for the lack of recognition may be simply the unexpectedly large size of fluvial megafans (N. Cameron, pers. comm. to MJW), compared with the well-known small alluvial fan with its voluminous geomorphic and geologic literature (Cooke et al. 1993). Furthermore, several examples could be quoted in which low-angle, extensive megafan surfaces have been interpreted as the more familiar pediment feature – which also boasts a voluminous literature (Cooke et al. 1993) – because of similarities if only in morphology and piedmont setting.

The criteria for identifying megafans were mainly designed to identify the most common megafan attributes, using experience from other continents. This experience suggests that overt proximal conicality is axiomatic. But derived megafans show little or no conicality (Wilkinson, Ch. 17). It is therefore possible that some areas of fluvial deposition of a fan-like, wide planform have not been identified, especially in those cases of secondary / derived megafans.

Another possible cause for undetected megafans on a continent the size of Africa are the insufficiencies of current digital terrain algorithms for identifying megafans (e.g., Weissmann et al. 2015). This derives from the fact that algorithms generally designed under the assumption that rivers lie within valleys as indicated by contour pattern. However, in cases where large rivers build megafans with positive topographic signatures, namely alluvial ridges and low-gradient cones, such features are exactly the opposite of topographic valleys. Authors who map river courses from source to sea thus obtain continuity across megafan surfaces by mapping stream courses incised into the fans (e.g., Paul et al. 2014), and may as a result fail to detect megafans entirely, especially since the areal extent of megafans is generally under-appreciated. Indeed, megafans are typically indicated by the GIS default pattern of a straight line that joins the apex of fans to the axial river where the valley form once again becomes apparent, reasserted in the contour pattern.

3.5.2 Distribution: Swell-Flank-Only and Swell-and-Rift Tectonic Settings

Competing tectonic controls of megafan location appear to be the active swells, inter-swell basins, and rifts. Swell flanks are the dominant depositional environment: Africa's megafans are almost exclusively associated with rivers flowing directly off the slopes of Africa's many swells (94%; Tables 3.2, 3.3, Fig. 3.4). This swell-related Group A is most commonly sourced from simple consequent rivers, the dominant swell-flank type A.a, with the minor axial type A.b occupying basins between the swells and fed by rivers from more than one swell (A.a: 85%; A.b: 13%; Table 3.2).

However, the observed distributions suggest a second factor operates because only 39% of this group is located on swell slopes, a monocausal group designated the 'swell-flank-only' subtype A.a.2. This more numerous and more complex hybrid subtype, designated the 'swell-and-rift' subtype A.a.1, accounts for 60% of the swell-flank A.a type. The swell-and-rift setting accounts for the three largest clusters in which megafans are notably contiguous: these are clusters in the Muglad–Melut, Anza and Salamat–Chad basins, which account for 44% of the swell-related type (41% of the entire African population). The factor explaining such clustering appears to be the existence of rift depressions, leading us to conclude that *a combination of swell flanks immediately adjacent to rift depressions* explains the difference between the dispersed swell-flank-only subtype A.a.2. and groups that comprise

the swell-and-rift subtype A.a.1 (adjacency as an issue in the geomorphic development of megafans is examined in Section 3.5.4.1). Thus, in northeast Africa we find concentrations of contiguous nested megafans in the rift zones, with the distribution of the dispersed swell-flank-only subtype A.a.2 further west in the Sahara and in southern Africa (Fig. 3.2a, c).

That the combination of factors is critical can be seen from the fact that rifts alone are not conducive to megafan development. This is illustrated by the fact that in the arms of the East African Rift system, thousands of km long, not one megafan is found. The small number of 'rift-only' megafans (3% – Table 3.2) is so low as to confirm this observation.

The Africa-wide survey of topographic roughness demonstrates the close correspondence almost everywhere (see SOD 3.5.2 for the Kalahari exception) of the flattest and smoothest surfaces of the continent with clusters of megafans. These low-declivity, smooth surfaces of subregional extent are shown as the darkest tones in Fig. 3.5. Even the smaller areas of megafan surfaces appear, as in the case of the Teghahart and neighbouring Tamanrasset North (N6–7) features.

The distinctiveness of the swell-and-rift setting is further shown in the greater relative thicknesses of Neogene sediments contained within them. Although depression fills in the basin–swell continent of Africa are usually ~300 m thick (Burke and Gunnell 2008; Burke and Wilkinson 2016), Miocene to Recent thicknesses can reach double this in rift troughs (Genik 1992; Guiraud and Bosworth 1997). Salama (1997) suggested that the modern megafan style of sedimentation in the Muglad basin might be an analogue for the style of basin filling.

3.5.3 Examples of the Larger Megafan Clusters – Swell-Flank-Only and Swell-and-Rift Subtypes

3.5.3.1 Northwest Kalahari Basin: Swell-Flank-Only Setting

Ten megafans appear in three separate groups of contiguous megafans, summarised here from Wilkinson et al., Ch. 4. Two swell flank-only clusters (subtype A.a.2) lie on the eastern and southern flanks of the Bié swell: group S1–4 in the upper Zambezi drainage (western Zambia, eastern Angola) displays mainly relict surfaces with small active lobes or wide drainage lines. These megafans are broadly triangular because they abut the axial Zambezi River. In the other group, the very large Cubango (S7) and contiguous Kunene (S5) are relict, their formative rivers having been diverted away from the inter-swell basin (now occupied by the Etosha Pan dry lake in northern Namibia). Both rivers are now also incised: the Okavango River flows in a narrow, 60 m-deep trench-like valley southeastward to Botswana, and the Kunene River flows southwestward in a major valley to the Atlantic Ocean. The resulting inactivity of their respective megafan surfaces has allowed the small, active Cuvelai megafan (S6) to build a sediment wedge that buries the original interfan zone between the Cubango and Kunene megafans.

The three Okavango Rift megafans (S8–10) occupy the long depression formed by the southwesterly extension of the East-African Rift in northwest Botswana and easternmost Namibia. These are the rift-only set (Table 3.2) that are unusual (i) in being apexed where low fault scarps are exposed at the surface, unlike all other rift-related megafans of subtype A.a.1; and (ii) because they lie hundreds of kilometres distant from the swells that are their sediment sources. In addition, (iii) each is fed by a combination of swells different from its contiguous neighbour: the Okavango megafan (S10) is fed from the Bié and Namibia swells; the Cuando megafan (S6) is fed only from the Bié swell; but the smallest megafan, the Zambezi (S5), is fed by the vast upper Zambezi drainage from both the Bié and North Zambia swells (Table 3.3 and Wilkinson et al., Ch. 4).

As the Okavango Rift megafans are formed by rivers that cross the rift orthogonally, the fans are relatively short and triangular. The Okavango River is unusual in having built not one but two megafans along its length, the Cubango (S7) and the Okavango 'inland delta' downstream (S10), their apexes lying 620 km apart – a configuration of great interest in understanding the history of sedimentation of the Etosha inter-swell basin (Wilkinson et al., Ch. 4). Rivers that enter the depression longitudinally have deposited the Grootlaagte Flats to the southwest and the Machili Flats to the northeast of the three Okavango Rift megafans.

3.5.3.2 Swell-and-Rift Setting

Despite the similarities between the three tectonic settings of the abovementioned extensive A.a.1 landscapes (Muglad–Melut, Anza and Salamat–Chad basins, Section 3.5.2), we compare the geomorphic variety that has evolved between them. All three depressions discussed contain Neogene sediment stacks (Genik 1992; Guiraud and Bosworth 1997) and appear to be active today.

(i) **South Sudan**. Twenty-three megafans (N7–29) comprise the largest cluster of contiguous megafans in Africa (273,000 km^2). This cluster stretches southeast 1,250 km from the Darfur swell, occupies the long inter-swell depression between the feeder Nile–Congo and Alleira swells, and extends to the north end of the East-African swell. The cluster then angles north for 700 km along the west flank of the Ethiopian swell (Fig. 3.2b). This configuration broadly follows the axes of the connected Muglad and Melut rifts (Fig. 3.1b). In all, five swells supply sediment to these interlocking depressions.

The Muglad basin (Fig. 3.1b) hosts 13 megafans, all but two of which are aligned transverse to the basin margins, resulting in relatively short and approximately triangular planforms. Where the Muglad and Melut troughs merge, the topographic basin is wide enough to accommodate one of the largest megafans, the Nile/Sudd (470 km) (E17), which is truncated where it abuts the Alleira swell and therefore displays an approximately triangular planform. The East-African swell feeds this megafan and three smaller neighbours. Five short, triangular megafans in the Melut trough are fed by rivers draining off the western flanks of the Ethiopian swell. The Nile/Sudd megafan (E17) is anomalously large (83,000 km^2), being fed from the extensive equatorial Nile basin. Excluding this megafan, the average area of the Muglad–Melut megafans is ~8,500 km^2.

(ii) **Southern Chad: Salamat–Dosseo–Bongor rifts**. In contrast to South Sudan, four contiguous megafans (N23–26) are oriented longitudinally, *parallel* with the axes of the major host troughs (Fig. 3.1b). Drainage today is dominantly towards the southern margin of modern Lake Chad. This concentration of megafans, with Lake Chad itself, occupy the long and complex depression defined by the connected C.A.S. and W.A.S. rift troughs. These are supplied by swell-related rivers flowing off the Darfur, Nile–Congo, Cameroon Line and Adamawa swells. Importantly, the longitudinal orientation of these large fans allows single fans to develop large surface areas, resulting in the development of a small number of very large megafans (N23–26) covering an area of ~173,000 km^2, the average being 46,000 km^2.

(iii) **Kenya: Anza Rift**. Africa's third largest cluster is a string of nine megafans (E30–35, E38–40) that occupies the long (>450 km) Anza Rift depression. This depression widens from ~120 km in Kenya's north to ~320 km in the south (Foster and Gleadow 1996). Sediment is derived from rivers flowing off the East-African and Ethiopian swells. Unlike the orthogonal drainage orientations in the Muglad zone, those in the Anza are a mix of transverse orientation along the rift margins, trending to longitudinal orientations towards the centre of the rift zone (especially larger megafans E30–31, E38), where they are correspondingly relatively long, with narrow, markedly curved axes. Most megafans here are still active, providing another parallel with megafans in the Muglad–Melut and Chad basins, but a long history is also evident from the existence of relict lobes on several megafans. The total area of the contiguous megafans of this second largest cluster in Africa is one fifth of the area of the South Sudan cluster (>57,000 km^2). This area is a minimum because margins of palaeo-megafans and imprecise coastal terminations are difficult to ascertain from remotely sensed data. The existence of a major rift zone seems critical in explaining the existence of the cluster. Average fan size in the Anza Rift is 7,500 km^2.

3.5.4 *Zones Devoid of Megafans*

3.5.4.1 *Swell Flanks and Long-Profile Hypotheses*

The existence of a swell does not imply the existence of a megafan: distribution maps (Fig. 3.2a–c) show that megafans are in fact rare along swell margins, and entirely absent on 10 of Africa's 30 swells. To explain such asymmetric distribution, especially in settings that appear so similar to those where megafans have been identified, we suggest that the following subtle geometries in river long profiles probably operate to preclude development (for general statements, see Wilkinson, Ch. 17).

(i) River long-profile altitude. Most swell flanks are incisional domains throughout their length, but some locations provide accommodation zones, depending on whether the long-term stream profile lies below or above the surrounding swell slopes – with megafans forming where the higher-lying profile operates (double arrows for base levels, Fig. 3.6). Small differences in altitude between these profiles – as little as a few tens of metres in Africa's relatively

flat landscapes – seem to determine the very different incisional versus megafan-forming geomorphic responses.

(ii) River slope. Because river discharge in the longer term is usually inversely related to slope (SOD 3.5.4), the lower gradients of larger rivers increase the likelihood of fan development. In contrast, the many smaller swells give rise to shorter, steeper rivers which are less likely to produce fans, or less likely to produce fans reaching megafan dimensions (compare Fig. 3.6a and b).

(iii) Adjacent location of swell and depression. The topographic step between a source upland and a megafan-hosting lowland is more pronounced in locations where a rift-related depression lies adjacent to the swell. Such depressions specifically raise the probability of megafan development (apexed where the river profile crosses the topographic profile; single arrow, Fig. 3.6c). Furthermore, the region-wide accommodation next to rift depressions results in clustering (laterally along the rift margin) instead of localised megafan development. However, positive and negative topographic 'adjacency' of this kind is not common in Africa, which appears to explain why megafans have not formed on many swell flanks.

These factors all provide hypotheses for future testing in specific locales.

3.5.4.2 *Smaller Swells and Incisional Regimes*

The continent-wide perspective indicates that very widespread incisional geomorphic regimes are associated with the lack of megafans in Africa because most swell flanks fail to provide the necessary combination of long profile and swell-flank slope. The reason relates to the fact that such regimes are a norm in tectonic environments of up-arched swells (Fig. 3.6a). Although most megafan-forming rivers correspond to the consequent type, the reverse is not true: despite the existence of numerous consequent rivers etching the slopes of swells, megafans are relatively rare along the flanks of swells.

River discharge probably plays a role as well (Section 3.5.4.1 (ii)). Although this hypothesis remains to be tested, we suggest that none of the smaller swells – Tripoli, Sawda, Aïr, Adrar, Haruj, or Red Sea Hills in the Saharan zone, or even the well-watered West-African swells (Guinea, South-West Nigeria, Cameroon Line) (Fig. 3.2a) – give rise to rivers of sufficient discharge to achieve the low-slope profiles

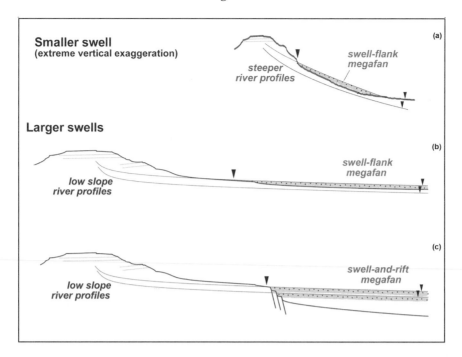

Figure 3.6 **Swell and river-profile geometry applied to megafan location**. The most common tectonic habitat for Africa's megafan is the swell flank. Smaller and larger swell flanks are diagrammed, showing in each case pairs of alternate long-term river profiles – related to base levels at slightly different altitudes, usually axial/trunk rivers (double arrows). Larger swells are associated with larger rivers and correspondingly lower channel slopes, a geometry that increases the probability of megafan development. Stipple indicates megafan sediment bodies. Single arrows indicate megafan apex locations.
(a) Smaller swells are associated with shorter, and therefore steeper fan-forming rivers due to smaller catchments; these rivers host shorter megafans with reduced horizontal reach. Upper profile lying above the swell flank leads to megafan aggradation; lower profile incises swell flank, preventing megafan formation. A small vertical change in local base levels (double arrows) determines whether a megafan forms or the river incises (preventing megafan formation, or incising an existing megafan).
(b) Larger swells associated with larger catchments give rise to lower-gradient rivers, correspondingly greater horizontal reach and hence megafans extending further into basins. River courses aligned along lower profiles incise the swell flank, preventing fan formation. Here too, a small vertical change in local base levels (double arrows) determines whether a megafan forms or the river incises (preventing megafan formation, or incising an existing megafan).
(c) The probability of the fan-forming river intersecting a depression and forming a megafan rises (i) on larger swells, and (ii) with adjacency to rift troughs. Compared to (a) and (b), river profiles aligned to lower base levels (lower parabola) also intersect basin depressions.

that are more conducive to generating large fans (compare Fig. 3.6a, b). This geometry apparently arises less often, as concluded earlier, and as evidenced by the fact that most swell flanks lack megafans of the 'swell-flank-only' subtype A.a.2 – which comprise only 39% of the swell-flank type.

3.5.4.3 Largest Swells

The perimeters of two of the three largest and highest swells, the East-African and the South-African, are almost devoid of megafans, despite a combined flank length of more than 9,500 km (we emphasise that smaller fans, less than 80 km long, are not mapped). This lack may result from the degree of uplift associated with their relatively large size and height, to produce what is undoubted region-wide incision. We note that no Cenozoic sediments are mapped anywhere along the inland-facing flanks (Petters 1991), and roughness signatures show that the Congo basin is mainly an eroded surface (Fig. 3.5b). Long profiles are everywhere incisional and demonstrably situated below the altitudinal threshold for megafan sedimentation (lower river profile, Fig. 3.6b).

This argument (swell size and megafan formation) appears to be supported by the existence of five megafans in the northwest quadrant of the Congo basin opposite the long East-African swell flank, where Neogene sedimentation is widely mapped (Kadima et al. 2011). These 'swell-flank-only' A.a.2 megafans (C1–5) lie on rivers draining two source swells, the Mayombe and Nile–Congo, i.e., swells that display intermediate areas and altitudes compared with the very large East-African and South-African swells. The flanks of these intermediate-sized swells apparently provide the narrow window in which megafan deposition, without rift depression assistance, occurs (upper profile, Fig. 3.6b).

A parallel can be seen in southern Africa. On the inland-facing slopes of the large South-African swell, the plateau drainage basins of the Limpopo and Orange rivers are devoid of megafans. However, the mid-size, mid-altitude swells nearby – the Nambia, Bié and North Zambia – account for seven megafans and four possible megafans.

In continental terms, the smaller number of megafans at the southern end of the continent, from the Congo basin southwards, seems to reflect the presence of the very large South African swell.

3.5.4.4 Coastal Lowlands

Africa's coastal zones, with sea-level base levels significantly lower than inland basin base levels, are devoid of megafans due to the presence of region-wide incisional regimes. The one exception is the Kenyan coast in the entire length of coast-facing lowland Africa, where a cluster of megafans has been described above. This is notably the sector in which the major Anza Rift zone intersects the coast (compare Fig. 3.2 panel b with panels a and c). Minor exceptions are found along the central coastline of Somalia, where the Shebelle River has laid down a megafan parallel to the coastline, so that the long runout of the river (its horizontal accommodation) has allowed deposition of a megafan body. Similar horizontal accommodation occurs in southern Mozambique, where the major Limpopo River, draining both the South African and Zimbabwe swells, has laid down the now relict Limpopo megafan (> 200 km long) in the setting of a wide continental shelf. This megafan ends 70 km from the coastline so that it is not a delta. The third zone concerns two possible megafans in central coastal Namibia, where several 'alluvial fans' have been interpreted to exist (Besler 1984), two of which are highly suggestive of megafans. We hypothesise that regional coastal incision was counteracted here by sufficiently large rivers to bring fluvial profiles into the megafan-forming 'window' (Fig. 3.6b).

3.5.4.5 Cratonic Blocks

Large cratonic blocks, especially those of the major Congo and Kalahari basins and of West Africa (stretching from the west-African coastline of Ghana–Sierra Leone north to Morocco), host only two swells interior to the vast area spanned by these cratons. These are the coastal Fouta Djalon and Guinea swells (Fig. 3.1a). The lack of swells corresponds to a lack of megafans on the cratons. The megafans that lie on these cratons are specifically peripheral and derive from swells situated along the craton margins. Thus, the vast area of the West African craton displays only five megafans (N6–8, N9, 10) four of which are sourced in the peripheral Hoggar and Fouta Djalon swells. The Congo craton displays no swells, and its five megafans (C1–5) are situated along the western margin, on the opposite side from the East African swell. All are fed from peripheral swells, four from the peripheral Mayombe swell and one from the Nile–Congo swell (Section 3.5.4.3). The Kalahari craton displays nine marginal megafans, all peripheral, fed from the Bié and Namibia swells (S1–7, S13–14).

3.5.4.6 Major Rivers

The fact that megafans are mainly associated with lower-order consequent rivers, and a few with larger axial rivers, raises the question of why the largest rivers of Africa show very few megafans. Along the many thousands of kilometres of the Nile, Niger, Senegal, Ubangi, Congo, Zambezi, Limpopo, and Orange rivers, we map only five megafans, the Zambezi (S8), Limpopo (S12), Nile/Sudd (N17), the 'inland delta' of the Niger (N10), and the Ubangi (C1). The Senegal, Congo, and Orange rivers are devoid of megafans.

As a general statement, incision is inimical to megafan development, and the fact that major rivers everywhere lie within valley-confined zones reduces the scope for megafan development, even where the major river is actively aggrading. The hypsometry of Africa offers an explanation that rehearses prior arguments. As the continent with the highest average altitude (Summerfield 1996), most rivers tend to be incised over a greater part of their courses. This includes the largest rivers, even where they approach the coasts.

Not all, but most of Africa's major rivers are also characterised by other incision-related patterns such as their much-remarked waterfalls (where they cross plateau margins, but inland also), and the lack of deltas. But even on lower-lying continents prone to lesser regional incision, major rivers mostly occupy valleys, albeit shallow valleys, and display no more than one or two small megafans, or none at all.

3.5.5 Morphology

Sizes vary from 2,700 km^2 (Chefu, S11) and 80 km long (Cuvelai, S6) at the low end, to 88,000/83,000 km^2 and 480/470 km long (Salamat, N23, Sudd N17 respectively) (SOD Table 3.1) at the high end. Average areas and longer radii are 14,700 km^2 and ~200 km.

Although it is beyond the scope of this study to examine megafan areas in relation to their feeder basin areas for the entire African continent, it becomes apparent that the relationship is poor; prime examples are the largest rivers such as the Nile and Zambezi (E17, S8) that have produced both the largest and smallest megafans, versus the Cuvelai with a small source basin that has nevertheless developed a megafan 80 km long. Unlike the radii of small alluvial fans whose short lengths are often unrestricted by the host basin (allowing meaningful relationships between fan area and feeder basin to be calculated), megafan length is truncated by distal barriers (usually structural) or the trunk river of the basin in 66% of cases, producing the well-known triangular planform. The remaining 34% of megafans are diamond shaped, a planform that is usually accompanied by parallel or (con)tributary drainage patterns distally. This is a significant finding considering that past methodologies (e.g., Hartley et al. 2010) have used radial drainage only as the criterion denoting the distal extent of the measured megafan.

Competition for space by contiguous megafans is a primary factor in determining planform and area. Factors affecting such competition appear to be discharge magnitude by the formative rivers, different fan slopes and exit altitudes from the upland, and fan orientation with respect to regional structural 'grain' – where drainage orthogonal to basin structure is more common and produces triangular planforms) (e.g., Melut group, E7–11). Longitudinal orientation is less

common but allows for extensive development. The southern Chad troughs are the prime example, where the mean area is five to six times larger than means for the other two major clusters.

The total area of megafans we identified in Africa is 1.2 million km^2. As such, megafans constitute 20 times the area of Africa's five major deltas, the Niger, Zambezi, Nile, Senegal and Tana combined (57,600 km^2). Indeed, Africa's megafans alone are broadly equivalent to the total area of 23 deltas of some of the largest rivers on the planet (SOD Table 3.5.6).

3.5.6 Megafans and Climate

Africa's megafans may well date back to the Pliocene or earlier and must therefore hold records of past climates. Varying climates have undoubtedly influenced river discharges repeatedly, especially for example in the Sahara Desert since establishment of desertic regimes ca. 2.7 Ma (deMenocal 1995; Leroux 1996; Rohling et al. 2014). The Sahara has also experienced wet episodes of varying intensity on a repeated basis during the more than twenty glacial–interglacial oscillations of the Pleistocene (deMenocal 1995). The fan-forming rivers have thus undoubtedly experienced repeated changes in discharge that would have affected the evolution of the megafans, especially their active and relict status.

3.6 Conclusions

All of Africa was examined for megafans > 80 km long, justifications being (i) to map the largest areas of fluvial sediment that are not floodplains of major rivers or the few deltas of Africa, and (ii) to limit the size of the study (there being numerous short-radius alluvial fans on the continent). We emphasise also that smaller fans, less than 80 km long, were not mapped, so that the entire area covered by fluvial sediment bodies is not rendered. Criteria by which megafan features were identified reflect our primary purpose of defining megafans as larger surficial lithological entities, rather than as morphological entities (e.g., Hartley et al. 2010; Weissmann et al. 2011).

On this basis, 87 megafans were mapped, amounting to > 1.2 Mkm^2 of Africa's land area and an order of magnitude more extensive than the total area of Africa's major subaerial deltas. This is a conservative estimate because minimum megafan areas were measured, and because six possible megafans and more than two dozen other features suggestive of megafans were encountered that do not meet all our criteria for designation as megafans. The unexpected scale of the megafan areas of Africa is an important conclusion of the study, suggesting that megafans ought to be considered as a normal component of continental landscapes, which in Africa have traditionally been thought of as characteristically erosional.

Almost all megafans – 82 of the 87 identified – are associated with relatively young tectonic swells. These are topographic edifices initiated by shallow-mantle convection beginning ~32 Ma and generating Afro-Arabia's globally unique, continent-wide topography dominated by basins and swells. This clear association is revealed by the fact that 94% of Africa's megafans are generated by rivers flowing off swells (the swell-related Group A). Almost all of these (85% of the swell-related Group) lie directly on the flanks of swell edifices (the swell-flank type A.a) formed by consequent rivers that follow the primary slope of these relatively young topographic features. The remainder lie in inter-swell depressions fed by rivers from more than one swell (the axial type A.b). This association demonstrates not only the critical importance of swell growth for the development of megafans, but it also suggests a date for swell-related megafan development no earlier than the initiation of the swells, ~30 Ma. This conclusion does not suggest that pre-swell landscapes in Africa were devoid of megafans, as evidenced by the tectonic settings of other continents where megafans are plentiful.

Although swell flanks demonstrably provide settings for megafan development, it is also apparent that swell flanks are an insufficient condition for the development of megafan clusters. The significant additional factor appears to be the existence of rift-related depressions immediately adjacent to swells. Such depressions increase the potential for megafan formation because they provide more ample accommodation space (Fig. 3.6c). The prime example is the region where three swells provide sediment for 21 contiguous megafans in the multi-rift Muglad–Melut basin of South Sudan. Clusters in the Anza and Chad basins illustrate the same dual factors of sediment supply and enhanced accommodation (Section 3.5.3.2).

We thus conclude that rift depressions are a critical factor for explaining the number, concentration, and location of Africa's largest single most numerous swell-and-rift subtype A.a.1 (61% of the swell-flank type). Factors determining this subtype appear to explain the existence of the only zone where megafans are numerous near an African coastline, namely the Anza Rift that extends to coastal Kenya.

Twenty of Africa's thirty swells are associated with the swell-related Group A. Of these, 15 are the sole source uplands of one or more megafans, accounting for 90% of swell-related megafans. Such swells thus reinforce the conclusion that swells are the most significant tectonic setting for the development of megafans in Africa.

Equally however, several topographic zones are devoid of megafans: these are (i) most swell-flank zones, especially margins of the largest swells, a setting that includes those ten swells that are entirely barren of megafans; (ii) the coastal lowland fringe of Africa; and (iii) the major cratonic blocks of Africa (Congo, Kalahari, West Africa). In these cases, mid-sized swells peripheral to the cratons explain the presence of isolated megafans located on the margins of these cratons. We hypothesise that altitudinal geometries of river long profiles (of the order that determine incisional versus aggradational stream behaviour) are undoubted explanations in the cases of the largest swells, and in Africa's coastal zones where incision is clearly a regional fluvial regime and megafans are absent. But such geometries are subtle on most swell flanks such that megafan occurrence appears almost random, with larger sectors devoid of megafans.

The group of three megafans within the Okavango Rift in southern Africa is unusual. The location of the apex of these is easily recognised as fixed at exposed fault scarps (see Wilkinson et al., Ch. 4). The setting is so straightforward and well-known from the intensive attention given the Okavango megafan in particular (e.g., McCarthy 2013) that it tends to be assumed that Okavango-style megafans are the norm. However, in the rest of Africa rift depression control is manifested in a more complex fashion, involving broad, rift-based, steer's head basins in which marginal faults are buried because the basins are overfilled.

Salama (1997) speculated that the thick continental fills of the Muglad troughs may be buried megafans, reflecting the modern megafan style of aggradation. The Termit troughs of southeastern Niger may well display the same palaeogeographies and depositional environments: these troughs contain one of the thickest Miocene to Recent continental sedimentary sequences (Genik 1992), which today occupy the interswell Ténéré depression flanked by the Adrar and Aïr swells, where many indications of megafans were encountered. Application in the subsurface of these findings by no means ought to be associated only with swells. The findings rather suggest that megafans are a depositional norm in any palaeogeographic setting where a topographic high – not necessarily a swell – was located adjacent to a rift depression.

Ultimately, it remains interesting that the significance of the megafan feature has remained unrecognised for so long. Africa's megafan population approaches one third of the global population of ~272 as identified by Wilkinson and Currit (Ch. 2). This result alone illustrates the uniqueness of Africa as a setting for megafan development, where foreland basins such as those of the Himalaya and Chaco Plains east of the Andes are far better known and thought to be the type regions for megafan development. Although the viability of Africa's swell environment for megafan formation appears to be comparatively unusual, it is nevertheless conducive, as this chapter has shown. In this context, the relict Loire megafan has also shown, however, that megafan features tend to become obscured by various processes over time (Gunnell, Ch. 9). Active fans stay mostly recognisable, usually from the presence of dark-toned wetland vegetation, but erosional destruction is dramatic in the almost unrecognisable remnants of megafans on the southern flanks of the Tibesti massif. These show numerous short, disconnected reaches of channel systems (e.g., N18, Fig. 3.2a) that are today largely destroyed by aeolian abrasion. The group of 'possible' megafans illustrates the trend because they no longer display all the criteria for secure identification.

These conclusions also suggest that more megafans will be identified as further detailed attention is given not only to other continents, but also to Africa (especially to the Ténéré and Iullemeden basins of Niger). Additional perspectives are offered in Chapters 17 and 18.

Acknowledgements

MJW expresses his sincerest appreciation to his late co-author Kevin Burke, for setting Africa in its modern tectonic context for this geomorphologist. Yanni Gunnell and two anonymous reviewers are thanked for insightful critiques.

References

Besler, H. (1984). The development of the Namib dune field according to sedimentological and geomorphological evidence. In J. C. Vogel, ed., *Late Cainozoic Palaeoclimates of the Southern Hemisphere*. Balkema, Rotterdam, 445–454.

Bosworth, W. and Morley, C. K. (1994). Structural and stratigraphic evolution of the Anza rift, Kenya. *Tectonophysics*, 236, 93–115.

Burke, K. (1976). The Chad basin: an active intracontinental basin. *Tectonophysics*, 36, 197–206.

Burke, K. (1996). The African plate. *South African Journal of Geology*, 99, 341–409.

Burke, K. and Gunnell, Y. (2008). *The African Erosion Surface: A Continental-scale Synthesis of Geomorphology, Tectonics, and Environmental Change over the Past 180 Million Years*. Geological Society of America Memoir, 201, 66 pp.

Burke, K. and Dewey, J. F. (1974). Two plates in Africa during the Cretaceous? *Nature*, 249, 313–316.

Burke, K. and Wilkinson, M. J. (2016). Landscape evolution in Africa during the Cenozoic and Quaternary–the legacy and limitations of Lester C. King. *Canadian Journal of Earth Sciences*, 53, 1089–1102.

Burke, K. and Wilson, J. T. (1972). Is the African plate stationary? *Nature*, 239, 387–390.

Cooke, R., Warren, A., and Goudie, A. (1993). *Desert Geomorphology*. University College London Press, London, 2nd edn, 526 pp.

DMAAC (Defense Mapping Agency Aerospace Center), various dates. Operational Navigation Charts (ONC), 1:1 million scale, St. Louis, Missouri, USA.

DeMenocal, P. B. (1995). Plio-Pleistocene African climate. *Science*, 270, 53–59.

Du Toit, A. (1933). Crustal movement as a factor in the geographical evolution of South Africa. *South African Geographical Journal*, 16, 3–20.

Earth Science and Remote Sensing Unit, NASA Johnson Space Center, http://eol.jsc.nasa.gov. <Accessed February 2022>

England, P. and Houseman, G. (1984). On the geodynamic setting of kimberlite genesis. *Earth and Planetary Science Letters*, 167, 89–104.

Foster, D. A. and Gleadow, A. J. W. (1996). Structural framework and denudation history of the flanks of the Kenya and Anza rifts, East Africa. *Tectonics*, 15, 258–271.

Genik, G. J. (1992). Regional framework, structural and petroleum aspects of rift basins in Niger, Chad and the Central African Republic (C.A.R.). *Tectonophysics*, 213, 169–185.

Guiraud, R. and Bosworth, W. (1997). Senonian basin inversion and rejuvenation of rifting in Africa and Arabia: synthesis and implications to plate-scale tectonics. *Tectonophysics*, 282, 39–82.

Hartley, A. J., Weissmann, G. S., Nichols, G. J., and Warwick, G. L. (2010). Large distributive fluvial systems: characteristics, distribution, and controls on development. *Journal of Sedimentary Research*, 80, 167–183.

Hofmann, C., Courtillot, V., Féraud, G., et al. (1997). Timing of the Ethiopian flood basalt event and implications for plume birth and global change. *Nature*, 389, 838–841.

Kadima, E., Delvaux, D., Sebagenzi, S. N., Tack, L., and Kabeya, S. M. (2011). Structure and geological history of the Congo Basin: an integrated interpretation of gravity, magnetic and reflection seismic data. *Basin Research*, 23, 499–527.

King, L. C. (1942). *South African Scenery*. Oliver and Boyd, London and Edinburgh, 340 pp.

King, L. C. (1967). *The Morphology of the Earth*. Oliver and Boyd, London and Edinburgh, 2nd edn, 726 pp.

Krenkel, E. (1922). *Die Bruchzonen Ostafrikas: Tektonik, Vulkanismus*. Erdheben und Schwereanomalien. Gebrüder Borntraeger Verlag, Berlin, 184 pp.

Kreslavsky, M. A. and Head, J. W. (1999). Kilometer-scale slopes on Mars and their correlations with geologic units: initial results from Mars Orbiter Laser Altimeter (MOLA) data. *Journal of Geophysical Research*, 104, 21,911–21,924.

Kreslavsky, M. A. and Head, J. W. (2002). Kilometer-scale roughness of Mars' surface: Results from MOLA data analysis. *Journal of Geophysical Research*, 105, 26,695–26,712.

Leroux, M. (1996). *La Dynamique du Temps et du Climat*. Masson, Paris, 310 pp.

McCarthy, T. S. (2013). The Okavango delta and its place in the geomorphological evolution of Southern Africa. *South African Journal of Geology*, 116: 1–54. Thirtieth Alex L. du Toit Memorial Lecture. doi:10.2113/gssajg.116.1.1

McKenzie, D. and Weiss, N. (1975). Speculations on the thermal and tectonic history of the earth. *Geophysical Journal of the Royal Astronomical Society*, 42, 131–174.

Miall, A. D. (1996). *The Geology of Fluvial Deposits*. Springer, New York, 582 pp.

Miall, A. D. (2014). *Fluvial Depositional Systems*. Springer, New York, 315 pp.

Paul, J. D., Roberts, G. G., and White, N. (2014). The African landscape through space and time. *Tectonics*, 33, 898–935.

Petters, S. W. (1991). *Regional Geology of Africa*. Springer-Verlag, Berlin, 722 pp.

Rohling, E. J., Foster, G. L., Grant, K. M., et al. (2014). Sea-level and deep-sea-temperature variability over the past 5.3 million years. *Nature*, 508, 477–482.

Salama, R. B. (1997). Rift basins of the Sudan. In R. C. Selley, ed., *African Basins, Sedimentary Basins of the World*. Elsevier, Amsterdam, 105–149.

Summerfield, M. A. (1996). Tectonics, geology and long-term landscape development. In W. M. Adams, A.S. Goudie, and A.R. Orme, eds., *The Physical Geography of Africa*. Oxford University Press, Oxford, 1–17.

Weissmann, G. S., Hartley, A. J., Nichols, G. J., et al. (2011). Alluvial facies distributions in continental sedimentary basins – distributive fluvial systems. In S. K. Davidson, S. Leleu, and C. P. North, eds., *From River to Rock Record: The Preservation of Fluvial Sediments and their Subsequent Interpretation*. SEPM Special Publication, 97, 327–355.

Weissmann, G. S., Hartley, A. J., Scuderi, L. A., et al. (2015). Fluvial geomorphic elements in modern sedimentary basins and their potential preservation in the rock record: a review. *Geomorphology*, 250, 187–219.

Wilkinson, M. J., Marshall, L. G., and Lundberg, J. G. (2006). River behavior on megafans and potential influences on diversification and distribution of aquatic organisms. *Journal of South American Earth Sciences*, 21, 151–172.

Wilkinson, M. J., Kreslavsky, M. H., and Miller, R. McG. (2008). Megafans of the Northern Kalahari Basin. *Third Southern Deserts Conference–Kalahari 2008* (University of Oxford, School of Geography), Molopo Lodge, Northern Cape, South Africa, 16–19 Sept 2008.

Wilkinson, M. J., Marshall, L. G., Lundberg, J. G., and Kreslavsky, M. H. (2010). Megafan environments in northern South America and their impact on Amazon Neogene aquatic ecosystems. In C. Hoorn and F. P. Wesselingh, eds., *Amazonia, Landscape and Species Evolution: A Look into the Past*. Blackwell, London, 162–184.

4

Megafans of the Northern Kalahari Basin
(Angola, Botswana, Namibia, Zambia)

M. JUSTIN WILKINSON, ROY MCG. MILLER, FRANK ECKARDT,
and MIKHAIL A. KRESLAVSKY

Abstract

Maps generated from various data sources reveal ten new megafans in the northern Kalahari region where, until now, the Okavango had been the only one recognised. Seven megafans were generated by rivers flowing off the Bié Swell of southern Angola, east to the Zambezi basin and south to the Owambo basin. Only three (Okavango, Cuando, Zambezi) are apexed at shoulders of the Okavango Rift (northern Botswana). Unusually, the Cubango/Okavango River has given rise to two megafans: the upstream Cubango megafan, and the well-known Okavango megafan downstream. Avulsion behaviour of three rivers has also demonstrably shifted discharge between major basins over time: the Cassai has, at times, flowed north into the Congo basin; the Cubango flowed into the Owambo basin (Etosha dry lake), but now discharges into the Makgadikgadi basin (via the Okavango megafan); and the Kunene, which now flows to the Atlantic Ocean, at one time discharged into the Etosha pan. Recognising the existence of so many more megafans than previously appreciated, as well as their autogenic, avulsive dynamics, is an invitation to reconsider the regime of sedimentary sequence deposition in these basins, which may have erroneously been interpreted as resulting from climatic or other external forcing factors.

4.1 Introduction

4.1.1 Study Area

We identify ten megafans, either active or inactive, each comprising large wedges of fluvial sediment (5×10^3–$10^4\,\mathrm{km}^2$), in the northern Kalahari Basin (SOD 4.1.1). The megafans cluster in three sub-basins that form an arc extending southwest to northeast from the Owambo Basin (OWB) of northern Namibia, through the Okavango Rift Basin (ORB) of northern Botswana, to the Upper Zambezi Basin (UZB) of western Zambia and easternmost Angola (Fig. 4.1). The large number of megafans contradicts a commonly expressed view, from the time of du Toit (1927), that the Okavango swampland is 'unusual in terminating in a major inland delta' (Wellington 1955; Moore and Larkin 2001:59; McCarthy et al. 2002). Fully 87 megafans are identified in Africa alone (the highest number for a continent, on a par with SouthAmerica) (Wilkinson and Currit, Ch. 2). Other landscapes of contiguous megafans, subregional in scale, are known from several parts of the world, most famously in the Indo-Gangetic Plains of northern India (Geddes 1960).

In the course of building megafan cones, the fan-forming rivers necessarily avulse to every radial position on the fan cone. This basic phenomenon is key to understanding the dominant Kalahari hydrological systems because a major finding is the existence of a relatively large number of *divide megafans*. Rivers on such megafans are critically situated such that they contribute the discharge of a large hinterland basin alternately to different major basins. The dominantly autogenic controls of river orientations on these fans, and hence sediment routing, must be taken into account in understanding hydrological dynamics and interpreting sedimentary successions in the region.

Various spellings of features in this chapter are used in the literature and on maps of the English-,

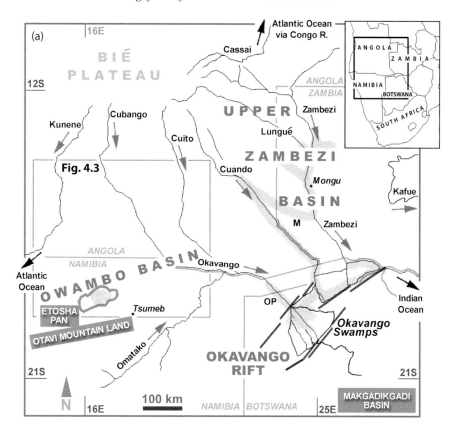

Figure 4.1 **Study area: major basins and rivers**.
(a) Three basins, the Owambo (OWB), the Okavango Rift (ORB), and the Upper Zambezi (UZB) are today linked by rivers. The largest rivers flowing off the Bié Highlands of Angola have all given rise to megafans. The Kunene and Cassai rivers flow to the Atlantic Ocean, the Zambezi and Cuando rivers to the Indian Ocean. The OWB is mainly an internal drainage basin centred on the Etosha Pan (dry lake). The Makgadikgadi Basin is fed by the regionally important Okavango and Cuito rivers. Major faults bounding the Okavango Rift depression are shown (dark straight lines). Modern wetlands (grayed) occupy parts of the Okavango rift zone and floodplains of rivers draining into this depression: Okavango 'panhandle' (OP), the Zambezi, Cuando, and Mulonga (M) floodplains.
(b) The well-known Okavango megafan ('inland delta'), radius 150 km, with wetland sectors dark green. Distal bounding faults are prominent. The 'panhandle' valley-confined reach of the Okavango River is ~100 km long. The Boteti River (right centre margin) drains trivial discharge from the megafan. A smoke plume rises from a fire, image centre. Astronaut image ISS036-E-7726, 12 June 2013, courtesy of the Earth Science and Remote Sensing Unit, NASA Johnson Space Center (http://eol.jsc.nasa.gov). A colour version of this image is available in the SOD for Chapter 4.
(c) Owambo Basin looking west, showing prior drainage orientations on the Kunene and Okavango megafans, and possible Omatako megafan, towards the inland drainage basin of the Etosha Pan. Drainage on the Kunene River is today oriented to the Atlantic Ocean, and towards the Makgadikgadi basin for the Okavango and Omatako rivers. Astronaut image STS101-706-25 courtesy of the Earth Science and Remote Sensing Unit, NASA Johnson Space Center (http://eol.jsc.nasa.gov). A colour version of this image is available in the SOD for Chapter 4.

French- and Portuguese-speaking countries in the study area. Our usage is to refer to the entire river as the *Okavango*, but distinguish two megafan features developed along its length as the *Cubango megafan* (Angola–Namibia border) and the *Okavango megafan* (Botswana). The designation *Okavango* River is also used throughout, although cognates are *Cubango* in Angola and *Kubango* and *Kavango* in Namibia (McCarthy 2013). Alternative spellings for other rivers

in the region are *Cunene/Kunene, Cuando/Kwando* (not to be confused with the smaller Caundo River), *Cuito/Quito, Cassai/Kasai,* and *Lungué-Bungo/ Lungwebunga* (abbreviated hereafter as *Lungué*).

4.1.2 The Kalahari Basin

The Kalahari Basin is one of the three largest tectonic basins of Africa's Neogene basin-and-swell topography

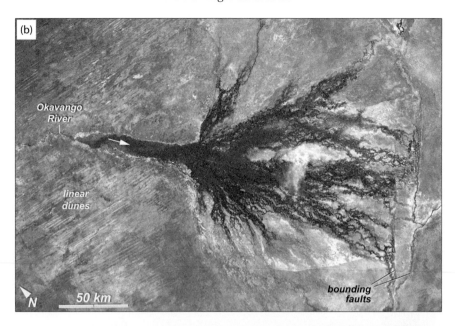

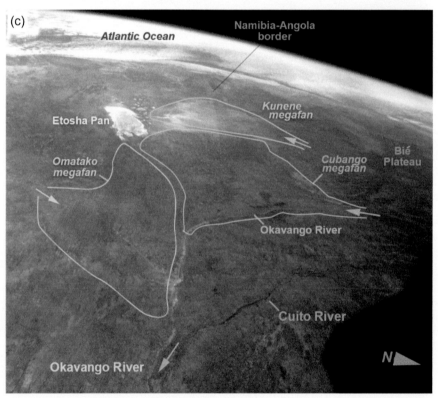

Figure 4.1 (*cont.*)

(Moore 1999; Burke and Gunnell 2008; Burke and Wilkinson 2016). Components of the Kalahari Basin are the OWB and UZB that lie on the flanks of the Bié Plateau in south-central Angola; the ORB appears as a distal extension of an arm of the East African Rift System (among others: Thomas and Shaw 1991;

Haddon and McCarthy 2005; Daly et al. 2020) (Fig. 4.1a).

The largest river in the study region, the Zambezi, receives flow from Angola and western Zambia, making its discharge equivalent to that of the River Nile (Haddon and McCarthy 2005). Other major rivers are the perennial Kunene, Okavango and Cuando, which rise in southern Angola. All have laid down megafan bodies. Both the Kunene and Zambezi rivers reach the oceans, but the Okavango drainage system ends in the extensive inland dry lake complex of the Makgadikgadi depression in central Botswana (Fig. 4.1a). The OWB is also an internal drainage basin, fed today only by the small, seasonal Cuvelai River system that has built the Cuvelai megafan (Fig. 4.2a). In the northwest Kalahari semidesert, the regionally important rivers receive runoff almost entirely from Angola's well-watered Bié Plateau, with negligible runoff from downstream tributaries or slopes (Mendelsohn and El Obeid 2004). The exception is the upper Zambezi drainage, which also receives discharge from east-bank tributaries in Zambia. The Okavango and Zambezi rivers have both developed modern courses that cross the depression of the ORB (Figs. 4.1a and 4.2a).

The Okavango Rift Basin (ORB), almost 160 km wide in the SW sector and narrowing to 50 km in the central sector (Fig. 4.1b), has been intensively studied (Hutchins et al. 1976a, b; Cooke 1980; Mallick et al. 1981; Thomas and Shaw 1991; McCarthy 2013). The Okavango River has given rise to a permanent wetland in the proximal sector of the Okavango megafan (McCarthy 2013). The flatness of the landscape allows the Thamalakane River at the foot of the Okavango fan to flow in either direction depending on whether the Okavango or Cuando basins receive greater rainfall in any one year (Debenham 1952; Wellington 1955).

The Owambo basin (OWB) (Fig. 4.1a, c) has also received significant attention in recent years, partly due to drilling projects undertaken to understand its aquifer potential. The upper Kalahari Group sediments that include all the geomorphic features treated in this study are described by Miller (Miller 1997; Miller et al. 2010; Miller et al., Ch. 15).

The UZB, defined as the Zambezi drainage basin above the Mambova Falls (at the southern rim of the ORB), is a large area in which surficial Kalahari Group sediments dominate and little geomorphic research has yet been conducted, the work of Cotterill and de Wit (2011) and Moore et al. (2007, 2012) being important exceptions. Fault lines coinciding with minor fault systems aligned NW–SE control the strikingly parallel courses of the lower Okavango, Cuando and Zambezi

rivers (Fig. 4.1a) (Thomas and Shaw 1991). Smaller, low-angle alluvial fans with associated wetlands comprise the Machili Plains along the north flank of the Zambezi River within the rift (SOD 4.1.2). They are not part of this analysis of the larger megafan features.

Pleistocene palaeolakes in the Makgadikgadi depression are revealed by long, curved shorelines (McFarlane and Eckardt 2008; review in Moore et al. 2012). The larger, late Pleistocene lakes inundated not only the Makgadikgadi pans but also lower sectors of the ORB megafans.

Annual precipitation of 400 mm in the OWB is classed as a BSh climate in the Köppen system (Kottek et al. 2006), lacking surface water except during the rainy season. Towards the northeast the study area is seasonally wet (Aw climate), with precipitation of 1,400 mm measured on the Congo divide. Between these extremes southwestern Zambia records an annual rainfall of 688 mm (Victoria Falls, Livingstone station: Moore et al. 2007).

4.1.3 River Behaviours and Divide Megafans

Construction of a megafan cone demands that the megafan-building river must shift repeatedly back and forth across the lowland surface. Avulsion behaviour ceases when the river cuts into the megafan, in incision mode. We describe these opposing modes (wide sheetlike deposition *vs.* vertical incision) and the phenomenon of divide megafans as background for the Discussion.

4.1.3.1 Avulsion and Incision Modes

Avulsion mode. The multi-channel avulsion style is well known because it is documented extensively on the modern Okavango megafan (e.g., McCarthy and Ellery 1998; McCarthy et al. 2002, reviewed in McCarthy 2013 – see SOD 4.1.3), where river discharge divides into a series of radiating channels (simultaneously active or not), the level of activity of individual channels and channel locations changing with time. Significant portions of the megafan are inundated in this way: of the 28,425 km^2 area of the entire megafan (Table 4.1), the maximum flooded zone fluctuates between 5,300 km^2 in the driest year on record (1996), and 15,500 km^2 in one of the wettest (1963) (Mendelsohn et al. 2010).

Table 4.1 *Megafans of the northern Kalahari Basin, named after their formative rivers, in three subsidiary Owambo (OWB), Okavango (Rift ORB), and Upper Zambezi (UZB) basins*

Basin	Megafan name	Megafan length[b] (km)	Megafan area (km²)	Fan slope in median sector[c]	Feeder basin area directed to Etosha (present)[d] (km²)	Feeder basin area directed to Etosha (past) (km²)	Gain / loss to downstream basin (%)	Status	Divide megafan	Weissmann/ Hartley[e]
OWB	Kunene	320	29,930	0.000146	0	66,900	−49[g]	Inactive	✓	—
	Cubango	310	50,070	0.00028[f]	0	51,625	−38[h]	Inactive	✓	—
	Cuvelai	84	2,540	0.000614	4,105	4,105		Active		—
	Other non-fan				14,160	14,160				
ORB	Okavango	150[j]	28,425	0.000360[j]	188,630[k]	137,005[m]	+26[n]	Active	✓	Botswana 1
	Cuando	108	5,090	0.000234+	113,393	113,393		Active	✓	Namibia 2
	Zambezi	102	8,250	0.000281	336,014	316,894[p]	−3[q]	Active	✓	Namibia 1
UZB	Cassai-Luena[r]	180	19,030	0.000316	3,150	12,400	−75[s]	Partly inactive	✓	Angola 1
	Lungué	170	15,070	0.000439	28,380	28,380		Partly inactive		Angola 2
	Lungué-Luio[t]	160	10,650	0.000424+	42,485	42,485		Partly inactive		Angola 3
	Luanginga	205	9,880	0.000471	10,365	10,365		Partly inactive		—

Notes. [a] Interaction between basins rendered as gains or losses of active catchment area resulting from river avulsion/incision on megafans. OWB losses are for the Etosha depression as a whole; gains/losses for the ORB and UZB are for individual megafans only;[b] longer radial lengths in asymmetric fans—see text;[c] apex-toe slope on smaller fans indicated (+);[d] drainage basin area above megafan apex;[e] Weissmann et al. (2010), Hartley et al. (2010);[f] distal half of fan;[g] loss of Kunene catchment to Etosha depression;[h] loss of Okavango catchment to Etosha depression;[j] length, slope from Stanistreet and McCarthy (1993);[k] 188,630 km² Okavango catchment above Panhandle from Wehberg and Weinzierl (2013);[m] Okavango catchment above Panhandle (Wehberg and Weinzierl 2013) less Cubango catchment above Cubango megafan apex (51,625 km²);[n] catchment gain (addition of Cubango megafan feeder basin) to Okavango megafan;[p] based on World Bank Report (http://siteresources.worldbank.org/ INTAFRICA/Resources/Zambezi_MSIOA_-_Vol_1_-_Summary_Report.pdf); excludes Cuando drainage as separate megafan-forming drainage;[q] catchment loss of Cassai to Zambezi megafan;[r] two rivers, the Cassai and Luena, built this megafan; Cassai now diverted;[s] loss due to diversion of Cassai to Congo basin (only 6% of entire Upper Zambezi Basin—see footnote p);[t] two rivers, the Lungué and Luio, built this megafan

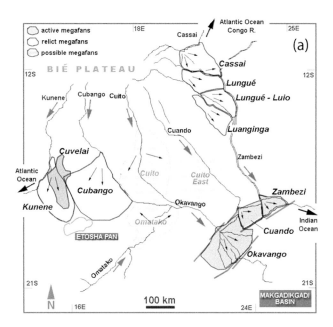

Figure 4.2 **Megafans of the northern Kalahari**.
(a) Megafans of the northern Kalahari Basin – named after their formative rivers (large arrows show drainage directions). *Owambo Basin:* Cubango and Kunene megafans (relict), overlain by small, active Cuvelai megafan (shaded). *Okavango Rift Basin (rift-margin bounding faults – straight gray lines):* Okavango, Cuando and Zambezi megafans are all active (shaded). *Upper Zambezi Basin:* Cassai, Lungué, Lungué-Luio, and Luanginga megafans. Small arrows show radial drainage associated with each megafan cone. Smooth, flat areas in the lower Cuito and Omatako drainages are possible megafans (hatched) (see text). From Burke and Wilkinson (2016).
(b) Topographic roughness signatures (scale identical to map a). Low-declivity, smooth surfaces of subregional extent (darkest tones) are almost everywhere aggradational zones, indicating megafans in most of the study area. *Owambo Basin megafans:* 1– Kunene, 2– Cuvelai, 3– Cubango, 12– Omatako (possible), 11– Cuito (possible). *Okavango Rift Basin megafans:* 4– Okavango, 5– Cuando, 6– Zambezi. *Upper Zambezi Basin megafans:* 7– Cassai, 8– Lungué, 9– Lungué–Luio, 10– Luanginga. Megafan landscapes here total nearly 200,000 km² (~ 300,000 km² including possible megafans). Other dark-toned aggradational areas are (i) Okavango panhandle floodplain (OP); (ii) lake floors (Etosha, Makgadikgadi dry lakes); (iii) calcrete-covered landscape: Etosha Calcrete Fm. (c). Narrow linear zones of fluvial incision appear rough (lighter tones) along the courses of the Okavango, Cuito, Omatako rivers. Linear patterns of tan and green are rough dune field landscapes.

The radial multi-channel style of river behaviour is significant to the present study because it is associated with extreme discharge loss: 98% of the discharge entering at the apex of the Okavango megafan fails to reach the drainage exit (the Boteti River) due to infiltration and evapotranspiration (McCarthy et al. 2002). However, megafan rivers commonly behave as single channels, with the river avulsing to different radial positions over time. The single-channel style is significant because a greater proportion of the discharge reaches the toe of the fan – and thence to sectors of the basin further downstream (see Discussion, Section 4.4). An example is the Bermejo River of northern Argentina that delivers high discharges to the trunk Paraguay River. Important for the present study is the observation that any one megafan can display both single- and multiple-channel styles of avulsion behaviours at different times. The

multi-channel drainage of the modern Okavango megafan itself is described in Section 4.3.3.

Incision mode. Deep, and therefore more permanent, incisions by major rivers are a key consideration in this study. The Okavango River provides a prime example from the study area where it cuts a 50-m-deep trench into the Cubango megafan that extends down the full length of the megafan (Figs. 4.2b, 4.3b, c, 4.4). This kind of incision is significant because it enables the river to deliver most of its discharge to the foot of the megafan zone, and beyond, with little loss because the discharge is not distributed across the megafan surface. In this case, the Okavango River not only maintains flow all the way to the toe of the Cubango megafan, but it delivers discharge to the apex of the modern Okavango megafan 250 km downstream in the ORB. In this sense incision is hydrologically similar to the single-channel

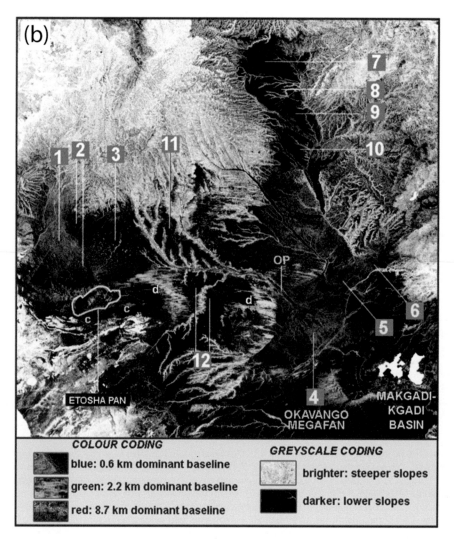

Figure 4.2 (*cont.*)

avulsion style, although the river is inhibited from moving across the megafan surface (we note that incised courses would likely have been impermanent when incision was on the order of a few metres only—because cut–fill cycles of erosion and sedimentation are able to 'repair' such incisions, thereby allowing the fan-forming river once again to avulse across the cone surface).

4.1.3.2 Divide Megafans: 'On–Off' Discharge Delivery

Megafans that occupy the drainage divides between major drainage basins occupy a unique status because of their influence on basin hydrology. In an under-appreciated phenomenon, rivers in avulsion mode on 'divide megafans' (see Wilkinson et al. 2006) deliver discharge of the entire upland feeder basin alternately to one and then to the other downstream basin. Examples are the Yamuna River in India, which can be shown to have delivered its discharge alternately to the Ganga and Indus drainage basins. Another example is the Parapetí River, which has drained alternately to the Amazon and the Paraná river basins (Wilkinson et al. 2006). Avulsions on divide megafans reorient river courses relatively quickly. In contrast, when the megafan river is in incision mode it remains in a fixed orientation in the longer term. Examples of the latter are the modern Kunene and Okavango rivers (in the sectors of the Kunene and Cubango megafans).

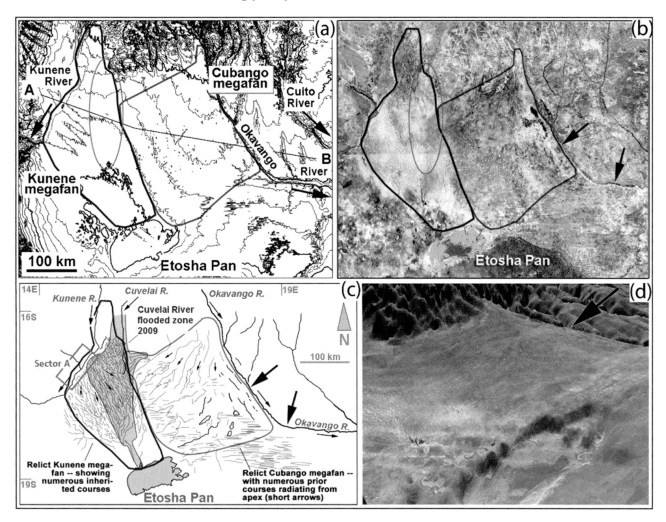

Figure 4.3 **Owambo Basin megafans**. (a) Low-gradient, conical form of Cubango and upper Kunene megafans. Ellipse indicates overlying active Cuvelai megafan. A–B transverse profile shown in Fig. 4.4. Map contour interval: 20 m. (b) Landsat false-colour infrared image of Cubango and Kunene megafans. Okavango trench – heavy arrows. (c) Surface features showing modern flow directions (arrows) of channels (dominantly plains-fed) of the relict megafans (Kunene-zone channels after Stengel 1963), and dune trends east of Etosha Pan. Flooded zone of 2009 (shaded) covers a greater area than the Cuvelai megafan. Okavango River trench: heavy arrows. (d) Cubango megafan cone, 320 km long – oblique DEM view with overlaid Landsat image, looking northeast from Etosha Pan (blue, foreground). Bié highlands (top): source region for Okavango R. discharge. Okavango River trench: heavy arrow. A colour version of this figure is available in the SOD for Chapter 4.

4.2 Methods and Criteria

4.2.1 Methods

Analysis of modern and ancient stream courses was based on remotely-sensed images from the astronaut handheld camera archive of colour images at the Johnson Space Center (Earth Science and Remote Sensing Unit 2022 – http://eol.jsc.nasa.gov), and from true- and false-colour Landsat imagery of the 1990 series and 2000 series. Field campaigns in the flat landscapes of northern Namibia have allowed confirmation of interpretations of ill-defined topographic features.

For detecting megafan surfaces and mapping their margins and dimensions, several georeferenced sources of data were used. GMTED2010 digital topographic data (https://earthexplorer.usgs.gov.) at a 3 arcsecond resolution were used to generate maps with 10-metre interpolated contour intervals, from digital elevation models constructed on a MICRODEM freeware platform (https://www.usna.edu/Users/oceano/pguth/website/microdem/microdem.htm). Topographic cross-sections, variously aligned, were constructed from the DEMs. Analysis of stream courses was based on remotely-sensed images from the image archive of

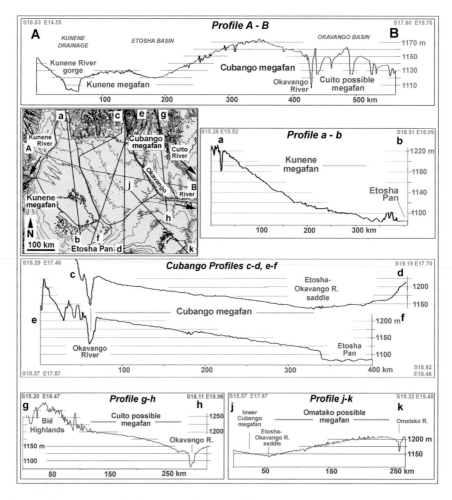

Figure 4.4 **Owambo Basin: Kunene and Cubango megafan profiles** (map contour interval: 20 m). Transverse profile showing positive, very low relief across cones of adjacent megafans (Kunene, Cubango) over a distance of 350 km (> 450 km including possible Cuito megafan). Note the narrow trench of the Okavango River. Thin Cuvelai megafan body is not apparent. Longitudinal profiles: Kunene and Cubango megafans, and Cuito and Omatako possible megafans; ragged Omatako profile indicates overlying linear dunes.

astronaut-handheld images at the Johnson Space Center (Earth Science and Remote Sensing Unit 2022). Google Earth imagery and true- and false-colour Landsat 4-5 Thematic Mapper images with a spatial resolution of 30 m in the VNIR wavelengths (https://earthexplorer.usgs.gov) were major data sources. Global ONC maps (1:100,000 scale – DMAAC, n.d.) gave further detailed information on dormant and discontinuous stream beds and some smaller river names.

Because of the relatively low rainfall of the study area's prevailing low topography, other small contributing areas were excluded as insignificant (the so-named *foothills fed* and *plains-fed*, i.e., megafan-fed, drainages). In the northwest Kalahari, the uplands generate almost all runoff – the 'active catchment' – compared with semiarid/arid lowland sectors, often dune-covered, here termed the inactive catchment zones (*sensu* Wehberg and Weinzierl 2013). Only the active catchment was used in the case of the OWB: these are the northern, Bié Plateau-related drainages, and exclude southern drainages which supply almost no surface flow. In contrast, the entire basin of the UZB is included in calculations of basin area because rainfall is significantly higher across the whole basin in terms of the *active catchment* designation.

We constructed a roughness map of the study area based on SRTM digital topographic data (available at ftp://e0srp01u.ecs.nasa.gov/). Topographic roughness is a measure of slope variability along baselines of different length (0.6 km, 2.2 km, and 8.7 km) (Wilkinson et al. 2010). The roughness algorithm was first applied on Mars by Kreslavsky and Head (1999, 2002) using a similar array of baselines (SOD 4.2.1). Lighter shades denote the higher roughness of hilly landscapes such as the Bié Plateau (Fig. 4.2b, top left), and darkest tones indicate flat and smooth plains landscapes. Shades of green indicate kilometre-scale topographic variability, such as dunefields.

Borehole logs from several wells in the OWB, which provided the context of the surface and subsurface data. Hillshading, available in DEM software, has been informative of megafan slopes and subtle margins, as have published reports in a few cases.

4.2.2 Criteria for Megafan Definition

In this and the Africa-wide survey (Wilkinson et al., Ch. 3), criteria were chosen primarily for the purpose of defining megafans *as lithological entities* rather than morphological entities, as employed, for example, by Hartley et al. (2010) and Weissmann et al. (2011). This approach allows for mapping of probable fluvial sedimentary bodies by remote means and it has the benefit of being testable: megafans supplied by a single major river can be expected to show a unique mineralogical signature, at least for the fluvial component. The criteria listed in SOD Table 4.2.2 are derived mainly from the approximately 25 well documented megafans, which include the Okavango megafan. Criteria also derive from experience gained in the world survey reported in Wilkinson and Currit (Ch. 2).

The most important criteria are the existence of a river as a sediment source, and a smooth, very low-gradient (usually $\ll 0.5°$) surface sloping away from the upland for at least 80 km. Radii were measured from the apex, taken as the point where the formative river exits the upland, and the toe at the lowest point of the continuous slope starting at the apex.

Partial-cone morphology is considered to be a universal criterion (SOD 4.2.2), at least for primary megafans (see Wilkinson, Ch. 17) in their proximal zones. We include zones of non-radial/contributory drainage on lower slopes, a pattern common on diamond-shaped fans – whereas Hartley et al. (2010) in their morphological approach exclude areas of non-radial patterns and terminate fan length and areal dimensions at the point where radial patterns give way to contributory

patterns, even if long-profile continuity of the land surface is maintained. Concave longitudinal profiles are common in small alluvial fans, but our preliminary survey of Africa reveals a range from slight concavity, to convexo-concave, to slight convexity, so that we do not designate long-profile concavity as a criterion. Continuity of the low-gradient, depositional surface is regarded as an indication of fluvial deposition in our lithology-oriented set of criteria.

Fan perimeters can be variably interpreted because distal portions of some megafans grade into lake environments (e.g., the well-known Lake Ngami south of the Okavango megafan) or extend beneath dunefields. More importantly, modern perimeters do not always represent the original fan margins. An example is the Cubango megafan where steeper slopes at the extreme distal margin in the Etosha sector of the megafan indicate erosional deepening of the Etosha Pan (Miller et al. 2010). Areas mapped are conservative as including only sectors that are demonstrably part of the low-gradient surface and can be lower than original areas by as much as 10%.

Topographic roughness, combining measures of relative smoothness and low declivity (see Wilkinson et al. 2010), is a distinct identifier of depositional landscapes (Fig. 4.2b) – especially for megafan landscapes but also for some few other flat landscapes (Fig. 4.2b). The vast megafan landscapes of the Andean foreland and Indo-Gangetic plains (Wilkinson et al. 2010) make roughness an attribute that we have employed as an added indicative feature (though not a criterion). Recognisable fluvial channel morphologies are often widespread on megafan surfaces. However, these can become so degraded and overprinted on relict surfaces as to be unrecognisable, so that they are not classed as criteria. The case of the large Cubango megafan, 310 km-long, and long unrecognised, is a prime case in point.

4.3 Results

4.3.1 Megafan Identification in the Northern Kalahari Basin

Ten megafans (Table 4.1 and Fig. 4.2a, b) are identified in the northern Kalahari (by criteria listed in SOD Table 4.2.2), of which only four are active today. The

Table 4.2 *'Possible megafans' (meet fewer criteria than megafans)*

Possible megafans (named after formative river)[a]	Length of low-angle slope (km)	Associated area (km^2)	Slope of proximal third	Area of modern drainage basin (total incised area, excluding inferred fan) (km^2)
Cuito (Owambo Basin)	~350	~22,400	0.000387	33,960
Omatako (Owambo Basin)	217	~43,350	0.00051	49,850
Cuito East [b] (Owambo Basin)	~390	~31,000	0.000527	Bisected by Cuando drainage?

Notes: [a] See Fig. 4.2a for locations. See text and SOD Table 4.2.2 for criteria; [b] Cuito East feature meets even fewer criteria than Omatako or Cuito.

others are relict, their formative rivers being incised and thus unable to avulse into alternate radial courses. Their fan-like planforms and surfaces are nevertheless recognisable in the landscape. The Okavango River displays the rare phenomenon of having deposited two major megafans along its course, namely the Cubango and Okavango fans (Fig. 4.2a) – the former relict and the latter active. Three possible megafans that meet fewer megafan criteria are also recognised (Table 4.2). Hartley et al. (2010) include six of these features in their global list (Table 4.1).

The Africa-wide survey of topographic roughness (Wilkinson et al., Ch. 3) suggests with little doubt that northern Kalahari lowland landscapes are almost everywhere relatively recent aggradational surfaces, modified to different degrees by fluvial erosion and/or aeolian processes. Apart from lake floors and floodplains such as that of the Zambezi River, these low-declivity, smooth surfaces of subregional extent (darkest tones in Fig. 4.2b) can be classified as megafans.

4.3.2 Owambo Basin (OWB)

Modern surface expressions of the OWB features are the slightly *convex*, sand-dominated Cubango megafan deposited by prior southward-flowing courses of the palaeo-Okavango River, and the flatter, clay-rich Kunene megafan (Fig. 4.2a, b). Palaeolake Etosha, in the vicinity of the present Etosha Pan (Figs. 4.1a, 4.2a, b), was the end point of both the Okavango and Kunene rivers for most radial orientations of these rivers on their megafan cones.

4.3.2.1 Cubango Megafan

The Cubango megafan has been described by Wilkinson et al. (2008), Miller (2010, 2013), Miller et al. (2010, 2016), Houben et al. (2020), and Miller et al. (Ch. 15). Here we place it in a wider context. The Cubango megafan is one of the longest in the study area (310 km), appearing as a vast, almost featureless plain north of the dry lakebed of the Etosha Pan. It displays distinct cone morphology (Fig. 4.3a, d), very low gradients, and extreme topographic smoothness (Fig. 4.2b). Despite its size it is the most cryptic of the Kalahari megafans. Its existence was not suspected until the setting of previously identified megafans was understood (Wilkinson et al. 2008). Palaeochannels are discontinuous and sufficiently widely spaced not to have been recognised. Vegetation patterns in the false-colour infrared imagery give little indication of the channels (Fig. 4.3b), and aeolian cover sands of the Andoni Formation and many associated small dry lakes (pans) on the distal slopes obscure channel configurations. Profiles show the downfan continuity of the surface (Fig. 4.4, profiles *c–d, e–f*). The Okavango River has cut a narrow, 30–50 m-deep, trench-like incision into the eastern margin of the fan surface.

The megafan units host three separate aquifers of wide extent (Miller et al., Ch. 15). Numerous drilled boreholes have demonstrated that the vast Cubango megafan is a fluvial sedimentary body, laid down by the Okavango River. Drilling just south of the Namibia–Angola border in the southern half of the megafan has established that the sediment body is 270 m thick (Houben et al. 2020). It hosts three aquifers, one of which is as much as 110 m thick. The architectural distinction implied by very large lateral dimensions (> 150 km) was duly recognised: aquifer dimensions apparently reflect their origin as sheet-like units laid down in a broad megafan depositional environment, rather than as units deposited as relatively narrow bodies of floodplain sediment confined within river-carved valleys.

4.3.2.2 Kunene Megafan

In the modern landscape, the morphologically more complex Kunene megafan meets the criteria of slope, roughness, and length, and is the longest in the study area at 320 km (Table 4.1). The apex is located on the basis of contour evidence showing conicality, low-angle slopes, and numerous radial drainage lines visible in remotely sensed imagery, in contrast with those of the neighbouring Cubango megafan. Repeated incision near the apex has removed the original apex entirely such that the younger, smaller Cassinga River fan now overrides the upper portions of the Kunene megafan (SOD 4.3). The marked change in slope midway down the fan (Figs. 4.3a, 4.4 profile *a–b*) occurs at a grain-size change from a sandier upper sector to the clay-rich lower sector (see also Miller et al., Ch. 15). The long profile may also represent a later erosive phase when the Kunene River avulsed towards the Etosha Pan from subapexes operating 55 km and more from the mapped apex (channels oriented broadly southeast, small arrows, sector A, Fig. 4.3c) (subapex activity is common on megafans – see Wilkinson, Ch. 17). The recency of such flooding is evident from the fact that the present bed of the Kunene River at the westernmost extremity of the fan (Sector A, Fig. 4.3c) lies only 13.5 m below the fan surface an altitudinal difference remarked upon by Wellington (1938, 1955) and Houben et al. (2020). However, the Kunene River no longer floods into the SE-oriented channels (Miller et al. 2010).

4.3.2.3 Cuvelai Megafan

This smaller megafan, almost 100 km long (Fig. 4.3a, c), is superimposed on the relict Kunene megafan (SOD 4.3). Discharge in wetter years continues beyond this megafan, and reaches the Etosha Pan (Fig. 4.3c).

4.3.2.4 Possible Megafans

The Omatako and Cuito features are both connected to major local rivers, but the continuity of their surfaces cannot be satisfactorily demonstrated, although some

low-gradient, long profiles (radii ~350 km, ~220 km) and smooth surfaces are strongly suggestive of mega-fan slopes (Fig. 4.2a, b; Fig. 4.4, profiles *g-h, j-k*; Table 4.2). Boreholes penetrating the Omatako feature show that this is also a sand-dominated system (Miller 2008) rather than an erosional, rock-cut feature. Circumstantial evidence for these features being degraded megafans is (i) the existence of other mega-fans in the study area (of demonstrated fluvial provenance), and (ii) the known characteristics of megafans in the region to cluster laterally and to display multi-phase aggradational and erosional histories over time. The Cuito East feature (Table 4.2 and Fig. 4.2a) straddles the modern Cuando River zone and is the least securely assigned as a megafan.

4.3.3 Okavango Rift Basin (ORB)

Three adjacent cones deposited by the Okavango, Cuando, and Zambezi rivers are identified as mega-fans (Table 4.1; Fig. 4.1a, b; 4.2a, b; 4.5) and show low, but distinct convex transverse profiles (Fig. 4.6, profile *A-B*). All are active in the sense that the forma-tive rivers lie at or within a few metres of at least a major part of the fan surfaces. Wetland zones are indicated in the infrared images by dark greens and show the radial channel pattern in the case of the Okavango fan, the largest of the three. Drainage patterns are less apparent on the other two fans but emphasise the relatively small presently active lobes (Fig. 4.5).

The Okavango megafan is demonstrably in multi-channel distributary mode today, as evidenced by the wide areas that are inundated during the annual flood [links to: https://eol.jsc.nasa.gov/Collections/ EarthObservatory/articles/OkavangoDelta.htm; https:// eol.jsc.nasa.gov/Collections/EarthObservatory/articles/ Okavango_Swamp_ISS028.htm] (McCarthy 2013). But in relatively recent times the megafan has displayed the contrasting single-channel style: large palaeo-courses in midfan and distal locations display meander wavelengths with dimensions similar to those of the Okavango River floodplain upstream of the fan apex. Indeed, McCarthy (2013) has reported palaeochannel dimensions in the midfan zone that are roughly com-parable to the present single-channel Okavango River wavelengths in the floodplain upstream of the fan

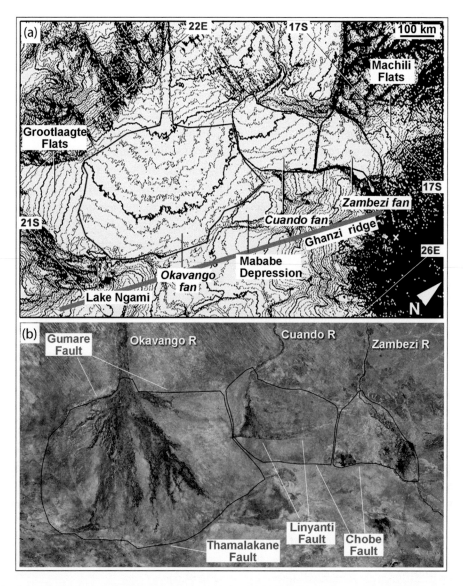

Figure 4.5 **Okavango Rift Basin megafans**. Okavango, Cuando and Zambezi megafans (outlined). (a) Contours (5 m interval) show cone morphology of the Okavango megafan. More complex morphologies show influence of faults (Cuando megafan) and partitioning of the surface by a wide, shallow incision zone (Zambezi megafan). (b) Linyanti and Chobe Faults act as distal margins for Cuando and Zambezi megafans, and form the Linyanti and Chobe swamps (darker green tones) (Landsat 1990 false-colour infrared image). A colour version of this figure is available in the SOD for Chapter 4.

apex – i.e., larger by orders of magnitude than those of modern channels in the same zone. Such dimensions indicate that the Okavango River has flowed on occasion *as a single large channel* to the toe of the megafan, and suggest with some certainty that the river has delivered much greater discharges to the toe of the fan than it does today – implying discharges closer to those of today (average measured at the gauge upstream of the megafan apex is 10,100 M m³/yr, McCarthy, 2013), as compared with the small fraction that flows off the

toe of the fan (to the Boteti River) which measures only 1–2% of the influx (McCarthy 2013).

The active smaller Cuando megafan lies adjacent to the eastern margin of the Okavango fan. It hosts not one, but two distal wetlands, dammed against the transverse Linyanti and Chobe faults (Figs. 4.5 and 4.6c, profile *E-F*). Both wetlands drain east towards the Zambezi megafan. However, fan gradients indicate, and local observation has documented (Mendelsohn et al. 2010, their chapter 5, note 11 therein) the fact that

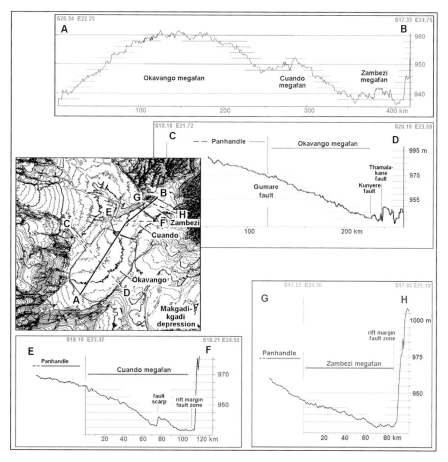

Figure 4.6 **Okavango Rift Basin: Okavango, Cuando, and Zambezi megafans** (key map: longitudinal profiles *C–D, E–F, G–H* as dashed lines; map contour interval: 20 m). Transverse profile *A–B* (strong vertical exaggeration): landscape of three adjacent megafan cones with a low relative relief of 25 m (along profile) over a linear distance of 400 km. Other longitudinal profiles (locations shown as dashed lines; horizontal and vertical scales approximately the same in each) show convexity and similarity of feeder stream slopes to megafan slopes. Linyanti fault scarp shown in *E–F*.

the Cuando megafan at times is connected hydrologically to the Okavango system. Flow can take place directly from the Cuando River (when it is oriented in southerly/southwesterly directions on the fan cone), or from the Linyanti wetland, into the common Cuando–Okavango fault scarp-defined depression. Flow can actually occur in either direction depending on which river delivers the higher discharge (Okavango discharge can flow east and enter the Linyanti wetland). The history of the Cuando megafan is complicated by course changes documented at the Mulonga Plain in Zambia (M, Fig. 4.1a), a confined plain occupied by the Cuando River when it flowed east directly into the Zambezi River (Moore et al. 2012). The river has subsequently changed course to its present southeasterly orientation (SOD 4.3).

The Zambezi megafan shows an active distal lobe in the southeast, inset 5–6 m below an older surface. The Chobe fault, continuing east, acts as the distal termination scarp (Fig. 4.5).

4.3.4 Upper Zambezi Basin (UZB)

Four megafans are identified (Table 4.1 and Figs. 4.7, 4.8), all prominent in infrared imagery (Fig. 4.7) due to water tables at shallow depth. Description of these fans has been almost non-existent, to our knowledge, except for the inclusion by Hartley et al. (2010) in their studies (Table 4.1). The UZB displays a more complex and unusual geomorphic history involving a combination of slight incision combined with on-fan sedimentation in wide floodplains that has resulted in a

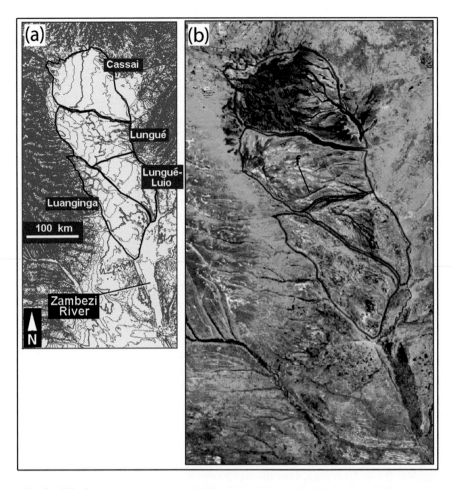

Figure 4.7 **Upper Zambezi Basin**.
Cassai, Lungué, Lungué–Luio, Luanginga megafans (outlined) stretch along strike in excess of 440 km. (a) Cone and incised cone morphology (map contour interval: 10 m). (b) Area shown in (a); infrared colour combinations (pinks, reds, purples, dark blues) show distinctive vegetation response to near-surface water tables, strongly associated with the areas of four megafans; note contrast with bright greens of woodlands of the backing highlands (left side of frame). (Landsat 1990, IR).

patchwork of surfaces being inactive in terms of mega-fan river avulsion. Minor faulting is evident in the linear control of some river courses, especially on the Lungué fan surface (Fig. 4.7b). A buried rift underlies the broad valley of the wide Barotse Plain (Daly et al. 2020). The control by this structure is apparent in the broadly linear N-S orientation of the main stem of the upper Zambezi River.

The Cassai River is today diverted to the Congo basin. Its megafan is inset ~ 15 m below the altitude of the Lungué, its neighbour to the south, suggesting a long history of geomorphic events (SOD 4.3). The Lungué–Luio megafan is a product of discharge by two megafan-forming rivers, the Lungué and Luio (Fig. 4.7), and is thus a 'derived megafan' (see Wilkinson, Ch. 17, Section 17.4.3), the only example

of this type in the study area. This unusual conform-ation appears to be a function of the greater proximity of exit points of the southern rivers.

4.3.5 Morphometric Properties

Table 4.1 shows megafan radii, slopes and areas. Feeder basin areas, and changes in these areas gener-ated by major avulsions, are also shown. The OWB megafan areas and lengths (excluding the small, locally sourced Cuvelai feature) are far larger than those of either the ORB or UZB, despite the fact that the OWB feeder areas are significantly smaller than those of the other two subbasins (Table 4.1). Very low fan gradients (Table 4.1) conform with those in other parts of the world.

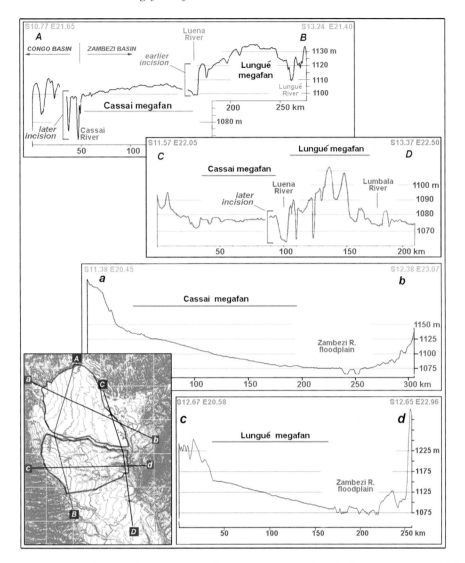

Figure 4.8 **Upper Zambezi Basin – northern megafans** (map contour interval: 5 m). Transverse profiles (*A–B*, *C–D*) of low relief over a linear distance in excess of 200 km across two juxtaposed megafans; note conical form of proximal sectors of Cassai and Lungué megafans, and abrupt incision of distal slopes of the Lungué megafan but shallow basin-like incision of the distal Cassai megafan (compare *A–B* and *C–D*). Longitudinal profiles *a–b* and *c–d* show very smooth, low-gradient, continuous slopes of the Cassai and Lungué megafans.

Where wide 'sweep angles' are available for radial drainage at the apex, approaching 180 degrees, megafans with semicircular planforms can evolve. The Okavango megafan is the most classic example. The Cassai and Lungué megafans in the northern UZB also display wide sweep angles compared with their southern neighbours that display sweep angles of < 45 degrees. Reduced sweep angles in turn translate into narrower, more triangular planforms because sweep angles are a function of distance between the exits of fan-forming rivers from the source upland. Exit distances are shorter in the southern UZB.

4.4 Discussion

4.4.1 Identification and Setting of Northern Kalahari Megafans

Ten features are identified as megafans (Fig. 4.2 and Table 4.1) in the northwest margin of the Kalahari Basin, nested in three subbasins. Four (Cuvelai, Okavango, Cuando, Zambezi) are demonstrably fluvial in origin because the formative rivers are active on the surfaces of the three ORB features, and the Cuvelai in the OWB (e.g., Miller et al., Ch. 15; McCarthy 2013; Podgorski et al. 2013). The similarity of the other six

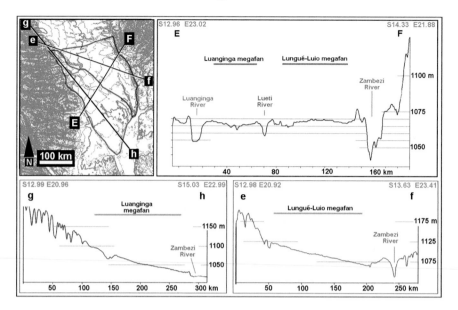

Figure 4.9 **Southern megafans of the Upper Zambezi Basin** (map contour interval: 5 m). (a) Transverse profile: landscape of adjacent megafans 130 km long. Note very low topography of both and slight incision into upper Luanginga megafan. (b), (c) Longitudinal profiles of the Lungué–Luio and Luanginga megafans at similar scales.

to these in both setting and conformity to megafan criteria (SOD Table 4.2.2) suggests with little doubt that they too are fluvial in origin. The Cubango and Kunene megafans are relict but recognisable in the modern landscape. The four UZB megafans are more complex, being partly active. Other surfaces that appear to have been generated under megafan fluvial dynamics (before significant erosional modification) are the Cuito, Cuito East, and Omatako possible fans (Fig. 4.2 and Table 4.2).

With broad altitudinal parameters imposed by the basin-and-swell geomorphology of the Kalahari Basin (Burke and Gunnell 2008), the OWB and UZB megafans are emplaced on the lower flanks of the Bié Plateau swell, reaching almost to the lowest altitudes of these sub-basins. Relative subsidence of the ORB (or uplift of the rift shoulders – McCarthy et al. 2002) has dropped intra-rift surfaces below the base profile of the three inflowing rivers such that megafan sedimentation is ongoing, and is indeed so effective that the Okavango megafan exports no fluvial sediment – i.e., it acts today as a closed sedimentary system (McCarthy 2013). The Cuando megafan may be of the same type. The Zambezi megafan, with the through-flowing Zambezi River, may be mainly a sediment-passing system.

Long profiles show characteristically low-gradient megafan slopes in all three subbasins. Profiles end abruptly at the structural margins of the ORB and UZB (Figs. 4.6, 4.8, 4.9). Incision by the UZB fan-forming rivers and the Okavango has rendered parts of the UZB fans inactive and the entire Cubango fan inactive. Capture by coastal drainages has diverted the upper Kunene River away from the OWB, and incision in the southern Congo system has diverted the Cassai River drainage northwards, out of the UZB.

The present linear style of incision is apparent in the roughness map (Fig. 4.2b) where it appears as narrow, light-toned ragged zones, most notably extending up courses of the Okavango, Zambezi, and Omatako rivers and their tributaries (Fig. 4.2b). Tectonic activity in the rift zone may be responsible for these phases of fluvial incision that have extended hundreds of kilo-metres upstream, fixing the present orientation of these major rivers.

Comparison of surface moisture signatures in the infrared wavelengths reveals that the more deeply incised megafans (Kunene and Cubango) are almost devoid of modern wetland vegetation, appearing arid compared with major wetland zones on the surfaces of the unincised ORB megafans and the slightly incised UZB megafans (compare Fig. 4.3b with Figs. 4.5 and 4.7b).

Hartley et al. (2010) and Weissmann et al. (2010) recognised six of the ten features identified in this study (Table 4.1). The disparity relates partly to the use of slightly different criteria, but most importantly to the inclusion of older, relict and variously overprinted forms in this study. The most notable example was the undetected relict Cubango fan, whose existence was predicted from patterns of fluvial sedimentation (hierarchical Groups 9 and 10 – see Wilkinson, Ch. 17) derived from the global study, and subsequently confirmed from borehole data (Miller et al., Ch. 15; Houben et al. 2020).

The ORB megafans have alternated between megafan and megafan-delta status during periods when the Makgadikgadi basin filled with extensive lakes that reached as far north as the ORB depression, inundating slopes of the Okavango and Cuando megafans, and all of the Zambezi megafan (Burrough et al. 2009; Moore et al. 2012). For the present analysis we emphasise river and megafan activity during times when distal inundation was not operating. It is possible that the UZB megafans also acted as megafan-deltas on occasion in the past during the tenure of a possible Palaeolake Bulozi (Moore et al. 2012). Megafan-deltas are discussed by Lane et al. (Ch. 12) and Wilkinson and Gunnell (Ch. 18).

4.4.2 Morphometric Characteristics

Table 4.1 shows megafan radii, slopes and areas. Feeder basin areas and changes in these areas, imposed by critical river avulsions, are also shown. We note in particular that basin areas likely do not have the same significance in the study area as do those in studies of classical alluvial fans. Well attested relationships between feeder basin area and fan area hold almost universally for alluvial fans. Reasons why this is not so for the study area megafans are several. It is commonly expected that drainage basins display precisely defined divides, are underlain by rocks of low infiltration capacity, and that extensive dune fields in areas of extremely low declivity are not complicating factors. But each one of these complications applies to one or another megafan in the study area. Drainage areas in Table 4.1 are supplied for comparative purposes rather than statistical analysis.

Comparisons of total fan area to total contributing drainage basin area (Table 4.1) are nevertheless revealing. First, the OWB shows that fan area totals

(83,000 km^2) are ~60% of the area of the feeder basin area total (~136,000 km^2). The same comparison in the ORB shows that the total fan area (~42,000 km^2) is only 7% of the total feeder basin area (600,000 km^2). Second, individual megafans within these two basins show the most extreme differentiation: the Cubango megafan (310 km long) is roughly the same area as its feeder basin (~50,000 km^2 and 52,000 km^2, respectively), whereas the Zambezi megafan in the Okavango Rift Basin amounts to only 2.4% of its vast feeder basin.

The major explanation for these discrepancies is primarily the horizontal accommodation available for lateral megafan expansion, but also a function of spacing between the fan apexes: the topographically unconfined OWB, where megafans have been able to extend across the wide Etosha Pan interswell basin and apexes are 230 km apart, has apparently allowed the fullest expansion of megafan sediment wedges. By comparison, megafan development in the ORB rift depression has been confined strictly to the width of the exposed rift (150–60 km), and apexes are only 130 km apart.

These results represent a reversal of the basic alluvial-fan tenet: the largest river in the region (Zambezi) has built one of the smallest megafans (Zambezi, 100 km) in the narrowest sector of the rift where fault scarps are exposed. The two large OWB fans, by contrast, are not restricted in this way. Similarly, the UZB megafans display longer radii than the ORB fans because the wider UZB depression allows more flexibility in spatial organisation of the fans.

A major conclusion of this study is therefore that basin dimension can act as a dominant control on megafan dimension, whereas this is seldom a control on smaller fans that are orders of magnitude smaller than their host basins (see Wilkinson, Ch. 17). Restated, it appears that where basin dimensions are sufficiently extensive and apex locations sufficiently spaced, fans will attain sizes more commensurate with relationships ascertained from studies of small alluvial fans.

The slopes of all megafan features accord with the findings of Hartley et al. (2010) who show that larger fans display gradients lower than 0.005. Although the known inverse relationship between fan slope and feeder basin area is poor in the study area, the Zambezi megafan displays the lowest slope, being formed by the largest river (basin size standing as a proxy for sediment grade and stream gradient – see Wilkinson, Ch. 17); likewise, the small Cuvelai mega-fan, fed by the smallest basin, is the steepest (Table 4.1).

Four of the ten megafans show overtly *convex-up* long profiles, and three others are straight or slightly convexo-concave. One reason may be the relationship that obtains in more desertic climates, that channel profiles steepen with the downstream reduction of flow (Cooke et al. 1993). This suggests that the generalisa-tion that megafans ought universally to show concave-up long profiles does not hold.

The similarity of feeder river and megafan gradients in the ORB (Fig. 4.6, profiles *C-D, E-F, G-H*) supports the idea that feeder river slope frequently determines fan slope (see Wilkinson, Ch. 17). Continuity of fan-surface slope is also a consistent feature of the Kalahari megafans. One exception is the Kunene megafan, where a distinct break of slope between the upper and lower sectors has been noted. The existence of a series of subapexes along the medial western margin of the megafan where the overspill from the Kunene River (at altitudes closest to that of the Etosha Pan) appears to explain the broad orientation of the numer-ous channels that trend southeast on the lower slopes of the megafan and drain towards the Etosha Pan (rather than trending directly southwest, downslope from apex to toe). The break of slope may derive from the activity of these channels in accentuating the dif-ference between the upper and lower slopes.

Surfaces of active megafans and some surfaces of inactive fans display consistently low-gradient, smooth surfaces (Fig. 4.2b), a prominent feature of megafan landscapes in South America (Wilkinson et al. 2010) and Africa (Wilkinson et al., Ch. 3). The older upper Kunene megafan surface and Okavango megafan sur-face display greater roughness than the other mega-fans, presumably as a result of erosional processes and sandy relict channel ridges which are a major compon-ent of the modern surface (McCarthy et al. 2002) respectively, and also in the latter case because of degraded longitudinal dunes in the distal sectors.

4.4.3 Interbasin Effects: Basin Hydrology and River Switching on Divide Megafans

In most parts of the world, the effects of river-course avulsions on megafans are restricted to the individual megafan cone and immediate fan-margin zones. However, the northern Kalahari is unusual in display-ing fully six 'divide megafans' (Section 4.1.3) – a megafan type situated astride the divides between major basins. These are the Kunene and Cubango megafans on the margins of the OWB, the Okavango, Cuando and Zambezi on the margins of the ORB, and the Cassai on the margin of the UZB. The location on divides gives avulsions on these megafans much wider, regional significance, that we examine fully below.

Sweep angles show the available orientations of radial river courses on these megafans (curved arrows, Fig. 4.10a, b) and demonstrate that six separate river diversions between major basins have occurred. In what may be a unique situation, these diversions involve two endoreic and three exoreic basins, and all except one are significant in terms of the loss, or addition, of major proportions of discharge to the basins involved.

Periodicity of the short-term diversions on divide megafans is probably irregular because a full-sweep cycle of the fan-forming river from one cone margin to the other and back again, varies widely (SOD 4.4.3). This has been referred to as the 'on–off' phenomenon that characterises divide megafans (Section 4.1.3.2), adding great complexity to discharge regimes in the receiving basins because it is geologically so fast (e.g., ~250 yr for the Kosi River to shift course on its fan from one margin to the other; Wells and Dorr 1987).

The following five interbasin connections are con-sidered in this section.

(i) Owambo Basin (OWB) connections to the Okavango Rift Basin (ORB) (via the Okavango R.).
(ii) Owambo Basin (OWB) connections to coastal (Atlantic) drainage (via the Kunene R.).
(iii) Okavango Rift Basin (ORB) connections to the Makgadikgadi Basin (via the Okavango, Cuando, and Zambezi rivers).
(iv) Okavango Rift Basin (ORB) connections to the Middle Zambezi Basin (UZB) and Indian Ocean (via the Okavango, Cuando, and Zambezi rivers)
(v) Upper Zambezi Basin (UZB) connections to the Congo Basin/Atlantic Ocean (via the Cassai R.).

We examine first-order implications of these diversions which are most significant hydrologically in the OWB and most complex in the ORB. Three sets of com-monly proposed controls of interbasin connections

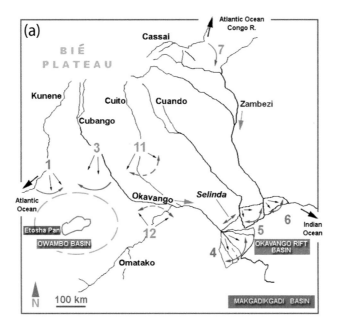

Figure 4.10 **Divide megafans, and endorheic and exorheic basins**.
(a) *Divide megafans* are located on drainage divides such that the fan-forming river flows alternately into neighbouring major drainage basins. Divide megafans are numbered as in Fig. 4.2b and show radial drainage (straight arrows) and sweep angles of prior river orientations (curved arrows). Most megafan rivers in the study area have displayed both endorheic and exorheic orientations – i.e., towards the interior drainage Etosha and Makgadikgadi basins and to the Atlantic and Indian Oceans (heavy arrows). Active and recently active megafans: Kunene R., presently oriented to the Atlantic Ocean, has emptied into the Etosha basin (1– Kunene megafan). Middle course of the Okavango R. (3– Cubango megafan), presently oriented to the Magkadigkadi inland basin, flowed episodically into the Etosha Pan basin at times when the Cubango megafan was active. Lower course of the Okavango R. (4– Okavango megafan), presently part of the interior Makgadikgadi basin. Okavango drainage can be oriented to the Cuando fan (straight red arrow) via east-draining courses on the Okavango megafan. The Cuando R., presently discharging mainly to the Zambezi R. (as shown by the swamplands), drains on occasion (5– Cuando megafan) to the Makgadikgadi basin via the Okavango system. Zambezi R. in the recent geological past will have flowed SW along the Linyanti Fault towards the lower Okavango fan (6– Zambezi megafan). The Cassai R., oriented today to the Congo Basin, flowed episodically to the Zambezi R. and ultimately the Indian Ocean (7– Cassai megafan). Possible megafans (dashed curved arrows): the Cuito R., presently oriented via the Okavango R. to the Makgadikgadi basin, on occasion would have been oriented to join the Cuando R. (11– Cuito possible megafan), thereby contributing to the Zambezi drainage. Omatako R. presently oriented via the Okavango R. to the Makgadikgadi basin, would have emptied on occasion into the closed Etosha Pan depression (12– Omatako possible megafan).
(b) Roughness map (same area as panel a) showing dark-toned, very low-gradient, smooth surfaces and the associated sweep angle of fan-forming rivers (curved arrows show past changes of orientation; dashed arrows show orientations on possible megafans). A colour version of this image is available in the SOD for Chapter 4.
(c) Palaeolake Makgadikgadi showing areas inundated (darker gray) to the 946 m level which includes local 'basin sumps' north and south of the Okavango megafan. From Burrough et al. (2009).

and their duration are (i) rapidly changing orientations of the megafan river on its cone (Figs. 4.11, 4.12), (ii) arid–humid climate alternations, and (iii) avulsion- *vs.* incision-mode dynamics of megafan rivers (Section 4.1.3). (iv) We include here a fourth dynamic, namely changes in avulsion style (multiple-channel *vs.* single-channel drainage patterns) on a single megafan (Section 4.1.3), because of their influence on discharge transmission downfan, evidence for which is seen on the Okavango megafan (Section 4.3.3). Discharge gain/loss percentages discussed are order-of-magnitude indications only.

4.4.3.1 Owambo Basin

River orientations on the Cubango and Kunene megafans involve the most important hydrological changes of any of the megafan-mediated effects in the study area basin.

The existence of the Okavango and Kunene megafans indicates clearly that the Okavango and Kunene rivers have been oriented to the Etosha Pan in the past. The scenarios below imply two axiomatic geometries: first, radial channel orientations on both megafans lead discharge mainly to the Etosha Pan

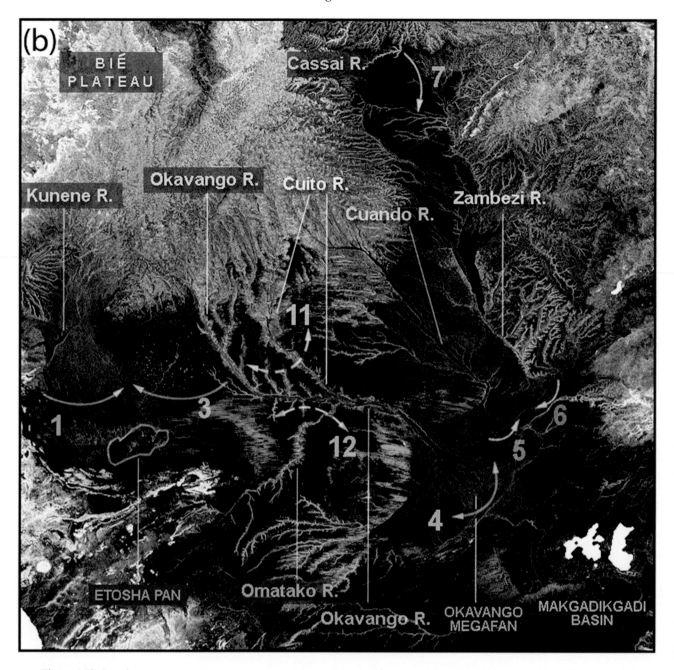

Figure 4.10 (*cont.*)

because these two megafans are situated within the OWB, versus modern orientations; second, avulsion across the entire surface of the megafans means that all orientations would have come into play repeatedly. This means that significant although infrequent orientations must have led discharge both east and west beyond the bounds of the OWB (before the modern orientations were established as permanent) because the megafans lie at the outer margins of the OWB subbasin.

The discussion below is based mainly on the known modern dynamics on the Okavango megafan wetland, and the waterless, incised Cubango megafan on which channels and small deflation pans are usually dry. We examine impacts of avulsions on the Cubango and more ancient Kunene megafans. Because of this more complex relation to other systems in the study area we render diagrammatically the effects of drainage style (avulsion and incision modes) and climate in both subbasins in Figs. 4.11 and 4.12.

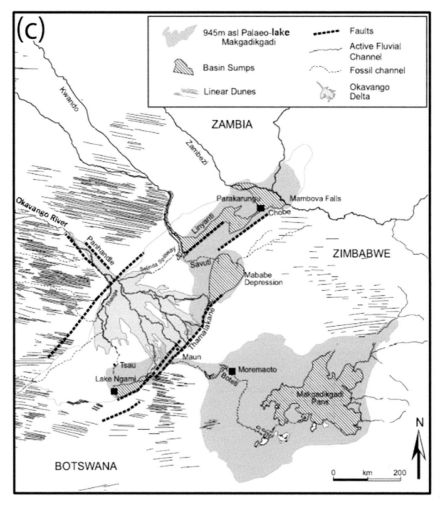

Figure 4.10 (*cont.*)

4.4.3.1.1 Avulsions on the Cubango Megafan: Impacts on Palaeolake Etosha

In the *multi-channel avulsion mode* (upper panels, Fig. 4.11), the Okavango River on the Cubango cone discharged mainly in the direction of the Etosha Pan, but assuming behaviour analogous to that of the modern Okavango megafan, very little discharge would have reached the pan, lost to infiltration and evapotranspiration (scenario *Av-1*, Fig. 4.11). In climates wetter than those of today, more water would probably have reached the pan, perhaps supporting a Palaeolake Etosha (climate alternative *Av-2*, Fig. 4.11). Alternation between long-duration wetter and semi-arid/arid climates (Dill et al. 2013) suggests pulsed phases of activity on the megafan and flooding of the palaeolake.

In cases of river behaviour in *single-channel avulsion mode* (as known from large palaeochannels on the Okavango megafan, and the most common type on world megafans), however, the Okavango River in most orientations would have been able to deliver the discharge of fully ~38% of the original active catchment to the pan (Table 4.1; scenarios *Av-3, Av-4*, Fig. 4.11). Such an increase likely supported the existence of a Palaeolake Etosha.

Diversions of this type impose major and continuing fluctuations in discharge directed to each basin, a regime of frequent autogenic forcing that has not been mentioned for the northern Kalahari hydrological patterns. Implications for interpretation of the sedimentary sequences in both basins therefore may be profound.

Implications of the Okavango River in *incision mode* on the Cubango megafan are shown in the upper panels of Fig, 4.12 (*In-1, In-2*). Incision down the full

Cubango Megafan Mode ⟍ Climate	ETOSHA Arid / semi-arid conditions (~as today)	DEPRESSION Humid conditions	OKAVANGO / MAKGADIKGADI DEPRESSIONS Role of Okavango R.
Avulsion – Multi-channel style Slow flood pulse (orientation mainly to Etosha depression)	*Av-1.* Wetland similar to that of today's Okavango megafan; negligible discharge from fan to Etosha depression (as with modern Okavango megafan) 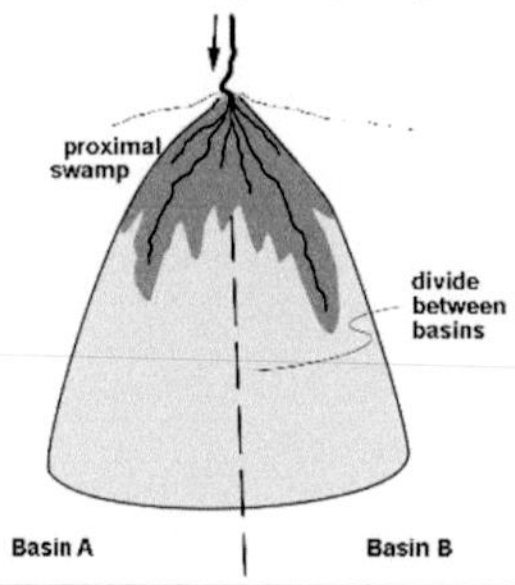	*Av-2.* Probable large wetland on Cubango megafan; and discharge to Etosha depression 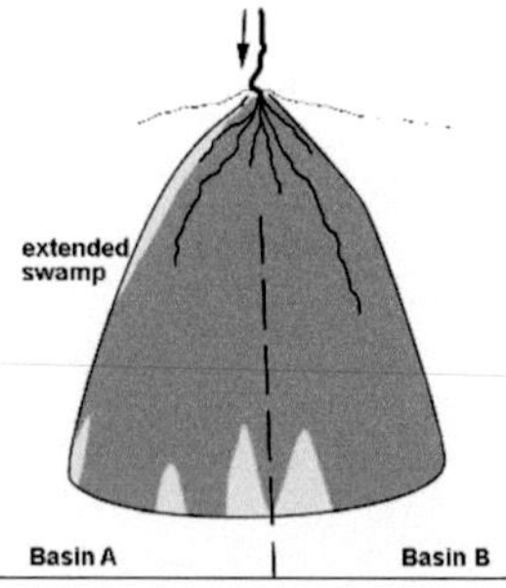	*Av-5.* Negligible discharge to rift (ORB) (or Makgadikgadi palaeolakes when these existed); reduced Okavango megafan wetland
Avulsion – Single-channel style Rapid, unrestricted avulsion of Okavango R. on megafan cone—orientation alternates between Etosha depression and Okavango rift	*Av-3.* Highly variable ("on-off") discharge (but reduced) alternately between Etosha depression and Okavango megafan — likely on shorter (< 10^{2-3} yr?) time scales, dependent on river orientation; Etosha depression dry or wet (Palaeolake Etosha) (El Niño cyclicity?) depending on Kunene River orientation	*Av-4.* Highly variable discharge continuously (but greater than in arid regime) to Etosha depression — likely on shorter (10^{2-3} yr?) time scales, dependent on river orientation; probable permanent water in Etosha depression (Palaeolake Etosha), with/without Kunene R. discharge; apical wetland on Cubango megafan	*Av-6.* Variable Okavango R. discharge to rift (ORB)— 'on–off' discharge as river is oriented to OWB or ORB

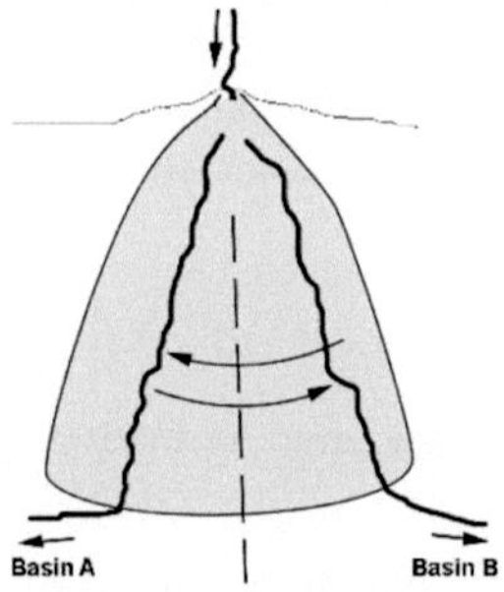

Figure 4.11 **Palaeoenvironmental scenarios – Avulsion modes (*Av-1–Av-6*)**. Based on the Okavango River reorientations on the Cubango 'divide megafan', showing effects in the Etosha basin (OWB-Owambo Basin A) and Okavango-Makgadikgadi system (ORB-Okavango Rift Basin B) under two climatic conditions. Heavy shading in the upper panel indicates flooded zone (*Av-1* typifies the modern limited Okavango megafan flood zone near the megafan apex). Modern geomorphic and climatic conditions shown as bold text in Fig. 4.12. Minimal discharge from interfan zones and megafans per se are not considered.

length of the megafan (as today) can be expected to have had similar effects to those described in the single-channel avulsion scenario above (*Av-3, Av-4*) – that is, leading most of the river's discharge all the way to the Etosha Pan, and thus probably supporting a Paleolake Etosha. This conclusion is based on the assumption of behaviour analogous to that of the modern Okavango River (discharge sufficiently large

to continue hundreds of km further downstream from the Cubango megafan). During climates wetter than those of today (*In-2*, Fig. 4.12), even more water would probably have reached the relatively nearby pan, and consequently with lower transmission loss, in this case also likely supporting a Palaeolake Etosha.

Wider regional impacts of Okavango River avulsions are also at least as important (*Av-5, Av-6,*

Climate / Cubango Megafan Mode	ETOSHA DEPRESSION — Arid / semi-arid conditions (~as today)	Humid conditions	OKAVANGO / MAKGADIKGADI SYSTEM — Role of Okavango R.
Incision – Okavango R. oriented to Etosha depression (OWB) in the longer term	*In-1.* Permanent but reduced discharge to Etosha depression (Palaeolake Etosha); no wetland on fan apex	*In-2.* Permanent higher discharge into Etosha depression (Palaeolake Etosha); no wetland on fan apex	*In-5.* Okavango River discharge removed from Okavango megafan hydrological budget; Okavango fan wetland reduced (fed only from the Cuito and Omatako rivers)
Incision – Okavango R. oriented to rift (ORB) in the longer term	***In-3.* Entire channelised discharge of Okavango R. removed from Etosha Pan; pan dry except in very wet years when fed by smaller rivers; no wetland on fan apex**	*In-4.* Entire channelised discharge of Okavango R. diverted from Etosha basin; pan probably dry depending on Kunene R. orientation (El Niño cyclicity?); no wetland on fan apex	***In-6.* Permanent (higher) discharge to Okavango rift, Okavango megafan wetland of modern size.** In wetter climates Okavango R. discharge may contribute to palaeolake hydrological budgets despite on-fan evaporative and infiltrational losses

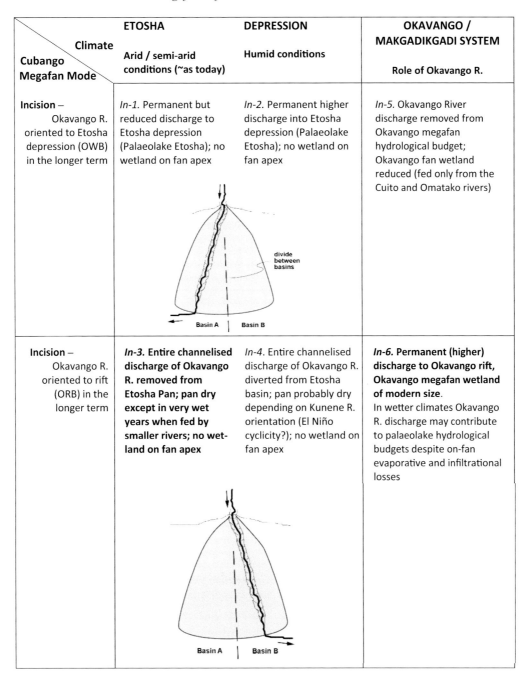

Figure 4.12 **Palaeoenvironmental scenarios – Incision mode (*In-1–In-6*).** Okavango River reorientations on the Cubango 'divide megafan', with two climatic conditions, showing effects in the Etosha and Okavango-Makgadikgadi systems. Bold text in scenarios *In-3* and *In-6* indicates modern geomorphic and climatic conditions (see text). Discharges from interfan zones and megafans *per se* are not considered.

Fig. 4.11, and *In-5, In-6*, Fig. 4.12). When the Okavango River was oriented along the extreme eastern margin of the Cubango megafan, discharge would have been oriented out of the OWB and into the ORB, thereby feeding not only the Okavango megafan downstream, but with overflow even gaining the Makgadikgadi depression (as occasionally happens today). In single-channel avulsion mode (discussed in Section 4.4.3.2), the orientation towards ORB would have been short-lived (the 'on–off' alternation of delivery); but when incised in this orientation the delivery of all discharge to the Okavango megafan would have continued significantly longer. A reorientation of the Okavango towards the ORB imposes a loss of discharge from 38% of the active upland source area to the Etosha Pan (Etosha incision scenario, *In-6*, Fig. 4.12).

4.4.3.1.2 Avulsions on the Kunene Megafan: Impacts on Palaeolake Etosha

Before its permanent diversion to the Atlantic Ocean, the Kunene River discharged mainly in the direction of the Etosha Pan. In terms of hydrological budgets, the Kunene River delivered discharge from an area increased greatly above that of today – that is, larger by 49% of the original (pre-diversion) total active catchment area of the pan (Table 4.1). This diversion has been recently dated to ~42 Ma (Houben et al. 2020), although relatively recent inundation of the lower slopes of the megafan by the Kunene River itself seems to have occurred from subapexes, in a zone where river bed lies little more than ten metres below the level of the lower surface of the megafan. Discharge to the pan would have varied depending on the mode of delivery (*avulsion mode* or *incision mode*) and on climate, perhaps at times supporting a Palaeolake Etosha. Despite the Paleogene age of the fan sediments, its contribution to the more recent history of the palaeolake needs to be considered because of the apparently more recent activity of the lower slope channels that appear to have led discharge from the Kunene River to the Etosha Pan.

Headward erosion by coastal Atlantic-oriented streams would have progressively encroached on the Kunene drainage basin, ultimately diverting fully half of the active catchment away from the pan (Table 4.1).

Today, with the Okavango and Kunene rivers both oriented away from the Owambo basin, fully ~118,525 km², or 87% of the original feeder area, are now removed from contributing to the Etosha Pan hydrological budget (Table 4.1). In this configuration discharge is so low that Etosha Pan is a dry lake (due to low discharges from the small Cuvelai and Omatako drainages or from groundwater). The resulting lowering of the pan floor and associated truncation of the fan toe, probably by long-continued wind action, is shown by the convex toe of the Cubango megafan (see Miller et al., Ch. 15). This kind of deflation is evidenced today by dust plumes emanating from the pan floor and extending hundreds of km offshore (SOD Fig. 4.4.3.1).

The complexity of the Etosha basin hydrology becomes apparent considering the gains and losses from one or both of the dominant feeder basins, the different styles of discharge delivery (avulsion *vs.* incision modes), and the synchronicity (or lack thereof) of river behaviours on their fans. Apart from these autogenic controls (fluvial avulsions on megafan surfaces), allogenic controls such as climatic fluctuations, activity on the rift faults and headward extension of coast-oriented streams further complicate the history of the Owambo basin. It deserves note that the Etosha depression under ample discharge would have become exoreic: i.e., under conditions of overflow, a Palaeolake Etosha with a depth of ~25 m, basin drainage would have been reversed such that the lake would have overflowed westwards into the present Kunene gorge.

It is not surprising, therefore, that highly variable conditions are recorded in the basin stratigraphy. For example, the Etosha Pan Clay Member records the long-term existence of a lake since the Oligocene, until ~6 Ma (Miller et al. 2010). Numerous lacunae are suggested by sand and salt horizons, suggesting repeated desiccation of the lake, with saline episodes. This history, at least latterly, can be explained by the demonstrably varied behaviours of major rivers on the two feeder megafans. We suggest that the Clay Mb. could well represent drainage configurations in which either or both the Kunene and Okavango rivers emptied into the pan. Periods of complete drying feasibly relate to times when both rivers were oriented away from the Owambo basin, in configurations like those of today. The Clay Mb. must also record major deflational lacunae (Miller et al. 2010), also analogous to present conditions. Desiccation starting at 6 Ma marks cessation of accumulation of the Clay Mb. (Miller et al. 2010), suggesting the permanent diversion of one or both of these rivers away from the Etosha depression. The evolution of the small Cuvelai megafan within the fan-margin groove between the Cubango and Kunene cones is undoubtedly connected to the complete abandonment by the much larger Okavango and Kunene rivers from flowing across of their megafan cones.

Although presently oriented to the ORB, the Omatako River is another divide-occupying feature (11, Fig. 4.10a, b; Table 4.2) that would have been oriented periodically towards the Etosha depression – further complicating the palaeoenvironmental reconstructions above.

4.4.3.2 Okavango Rift Basin (ORB) Rivers and Makgadikgadi Basin Palaeolakes

The ORB is the most complex of the three basins. Rivers entering the ORB drain two different highlands, flow towards two different basins, and the component megafans adjoin one another so that their hydrological behavior can be connected, especially during flood events. Megafans have provided significant discharge to the Makgadikgadi basin via the Boteti River outlet. But it is also true that all three ORB megafans have flowed on occasion towards the middle Zambezi drainage to the east.

4.4.3.2.1 Variable Discharge in the Okavango River: Impacts on the ORB

When the Okavango River on the Cubango divide megafan was oriented to the Etosha Pan, fully 26% of its active source area was lost to the Okavango megafan (Table 4.1) downstream in the ORB. Loss of volumes of this magnitude at times when the Cubango megafan was active would have reduced significantly the hydrological activity on the Okavango megafan, and it suggests a significant reduction the size of its present wetland. In turn, it seems probable that essentially no discharge would have reached the Makgadikgadi sump from the Okavango drainage system because, under the present far higher discharge regime, only 2% of the influx at the head of the fan exits the toe towards the Makgadikgadi depression (McCarthy 2013).

However, when the Okavango River in the OWB was diverted towards the ORB (as at present) the active source area of the Okavango megafan of 26% (Table 4.1) translates into fully 57% of the present discharge to the Okavango megafan (McCarthy and Ellery 1998) – i.e., effectively doubling the discharge at the apex. In turn, in the *single-channel mode*, far larger proportions of the Okavango River flow would have reached the toe of the fan (Section 4.1.3.1). Increased delivery of water downstream to the Makgadikgadi depression is thus a likely result. Such longer-term conditions may have lasted as long as millions of years, if dating of the present incised

configuration of the Okavango River is an indication – i.e., since 4 Ma (Miller et al. 2010).

4.4.3.2.2 Avulsions on the Okavango, Zambezi, and Cuando Megafans

Sweep-angles of the fan-forming rivers emphasise the variety of possible flow orientations and interactions that can occur on the three ORB megafans (curved arrows on fans 4, 5, 6, Fig. 4.10a, b) (SOD 4.4.3.2). The terminal fault-controlled depressions at the foot of all three fans are connected by waterways. Okavango discharge would have been fed eastwards along this depression to the Cuando megafan (Linyanti and Chobe swamps), as today, as well as along radially oriented channels such as the modern Selinda Spillway/Savuti channel (Fig. 4.10c), especially under conditions of single channel discharge (Section 4.1.3.1). Avulsions on the Zambezi megafan may have had effects on its neighbours. When oriented southwest, as opposed to the present southeast orientation, some of the Zambezi discharge would flood the fault-related depressions of the Cuando megafan. Such flooding would be more pronounced at times when the entire Zambezi discharge was directed southwest radially down its small megafan. The far higher discharge of the Zambezi River on this smallest megafan, even the multi-channel mode, would have been attenuated very little – unlike the almost complete attenuation currently of this mode on the large Okavango megafan with the Okavango River's far smaller discharge.

The regional and longer-term significance of these demonstrable river orientations is, however, likely to remain moot until more is known about the sequence of controls of tectonic movements and drainage captures in deeper time (at least as far back as the Late Miocene), especially in Zambia and the southeastern Congo basin (Moore et al. 2012). Such controls would outweigh autogenic megafan dynamics, with activity on rift faults acting as major controls responsible for the diversion of the entire discharge of the Zambezi River (36.73 km^3/yr, at the Mosi oa Tunya/Victoria Falls; Moore et al. 2007) towards the Makgadikgadi depression, prior to the Late Pliocene (Moore et al. 2012, their fig. 7A). This massive discharge from a catchment fully 336,000 km^2 in area (Table 4.1) is seen as one source large enough to have fed the vast Palaeolake Makgadikgadi (66,000 km^2; Burrough et al. 2009; Moore et al. 2012), which inundated the now-dry pans of the Makgadikgadi depression and included Lake Ngami (south of the Okavango mega-fan) and distal parts of the Cuando and Zambezi mega-fans (Fig. 4.10c) (Grove 1969; Thomas and Shaw 1991; Moore et al. 2012). Modern seismic activity in the wider region is well attested (Haddon and McCarthy et al. 2005), and recent movement on the faults is graphically demonstrated on the Cuando megafan (Fig. 4.6, profile *E-F*). Compared with these longer-term dynamics, the shorter term climatic shifts (mid to upper Pleistocene) based on extensive dating of beach ridges that encircle the Makgadikgadi depression, and on evidence for lake high-stands in central Africa, Burrough et al. (2009) have suggested that wetter climatic phases can account for the repeated establishment of permanent waterbodies in the basin.

The connections between the three contiguous ORB megafans suggests that separately or in concert they are able to direct flow to the middle Zambezi basin or alternately to the Makgadikgadi depression. Hydrological conditions related to megafan behaviours – incision/no incision and single/multi-channel river styles – probably operated regionally against the backdrop of swell uplift, the extension of rifting into the ORB region, and repeated climatic fluctuations of the Quaternary. Headward advance of the Mambova Falls knickpoint on the Zambezi River (presently at the distal eastern tip of the megafan) will slowly divert all inland (Makgadikgadi) directed flow on all three megafans to the exorheic Zambezi River (Moore et al. 2012).

4.4.3.3 Upper Zambezi Basin (UZB) Connections to the Congo Basin

The Cassai River on its megafan (7, Fig. 4.10a, b) has demonstrably flowed alternately into the upper Zambezi and upper Congo basins. The Cassai discharge to these very large basins, however, is trivial (Table 4.1). Its interest lies more in comparison with the other two major diversions across the Zambezi–Congo divide that have been documented further east in Zambia – namely that of the Chambeshi River diverted into the Congo system from an earlier permanent orientation towards the Makgadikgadi, and the Kafue River that was diverted from the Makgadikgadi

basin to the Indian Ocean, both at times in the Mid-Pleistocene (Moore et al. 2012). By comparison to these permanent diversions, the Cassai River on its divide megafan has straddled the Zambezi-Congo divide, connecting these basins hydrologically. Interestingly, this repeated connection appears to have operated on two occasions widely separated in time (Upper Zambezi Basin stages 1 and 3, SOD 4.3).

4.5 Conclusions

4.5.1 Identification and Location of Megafans

Using remotely sensed data to ascertain morphologic and morphometric attributes, ten landforms of partial cone morphology and very low gradient were identified according to a set of criteria (SOD Table 4.2.2) as active or inactive megafans, i.e., as probable fluvial sedimentary masses. These cones lie in three subbasins and stretch in an arc from northern Namibia (Owambo Basin) via the Okavango rift (Okavango Rift Basin) in Botswana, to northwestern Zambia/eastern Angola (Upper Zambezi Basin). The Okavango River is highly unusual in global terms by having given to rise to *two* megafans along its course, the downstream feature being the well-known Okavango 'delta', long considered the only megafan in the region. The Lungué–Luio fan in the Zambezi-basin is possibly of the derived type (Wilkinson, Ch. 17), another relatively uncommon geomorphic feature.

Although megafans comprise only 24% of larger fans (i.e., >30 km-long in a world-wide study; Hartley et al. 2010), their size confers on them a major significance in the landscapes of continental basins. In the northern Kalahari Basin, they measure collectively 180,000 km^2. The total is closer to 300,000 km^2 if the 'possible fans' are included. The study also demonstrates that such features are a normal component of the Kalahari landscape. Eight of the Kalahari megafans have formed in open topographic basins (demonstrating also that megafans are not deltas).

4.5.2 Active and Inactive Megafans

In the northern Kalahari region, subsidence in the Okavango Rift appears to be the main cause for active megafan sedimentation, the basin with dominant, recent fanlike sedimentation. The Okavango megafan captures virtually all sediment supplied to it, both clastic and solute. Tectonic activity associated with the rift in particular, may explain regional incision that has entrenched those rivers that flow into the rift depression. Such incision has in turn rendered large areas of the megafan surfaces inactive in the Owambo and Upper Zambezi basins.

4.5.3 Morphometric Analysis

The usual relationships seen in smaller alluvial fans – correlation between fan area and feeder basin area – are eclipsed in the Okavango Rift Basin due to the constriction imposed by fault scarps on the horizontal extent of the potentially far larger megafan sediment bodies. More normal relationships appear to exist in the other component basins, but the study area population is too small to fully examine the hypothesis. Expected slope relationships are retained, with the largest rivers displaying the lowest slopes and vice versa. Roughness maps confirm that megafan surfaces in the study area follow the global pattern of very low declivity and great smoothness.

4.5.4 Divide Megafans

Six of the ten megafans are located on divides between major drainage basins, a configuration of drainage and depressions that is another unusual characteristic in world terms. Two of the possible megafans are also located on divides. Divide megafans deliver water and sediment alternately to neighbouring basins. Short- and long-duration avulsions must have occurred repeatedly in the study area in the Neogene. Long-duration orientations must have affected hydrology of the Owambo, Okavango, and Makgadikgadi basins in particular. A matrix of controls common in the study area – river orientation on divide megafans, feasible precipitation ranges, avulsion styles, and incision/non-incision river dynamics – showed the complexity of hydrological response that can arise in basins in which megafans are located, demonstrating extremes of gain/loss of 25–50% of upstream catchment areas to the downstream basins. The most striking examples are the existence of Palaeolake Etosha as a permanent lake at times, and the likely repeated and significant reduction

of the discharge that supports the modern wetland on the Okavango megafan.

Major reorientations of regional drainage on six megafan cones amount to significant and continuing fluctuations in discharge to each basin under a regime of frequent (in geological terms) autogenic forcing that has not been mentioned heretofore for the northern Kalahari hydrological systems.

4.5.5 Interpretation of Continental Sedimentary Sequences

Ventra and Clarke (2018) have noted that megafan dynamics involve both auto- and allogenic controls, complicating interpretation of sedimentary records in particular. Sediment supply to basins will be determined at least partly by autogenic controls of megafan river behaviour, and consequently ought to be included with more commonly invoked controls such as tectonics and climate fluctuations, especially in basins where divide megafan dynamics may have operated. It can be argued that the effects induced by megafan avulsions in some circumstances even override the effects of climate and tectonics, especially for sedimentary sequences in two of the three basins in this study.

Acknowledgements

MJW gratefully acknowledges Initiative Funding for field work from Jacobs Technology, Houston. Yanni Gunnell provided important discussion and critical assistance in simplifying the exposition.

References

Burke, K. and Gunnell, Y. (2008). *The African Erosion Surface: A Continental-scale Synthesis of Geomorphology, Tectonics, and Environmental Change over the Past 180 Million Years*. Geological Society of America Memoir, 201, 66 pp.

Burke, K. and Wilkinson, M. J. (2016). Landscape evolution in Africa during the Cenozoic and Quaternary–the legacy and limitations of Lester C. King. *Canadian Journal of Earth Sciences*, 53, 1089–1102.

Burrough, S. L., Thomas, D. S. G., and Bailey, R. M. (2009). Mega-Lake in the Kalahari: A Late Pleistocene record of the Palaeolake Makgadikgadi system. *Quaternary Science Reviews*, 28, 1392–1411.

Cooke, H. J. (1980). Landform evolution in the context of climate change and neo-tectonism in the middle Kalahari of north-central Botswana. *Transactions of the Institute of British Geographers,* 5, 80–99.

Cooke, R. U., Warren, A., and Goudie, A. S. (1993). *Desert Geomorphology.* University College London Press, London, 2nd edn, 526 pp.

Cotterill, F. P. D. and de Wit, M. J. (2011). Geoecodynamics and the Kalahari epeirogeny: linking its genomic record, tree of life and palimpsest into a unified narrative of landscape evolution. *South African Journal of Geology,* 114, 489–514.

Daly, M. C., Green, P., Watts, A. B., et al. (2020). Tectonics and landscape of the Central African Plateau and their implications for a propagating Southwestern Rift in Africa. *Geochemistry, Geophysics, Geosystems,* 21, e2019GC008746. https://doi.org/10.1029/2019GC008746

Debenham, F. (1952). The Kalahari today. *Geographical Journal,* 118, 12–23.

Dill, H., Kaufhold, S., Lindenmaier, F., et al. (2012). Joint clay-heavy-light mineral analysis: a tool to investigate the hydrographic-hydraulic regime of Late Cenozoic deltaic inland fans under changing climatic conditions (Cuvelai-Etosha Basin, Namibia). International Journal of Earth Science, 102, 1–40.

DMAAC (Defense Mapping Agency Aerospace Center), various dates. Operational Navigation Charts (ONC), 1:1 million scale, St. Louis, Missouri, USA.

Du Toit, A. L. (1927). The Kalahari. *South African Journal of Science,* 24, 88–101.

Earth Sciences and Remote Sensing Unit—http://eol.jsc.nasa.gov, accessed February 2022.

Grove, A. T. (1969). Landforms and climatic change in the Kalahari and Ngamiland. *Geographical Journal,* 135, 191–212.

Haddon, I. G. and McCarthy, T. S. (2005). The Mesozoic–Cenozoic interior sag basins of Central Africa: the Late-Cretaceous–Cenozoic Kalahari and Okavango basins. *Journal of African Earth Sciences,* 43, 316–333.

Hartley, A. J., Weissmann, G. S., Nichols, G. J., and Warwick, G. L. (2010). Large distributive fluvial systems: characteristics, distribution, and controls on development. *Journal of Sedimentary Research,* 80, 167–183.

Houben, G. J., Kaufhold, S., and Miller, R. McG., et al. (2020). Stacked megafans of the Kalahari Basin as archives of paleogeography, river capture and Cenozoic paleoclimate of southwestern Africa. *Journal of Sedimentary Research,* 90, 980–1010.

Hutchins, D. G., Hutton, S. M., and Jones, C. R. (1976a). The geology of the Okavango Delta. *Proceedings of the Symposium on the Okavango Delta and its future utilization.* National Museum, Botswana, 13–19.

Hutchins, D. G., Hutton, L. D., Hutton, S. M., Jones, C. R., and Leonhert, E. P. (1976b). A summary of the geology, seismicity, geomorphology and hydrogeology of the Okavango Delta. *Geological Survey of Botswana, Bulletin* 7.

Kottek, M., Grieser, J., Beck, C., Rudolf, B., and Rubel, F. (2006). World map of Köppen–Geiger climate classification. *Meteorologische Zeitschrift,* 15, 259–263.

Kreslavsky, M. A. and Head, J. W. (1999). Kilometer-scale slopes on Mars and their correlations with geologic units: initial results from Mars Orbiter Laser Altimeter (MOLA) data. *Journal of Geophysical Research,* 104, 21,911–21,924.

Kreslavsky, M. A. and Head, J. W. (2002). Kilometer-scale roughness of Mars' surface: results from MOLA data analysis. *Journal of Geophysical Research,* 105, 26,695–26,712.

Mallick, D. I. J., Habgood, F., and Skinner, A. C. (1981). *Geological interpretation of Landsat imagery and air photography of Botswana.* Overseas Geology and Mineral Resources, Institute of Geological Sciences, National Environmental Research Council, London.

McCarthy, T. S., and Ellery, W. M. (1998). The Okavango Delta. *Royal Society of South Africa, Transactions,* 53, 115–126.

McCarthy, T. S., Smith, N. D., Ellery, W. M., and Gumbricht, T. (2002). The Okavango Delta—semiarid alluvial-fan sedimentation related to incipient rifting. In M. W. Renaut and G. M. Ashley, eds., *Sedimentation in Continental Rifts.* SEPM Special Publication, 73, 179–193.

McCarthy, T. S. (2013). The Okavango Delta and its place in the geomorphological evolution of southern Africa. *South African Journal of Geology,* 116, 1–54.

McFarlane, M. J. and Eckardt, F. D. (2008). Lake Deception: a new Makgadikgadi palaeolake. *Botswana Notes and Records,* 38, 195–201.

Mendelsohn, J. and El Obeid, S. (2004). *The Okavango River.* Struik Publishers, Cape Town.

Mendelsohn, J. M., van der Post, C., Ramberg, L., et al., 2010. *Okavango Delta: Floods of Life.* RAISON, Windhoek, Namibia, 144 pp.

Miller, R. McG. (1997). The Owambo Basin of northern Namibia. In R.C. Selley, ed., *African Basins, Sedimentary Basins of the World.* Elsevier, Amsterdam, 237–268.

Miller, R. McG. (2008). *The Geology of Namibia, Vol. 3: Palaeozoic to Cenozoic.* Ministry of Mines and Energy, Geological Survey, Windhoek, Namibia.

Miller, R. McG. (2010). Lithology of *Boreholes* WW 201216 and WW 201217, Ohangwena region, Namibia. Dept. of Water Affairs and Forestry, Windhoek, Namibia.

Miller, R. McG., Pickford, M., and Senut, B. (2010). The geology, palaeontology and evolution of the Etosha Pan, Namibia: implications for terminal Kalahari deposition. *South African Journal of Geology,* 113, 307–334.

Miller, R. McG. (2013). *Groundwater for the North of Namibia: Technical Note no. 1—drill-log interpretation and evaluation of drillings KOH I and KOH II aquifers,* Ohangwena Region, Cuvelai–Etosha Basin. BGR report 05-2345, DWAF, Windhoek, Namibia and BGR, Hanover, Germany.

Miller, R. McG., Lohe, C., Hasiotis, S. T., et al. (2016). The Kalahari Group in the 400-m deep core borehole WW 203302, northern Owambo Basin. *Communications of the Geological Survey of Namibia*, 17, 143–238.

Moore, A. E. (1999). A reappraisal of epeirogenic flexure axes in Southern Africa. *South African Journal of Geology*, 102, 363–376.

Moore, A. E. and Larkin, P. A. (2001). Drainage evolution in south-central Africa since the break-up of Gondwana. *South African Journal of Geology*, 104, 47–68.

Moore, A. E., Cotterill, F. P. D., Main, M. P. L., and Williams, H. B. (2007). The Zambezi River. In A. Gupta, ed., *Large Rivers: Geomorphology and Management*. Wiley, Chichester, 311–332.

Moore, A. E., Cotterill, F. P. D., and Eckardt, F. D. (2012). The evolution and ages of Makgadikgadi palaeo-lakes: Consilient evidence from Kalahari drainage evolution. *South African Journal of Geology*, 115, 385–413.

Podgorski, J. E., Green, A. J., Kgotlhang, L., et al. (2013). Paleo-megalake and paleo-megafan in southern Africa. *Geology*, 41, 1155–1158.

Stengel, H. W. (1963). *Wasserwirtschaft, Waterwese, Water Affairs in S. W. Africa*. Verlag Der Kreis, Windhoek, Namibia.

Thomas, D. S. G. and Shaw, P. A. (1991). *The Kalahari Environment*. Cambridge University Press, Cambridge, 284 pp.

Ventra, D. and Clarke, L. E. (2018). Geology and geomorphology of alluvial and fluvial fans: current progress and research perspectives. In D. Ventra and L. E. Clarke, eds., *Geology and Geomorphology of Alluvial and Fluvial Fans: Terrestrial and Planetary Perspectives*. Geological Society of London, Special Publication, 440, 1–21.

Wehberg, J. and Weinzierl, T. (2013). The Okavango Basin – physical-geographical settings. *Biodiversity & Ecology*, 5, 11–13.

Weissmann, G. S., Hartley, A. J., Nichols, G. J., et al. (2010). Fluvial form in modern continental sedimentary basins: distributive fluvial systems. Geology 38, 39–42.

Weissmann, G. S., Hartley, A. J., Nichols, G. J., et al. (2011). Alluvial facies distributions in continental sedimentary basins – distributive fluvial systems. In S. K. Davidson, S. Leleu, and C. P. North, eds., *From River to Rock Record: The Preservation of Fluvial Sediments and their Subsequent Interpretation*. SEPM Special Publication, 97, 327–355.

Wellington, J. H. (1938). The Kunene River and the Etosha Plain. *South African Geographical Journal*, 20, 21–32.

Wellington, J. H. (1955). *Southern Africa, A Geographical Study, Vol. 1: Physical Geography*. Cambridge University Press, Cambridge, 528 pp.

Wells, N. A. and Dorr, J. A. (1987). Shifting of the Kosi River, northern India. *Geology*, 15, 204–207.

Wilkinson, M. J., Marshall, L. G., and Lundberg, J. G. (2006). River behavior on megafans and potential influences on diversification and distribution of aquatic organisms. *Journal of South American Earth Sciences*, 21, 151–172.

Wilkinson, M. J., Kreslavsky, M. H., and Miller, R.McG. (2008). *Megafans of the Northern Kalahari Basin. Third Southern Deserts Conference – Kalahari 2008*. University of Oxford, School of Geography, Molopo Lodge, Northern Cape, South Africa, 16–19 Sept 2008.

Wilkinson, M. J., Marshall, L. G., Lundberg, J. G., and Kreslavsky, M. H. (2010). *Megafan environments in northern South America and their impact on Amazon Neogene ecosystems*. In C. Hoorn and F. P. Wesselingh, eds., *Amazonia, Landscape and Species Evolution: A Look into the Past*. Blackwell, London, 162–184.

5

The Chaco Megafans, South America

EDGARDO M. LATRUBESSE, EDWARD PARK, CARLOS RAMONELL,
M. ANWAR SOUNNY-SLITINE, and EDGARDO CAFARO

Abstract

The Chaco plain, covering $> 800,000 \, \mathrm{km}^2$ in Bolivia, Argentina, and Paraguay, comprises six megafans. These are fed mainly from Subandean basins that are among the highest sediment-yielding basins of the Andes: of the total supplied ($\sim 325 \, \mathrm{Mt \cdot yr}^{-1}$) $\sim 68\%$ is trapped on the megafans – because all the rivers except one end on the megafans, reaching at most $\sim 50\%$ of the distance of their late Pleistocene ancestors. As such, the Chaco plain is one of the largest active continental sedimentary sinks of the planet, and includes the longest known megafan. The rivers terminate in the largest area of seasonal wetlands in South America, a product of (i) extremely flat megafan surfaces, (ii) the mosaic of palaeolandforms and present fluvial and lacustrine patterns, and (iii) the hydrogeomorphological dynamics under the current Holocene humid climate. Calculations of specific power appear to explain the effectiveness of these rivers in transporting the current inputs of water and sediment. However, during part of the Late Pleistocene the hydrological and sedimentological regimes allowed the fan-forming rivers to deliver sediment to the regional Paraguay and Paraná trunk rivers and thence to the ocean. New morphometric data describe relationships between feeder-basin area and megafans area, slope and circularity.

5.1 Introduction

The Chaco, in South America, is a vast plain that extends over more than $800,000 \, \mathrm{km}^2$ through Bolivia, Argentina, and Paraguay. From a geologic point of view, the Chaco plain is an Andean retroarc foreland basin that merges in the east and south with the Chaco–Pampean platform, and in the north transitions into the Beni plains of Bolivia (Moxos plains) (Fig. 5.1). The landscape is characterised by an extremely flat plain, ranging in elevation between 600 and 40 m above sea level, and covered by an almost continuous layer of Quaternary sediments. Owing to favorable tectonic and climatic factors, the Chaco is the largest surface area of the planet covered by Quaternary megafans. These gigantic landforms were first described by Iriondo (1993), with subsequent studies investigating their geomorphology, tectonics, and sedimentary basins (Barnes and Heins 2009; Hartley et al. 2010; Horton and DeCelles 2001; Latrubesse 2015; Latrubesse et al. 2012; May et al. 2008a, b; Wilkinson et al. 2006). Here we present a state-of-the-art overview of the Chaco megafans; introduce new morphometric data; characterise the hydro-geomorphologic dynamics of the generator rivers; analyse the relationship between megafans and wetlands; and discuss the role of the Chaco on a continental scale as an active sediment sink.

5.2 Methods

We compiled information from the literature and also produced new results after conducting multiple field and boat surveys in Bolivia, Argentina, and Paraguay in 1995, 2005, 2006, 2007, 2008, 2009, 2010, 2015, and 2016. We used remote sensing products (Shuttle Radar Topographic Mission – SRTM – digital elevation data, and optical satellite images), and thematic and historical maps and the global HYDROSHED dataset was utilised to extract stream networks (Lehner et al. 2006) and thus generate catchment maps for each fan system. We then measured a selection of morphometric attributes of the drainage basins, megafans, and river channels such as area, relief, hypsometry, gradients, perimeters, river widths, lengths, sinuosity, and various shape indices (Table 5.1).

The hydrological record and sediment yields of the Chaco fluvial systems were obtained from gauge stations and the literature (Cochonneau et al. 2006; Guyot et al. 1994; Guyot et al. 1996; Latrubesse and Restrepo 2014; SSRH 2004).

The hypsometry of the drainage basins feeding into the Chaco megafans was calculated from the SRTM GL1 DEM product at 30 m ground resolution (downloaded from USGS Earth Explorer). Individual drainage basin DEMs were extracted using the catchment delineation method advocated by Lehner et al. (2006). The pixel elevation values of the DEM were ranked by frequency, and total cumulative areas normalised to 100% were produced on that basis for inter-basin comparison (Fig. 5.2).

To map the typically flooded area on the Chaco fan during the high-water season, we used the Enhanced Vegetation Index (EVI) layer of the MODIS-Terra MOD13Q1 product (v. 6) supplied from the Land Processes Distributed Active Archive Centre (LPDAAC). The EVI is calculated as:

$$EVI = G * \frac{(NIR - RED)}{(NIR + C1 * RED - C2 * Blue + L)} \tag{5.1}$$

where $C1$, $C2$, L, and G are the coefficients of the MODIS-EVI algorithm. This product is a 13-day composite driven from the daily product (processing Level-3 data), which yields vegetation indices at 250 m resolution. We extracted EVI layers from March (i.e., peak flood month) between 2015 and 2018 and averaged them to highlight the flooding patterns. Since the Chaco plain is a vast flat area with relatively dense vegetation and grass, we decided to use the EVI to map the flooded vegetation area during the flood season, which is more sensitive to the water content than the Normalised Difference Vegetation Index (NDVI). Given the vegetation cover over the flat area, a Landsat-derived flood frequency map (such as Pekel et al. 2016) largely underestimates the extent of the

flooded area. The EVI has been used before in the Amazonian wetlands (*várzea*) to map the flooded vegetation area during the high water season (Park and Latrubesse 2017). Here, we applied two-step filters to retrieve the flooded area. The method is based on thresholds identified using a density slicing method. We used EVI = 0.36 as a first filter to obtain open-water pixels. Open-water features are efficiently extracted using EVI because most of the spectra from the Near Infrared (NIR) band are absorbed by the water. Then EVI = 0.45 was used to retrieve flooded vegetation pixels. We overlaid on the EVI map a flooded vegetation boundary derived from RapidEye imagery at several locations along the Pilcomayo River during the wet season to test and validate this threshold. Also, several DEM-derived elevation profiles were drawn to map the distribution of local topographic hollows where water is likely to be stored during the flood season. Finally, the open water and flooded vegetation pixels were combined to generate the flood map. Our inundation map was visually inspected by comparing it with the results from the available literature on Chaco wetlands (Marchetti 2017).

5.3 The Modern Chaco Environment

The Chaco plain is dominated by subtropical semi-deciduous vegetation. Because of its biogeographic imprint, the Chaco is also considered the third largest biome of South America, after the Amazon and the Cerrado. The Chaco climate is driven by a monsoonal circulation system and ranges from seasonally dry tropical to semiarid. During austral summer (January, February, March – JFM), the South American Summer Monsoon (SASM) (Nogués-Paegle et al. 2002; Zhou and Lau 1998) generates intense convective rainfall and causes the summer precipitation maximum. The Andes trigger heavy orographic convective rainfall, while the moisture-laden northerly low-level flow travels along the eastern slopes of the Andes (e.g., Garreaud et al. 2003; Vizy and Cook 2007; Vuille et al. 1998). The Andean morphotectonic unit that concentrates a large proportion of the orographic precipitation is the Subandean Belt. Precipitation and cloudiness sustain the highly biodiverse montane and submontane forests (Yungas), which reach as far as ~28° S into Tucuman Province, Argentina.

During winter, the Inter-Tropical Convergence Zone migrates to the north and the Chaco experiences a long dry season. The E–W Chaco annual rainfall gradient is strong, changing from 1,300 mm/yr in the east to near 400 mm/yr at the Paraguay–Bolivia border, then increasing once again to more than 1,200 mm/yr at the Subandean Mountains and footslopes. During winter (June, July, August – JJA), cold air incursions from the south polar advection (known as *surazos* in Bolivia, or *friagems* in Brazil) can episodically cause severe temperature drops and additional rainfall throughout the year (Garreaud 2000). The dominant winds nonetheless come from the northern quadrant. They persist during a good part of the year, blowing at mean velocities above 4.5 m/s, producing significant deflation in the Chaco plain, and generating dust transport and aeolian dunes.

5.4 The Quaternary Chaco Megafans

5.4.1 General Characteristics and Metrics

The main Chaco rivers draining the Andes and debouching in the Chaco plain have formed fans that range from typical piedmont alluvial fans to giant fluvial fans (Horton and DeCelles 2001; Iriondo 1993; Latrubesse et al. 2005, 2012). They form the flat, aggradational alluvial–aeolian terrain that extends from the Subandean piedmont towards the east and south. The megafans are fed by large drainage basins, such as the Pilcomayo and the Grande, which flow through diverse geotectonic units such as the Puna–Altiplano plateau, the Eastern Cordillera, and the Subandean zone. The Bermejo, Juramento, and Dulce watersheds in the south descend from the Santa Barbara system and the Pampean Ranges (Fig. 5.1).

The main northern Chaco megafans are generated by the Grande and Parapetí rivers. The drainage area of the Parapetí in the catchment upstream of the fan apex is $\sim 8,000\,\mathrm{km^2}$, whereas the Grande River drainage basin is $59,800\,\mathrm{km^2}$ (Guyot et al. 1994) (Fig. 5.1). The Grande megafan spreads across an area of $\sim 58,140\,\mathrm{km^2}$ (Table 5.4). Presently, the Grande fluvial system flows towards the Amazon basin; but in the past, the system coalesced with the Parapetí megafan ($\sim 60,000\,\mathrm{km^2}$) located to the south (Fig. 5.1).

The Parapetí River currently lacks the discharge to reach the regional base level and ends in the Izozog swamps, a remarkable inland wetland system of the Bolivian Chaco located $\sim 130\,\mathrm{km}$ from the Andean mountain front. As indicated by palaeochannels and topography, the Parapetí River shifted laterally across a broad area (alternately discharging into the Paraguay River in the Pantanal Basin, or crossing the territory of Paraguay and coalescing with the Pilcomayo megafan), or flowing north towards the Amazon watershed (Iriondo 1993; Latrubesse et al. 2012) (Fig. 5.1).

The major southern Chaco megafans (Pilcomayo, Bermejo, Salado–Juramento, and Dulce) cover an area of approximately $400,000\,\mathrm{km^2}$, 570 km long (from 21 to 27° S) and 700 km wide, between the Subandean Ranges and the Paraguay and Paraná rivers (Fig. 5.1). The largest megafan is the Pilcomayo, with an area of $223,000\,\mathrm{km^2}$, a $\sim 610\,\mathrm{km}$ mean radius and a maximum length of $\sim 720\,\mathrm{km}$. The megafan's apex has been suggested to occur near Villamontes in Bolivia (Iriondo 1993), i.e., at the mountain front of the Subandean Ranges; but because of low structural ridges of Cenozoic rock striking across the foreland basin east of Villamontes, we suggest instead that the apex of the Quaternary Pilcomayo megafan is situated $\sim 60\,\mathrm{km}$ further downstream, near Ibibobo. From there, the Pilcomayo River enters the Chaco plain without further obstacles or lateral geological constraints, forming palaeochannels and the present fluvial lobe in a radial pattern.

The Bermejo megafan is a narrower system constrained between the Pilcomayo and Salado megafans. It extends for $\sim 650\,\mathrm{km}$ from the apex, near the town of Embarcación to the confluence with the Paraguay River. This is currently the only southern Chaco river that reaches the regional base level, i.e., the Paraguay–Paraná rivers.

The Juramento–Salado megafan constrains the Bermejo megafan in the south. Its oval shape is nearly 650 km long and 150 km wide. It coalesces in the north with the Bermejo, in the east with older Quaternary units of the Paraná River fluvial belt, and in the south with the Dulce megafan.

Despite their large differences in size, the Pilcomayo Af/Ad ratio (fan area/source-catchment area) is 2.5, while the ratio in the relatively small Parapetí is 9.3 (Table 5.1), meaning that a smaller

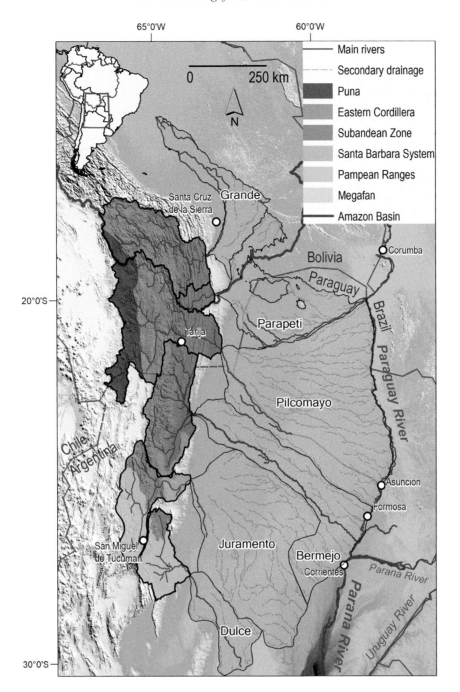

Figure 5.1 **Major megafans of the Chaco, drainage basin areas, and the main Andean geotectonic domains**. Solid blue lines indicate the active channel of the major rivers. Dashed lines represent palaeochannels and underfit streams within palaeochannels.

drainage area in the Parapetí generated a proportionally larger megafan than did the Pilcomayo. The Grande ratio of 0.8 is even less efficient, which implies a large basin with a relatively small fan.

The Parapetí watershed shows the lowest range in basin relief (BR = Basin Max. Elevation − Basin Min. Elevation), and the larger proportion of the hypsometry is concentrated from 700 to 2,000 m a.s.l. in the Subandean Zone (Fig. 5.2). The largest drainage basins, such as the Salado–Juramento, Grande, and Pilcomayo, show wider hypsometric curves because their watersheds spread across a larger range of elevations and geologic provinces above the Subandean Zone, such as the Eastern Cordillera, and

Table 5.1 *Metrics of the megafans (Af) and their contributing watershed areas (Ad)*

Megafan name	Basin area (km²)	Subandean area (km²)	Subandean/Basin area ratio (%)	Basin elevation attributes[a] (m)						Fan size, shape and elevation attributes									Fan Area/Basin Area (km²/km²)	Fan Length–FL[c] (km)	River Length–RL (km)	RL/FL ratio (km/km)
				Min.	Max.	Mean	STD	Median	Relief–BR (Max–Min)	Area (km²)	Perimeter (km)	Circularity index[b]	Min. elev. (m)	Max. elev. (m)	Mean elev. (m)	STD (m)	Relief–RI (Hmax–hmin)[c] (m)	Slope (m/m)				
Grande	59,378	18,660	31.43	431	5,135	2,521	973	2,488	4,704	58,140	1,748	0.20	154	547	267	66	393	0.000823899	0.8	477	346	0.73
Parapetí	7,456	7,235	97.04	597	3,539	1,312	445	1,193	2,942	59,656	2,823	0.11	82	658	253	89	576	0.00117551	9.3	490	174	0.36
Pilcomayo	87,349	20,372	23.32	315	5,904	3,006	1,213	3,372	5,589	216,216	2,054	0.64	60	350	134	54	290	0.000411932	2.5	704	180	0.26
Bermejo	51,949	26,051	50.15	279	5,911	1,771	1,224	1,384	5,632	111,889	1,970	0.36	50	285	124	58	235	0.000335714	2.2	700	700	1.00
Juramento	41,159	5,408	13.14	387	6,342	2,643	1,242	2,615	5,955	150,575	1,661	0.69	53	397	148	62	344	0.000577181	3.7	596	200	0.34
Dulce	27,557			188	5,534	915	852	496	5,346	37,339	949	0.52	68	203	102	26	135	0.000166256	1.4	812	240	0.30

Notes: [a] the hypsometric curves for each drainage basin are presented in Fig. 5.2. [b] Circularity index given by: $4\pi A/P^2$. [c] Relief and fan length are calculated based on the SRTM DEM GL1 (30 m)

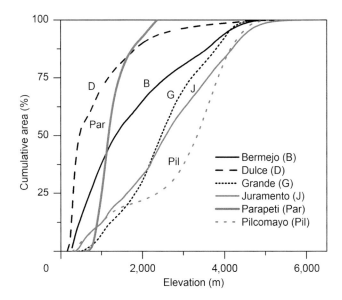

Figure 5.2 Hypsometric curves of the watershed areas feeding the Chaco megafans.

in the case of the Pilcomayo, even the Puna–Altiplano plateau (Fig. 5.2). The Pilcomayo also exhibits the largest basin relief and the highest elevation median, wheras the Parapetí presents the lowest basin relief. Although the relief of the Dulce watershed attains ~5,334 m, the median elevation is only 496 m because the basin has developed on older Lower Paleozoic and Precambrian terrain in the Pampean Ranges which were reactivated and uplifted by the Andean tectonics (Figs. 5.1–5.2)

The megafans also display a diversity of shapes. Megafan relief decreases from the Parapetí to the south, with the largest value for the Parapetí (FR = 396 m) and the smallest for the southernmost megafan, the Dulce (FR = 131 m) (Table 5.1). Megafan planform also changes from north to south. The circularity index is lowest in the Parapetí and Grande, and increases in the Pilcomayo, Bermejo, and Dulce. The configuration of the foreland and the proximity of the Brazilian Shield and other ancient geological outcrops in eastern Bolivia and along the Brazil–Bolivia border constrain and distort the shape of the Grande and Parapetí towards more elongate and irregular shapes, with circularity indices (CI) of 0.20 and 0.11, respectively. In contrast, the vast Chaco–Pampean plain provides more lateral space to accommodate the development of more rounded megafans such as the Pilcomayo, Juramento–Salado, and Dulce, which – without ever attaining the

highly circular pattern of the Taquari megafan in the Pantanal (see Santos et al., Ch. 6) exhibit triangular to oval shapes. A special case is the Bermejo, which is laterally constrained by the Pilcomayo and Juramento–Salado megafans, and as a consequence, displays a more elongated shape than its neighbours with a circularity index of 0.36 (Table 5.1).

Correlations between drainage and fan areas, and relationships between fan area and fan slope and channel slope, have been used by several authors since Bull's pioneering studies (1964, 1968) to characterise alluvial fans. For example, a power relationship between the surface areas of the fan and the feeder watershed (Ad) has been postulated, and in many areas of the world exhibits a strong positive correlation (Bowman 2019). The limitation of this mathematical relationship is that it does not consider the volume of the fan. It uses instead the planimetric fan area (Af) as a proxy for sediment yield, which introduces uncertainties when applied to subsiding areas (Giles 2010). For the Chaco fans, the Ad–Af power-law correlation is not statistically strong (Fig. 5.3). The reasons for the weak correlation are not discussed here, but we suggest a possible link to factors such as spatial variability of precipitation regimes in the watersheds, lithological complexity in the watersheds, basin factors (elevation, relief roughness, among others), differences in sediment yield, and the available lateral and vertical accommodation space in the foreland and Chaco–Pampean platform.

Correlations between alluvial-fan mean slope and watershed area have also been interpreted in terms of alluvial fan dynamics. Mean fan slope has been assumed to be inversely proportional to fan size (as a proxy for discharge), with fans in humid regions displaying lower slopes than in arid regions (Milana and Ruzycki 1999). However, in the case of the Chaco megafans, slope gradients are one order of magnitude lower than those of the alluvial fans plotted by those authors. We find a correlation between Ad and mean fan gradient, with correlation strength increasing when watershed areas outside the Subandean Zone – regarded as the dominant source of sediment load and water discharge – are excluded. However, the correlation between fan area (Af) and mean fan slope is difficult to assess as all the megafans slopes are extremely low (Fig. 5.3). Also, a logarithmic

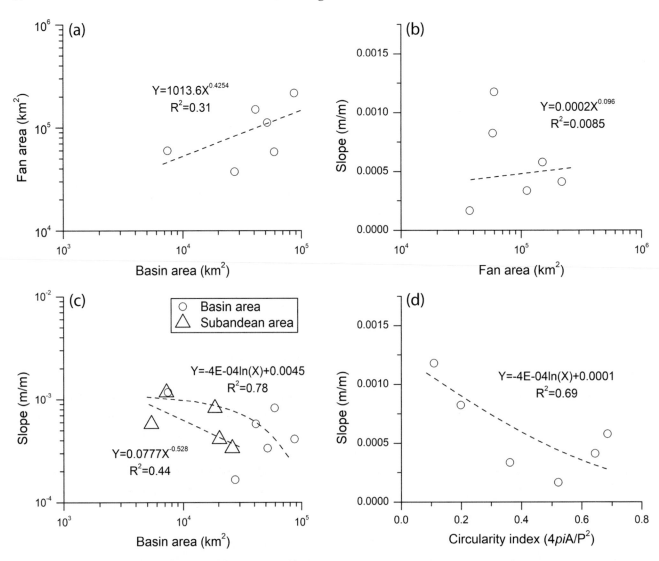

Figure 5.3 **Statistical correlations between Chaco morphometric attributes.**
(a) power function relationship between drainage area (Ad) and megafan area (Af). (b) plot of fan area (Af) *vs.* fan gradient. (c) fan gradients and drainage areas for the total basins of the six *megafans*, and for the portions of the basins dominantly in the Subandean zone (excluding the Dulce River which drains other geological units as shown in Fig. 5.1). (d) plot of fan circularity *vs.* fan gradient.

correlation is suggested between fan slope and circularity, although the reasons behind the correlation are still inconclusive (Fig. 5.3).

5.4.2 *Quaternary Record*

The Chaco megafans are complex landforms that developed during the Quaternary and underwent climatic changes that influenced their hydro-sedimentological regimes, geomorphologic dynamics, and triggered interactions between fluvial and aeolian systems. Gravelly and sandy sediments were, and still

are, deposited in the most proximal areas of the Andean foothills, and sandy and muddy deposits have been dominant from the proximal to the distal areas of the megafans.

The surface morpho-sedimentary units composing the megafans range in age from the Late Pleistocene to the present (e.g., Iriondo 1993; Latrubesse et al. 2012). The coalescence between the fan sediments and palaeochannels of the largest systems has been recognised by geomorphological mapping and, in some areas, delineating precise limits between fans is diffi-cult (Fig. 5.1). For example, the Parapetí River flowed to the north in the direction of the Grande–Mamoré drainage basin, but at other times coalesced and interacted with the Pilcomayo megafan to the south. Similarly, the southern limit of the Bermejo megafan overlaps and coalesces with palaeochannels of the Juramento–Salado megafan, making it difficult to identify the boundary near the zone of coalescence, i.e., where the palaeobelts approach the Paraná River fluvial belt.

Iriondo (1993) suggested that the megafans were formed by alternate dry and wet periods, with stable fluvial belts during humid climates such as at present, and generalised sedimentation by spillouts and ephemeral channels during drier climates such as the Last Glacial Maximum (LGM). We suggest an alternative explanation. With the exception of the Bermejo River, the present-day channel lengths of the Chaco rivers are shorter than the radius of the older alluvial fans, and the rivers do not reach either the regional Paraguay–Paraná base level to the south or the Amazon basin (i.e., the Mamoré–Madeira Amazon tributary) to the north (Table 5.1, see also Discussion). However, during a large part of the Last Glacial Period (LGP), the Chaco rivers were capable of reaching those regional base levels. The lengths of these larger LGP rivers were double or more than they are today. On a regional scale, many of the palaeochannels were wide and generated nodal points where lobes, with radial systems, recreated fan-in-fan morphologies because of longitudinal autogenic adjustments through the sediment transfer system, e.g., in the Parapetí system (Latrubesse et al. 2012).

The widespread palaeochannels imply that sediment supply and rainfall sourced from the Andes were at one time more abundant than today, and water discharges were powerful enough to move an augmented sandy load further from the Andean foothills than at present. The tendency for these systems was aggradation, avulsion, and spillway systems redirecting discharge into local flood basins, locally known as *derrames*. The older abandoned flooded areas were similar to the presently active flood basins of the southern Chaco generated as a consequence of modern avulsion in the Pilcomayo and Bermejo rivers. Seasonal water discharge in the Andean source area was substantially larger than now. In the case of the Parapetí, several fan palaeochannels reached the Paraguay River more than

500 km eastwards of the fan apex, which means the river flowed more than three times farther than it does in the present-day system. In megafans such as the Parapetí, channels were significantly wider, but they narrowed downstream – as expected in this kind of fluvial system because avulsion at nodal points produces channel planform adjustments, decreasing discharge downstream and in distributary branches.

The Pilcomayo palaeochannels also reached the Paraguay River. During the LGP, the Juramento–Salado megafan also expanded and large, late Pleistocene sinuous paleochannels reached the distal part of the fan in the Santa Fe province (Iriondo 1993). Low-sinuosity palaeochannels 1–3 km in width were also identified in the Juramento–Salado megafan, flowing in a regional NW–SE direction (Peri and Rossello 2010).

Although absolute chronologies of the morphosedimentary units are scarce, the available chronological and biostratigraphic information allows us to postulate a general chronological and palaeoenvironmental framework. Morphostratigraphic analysis, geochronological biostratigraphic data, and regional correlations between these datasets suggest that the surfaces of the Chaco megafans and minor but large piedmont fans probably started to be created during MIS 4, and reached maximum development during the Early to the late Middle Pleniglacial (ca. 60–28 ka). Levels of Pleistocene and Holocene terraces and outcrops have been described in the megafans (Iriondo 1993; Latrubesse et al. 2012; May et al., Ch. 7). In the Bolivian megafans, optically stimulated luminescence (OSL) dating of alluvial and aeolian materials indicates strong simultaneous fluvial and aeolian activity during the LGM, particularly during MIS 3 and early MIS 2 (Latrubesse et al. 2012). Also, Pleistocene alluvial units from several localities in the Chaco provided megafauna from the late Pleistocene to the early Holocene (Lujanian mammal age). The recorded Pleistocene faunas are similar in the Formosa and Chaco provinces (Bermejo and Juramento–Salado megafans) (Tonni and Scillato-Yané 1997; Zurita et al. 2014), at localities of the Pilcomayo megafan in Paraguay (Carlini and Tonni 2000; Hoffstetter 1978), and in alluvial deposits in the Andean foothills of southeast Bolivia (Hoffstetter 1968). Three OSL ages in the Pleistocene alluvial units of the Bermejo megafan (Villa Escolar) containing palaeofauna have provided ages ranging from 92,200 ($\pm$ 9,650) to 58,160 ($\pm$ 4,390) years BP (Zurita et al. 2009, 2014). The faunistic association consist of *Chaetophractus* sp., *Propraopus* sp., *Pampatherium typum*, *Holmesina paulacoutoi*, *Glyptodon* sp., *Neosclerocalyptus* sp., *N*. cf. *paskoensis*, *Panochthus* cf. *tuberculatus*, *Glossotheirum* cf. *robustum*, *Scelidotherium leptocephalum*, *Megatherium* sp., cf. *Morenelaphus*, *Antifer* sp., *Hemiauchenia paradoxa*, *Toxodon* sp., *Macrauchenia* sp., *Procyon cancrivorous*, *Protocyon* cf. *troglodytes* (Zurita et al. 2014). The presence of *Equus* (*A.*) *neogeus* Lund in two other fossiliferous localities of the Bermejo megafan (Barranqueras and General San Martín) also constrains the age of the sediments to the Lujanian *s.st.* (late Pleistocene–early Holocene) (Zurita et al. 2009, 2014).

Based on the available chronologies in Bolivia and Argentina and the palaeofauna described above, we argue that though the megafans evolved during the whole of the Last Glacial Period (LGP), they reached their peak extent in MIS 3 and early MIS 2. During this period, aeolian activity was concomitant with the existence of dynamic fluvial belts. The interactions between aeolian and fluvial processes are still characteristic, particularly in the Bolivian Chaco, where large dune fields have been intermittently active since the late Pleistocene because of the deflation of alluvial sediments. The main dunes cover ~ 22,349 km^2 of the Bolivian Chaco. Older aeolian sands dominate the substrate of these areas, frequently reddish due to oxidation, with linear features striking N–S. They have been interpreted as long arms of parabolic dunes, or perhaps even linear dunes. The older Pleistocene units covered an area larger than that occupied by the Holocene coversands. Whether Pleistocene or Holocene, all dunes are oriented N–S to NW–SE, suggesting unchanging dominant wind directions in the past as in the present (Latrubesse et al. 2012).

A peak in aridity was reached during the LGM and during the transition from the Lateglacial to the early Holocene (dates oscillating between 20 ka and 10 ka) (Latrubesse et al. 2012). Loess deposits of the Urundel Formation (Iriondo 1993) partly covered the Bolivian Chaco, the eastern valleys of the Subandean Ranges, extending 900 km from Santa Cruz de la Sierra to Tucumán (Argentina). As suggested by Iriondo (1993), the loess deposits could have been related to the north-south wind circulation more than to the Pampean loess aeolian system located much further to the south in Argentina. However, loess-like deposits sourced from the south by the Pampean aeolian system covered the southern-most area of the Chaco (Iriondo 1990). 'Red beds' (displaying Bw horizons) also developed in the northern Chaco piedmont area of Bolivia, indicating dominant dry conditions at some time before ~ 18 cal ka BP (May and Veit 2009). Charcoal radiocarbon dates from fluvial sands around Santa Cruz (15.14 $\pm$ 0.07 ^{14}C ka BP) and the Grande River area (18.7 $\pm$ 0.09 ^{14}C ka BP) indicate aridity and decreasing fluvial activity in eastern Bolivia during the LGM and early Lateglacial (14–10 ka) (May and Veit 2009; May et al. 2008).

The regional environmental setting in the Chaco during MIS 2 (i.e., the LGM) was one of increasing aridity. During the LGM, the rivers had smaller discharges than during MIS 3 and early MIS 2, and aeolian morphogenesis (dunes and loess deposits) was more intense on the Andean foot-slopes, with loess and loess-like deposits widespread regionally.

The megafans were more stable because of decreasing discharge availability which probably maintained more stable fluvial belts. Minor underfit channels slightly reworked the surface of the megafans. Stronger aridity during MIS 2 in the Chaco, and reduced annual and/or seasonal monsoonal rainfall, may also be explained by the increased effect of cold air masses coming from the south in winter during the LGM (Latrubesse and Ramonell 1994). In this scenario, the Chaco was highly susceptible to the friagem/surazo effect.

The Lateglacial (14–10 ka) and early Holocene are also poorly recorded regionally. The record of organic materials from fluvial sediments containing phytoliths and megafauna in the Bermejo megafan (close to the Villa Escolar area) has provided a ^{14}C age of 9570 $\pm$ 90 yr BP, suggesting the continuation of a relatively dry and cold climate during the late Pleistocene–Holocene transition for the eastern Chaco (Contreras and Zucol 2019; Zurita et al. 2014). However, the chronology of the Villa Escolar locality is partly controversial because the radiocarbon-dated stratigraphic units seem to be the same as those that provided older MIS 4 and MIS 3 OSL ages. Early Holocene palaeofaunal remains were also documented, and radiocarbon dating of charcoal in alluvial deposits of smaller proximal channels in the Bolivian foothills suggests that that the megafauna survived at least up to 5.5 ka BP (Coltorti et al. 2012).

The Holocene was characterised by an increase in fluvial activity, but rivers were incapable of entirely reworking or covering with younger sediments the older and larger late Pleistocene megafans. During the entire Holocene, fluvial dynamics have been remarkably less energetic than during a good part of the LGP (MIS 4 to early MIS 2). Younger fan lobes are small and restricted to areas more proximal to the Andes.

In the proximal areas, Holocene belts are incised into older megafan surfaces, generating inset terraces. However, there is an intersection point among the concave longitudinal profiles where the present rivers almost asymptotically join the megafan surface and become slightly perched above the surface of the fan downstream. The autogenic adjustments of the longitudinal profiles of modern rivers are nevertheless complex and dynamic (see Section 5.5), with the reworking of older surfaces being more intense towards the southern and northern Chaco. For example, on the Parapetí megafan an extensive area has undergone sedimentation processes and reworking in the last 6–7 ka (May et al., Ch. 7; note that the Quaternary Parapetí megafan covered ~50,000 km^2, a significantly larger area than the one postulated by May et al.). Despite the high degree of reworking, however, the modern active megafan does not reach even half of the extent of the Pleistocene megafan. Regarding the recent megafan dynamics, May et al. (see Ch. 7) inferred that the modern avulsive processes on the modern lobe are mostly restricted to a ~2,000 km^2 depositional area on the distal megafan. Although aeolian activity reached maximum activity during the LGP, the deflation of alluvial sediments from the megafans continued during part of the Holocene (ca. 10 ka and 3.5 ka). The widespread Holocene parabolic dunes covering thousands of kilometres indicate the persistence of semi-arid to sub-humid conditions (tropical wet–dry) during the lower and middle Holocene (Latrubesse et al.

2012). From 3.5 to 1.5 ka the climate was probably slightly wetter, with soil development (May 2007). Recent parabolic dunes seem to be related to a more arid climatic episode starting ca. 1.5 ka (Latrubesse et al. 2012).

Reworking and incision were of a smaller magnitude on the Pilcomayo and Bermejo. On these megafans, flood deposits and avulsive processes have blanketed the older megafan surface with a few metres of predominantly muddy materials. Quaternary and recent paludal deposits and swamps on those megafans cover more than 125,000 km^2 (Iriondo 1993). Despite a lack of chronological data, the uppermost fan of the Juramento–Salado megafans is also postulated to have developed during the Holocene (Iriondo 1993). Palaeofluvial belts 1–2 km in width and sinuous younger and smaller palaeochannels 90–300 m wide in the Juramento–Salado megafan indicate reworking and blanketing of the old Pleistocene surfaces during the Holocene in the eastern and southern areas (Peri and Rossello 2010).

5.5 Modern Dynamics of the Chaco Rivers

As detailed above, the apexes of megafans and more minor alluvial fans occur at the foot of the Subandean Ranges, and recent and modern lobes, with radial distributary patterns of active and inactive alluvial belts at different elevations across the surfaces of those fans. Moreover, in rivers such as the Pilcomayo, outside of apexes and proximal zones the present sedimentation lobes have developed over those of the late Pleistocene megafan; channel bottoms lie characteristically ~2 m above the fan surfaces. As previously described, modern river channels do not reach the distal parts of the fans, and typical morphological features in these zones are mosaics of spills, ponds and elongated swamps on the alluvial belts abandoned by avulsion.

The mean annual discharges of the southern megafan rivers range from 42 to 356 m^3/s (Tables 5.2 and 5.3), but mean annual discharge fails to show a linear correlation with watershed size (Fig. 5.4), i.e., a larger drainage area does not scale with greater discharge. One reason for this is that while rainfall concentrates in the Subandean Zone (Sections 5.3 and 5.4), larger drainage basins such as the Pilcomayo, Bermejo,

Table 5.2 *Data from gauge stations closest to megafan apexes*

Station	Area (km²)	Q (m³/s)	Q yield (m³/s/km²)	Sed. load (Mt/yr)	Sed. yield (t/km²/yr)
Northern Chaco					
Piray	1,420	11	0.0077	2.9	2,040
Piray	2,880	12	0.0042	2.3	799
Grande	23,700	130	0.0055	154.3	6,510
Grande	31,200	230	0.0074	206.9	6,630
Parapetí	7,500	79	0.0105	19.4	2,590
Southern Chaco					
Grande de Tarija	10,460	48	0.0046	14	1,338
Pilcomayo	96,000	204	0.0021	141	1,469
Pilcomayo-Talula	6,490	19.6	0.0030	10.8	1,664
Pilcomayo-Villa Quemada	13,500	49.9	0.0037	24.5	1,822
Bermejo	25,000	356	0.0142	120	4,800
Bermejo	2,260	22.9	0.0101	4.9	2,168
Bermejo	4,850	89.5	0.0185	15.7	3,237
Pescado	1,700	50.6	0.0298	24	14,117
Iruya	2,120	24	0.0113	17.7	8,349
Juramento	31,900	29	0.0009	34	1,066
Salí	4,700	15	0.0032	4.9	1,043
Dulce	15,000	98	0.0065	23.7	1,580

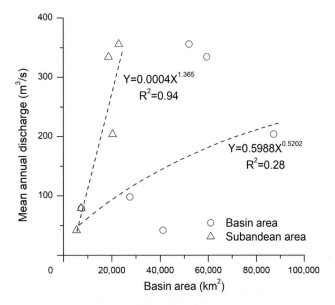

Figure 5.4 **Plots of mean annual discharge and drainage basin area**. (a) Results for the whole basin of each of the six megafans. (b) Results with drainage area restricted to the wetter Subandean Zone (Dulce River excepted).

Juramento–Salado and Grande extend across more arid high mountain areas of the Eastern Cordillera and the Puna Altiplano as well. Thus, the positive correlation between mean annual discharge and drainage area only

increases significantly when (drier) watershed areas lying outside the Subandean Zone are discounted (Fig. 5.4).

The relationship between low and high discharges is among the most extreme for perennial rivers in the tropics. High water discharge variability (maximum daily discharge/minimum daily discharge, i.e., Q_{max}/Q_{min}) related to mean annual water discharge (Q_{mean}) is as high as 150 in the Pilcomayo and 190 in the Bermejo (Latrubesse et al. 2005). Sediment yields are very high in all the basins, ranging from $\sim 1,000$ to more than 6,600 t/km²/yr.

There is no good correlation between basin relief (BR) and sediment yield (Fig. 5.5). However, as in the case of water discharge, the correlation between sediment yield and watershed area increases when restricting the watershed area to the wetter Subandean-Zone sub-catchments (Fig. 5.5). Likewise, the correlation suggests that the Subandean Zone provides the largest amount of suspended sediment load. It also indicates that, despite the presence in some watersheds of high mountains (of limited extent in the hypsometric curve: see Fig. 5.2), these are not the most substantial sources of sediment. An extreme case is the Dulce River catchment where the major proportion of the drainage area

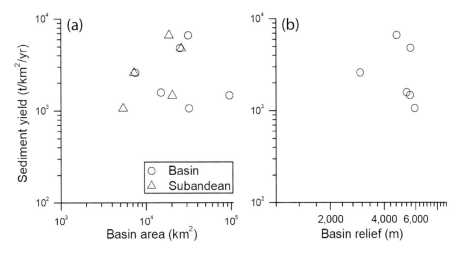

Figure 5.5 **Relationships between drainage area, sediment yield and basin relief (BR).** (a) Plot of the drainage area and sediment yield for the whole basin of each of the six megafans, and a plot of the drainage basins located in the Subandean zone (excluding the Dulce River) *vs.* sediment yield. (b) Plot of sediment yield *vs.* basin relief index (BR).

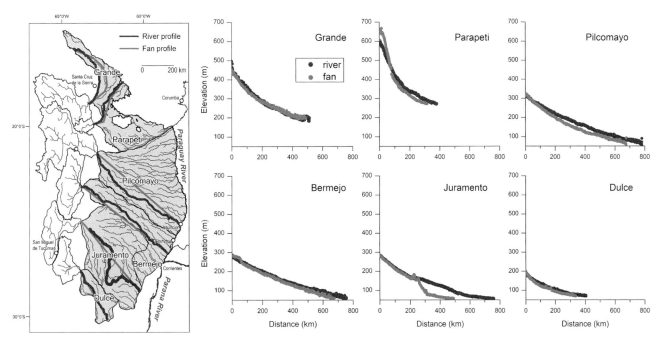

Figure 5.6 **Longitudinal profiles of surfaces of the Chaco megafans and their modern river channels.** A colour version of this figure is available in the SOD for Chapter 5.

is low-lying. The situation is reversed in the Pilcomayo watershed where large drainage areas occur at high elevations (median elevation 3,372 m); here relatively flat areas of the Altiplano include intermontane tectonic basins act as local sediment sinks (Table 5.1; Fig. 5.2).

River channel patterns on the fans vary according to flow direction and distance from the mountain front. On the southern Chaco megafans, channels near the apexes are dominantly braided but become meandering in distal zones. Pattern changes are gradual, accompanied by a reduction in channel widths. There are also transitional reaches where meandering and braided patterns co-exist, such as a non-regular meandering pattern with mid-channel bars (i.e., 'wandering', *sensu* Schumm 2007) (Fig. 5.6).

Fan slope is an important variable controlling channel dynamics, as variations in slope force rivers to

Table 5.3 *Hydro-geomorphologic data and metrics.*

River name	Q_{mean} (mean annual discharge)	Q_m (Dec–Apr) (m^3/s)	Q_m (May–Nov) (m^3/s)	Seasonal Q_m ratio	Channel gradient (m/m)	Mean channel width (m)	Channel pattern	Channel sinuosity	Stream power (W/m)	Specific stream power (W/m^2)
Pilcomayo	204	415	53	7.83	0.00057	1,443	Braided	1.04	2,318	1.61
					0.00052	642	Wandering	1.51	2,114	3.29
					0.00044	321	Meandering	1.71	1,789	5.57
Bermejo	356	717.7	103	6.97	0.00049	1,981	Braided	1.09	3,446	1.74
					0.00037	768	Wandering	1.57	2,602	3.39
					0.00034	410	Meandering	1.57	2,391	5.83
Juramento	42	71	22	3.23	0.00176	88	Meandering	1.54	1,224	13.92
					0.00112	154	Braided	1.19	779	5.06
					0.00101	66	Meandering	1.96	702	10.65

adjust some of their morphological parameters to fixed input variables (Bull 1968). However, for the Chaco megafans further systematic analyses of longitudinal profiles using higher-resolution digital elevation models are required as slight incision and very gentle knickpoint breaks are detected (Fig. 5.6). The explanation for these knickpoints lies in the reactivation of the longitudinal profile along the largest megafans, whereby the systems autogenically adjust their longitudinal profiles by locally increasing slopes to improve the efficiency of water and sediment transport to reach the base level, and on tectonic activity in some sections of the Chaco plain.

In the case of the Pilcomayo megafan, at a point situated about one-third of a channel's length measured from the fan apex, the river channel loses its entrenchment. Such breaks in slope may promote bed sedimentation, and the active channels become perched on the older fan surface (Fig. 5.6). A special case is the Juramento–Salado megafan, which exhibits an anomaly in the middle reach (Fig. 6). Thiswas produced by neotectonic activity along a NNE-elongated structural high, ~200 km in length, and 20–80 km wide locally known as Lomadas of Otumpa (Peri and Rosello 2010).

Loss of entrenchment is geomorphologically expressed also as a change in channel pattern and as a reduction of mean channel width (Table 5.3). The majority of the sediment transport, peak discharges, and work by the river on its channel happens during the rainy season between December and April. We calculated several parameters relevant to rainy season conditions:

- P: channel sinuosity (dimensionless)
- Ω: stream power (W/m)
- ω: specific stream power (W/m^2)
- Channel: mean channel slope (m/m)
- The mean discharge during the rainy season was assumed to be representative of the channel-forming (dominant) discharge for the Chaco rivers.

The expression for specific stream power is:

$$\omega = \frac{\gamma.Q.S}{B} \tag{5.2}$$

where γ is specific fluid weight, Q is mean annual discharge, B is mean channel width, and S is the energy slope.

Given that:

$$Q = v.B.h \tag{5.3}$$

and substituting (5.3) into (5.2), we obtain:

$$\omega = \gamma.h.S.v = \tau_o.v \tag{5.4}$$

where h is mean flow depth, v mean flow velocity, and τ_o shear stress.

These relations suggest that an increase in specific stream power is the result of a direct variation in shear stress, mean velocity, or both. In other words, these rivers change their slopes and patterns in order to attain a new state of equilibrium that maintains water flow and sediment transport. Thus, the loss of gross power in wider channels is compensated for by an increase in specific stream power in a narrower meandering channel further downstream.

The suspended sediment load (predominantly silt and clay) in all the Chaco rivers is extremely high.

For example, the Bermejo River, which today is the only river crossing the entire Chaco plain and reaching the Paraguay River collector system, transports ~95 Mt/yr of suspended load. It is currently the main source of fine sediments to the Paraná River and La Plata estuary (Amsler and Prendes 2000). In its middle reach, the suspended sediment concentration during flood stages averages 15,000 ppm (e.g., at Lavalle Port; 25°40' S; 60°10' W). The high flows take place during the austral summer, and the hydro-sedimentary pulse reaches the Paraguay River and propagates along the right margins of the Paraná River between 1 and 3 months later (Amsler and Drago 2009). In its middle to lower reach, the river channel has a meandering pattern with slopes of ~0.2 m/km, a mean bankfull width of 240 m, and a fine to very fine sandy bed. The mean discharges for low and flood stages are 60 m^3/s and 1,200 m^3/s, respectively.

Channel stability of the Bermejo river since 1945 was analysed along a 145-km-long valley reach around Lavalle Port using historical cartography and fieldwork. All known meander migration processes and landforms were identified in those reaches (i.e.,

meander expansion and translation, neck and chute cut-offs, etc.), and reactivations of abandoned meanders were also observed. The mean width of the meander belt built from 1945 to 2004 was 1,820 m, with a localised maximum lateral shift of 2,970 m. The maximum recorded rate of channel shift was 600 m/yr (Fig. 5.7).

The main Chaco rivers are still prone to avulsion, but historically only a few major episodes have been recognised: the Bermejo shifted to the north in the middle of the nineteenth century and abandoned a ~200-km-long tract called the Bermejito River, which is partly functional during big floods. Another river prone to lateral shifting and avulsion is the Pilcomayo: lateral shifting of the meandering reach 150 km long, built a meander belt ~1,000 m wide from 1976 to 2008. For that period, the local maximum width was 2,810 m, with a maximum channel shift rate locally attaining 80 m/yr (Fig. 5.7).

Two main mechanisms control the flood dynamics and the shifting of the Pilcomayo River. One is the sequence of palaeochannels that compose the old megafan and the recent lobe. When the river lacks

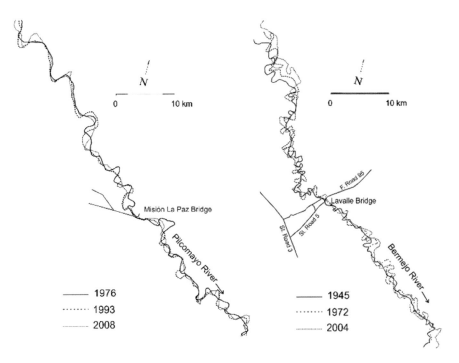

Figure 5.7 **Multi-decadal evolution of the Pilcomayo and Bermejo river channels**. Left: braided, wandering and meandering patterns of the Pilcomayo River. Right: braided and wandering sinuous patterns of the Bermejo River. Data sources: aerial photographs and Landsat images from the years 1976, 1980, 1989, 1993, 1996, 1999, 2006 and 2008 for the Pilcomayo River (Mision La Paz bridge: 22°22.69' S, 62°31.12' W); and 1945, 1972, 1989, 1998 and 2004 for the Bermejo River (Lavalle bridge: 25°39.248' S; 60°7.734' W).

entrenchment capacity, it becomes perched a couple of metres above the megafan surface. Thus, during floods the water flows towards palaeobelts 1.7–3 km in width and follows the megafan's surface slopes, which can be one order of magnitude steeper than the channel slope. Lateral overspill and migration can capture and reactivate old channel morphologies, thereby revitalising temporarily inactive landforms. Both situations favour the development of splays, concentrated runoff on the palaeofan surface, and potential avulsions. At the same time, however, the high abundance of sediment can block and abort the process by plugging the mouths of palaeochannels and flood channels.

The other major mechanisms triggering avulsion are the high accumulation rates of fine sediment and of woody debris in the channel, which raise the channel bed and generates log jams, respectively. This results in water spilling over the megafan surface and feeding large, permanent or temporary wetlands with sediment. As a result, local palaeochannels are frequently occupied by underfit streams.

At vegetation debris-jam locations, dome-like deposits accumulate and expand downstream, upstream, and laterally for several kilometres. In the Bermejo and Pilcomayo, the most common woody debris consists of *Tessaria integrifolia*, which is removed by bank erosion but also plays a major ecogeomorphological role as a pioneer species colonising abandoned channels and other fluvial landforms. The density and growth rate of *T. intergrifolia* can reach 10,000 trees per hectare in five years (J.J. Neiff, pers. comm. to Latrubesse). On the Pilcomayo the point of blockage has been moving hundreds of kilometres upstream throughout the 20th century (Iriondo 1993; Martín-Vide et al. 2014).

The Pilcomayo has been suggested to have flowed in the channel of today's Confuso River (Paraguay) during the early times of the Spanish conquest (Iriondo and Orfeo 2014), shifting in the 18th century to the zone where the Araguaymini River (Formosa, Argentina) currently flows. By the middle of the 19th century, another avulsion was in progress, with the river shifting towards the newly formed northern channel of the Pilcomayo (then known as the Itica or Araguay Guazu), and abandoning the southern Araguay Mini channel. Both channels at the time flowed towards the Paraguay River and created a triangular island identified as Father Patiño Island in historical manuscripts (Guzmán 2014).

After that, the Pilcomayo channel came to occupy its current axial position along the Paraguay–Argentina border. In the beginning of the 20th century the river kept its channel functioning as far downstream as about 400 km from Asunción, at which point a large, shallow swamp existed. Probably for several centuries, the Pilcomayo had also discharged water into a 15,000 km² swampy area located 250 km upstream of the Paraguay River confluence at the Paraguay–Argentina border (Esteros Patiño), filling the swampy depression with sediment (Cordini 1947; Iriondo 1993). In the period 1947–1976, 150 km of the channel were clogged by mud and organic debris, the point of blockage progressing upstream at a rate of 12 km/yr between 1968 and 1976, the rate rising to 22 km in 1976 alone. From 1975 to 1990 another 150 km became clogged (the maximum annual rate of blockage of 45 km was attained in 1984), thereby accelerating the disequilibrium process after that date (Martín-Vide et al. 2014).

The Juramento–Salado River has also experienced several avulsions. Historical maps from the 17th and mid-18th centuries indicate that the Juramento–Salado discharged into the Paraná. In the second half of the 18th century it shifted and joined the Dulce to the south, reaching Lake Porongos near the Santiago del Estero and Cordoba province borders in Argentina (Fabbian et al. 1979).

The Chaco plain also contains many small creeks and streams, local systems that drain or feed large swamps and marshy areas. The majority of these autochthonous streams are underfit rivers occupying larger palaeochannels of the Pilcomayo and Bermejo megafans (Iriondo 1974). They are fed by rain- and groundwater, and either discharge into or drain temporary (*bañados*) and permanent (*esteros*) swamps, carrying a big proportion of their sediment load as colloidal materials consisting of clay minerals and organic matter (Iriondo 1993; Orfeo 1986).

Fluvial-aeolian interactions are still active in the Bolivian Chaco. The climatic regime is characterised by very intense rainfall during the summer monsoon, followed by a long dry season with strong and persistent north winds. These environmental conditions allow the coexistence of fluvial and aeolian processes. The abundant river bedload is a major source of sand that becomes available to seasonal aeolian remobilisation and to the generation of coetaneous dune fields and blowouts towards the south. The current river beds of the Parapetí and Grande rivers are dominantly covered by aeolian landforms such as aeolian ripples, small dunes, aeolian lag deposits, and nebkhas. Where the sand supply is large, the deflation of fluvial sand creates lobes of barchanoid transverse dunes; parabolic dunes and blowouts develop in areas with smaller supply.

5.6 Megafans and Wetlands

The Chaco is one of the most important wetland regions in South America. The whole Chaco plain is a mosaic of permanent and seasonal wetlands related to relict fluvial landforms, widespread swamps and water bodies on the plain's flat surfaces, and areas that suffer the direct influence of river floods and water-table saturation (Fig. 5.8). The Chaco surface sediments and soils are composed predominantly of several metres of thick impervious paludal clays. Neotectonic movements affecting Quaternary deposits have also been postulated to promote the generation of water ponding in depression areas and swamps in the eastern Chaco and on the Juramento–Salado megafan (Iriondo 1986, 1993; Peri and Rossello 2010). The major swamps and poorly drained areas are very shallow. The vegetation of the Chaco is represented by semideciduous forests coexisting with hydrophilic forests, savannas, and spiny shrublands with xerophytic morphological features (Placci and Holz 2004). A mosaic of paludal vegetation, floating plants, grasslands and palms (*Copernicia alba*) is a common feature of wetlands. Such vegetation diversity is a function of bioclimatic conditions and a complex assemblage of macro- and mesoscale landforms. Detailed classification and analyses of wetlands are beyond the scope of this chapter, but further ecogeographic and vegetation syntheses appear in Ginzburg et al. (2005), Kandus et al. (2017), Marchetti (2017), Minotti (2017), and Navarro et al. (2006).

Major wetlands occupy abandoned paleochannels and local depressions, as on the interfluve between the abandoned Bermejito River and the Teuco–Bermejo depression (total of 14,000 km²; Ginzburg

et al. 2005). Other major wetlands are the Bañados La Estrella in the Pilcomayo system (5,600 km²), and a large area of 8,300 km² in the Dulce megafan extending from the south of Santiago del Estero province as far as the terminal delta of the megafan in Mar Chiquita lake. In the northern Chaco, the most important wetland is the ~6,000 km² Izozog swamp (a Ramsar protected area) on the Parapetí megafan in Bolivia.

The Juramento–Salado has also developed large swampy areas such as Bañados de Figueroa (3,000 km²). The Juramento–Salado in its middle reach takes an NNW–SSE direction and alternates between sections with well-defined channels and sections with more anarchic channel patterns associated with swamps, shallow lakes, and flooded areas such as the 'Bañado del Copo', which acts as a sink for sediment and water (Latrubesse and Brea 2009). In the distal part of the megafan, a depressed flat area of ~5,000 km² known as the Sub-meridional lowlands (*Bajos sub-meridionales*), is now inundated by local rainfall and watertable saturation. It connects to the south with Quaternary units of the Paraná River alluvial belt.

The wetlands of the Chaco are either freshwater or saline lakes and swamps. Somewhat permanent wetlands occupy the eastern Chaco, known as the Wet Chaco. The saline lakes occur along the western Chaco, or Dry Chaco, from Bolivia in the north to the Dulce megafan in the south. Because of the sharp climatic seasonality, the western Chaco is characterised by seasonal wetlands that become hydrologically active during the main circulation of the SASM. The seasonal areal contraction and expansion of the wetlands (elasticity coefficient) between the wet and the dry season is high, reaching an average ratio of 12.5 in the eastern (wet) Chaco (Neiff 1999).

We estimated the distribution and extension of wet areas for each megafan by photo-interpretation using remote sensing products at the end of the rainy season when the rivers reach their flood peaks (Fig. 5.8). Table 5.4 synthesises the flooded areas for each megafan. A total of ~306,000 km², i.e., 48% of the total surface of the megafans in this study, are seasonally flooded. The largest wet areas lie on the Pilcomayo and Juramento–Salado megafans, whereas the Dulce has the largest ratio of wetland to total megafan area (~64%), and the Bermejo has the smallest (~32%).

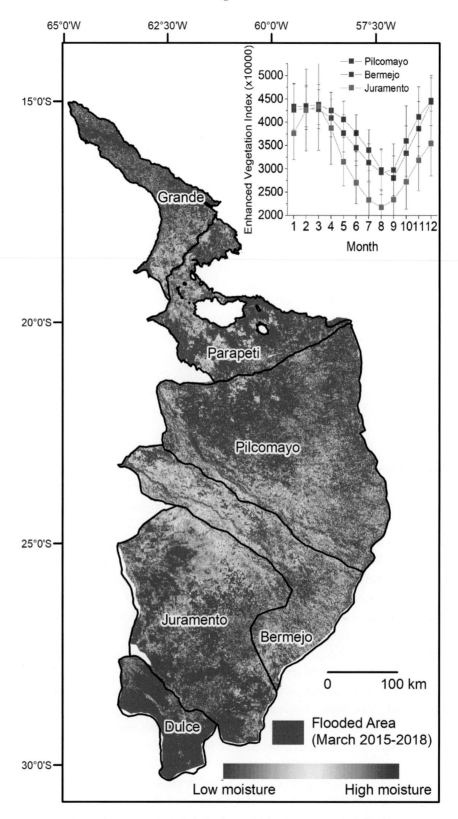

Figure 5.8 **Average flooded area (grey pixels) of the Chaco megafans in March between 2015 and 2018, assessed by MODIS EVI.** EVI is used to map open water (< 0.36) and flooded vegetation (> 0.47). Inset graph shows the interannual average (2015–2018) monthly EVI values for the three largest fans of the Chaco, the Pilcomayo, Bermejo, and Juramento–Salado. The graph shows that, on average, the highest EVI is attained during March for the three fans, the month shown on the map. The results from the map are summarised in Table 5.4.

Table 5.4 *Total flooded area (and its proportion of the total fan area) for each of the Chaco fans during the flood peak month (March) assessed by EVI*

Fans	Total fan area (km^2)	Total flooded area (within each fan)	
		Defined as MODIS EVI > 0.47[a] or EVI < 0.36[b] (km^2)	Portion of flooded area out of total fan area (%)
Grande	48,026	24,169	50.3
Parapeti	69,126	29,870	43.2
Pilcomayo	216,216	117,699	54.4
Bermejo	111,889	35,349	31.6
Juramento-Salado	150,575	74,894	49.7
Dulce	37,339	24,008	64.3

Notes: [a] This represents high moisture area, i.e., mostly flooded forests; [b] This represents open water area, where water body could be directly observed from satellite imagery.

Although massive deforestation is occurring across the Paraguayan Chaco in particular, several of the large wetlands mentioned above are currently protected by different conservation areas such as Biosphere Reserves, Natural and Provincial Reserves, National and Provincial Parks and Ramsar Sites.

5.7 The Chaco as a Sediment Sink

Notwithstanding landscape complexities, erosion is generally enhanced in drainage basins of the seasonally sub-humid and semiarid Tropics with an intense rainfall season, but is more subdued in catchments experiencing high rainfall rates and constant wet climates (Latrubesse and Restrepo 2014). Climate thus plays a complex role in erosion control, mainly because it influences other controls like vegetation cover. In the case of the Chaco rivers, high sediment loads are also related to the erosion of the susceptible sedimentary rocks of the Subandean Zone.

Being located in the subtropical belt (~15° S–30° S), the Chaco catchments exhibit high seasonal discharge variability (Latrubesse et al. 2005), high sediment yields, and relatively low values of specific runoff.

The mean elevation of the Andes in the Chaco hinterland is ~2,700 m, and water availability for the Chaco rivers for erosion and transport processes is ten times smaller compared to rivers in the humid tropics of the northern Andes of Colombia. However, sediment production of the central Andean rivers in the Chaco is two orders of magnitude greater than that of high-runoff rivers draining the northern Caribbean and

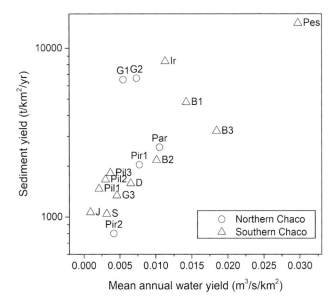

Figure 5.9 **Sediment yields in Andean rivers and mean annual water delivery per catchment** (data from Table 5.2). Northern Chaco: Pir1–2 (Piray), G1–2 (Grande), Par (Parapetí); Southern Chaco: Pil1 (Pilcomayo), B1–3 (Bermejo), Ir (Iruya), Pil2 (Pilcomayo–Talula), Pil3 (Pilcomayo-Villa Quemada), G3 (Grande de Tarija), J (Juramento), S (Sali), D (Dulce), and Pes (Pescado).

Pacific Andes (Latrubesse and Restrepo 2014). The relationship between specific runoff and sediment yield (Table 5.3 and Fig. 5.9) shows, to some extent, that transport capacity and the amount of water available for fluvial erosion in the Chaco are high.

Sediment yields of the fluvial basins in the central Andes of Bolivia and northern Argentina are among the highest in the world. The central Andes of Bolivia

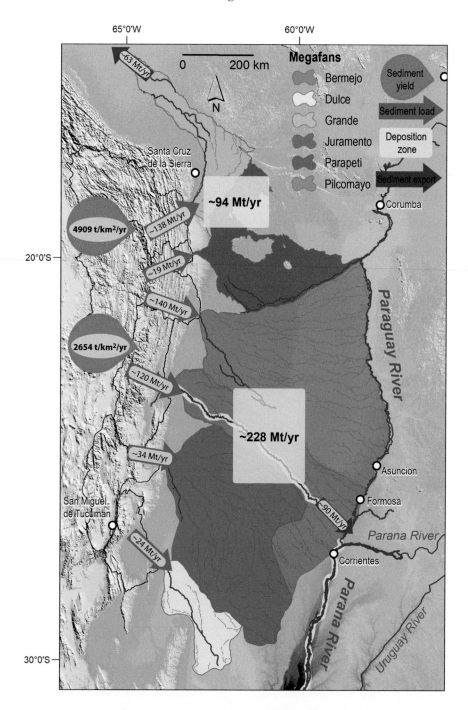

Figure 5.10 **Sediment budget for the Chaco megafans (sources and sinks).** Brown balloons: sediment yields of the Andes of Bolivia and southern Chaco. Blue arrows: annual fluxes of suspended sediment load at the apex of the major megafans. Values in yellow boxes: annual sediment trapping (sink) in the Chaco plain; red arrows: sediment export towards the Amazon and Paraná basins.

record the higher sediment yield values, with an upper quartile value around 2,070 t/km²/yr (Latrubesse and Restrepo 2014) (Fig. 5.9). In the southern Chaco, 25% of the available data from gauge stations document sediment yields of ~3,237 t/km²/yr, and smaller tributary catchments of the Bermejo River such as the Pescado and Iruya rivers have among the highest known sediment yields documented in South America (14,118 t/km²/yr and 8,349 t/km²/yr, respectively) (Latrubesse and Restrepo 2014).

The pronounced seasonal moisture pattern is one of the controls on erosion rates. From December to March during the rainy season, 85% of the total suspended sediment is transported in the Pilcomayo (Martín-Vide et al. 2014), and 75–90% in the Grande River (Guyot et al. 1994). The budget presented below is restricted to the megafan river systems. A substantial additional amount of sediment generated in the Andes is also trapped in intermontane basins, mountain floodplains, and the proximal alluvial fans of smaller catchments.

Approximately ~97 Mt/yr are stored by the Parapetí, Grande, and some minor rivers in the northern Chaco plain, and the Grande River exports an additional ~63 Mt/yr to the Mamoré River, and onwards to the Amazon basin (Guyot et al. 1996). The three largest southern Chaco megafans (Pilcomayo, Bermejo, and Salado–Juramento) store ~228 Mt/yr in the plain as far as ~500–700 km from the Andean foothills (Latrubesse 2015; Latrubesse and Restrepo 2014). The Pilcomayo River carries about 140 Mt/yr of fine sediments, but almost all of it is stored in the Chaco plain before it meets the Paraguay–Paraná systems. The Bermejo River contributes ~120 Mt/yr to the Chaco plain, but ~90 Mt/yr bypass the plain and are exported to the Paraná River via the Paraguay River.

From the total amount of sediment exported to the Andean piedmont by the major fluvial systems, ~68% is trapped in the Chaco plain by the six megafans, instead of being transferred to the ocean through the Paraná or the Amazon basins. Thus, the Chaco plain and its megafans are one of the largest (and perhaps the largest) active continental sedimentary basins (sediment sinks) on the planet (Fig. 5.10) (Latrubesse 2015).

5.8 Conclusions

The Chaco megafans are complex landforms that developed during the last glaciation of the Pleistocene and have continued to evolve under various climatic conditions until now. During the Last Glacial Period, more extreme climatic conditions favoured the development of the gigantic megafans that can be reconstructed and mapped from surface geology and geomorphology. Particularly from MIS 4 to early MIS 2, the hydrological and sedimentological regimes of these fluvial systems differed from those

of the present. Larger, but also more extreme seasonal water discharges allowed the Chaco rivers to deliver sediment beyond their fan limits, thereby reaching their regional base levels – i.e., the Paraguay and Paraná collector rivers connected to the ocean.

Nowadays the rivers of the Chaco display similar morphological characteristics and behaviours, changing their channel patterns according to position within the alluvial fan. However, the current climate sets constraints on the generation of smaller lobes. The rivers extend at most only ~50% of the distance down the length of their megafans compared with their late Pleistocene ancestors, and terminate in interior wetlands on the Chaco plain. Moreover, the sinuosity and depth of current river channels change in response to slope reductions on the palaeofan surfaces. Because of these regional changes, an increase in specific power seems to be an effective way for these rivers of transporting their current inputs of Andean water and sediment.

The Chaco basins are among the highest sediment-yielding basins of the entire Andes. Currently, the Chaco plain is also one of the largest active continental sedimentary sinks on the planet. Approximately ~325 Mt/yr of sediment transported by the six rivers generating the largest megafans are trapped on the Chaco plain. The combination of the extremely flat surface, the mosaic of palaeolandforms and present fluvial and lacustrine patterns, and the hydrogeomorphological dynamics of those systems, have generated the largest area of seasonal wetlands in South America under current humid Holocene climates.

Acknowledgements

We especially thank Jose C. Stevaux, Yanni Gunnell, Alfredo Carlini, and Pavel Adamek for their critical reviews and suggestions that helped us to improve the manuscript.

References

Amsler, M. L. and Prendes, H. H. (2000). *Transporte de sedimentos y procesos fluviales asociados. El río Paraná en su tramo medio* (pp. 233-306). Santa Fe, Argentina: Centro de Publicaciones, Universidad Nacional del Litoral.

Amsler, M. L. and Drago, E. C. (2009). A review of the suspended sediment budget at the confluence of the

Paraná and Paraguay Rivers. *Hydrological Processes: An International Journal*, 23, 3230–3235.

Barnes, J. and Heins, W. (2009). Plio-Quaternary sediment budget between thrust belt erosion and foreland deposition in the central Andes, southern Bolivia. *Basin Research*, 21, 91–109.

Bowman, D. (2019). *Principles of Alluvial Fan Morphology*. Springer, Berlin.

Bull, W. B. (1964). *Geomorphology of Segmented Alluvial Fans in Western Fresno County*, California. US Government Printing Office.

Bull, W. B. (1968). Alluvial Fan, Cone. In R.W. Fairbridge, ed., *Geomorphology, Encyclopedia of Earth Sciences Series*. Springer, Berlin, 7–10.

Carlini, A. A. and Tonni, E. P. (2000). *Mamíferos fósiles del Paraguay*. Departamento Científico Paleontología Vertebratos Museo de La Plata.

Cochonneau, G., Sondag, F., Guyot, J. L., et al. (2006). The environmental observation and research project, ORE HYBAM, and the rivers of the Amazon basin. *Climate Variability and Change – Hydrological Impacts*, 308, 44–50.

Coltorti, M., Della Fazia, J., Rios, F. P., and Tito, G. (2012). Nuagapua (Chaco, Bolivia): evidence for the latest occurrence of megafauna in association with human remains in South America. *Journal of South American Earth Sciences*, 33, 56–67.

Contreras, S. A. and Zucol, A. F. (2019). Late Quaternary vegetation history based on phytolith records in the eastern Chaco (Argentina). *Quaternary International*, 505, 21–33.

Cordini, I. R. (1947). *Los ríos Pilcomayo en la región del Patiño*. Anales I, Dirección de Minas y Geología (Buenos Aires), 82 pp.

Fabbian, T., Ferreyro, V., De Felippi, R., Bernal, W., and Sanches, M. (1979). Estudio Geomorfológico en la zona del Bañado de Copo. Area: Rio Salado, Provincia de Santiago del Estero. In, *Consejo Federal de Inversiones-CFI*.

Garreaud, R. (2000). Cold air incursions over subtropical South America: mean structure and dynamics. *Monthly Weather Review*, 128, 2544–2559.

Garreaud, R., Vuille, M., and Clement, A. C. (2003). The climate of the Altiplano: observed current conditions and mechanisms of past changes. *Palaeogeography, Palaeoclimatology, Palaeoecology*, 194, 5–22.

Giles, P. T. (2010). Investigating the use of alluvial fan volume to represent fan size in morphometric studies. *Geomorphology*, 121, 317–328.

Ginzburg, R., Adámoli, J., Herrera, P., and Torrella, S. (2005). Los Humedales del Chaco: clasificación, inventario y mapeo a escala regional. *Miscelánea*, 14, 121–138.

Guyot, J. L., Bourges, J., and Cortez, J. (1994). Sediment transport in the Rio Grande, an Andean river of the Bolivian Amazon drainage basin. *IAHS Publications-Series of Proceedings and Reports–International Association of Hydrological Sciences*, 224, 223–232.

Guyot, J. L., Filizola, N., Quintanilla, J., and Cortez, J. (1996). Dissolved solids and suspended sediment yields in the Rio Madeira basin, from the Bolivian Andes to the Amazon. *Erosion and Sediment Yield: Global and Regional Perspectives*, 55–63.

Guzmán, S. V. (2014). El esplorador J. Crevaux i el rio Pilcomayo. *Conferencia de la Sociedad Geográfica Argentina*, 83.

Hartley, A. J., Weissmann, G. S., Nichols, G. J., and Warwick, G. L. (2010). Large distributive fluvial systems: characteristics, distribution, and controls on development. *Journal of Sedimentary Research*, 80, 167–183

Hoffstetter, R. (1968). Ñuapua, un gisement de vertébrés pléistocènes dans le Chaco Bolivien. *Bulletin du Museum National d'Histoire Naturelle (2 Série)*, 40, 823–836.

Hoffstetter, R. (1978). Une faune de Mammifères pléistocènes au Paraguay. *Comptes Rendus Sommaires des Sciences de la Societé Géologique de France*, 1, 32–33.

Horton, B. and DeCelles, P. G. (2001). Modern and ancient fluvial megafans in the foreland basin system of the central Andes, southern Bolivia: Implications for drainage network evolution in fold-thrust belts. *Basin Research*, 13, 43–63.

Iriondo, M. (1974). Los ríos desajustados de Formosa. Una hipótesis alternativa. *Revista de la Asociación Geológica Argentina*, 29, 136–137.

Iriondo, M. (1990). Map of the South American plains-Its present state. *Quaternary of South America and Antarctic Peninsula*, 6, 297–308.

Iriondo, M. (1993). Geomorphology and late quaternary of the Chaco (South America). *Geomorphology*, 7, 289–303.

Iriondo, M. H. (1986). Dinámica fluvial y transporte de sedimentos en el arroyo Los Amores (Chaco-Santa Fe). *Actas, Primera Reunión*, Asociación Argentina de Sedimentología. La Plata: Asociación Argentina de Sedimentología, 19–21.

Iriondo, M. H. and Orfeo, O. (2014). Esquema hidrosedimentario de la cuenca del río Pilcomayo. In, *14 Reunion*, Asociación *Argentina de Sedimentología*. Puerto Madryn, Argentina: Asociación Argentina de Sedimentología, 140-141.

Kandus, P., Minotti, P., Fabricante, I., and Ramonell, C. (2017). Identificación y Delimitación de Regiones de Humedales de Argentina. In L. Benzaquen, R. D. E. Blanco, P. Bo, et al. eds., *Regiones de Humedales de la Argentina*. Universidad Nacional de San Martín y Universidad de Buenos Aires: Ministerio de Ambiente y Desarrollo Sustentable, Fundación Humedales/ Wetlands International, 31–48.

Latrubesse, E. M. and Ramonell, C. G. (1994). A climatic model for southwestern Amazonia in last glacial times. *Quaternary International*, 21, 163–169.

Latrubesse, E. M., Stevaux, J. C., and Sinha, R. (2005). Tropical rivers. *Geomorphology*, 70, 187–206.

Latrubesse, E. M. and Brea, D. (2009). Floods in Argentina. *Developments in Earth Surface Processes*, 13, 333–349.

Latrubesse, E. M., Stevaux, J. C., Cremon, E. H., et al. (2012). Late Quaternary megafans, fans and fluvio-aeolian interactions in the Bolivian Chaco, Tropical South America. *Palaeogeography, Palaeoclimatology, Palaeoecology*, 356, 75–88.

Latrubesse, E. M. and Restrepo, J. D. (2014). Sediment yield along the Andes: continental budget, regional variations, and comparisons with other basins from orogenic mountain belts. *Geomorphology*, 216, 225–233.

Latrubesse, E. M. (2015). Large rivers, megafans and other Quaternary avulsive fluvial systems: A potential "who's who" in the geological record. *Earth-Science Reviews*, 146, 1–30.

Lehner, B., Verdin, K., and Jarvis, A. (2006). HydroSHEDS technical documentation, version 1.0. *World Wildlife Fund US*, Washington, DC, 1–27.

Marchetti, Z. (2017). Region Humedales del Chaco. In L. Benzaquen, R. D. E. Blanco, P. Bo, et al., eds., *Regiones de Humedales de la Argentina*. Universidad Nacional de San Martín y Universidad de Buenos Aires: Ministerio de Ambiente y Desarrollo Sustentable, Fundación Humedales/Wetlands International, 59–72.

Martín-Vide, J. P., Amarilla, M., and Zárate, F. J. (2014). Collapse of the Pilcomayo River. *Geomorphology*, 205, 155–163.

May, J.-H., Argollo, J., and Veit, H. (2008a). Holocene landscape evolution along the Andean piedmont, Bolivian Chaco. *Palaeogeography, Palaeoclimatology, Palaeoecology*, 260, 505–520.

May, J.-H., Zech, R., and Veit, H. (2008b). Late Quaternary paleosol–sediment-sequences and landscape evolution along the Andean piedmont, Bolivian Chaco. *Geomorphology*, 98, 34–54.

May, J.-H. and Veit, H. (2009). Late Quaternary paleosols and their paleoenvironmental significance along the Andean piedmont, Eastern Bolivia. *Catena*, 78, 100–116.

Milana, J. P. and Ruzycki, L. (1999). Alluvial-fan slope as a function of sediment transport efficiency. *Journal of Sedimentary Research*, 69, 553–562.

Minotti , P. (2017). Subregion Riachos y Esteros del Chaco Húmedo. In L. Benzaquen, R.D.E. Blanco, P. Bo, et al., eds., *Regiones de Humedales de la Argentina*. Universidad Nacional de San Martín y Universidad de Buenos Aires: Ministerio de Ambiente y Desarrollo Sustentable, Fundación Humedales/ Wetlands International, 149–162.

Navarro, G., Molina, J. A., and de Molas, L. P. (2006). Classification of the forests of the northern Paraguayan Chaco. *Phytocoenologia*, 36, 473–508.

Neiff, J. J. (1999). El régimen de pulsos en ríos y grandes humedales de Sudamérica. *Tópicos Sobre Humedales Subtropicales y Templados de Sudamérica*, 229, 99–103.

Nogués-Paegle, J., Mechoso, C. R., Fu, R., et al. (2002). Progress in Pan American CLIVAR research: understanding the South American monsoon. *Meteorologica*, 27, 1–30.

Orfeo, O. (1986). Estudio sedimentológico de ambientes fluviales del Chaco Oriental. *Rev. Ambiente Subtropical*, 1, 60–72.

Park, E. and Latrubesse, E. M. (2017). The hydrogeomorphologic complexity of the lower Amazon River floodplain and hydrological connectivity assessed by remote sensing and field control. *Remote Sensing of Environment*, 198, 321–332.

Pekel, J.-F., Cottam, A., Gorelick, N., and Belward, A. S. (2016). High-resolution mapping of global surface water and its long-term changes. *Nature*, 540, 418–422.

Peri, V. G., Rossello, E. A. (2010). Anomalías morfoestructurales del drenaje del río Salado sobre las Lomadas de Otumpa (Santiago del Estero y Chaco) detectadas por procesamiento digital. *Revista de la Asociación Geológica Argentina*, 66, 636–648.

Placci, G. and Holz, S. (2004). Patrón de paisaje de bosques del Chaco Oriental. *Ecología y Manejo de los Bosques de Argentina. Editorial de la Universidad Nacional de La Plata*.

Schumm, S. A. (2007). *River Variability and Complexity*. Cambridge University Press, Cambridge, UK.

SSRH, S.d.R.H. (2004). Estadística Hidrológica de la República Argentina. *Presidencia de la Nación, Rep. Argentina*.

Tonni, E., and Scillato-Yané, G. (1997). Una nueva localidad con mamíferos pleistocenos en el Norte de la Argentina. Aspectos paleozoogeográficos. In *VI Congresso da Associação Brasileira de Estudos do Quaternário e Reunião sobre o Quaternário da América do Sul, Curitiba, Brasil, Anais*, 345–348.

Vizy, E. K. and Cook, K. H. (2007). Relationship between Amazon and high Andes rainfall. *Journal of Geophysical Research: Atmospheres*, 112. doi.org/10 .1029/2006JD007980

Vuille, M., Hardy, D. R., Braun, C., Keimig, F., and Bradley, R. S. (1998). Atmospheric circulation anomalies associated with 1996/1997 summer precipitation events on Sajama Ice Cap, Bolivia. *Journal of Geophysical Research: Atmospheres*, 103, 11191–11204.

Wilkinson, M. J., Marshall, L. G., and Lundberg, J. G. (2006). River behavior on megafans and potential influences on diversification and distribution of aquatic organisms. *Journal of South American Earth Sciences*, 21, 151–172.

Zhou, J. and Lau, K. (1998). Does a monsoon climate exist over South America? *Journal of Climate*, 11, 1020–1040.

Zurita, A. E., Miño-Boilini, Á. R., Carlini, A. A., Iriondo, M., and Alcaraz, M. A. (2009). Paleontología del Chaco Oriental: Una nueva localidad con mamíferos fósiles pleistocenos en el río Bermejo (Formosa, Argentina). *Revista Mexicana de Ciencias Geológicas*, 26, 277–288.

Zurita, A. E., Miño-Boilini, A., Francia, A., et al. (2014). Paleontología y cronología del Cuaternario de las provincias de Corrientes y Formosa, Argentina. *Acta Geológica Lilloana*, 26, 75–86.

6

Megafans of the Pantanal Basin, Brazil

MAURÍCIO G. M. SANTOS, JOSÉ C. STEVAUX, and MARIO L. ASSINE

Abstract

The Pantanal Basin (west-central Brazil) is one of the largest alluvial wetlands in the world ($> 150{,}000\,\mathrm{km}^2$) formed dominantly by coalescing Quaternary megafans. The Pantanal Basin is an efficient sediment trap: of $25\,\mathrm{Mt\,yr}^{-1}$ of suspended load that enters by the main river systems, only $10\,\mathrm{Mt\,yr}^{-1}$ is exported by the trunk river. Sediments are sourced by multiple rivers draining Precambrian lowlands and Paleozoic uplands. The eastern border displays tablelands of Paleozoic rocks of the Paraná Sedimentary Basin, with lowlands of Precambrian rocks on the northern, southern and western borders. The Taquari, Cuiabá, and São Lourenço megafans, tributaries to the Paraguay trunk-river system, are the largest fluvial fans in the Pantanal. The Paraguay River itself has produced two relatively small megafans. The megafans display four main landform assemblages: incised meander belts proximally, active aggradational lobes, abandoned degradational lobes, and mixed-process floodplains. Megafan surfaces display palaeo-drainage networks ranging from braided channel planforms to the current meandering and ana-branching planforms. Megafan areas seem to be a function of both feeder-basin area and catchment geology: those fed from sedimentary rock outcrops are larger, with more complex barform development than those supplied from Precambrian basement catchments.

6.1 Introduction

The Pantanal region in west-central Brazil is one of the largest alluvial wetlands in the world, formed dominantly by coalescing Quaternary megafans. The Pantanal Basin is located in the upper Paraguay drainage basin and is seasonally flooded by several rivers. It is home to a wide variety of wildlife and hosts more bird species than North America. Land-use change in the catchment areas is considered by local inhabitants as a major cause of river channel shallowing, in-channel bar development, and channel avulsions. Studies on the natural causes and mechanisms of such phenomena are thus paramount to the management of the area, and the understanding of the fluvial dynamics is also of crucial importance for resource management and wildlife conservation programmes.

Sediments of the Pantanal plain are sourced by multiple rivers draining Precambrian lowlands and Paleozoic uplands (Fig. 6.1), and these rivers have constructed alluvial fans of various sizes as they enter the plain (Assine 2005, 2015; Assine et al. 2014, 2015a,b,c). The study of megafans and distributive fluvial systems has been of key interest to many researchers in recent years (Wilkinson 2006; Weissmann et al. 2010; Hartley et al. 2010, Latrubesse 2015). Numerous studies have concentrated on the foreland basins of active orogens (e.g., Sinha and Friend 1994; Horton and DeCelles 2001; Latrubesse et al. 2012 among others), with few concentrating on active megafans in intracratonic settings (Assine et al. 2005; Buehler et al. 2011; Zani et al. 2012; Ch. 17).

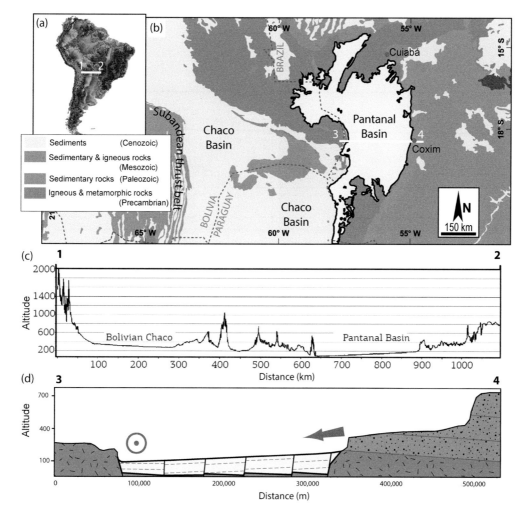

Figure 6.1 **The Pantanal in its regional context**. (a) SRTM (~ 90 m spatial resolution) digital elevation model showing South America and location of section 1–2 in panel (c). (b) Geological map of the surrounding area; 3–4 is location of profile in panel (d). (c) Topographic W–E profile from the outer Andes to the Paleozoic scarpland topography of the Paraná Basin, highlighting the Chaco and Pantanal basins. Location shown in a (1–2). (d) Topographic and geological profiles of the Pantanal Basin. Location shown in (b) (3–4). A colour version of this figure is available in the SOD for Chapter 6.

The Pantanal megafan surfaces display a record of palaeodrainage networks ranging from braided channel planforms to the currently prevailing meandering and anabranching planforms, and thus constitute an archive of changes in fluvial activity and palaeoclimate (Assine 2003; Assine and Soares 2004; Zani and Assine 2011; Zani et al. 2012; Assine et al. 2015a,b,c). Our understanding of the sedimentary architecture of the Pantanal is still in its early stages, but comprehensive studies of the basin fill and of processes currently operating across its surface are expected to provide analogues for fluvial sequences preserved in the rock record. This fluvial archive can also provide substantial evidence concerning the palaeoenvironmental evolution of the region.

Here, we provide a state-of-the art summary of the Quaternary environmental record of the Pantanal and the major geomorphological styles that characterise the main megafans.

Wetlands of the modern Pantanal cover > 150,000 km^2, i.e., approximately the combined area of England and Wales. Various fluvial systems develop and interact, forming very large fluvial-dominated alluvial fans, or megafans, on its alluvial plains (Fig. 6.2). The Pantanal Basin is an active interior sedimentary basin with a continuous record of up to M_L 5.5 earthquakes resulting from subsidence of fault-bounded basin blocks (Assumpção and Sacek 2013). The tectonic drivers, however, are disputed, and although these are commonly associated with

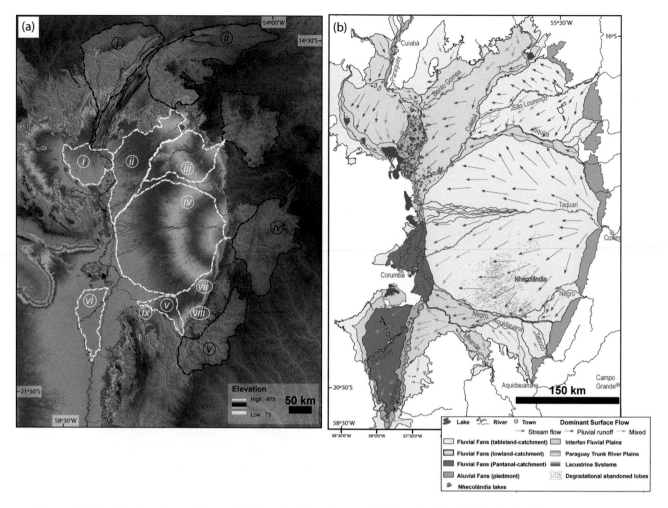

Figure 6.2 **The Panatanal megafans and their upland catchments**. (a) Megafans (i–vi, dashed white lines) and unnamed smaller fans (numbered vii–ix) of the Pantanal Basin and their respective contributing upland basins (dashed black lines, for larger megafans i–vi). Key to names: i– Paraguay Megafan; ii– Cuiabá Megafan; iii– São Lourenço Megafan; iv– Taquari Megafan; v– Aquidauana Megafan; vi– Nabileque Megafan; vii– Negro Fluvial Fan; viii– Taboco Fluvial Fan; ix– Miranda Fluvial Fan. (b) Depositional tract systems of the Pantanal Basin (after Assine et al. 2015c).

Andean tectonics (Almeida 1959; Shiraiwa 1994; Ussami et al. 1999; Horton and DeCelles 1997) the basin is characterised by a geometry of grabens and horsts with basin boundary faults displaying throws of hundreds of metres (Assine et al. 2015a). Preserved deposits up to 500 m thick record the climatic and tectonic evolution of central South America during the Pleistocene and Holocene (Ab'Sáber 1988; Tricart 1982; Klammer 1992; Clapperton 1993; May et al. 2008a,b; Whitney et al. 2011; McGlue et al. 2012). The area is currently under a savanna climate (Alho 2005), with mean annual values of 20–27 °C, 800–1,300 mm, and 1,300–1,600 mm for temperature, precipitation, and evapotranspiration respectively. Seasonal changes in rainfall in the catchment

areas of the Pantanal plain generate flood pulses from north to south (Assine et al. 2015b).

The basin is roughly elliptical, elongated in a north–south direction, and is bordered by faults to the east and west. These and other faults have also been mapped in the subsurface and divide the basin into a series of blocks. The eastern border displays tablelands of Paleozoic rocks of the Paraná Sedimentary Basin, with lowlands of Precambrian rocks on the northern, southern, and western borders. The Pantanal basin coalesces with the adjacent Chaco Basin to the west and south, limited by the NW-striking Bodoquena footwall upland. However, it is currently an open, externally drained basin, and part of the water and sediments of the basin are transferred via the

Paraguay River towards the main, continental-scale trunk river – the Paraná (Fig. 6.2).

The Taquari, Cuiabá, and São Lourenço megafans are the largest fluvial fans in the Pantanal, and are tributaries to the Paraguay trunk-river system. The Paraguay River has also produced a fluvial fan in the proximal area where it enters the basin (Paraguay fluvial fan), and it leaves the basin via the Nabileque fan. The Paraguay River flows from north to south along the western border of the Pantanal, connecting to the Chaco basin near the city of Corumbá. It develops complex river plains confined to the east by the megafan systems, e.g., Taquari and Cuiabá, and to the west by the basin's border fault. Smaller-scale fluvial fans include the Taboco, Aquidauana, Itiquira, and Negro river systems. Some rivers, including the latter two, develop fan systems at their entrance into the basin, but develop into interfan systems downstream where they are confined by flanking, larger-scale fans. The megafans lie across the basin structure, and interact with interfan systems and the trunk river system, which flows along the N–S basin axis. Satellite imagery provides evidence of the activity of heterogeneous fluvial environments evolving on different scales and generating planform patterns driven by tectonics and climate. Image analysis also reveal past fluvial dynamics, with channel planforms preserved on abandoned lobes of the fluvial fans which contrast markedly with modern river dynamics.

We describe different interactions between adjacent megafans, and describe and classify their distinct depositional dynamics. The latter result mainly from the contrasting geological characteristics of the respective catchment areas. Avulsions and other fluvial processes on megafans are investigated. Depositional and fluvial dynamics are discussed in light of the distinctive geology and relief of the catchment areas. We demonstrate that the size and fluvial style of each megafan is controlled by catchment-area geology and by the interaction with other alluvial fans on either side. We also show that distinct styles of megafan deposition can take place under exactly similar catchment and climatic conditions.

6.2 Methods

The results synthesised in this chapter are based on a combination of field studies and remote sensing analysis from satellite imagery and aerial photography. Multitemporal orbital data included high- to medium-resolution spatial and radiometric imagery such as Landsat, GeoCover, ASTER, SPOT, MODIS/Terra, and the 3 arcsecond SRTM digital elevation model. Sediment yield and river discharge information were collected from datasets of the ANA (Brazilian Water Agency; https://www.ana.gov.br/) and used in combination with geological and geomorphological maps to build up a comprehensive review of the previously proposed models for the geomorphological delimitation of the Pantanal alluvial plains. We used all the above to identify, map, and classify depositional systems in terms of their present-day morphology, landforms, and preserved palaeo-landforms.

6.3 Provenance-Based Classification of Pantanal Megafans

The numerous fluvial systems in the Pantanal Basin are all tributaries of the Upper Paraguay River drainage basin (Fig. 6.2a). However, when entering the plain, they form complex arrangements of floodplains, lakes, and fluvial fans. The basin is an efficient sediment trap: whereas 25 Mt/yr of suspended load enters the Pantanal Basin by the main systems (e.g., the Paraguay, Cuiabá, São Lourenço, Piquiri, Taquari, Aquidauana, and Miranda rivers), only 10 Mt/yr are exported by the trunk river (ANA 2004). Most of the tributaries of the Paraguay River and related megafans (Taquari, São Lourenço, and Cuiabá) are transverse to the basin axis (Fig. 6.2b). Megafans are classified here according to their sediment source area.

6.3.1 Megafans Fed by a Sedimentary Hinterland (Paraná Scarplands)

Three of the four larger megafans of the Pantanal Basin, i.e., the Taquari, São Lourenço (Fig. 6.3), and Aquidauana, are produced by rivers rising in the sedimentary tablelands composed of Paleozoic marine and continental rocks of the Paleozoic–Mesozoic Paraná Basin, on the eastern confines of the basin. They transport 22.5 Mt/yr of suspended sediment load (Macedo et al. 2014), and the Taquari River alone contributes 72% of the sediment load entering the basin. Of this total, almost the entire suspended load (90%) is

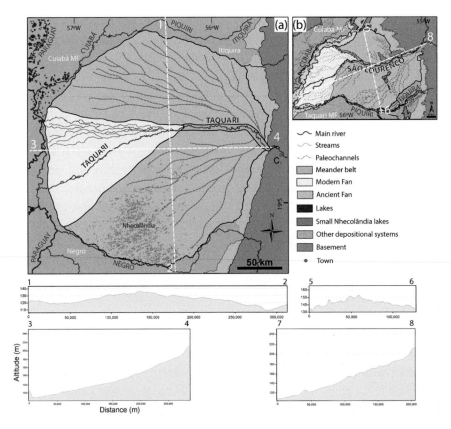

Figure 6.3 **Land system maps of the megafans produced by the Paleozoic sedimentary uplands**. (a) Taquari Megafan, C: town of Coxim. (b) São Lourenço Megafan. A colour version of this figure is available in the SOD for Chapter 6.

trapped in the basin (Assine et al. 2015c). Other, smaller-scale megafans sourced from the tablelands include the Taboco and Negro megafans.

6.3.1.1 The Taquari Megafan

The Taquari river (drainage area: 27,000 km²) flows for more than 200 km through Paleozoic terrain (Jongman 2006; Zani et al. 2012). The Taquari megafan occupies an area of 49,040 km², or 37% of the Pantanal Basin. With an apex-to-toe distance of 250 km, the Taquari is the largest fan on the Pantanal Basin (Assine 2003; Zani et al. 2012), and is among the largest distributive fluvial systems in the world (Hartley et al. 2010). It is limited to the north by the Itiquira interfan fluvial system, to the east by the sedimentary tablelands of the Paraná sedimentary basin, to the south by the Negro River fluvial fan and interfan systems, to the west by its collector – the Paraguay River – and to the northwest by the Cuiabá Megafan (Fig. 6.3a). It currently accumulates ~16 Mt/yr on its

floodplains and lobes, and is the most important sediment routing system of the Pantanal Basin (Assine et al. 2015c). The river flows from ~190 m at its apex to 80 m at its toe, with an average gradient of 0.36 m/km (Zani et al. 2012). The mean river discharge of the Taquari River at its entrance to the basin is around 400 m³/s, with 4.1 Mt/day of suspended sediment load (Padovani et al. 2010; Carvalho et al. 2005). Nearly 12% of the megafan area is permanently inundated, and during the flood season it can cover as much as 37% of the megafan surface (Jongman 2005).

Where the river enters the Pantanal Basin near the town of Coxim, its modern floodplain is only ~5 km wide and continues downstream as an active meander belt 110 km long, entrenched in older deposits of the megafan (Fig. 6.3a). The river displays a single-channel meandering planform (sinuosity index: 1.6), with common, well-preserved scroll bars, channel cut-offs, and oxbow lakes (Fig. 6.3a) (Assine 2005). Bar development is the most prominent in-channel feature, with many examples of longitudinal bars, unit bars, compound bars, and bank-attached bars. Although vegetation-stabilised bars are more common in the proximal areas of the meander belt, large areas of sand are permanently exposed on bars and these can provide a plentiful supply of sedimentary material for reworking during floods. This situation is not common in most of the other megafans (see below). River confinement in this part of the megafan currently hinders avulsion and channel crevassing, promoting overall sediment bypass to more distal areas. Incision of the river into older sediments of this system has been driven by base-level changes along the Paraguay trunk river (Assine 2005). Climatic changes are another possible mechanism, with the Taquari megafan having recording changes from arid to semiarid conditions during the late Pleistocene, followed by a more humid environment in early Holocene time (Braun 1977; Tricart 1982; Klammer 1982).

Downstream, at the end of the confined meander belt, the river loses confinement and builds numerous depositional lobes which are initiated by avulsion processes. The characteristically distributive fluvial system accordingly displays a radial drainage pattern (Hartley et al. 2010; Weissmann et al. 2010). Levees along the anastomosing distributaries are distinctly elevated above the surrounding floodplain. Channel sinuosity on the lobes is 1.3 (80% of confined meander belt). Channel gradients increase from 0.38 m/km in the meander belt to 0.44 m/km near the axis of the modern lobes and progressively decrease downstream of this point (Assine 2005). The number of barforms greatly diminishes in this more distal sector of the river, with only the development of small unit bars and scattered, vegetation-stabilised, in-channel compound bars and bank-attached bars (Table 6.1).

Table 6.1 *Landform assemblages on megafans of the Pantanal: a basis for classification*

Megafan	Fan area (km²)	Drainage basin area (km²)	Land unit	Assemblage of landforms	
Taquari	49,040	27,000	*Incised meander belt*	- single, sinuous channel - laterally confined meander belt - well-preserved scroll bars - longitudinal bars, unit bars, compound bars, and bank-attached bars - oxbow lakes - channel cutoffs	Sedimentary hinterland
			Modern depositional lobe	- distributary and anastomosing channels - downstream-decreasing sinuosity - elevated levees - number of barforms greatly diminishes - small unit bars - rare bank-attached bars - scattered, vegetation-stabilised, in-channel compound bars	
Sãolourenço	12,700	24,000	*Incised meander belt*	- single, highly sinuous channel - laterally confined meander belt - barforms mostly restricted to point bars - scattered alternate bars - rare cutoffs and crevasse splays - incipient scroll preservation - common oxbow lakes.	
			Modern depositional lobe	- distributary and anastomosing channels - downstream-decreasing sinuosity - conspicuous levee ridges - rare barforms - common crevasse splays	

Table 6.1 (*cont.*)

Megafan	Fan area (km^2)	Drainage basin area (km^2)	Land unit	Assemblage of landforms	
Aquidauana	3,129	15,675	*Incised meander belt*	- single, highly sinuous channel - laterally confined meander belt - very rare in-channel barforms - rare bank-attached, non-vegetated barforms - common oxbow lakes	
			Modern depositional lobe	- highly sinuous channels - downstream-decreasing sinuosity - very isolated, small unit bars; - elevated channels - downstream-decreasing levee height - exposed sand nearly absent	
Cuiabá	15,420	22,000	*Incised meander belt*	- single, highly sinuous channel - well-preserved scroll features - rare bank-attached unit bars - common crevassing - rare non-vegetated sand deposits at meander apexes - common oxbow lakes	Precambrian shield hinterland
			Modern depositional lobe	- unconfined meandering - anabranching distributive channels - net of levees - numerous small crevasse deltas - absence of in-channel bars - sparse crevasse splays - sparse channel cutoffs - sparse oxbow lakes	
Paraguay	7,182	33,860	*Incised meander belt*	- single, highly sinuous channel - compound barforms and point bars - common scroll preservation - common crevassing - vegetated longitudinal barforms - common channel cut-off	
			Modern depositional lobe	- anastomosing channel network to the east - a distributary pattern to the west - 30-km-long fluvial island - common scrolls - floodplain lakes - crevasse splays - oxbow lakes - rare to absent in-channel barforms	
Nabileque	9,100	> 15,000	*Incised meander belt*	- low sinuosity, single channel - common channel cutoffs - common scroll preservation - vegetated islands - floodplain lakes - rare in-channel barforms - common oxbow lakes	Pantanal Basin

Avulsion processes begin with splay progradation onto floodplains and flood basins, followed by construction of primitive levees and progressive diversion of flow from the parent channel, constructing new avulsion belts (Assine 2005; Buehler et al. 2011). Floodplain sediments on the Taquari megafan typically consist of coarse-grained sand (up to 60%) (Souza et al. 2002). The main channel of the Taquari river joins the Paraguay trunk river near the city of Corumbá, forming a complex arrangement of floodplain and anabranching channels. Part of the Taquari discharge is currently being diverted through the Caronal Avulsion (Assine 2005) to a point ~75 km north of its mouth, where it also meets the Paraguay River. The southern area of the Taquari megafan is an abandoned lobe known as the Nhecolândia region. Its main feature is a population of > 10,000 small, shallow, elliptical lakes (*baías*) which are separated from each other by ~5-m-high sandy ridges (*cordilheiras*) with dense vegetation. The geochemistry of these lakes is complex, with many examples of brackish to saline water, some of them with a pH of up to 10, and diverse colonisation patterns by macrophytes (Barbiéro et al. 2002).

6.3.1.2 The São Lourenço Megafan

In the northeastern part of the Pantanal Basin, the São Lourenço River flows westwards to join the Cuiabá River (Fig. 6.3B). It is the third largest megafan of the basin, with an approximate area of 12,700 km² and delivering 4.2 Mt/yr of suspended sediment load (Assine et al. 2015c; Carvalho et al. 2005). The 24,000 km² drainage basin erodes Paleozoic and Mesozoic siliciclastic deposits of the Paraná sedimentary basin. This sedimentary protolith is thus conveyed to the Pantanal wetlands as sediment already comminuted and well sorted as a result of its earlier geological history (Corradini and Assine 2012). The São Lourenço flows for ~110 km before being joined by the Vermelho River at the point where they both enter the Pantanal Basin. The megafan is laterally constrained by the Cuiabá megafan to the west, debris-flow dominated alluvial fans to the east, and the interfan fluvial system of the Piquiri River to the south, which separates this megafan from the Taquari

megafan (Fig. 6.3a). Immediately below its confluence with the Vermelho, the São Lourenço abruptly changes direction to flow parallel with the N65E-striking São Lourenço fault system (Corradini and Assine 2012). The São Lourenço River enters the Pantanal Basin at about 200 m a.s.l. at its apex, then descending to around 110 m a.s.l. where it encounters the Cuiabá River. The overall gradient is thus 0.4 m/km (Assine et al. 2014).

The river is currently incising older sediments near the fan apex, where the channel is confined by a meander belt similar to the Taquari River situation. Barforms at this reach of the river are mostly restricted to point bars and some scattered alternate bars. Channel cutoffs and crevasse splays are rare to absent, scroll preservation is incipient, but oxbow lakes are very common (Table 6.1). Relatively large areas of unstable sand deposits are stored on unvegetated, point-bar apexes across this reach of the megafan. This situation provides loose sedimentary material easily reworked during floods, with a direct impact on the river dynamics. Abandoned lobes on the middle and upper parts of the megafan preserve braided palaeochannel planforms indicative of contrasting fluvial dynamics which prevailed during the Pleistocene on this megafan (Assine et al. 2014; Pupim 2014).

The river continues to flow within an incised meander belt and flows downstream to the modern depositional lobe, where distributary channels have constructed conspicuous levee ridges. This buffers channelised flow from the floodplain and allows the São Lourenço to cross the distal floodplains with minimal loss of discharge and reach the Cuiabá River (Assine et al. 2005). Barforms are rare along the distal part of the São Lourenço River, but crevasse splays are common. The abandoned lobes of the São Lourenço megafan display a diverse gallery of remarkably well preserved palaeochannel planforms, collectively revealing a dynamic evolution of fluvial regimes most likely linked to variations in climate during the Pleistocene and Holocene (Assine et al. 2014). River planform analysis suggests that the São Lourenço River evolved over time from a multi-channel braided fluvial system to its modern meander-belt/distributary arrangement. Sedimentological analysis reveals a marked change in the sediment transport capability of this river system, as recorded by a fining-upwards trend from coarse-grained sand associated with braided palaeochannels, to much finer-grained silts and fine sand associated with the more recent meandering channel dynamics (Assine et al. 2014).

6.3.1.3 The Aquidauana Megafan

The Aquidauana River is 650 km long, 330 km of which occupies the alluvial plains of the Pantanal (Fig. 6.2). The river incises Paleozoic tablelands in a NE–SW direction as a result of drainage control and capture guided by the fracture net, and where higher channel gradients (~1 m/km) promote higher erosion rates. This situation is reflected in the high suspended sediment yield (1.4 Mt/yr; Carvalho et al. 2005) and in the presence of rapids. The drainage area is 15,675 km^2 (Gregório et al. 2016) and situated among south-dipping Paleozoic rocks of the Paraná Basin.

The Aquidauana megafan (surface area: 3,129 km^2; Merino 2017) lies in the southern part of the Pantanal Basin and is limited to the north by the Negro Interfan system, to the east by the Taboco fluvial fan, to the south by the Paraná tablelands, and to the west by the Miranda fluvial fan. The Aquidauana River flows in a NNW direction from the southern limits of the basin and, like the megafans described earlier, also currently displays an incised meander belt in its proximal area and a complex arrangement of prograding, multiple levee constructions on older floodplain deposits. The meander belt is incised by up to 6 m into Pleistocene deposits, and the height of the incision decreases downstream (Facincani and Assine 2010). The river channel is 50 m wide as it enters the basin, narrowing to 40 m where it joins the Miranda River. The river presents very limited in-channel bar development, abundant oxbow lakes, few scrollbars, and rare crevasses. Vegetation on the meander belt is dense, with rare exposures of bare sand (a few occurrences are limited to point-bar cusps) and an absence of river bar development other than point bars stabilised by dense vegetation.

This megafan can be subdivided into three geomorphological units: (i) a 2,200 km^2 Pleistocene palaeolobe characterised by distributary palaeochannels; (ii) an 80-km-long, narrow (2.5 km), 150 km^2 modern meander belt with a 0.15 m/km stream gradient, channel sinuosity index of 1.5, and abundant oxbow lakes; and (iii) a 750 km^2 modern depositional lobe displaying highly sinuous distributary channels (0.16 m/km stream gradient) which are elevated in relation to the surrounding floodplains (Facincani and Assine 2010; Merino 2017). These levee complexes diminish in height downstream, finally coalescing with the Miranda and Negro rivers in a complex arrangement of small channels and floodplain wetlands. As on the São Lourenço Megafan, downstream of the meander belt the levee systems develop and maintain the river channels in a raised position above the floodplain; channel sinuosity decreases to < 1.5. Crevassing and crevasse splay development are common. Channel bars are almost absent, with very rare examples of small unit bars. Exposures of bare sand are also nearly absent, with vegetation covering and potentially stabilising the entire river course (Table 6.1).

6.3.2 Megafans Fed from a Hinterland of Basement Rocks (Precambrian Shield)

6.3.2.1 The Cuiabá Megafan

The Cuiabá River flows 1,080 km from its source near the town of Rosário Oeste (near the northern tip of the Paraguay watershed) to its junction with the Paraguay River near the town of Puerto Quijarro, in Bolivia. It is one of the main tributaries of the Upper Paraguay River, flowing 550 km on Precambrian rocks in an estimated drainage-basin area of 22,000 km^2 before entering the Pantanal's alluvial plains (Pupim 2014; Pupim et al. 2014). The river hydrology is similar to other rivers of the Pantanal, i.e., seasonal, with 72% of the flow during the rainy season. Mean monthly discharge ranges from 200 to 800 m^3/s. The Cuiabá River loses discharge downstream unless supplemented by tributaries, but this trend reverses after its confluence with the São Lourenço and Piquiri rivers.

The Cuiabá River enters the Pantanal Basin near the town of Santo Antônio do Leverger (Fig. 6.4a), forming downstream of this point the largest of the Pantanal megafans generated by a lowland catchment in Precambrian rocks. The river has been building – at least since the late Pleistocene (Pupim 2014) – a 15,420 km^2 fluvial fan. It was named Cuiabá River

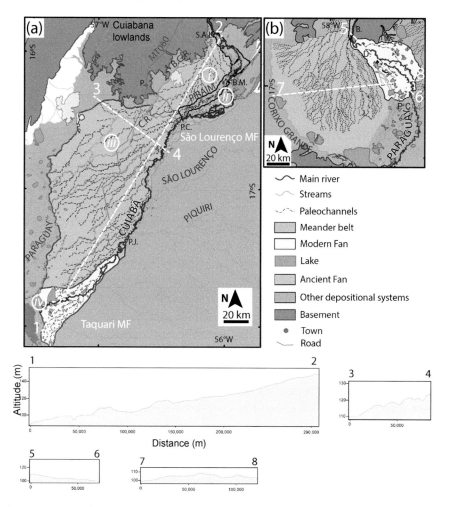

Figure 6.4 **Land system maps of the megafans sourced from the Precambrian uplands**. (a) Cuiabá Megafan; i) Ancient Cuiabá Fan; ii) Cuiabá Alluvial Plain; iii) Lower Fan; iv) Active Lobe. Place names. B.M.- Barão de Melgaço, P.- Poconé, P.C.- Porto Cerrado, P.J.- Porto Jofre; S.A.L.- Santo Antônio do Lebeger. River acronyms: B.G.R.: Bento Gomes River; C.R.: Claro River. (b) Paraguay Megafan. Place names. B.- Baiazinha, D.- Descalvado, T.- Taiamã Island, P.C.- Porto Cerrado. A colour version of this figure is available in the SOD for Chapter 6.

Megafan by Assine et al. (2015c). The construction of this megafan consumes 57% of an annual sedimentary load of 12.6 Mt (PCBAP 1997). The Cuiabá Megafan displays a smooth surface, with a general slope of ~0.2 m/km (apex-to-toe elevations fall from 155 to 90 m over a horizontal distance of 300 km). The megafan is divided into four geomorphological land units defined according to age and nature of the acting hydro-sedimentological processes (Pupim 2014): (i) the Ancient Cuiabá Fan, (ii) the Cuiabá Alluvial Plain, (iii) the Lower Fan, and (iv) the Active Lobe (Fig. 6.4a).

The Ancient Cuiabá Fan (878 km^2) occurs at the apex of the megafan. It is limited by the Cuiabana Lowlands to the west, the Cuiabá Alluvial Plain to the east, and the Lower Fan to the south. Probably at the end of Pleistocene, an avulsion repositioned the

Cuiabá River to the southeast, abandoning its now inactive original lobe. Meira et al. (2019) report that channel shifting processes continued until the early to middle Holocene. The Ancient Cuiabá Fan thus constitutes a raised surface relative to the adjacent land units. It is crossed by a distributary anabranching system of palaeochannels that still operate as conductors of rain- and groundwater-fed streams, but remains above the Cuiabá River floodwater line. The fan deposits are homogeneous, consisting of massive fine to very fine sand containing iron-oxide nodules. Pupim (2014) obtained OSL depositional ages between 19 and 48 ka.

The Cuiabá Alluvial Plain (1,650 km^2) is structurally controlled by the NE–SW lineaments of the Morros and Serras Baixas Domain and by NW–SE lineaments (Fig. 6.4a). The more obvious expression of this structural control is the abrupt 90° elbow in the river course from SE to SW near Barão de Melgaço (Fig. 6.4a). To the SW, the alluvial meander belt is confined between the Ancient Cuiabá Fan and the Lower Fan to the west, and by the São Lourenço Fan to the east. The fluvial system is a single meandering channel up to the Barão elbow. In this upper reach the river displays well-preserved scroll features, rare bank-attached unit bars, oxbow lakes, sand deposits at meander apexes, and common crevassing (Table 6.1). Vegetation is very densely distributed, similar to the Aquidauana, with very rare bare sand exposures.

The river at first constructed a typical 'raised channel system' along to the border of Ancient Cuiabá Fan border. This system operated until 8.5 ka when the river avulsed, directing its discharge into a 20-km-wide plain and developing an anabranching pattern that is still active (Meira 2019). The alluvial plain narrows again near Porto Cercado and is inset for 70 km in an incised belt displaying meanders until its reaches an unconfined zone where it forms the Active Lobe. The palaeochannel of the Cuiabá River is today occupied by the underfit Piraim River (Fig. 6.4a).

The Lower Fan (8,700 km^2) occupies the central and distal portion of the Cuiabá Megafan. It consists of a series of small lobes that migrated from north to south, ending up today in the present Active Lobe (Fig. 6.4). The distributive system of the Lower Fan is atypical: instead of a unique apical feeder channel, a succession of distributary channels deliver water through a series of crevasses cutting through the right banks of the Cuiabá River, or even from other rivers entering the fan separately such as the Bento Gomes and Claro rivers (B.G.R. and C.R. in Fig. 6.4a). Although there are no dates for the sediments of this geomorphic zone, it is reasonable to suppose that the fan began to aggrade after the avulsion of the Cuiabá River and abandonment of the Ancient Cuiabá Fan.

The Active Lobe appears downstream of Porto Jofre, where the Cuiabá River acquires an unconfined meandering pattern for 80 km before joining the Paraguay River. Along its course, channel crevasses give rise to a net of anabranching distributive channels that cover a triangular area (2,000 km^2). As a result, in the last 40 km before it reaches the Paraguay, the Cuiabá River enters a wide floodbasin where a net of

levees is associated with a population of small crevasse deltas leading sediment away from the Cuiabá channel to the floodplain. The interference between trunk and tributary flow dynamics near the Paraguay generates complex processes around the channel junction. Although the Paraguay trunk river has a larger water discharge, it reaches the distal part of the Cuiabá Megafan with a considerably reduced sedimentary load, whereas the Cuiabá River delivers a substantial load of suspended sediment to the Paraguay (2.3 Mt/yr) – thus practically being the only purveyor of floodplain construction material in this area even though it has already been losing water and sediment load to the Lower Fan. In this anabranching domain, the main landform assemblage is characterised by an absence of channel bars, some small patches of sand exposure, widespread scrollbars, sparse oxbow lakes, sparse channel cutoffs, and sparse crevasse splays (Table 6.1).

6.3.2.2 The Paraguay Megafan

The Paraguay River (Fig. 6.4b) collects water from all the tributaries of the basin and forms its own extensive floodplains across the Pantanal. The Paraguay River catchment covers 33,860 km^2 north of the Pantanal (Carvalho et al. 2005). Before entering the Pantanal Basin, the river flows on a 5 km-wide aggradational fluvial valley perhaps guided by NNE–SSW faults in the Precambrian basement, and with its channel cutting up to 10 m into Pleistocene deposits. Fluvial forms along this reach of the river include compound barforms and point bars, vegetated longitudinal bars, bank-attached bars, widespread scroll and crevasse features, intermediate-sized sand exposure areas compared to the other rivers in the region (the bare sand is typically exposed at meander apexes), and common channel cutoffs. The river flows in a SSW direction before entering the basin, where it abruptly changes direction towards the SE and forms the 7,182 km^2 Paraguay River Megafan (Assine and Silva 2009).

The megafan can be divided into two geomorphic zones: the Ancient and Modern Paraguay lobes. The Ancient Lobe has practically never been studied. It consists of a smooth surface with a network of many palaeochannels and avulsion belts (Assine and Silva 2009). These undergo considerable reworking during seasonal flood events, when sheetflow generated by overbank spill from the main channel degrades older deposits. Despite the present erosional and depositional activity over the Ancient Lobe, many of its Pleistocene palaeochannels with characteristic distributary networks and nodal points are nonetheless preserved (Silva 2010), emphasising the low energy of the floodwaters. Analysis of these palaeochannels reveals a contrast in fluvial styles with those prevalent during the late Pleistocene and early Holocene. Although this has not been studied, it is possible that the abandonment of the Ancient Lobe occurred by abrupt river avulsion to the SE.

The Modern Lobe presents diminished topographic gradients because of the loss of confinement and system progradation onto older deposits. The complex assemblage of fluvial forms includes crevasse splays, abandoned channels, and levee complexes. The proximal domain of the megafan features a 50 km long meander belt with a 250 m wide channel (sinuosity index: 1.6) and many crevasses. Downstream, it develops into an anastomosing channel network to the east and a distributary pattern to the west. A 30-km-long island, the Taiamã Island, has formed in the anastomosing reach, downstream of which a sudden loss of gradient is observed and avulsion processes are currently occurring, with flow diversion onto the floodplain and further development of anastomosing reaches (Assine and Silva 2009). Fluvial forms on the modern lobe are dominated by scrolls, floodplain lakes, crevasse splays, oxbow lakes, and rare to absent in-channel barforms (Table 6.1). Existing bars are densely vegetated with rare, small areas of bare sand.

6.3.3 Megafan Fed by the Pantanal Basin: The Nabileque Megafan

Where the Paraguay River exits the Pantanal and flows into the Chaco Basin, the Nabileque Megafan abuts the Tucuvaca (Bolivia) and Pilcomayo (Paraguay) megafans. Compared to all other Pantanal megafans, the Nabileque Megafan is unique in having its catchment area located entirely within the Pantanal Basin. It covers an area of approximately 9,100 km^2 (Fig. 6.5), with very low stream gradients, altitudes varying from 90 m at its apex to 70 m at its southern exit. Most of this megafan is composed of intensely reworked and eroded older deposits. Floods here are delayed by up to

four months in comparison with floods experienced by fans to the north (Stevaux et al. 2020). Although this is not the largest megafan of the basin, it is fed by the river with the largest discharge – the Paraguay. Here the river is confined to a meander belt with channels much smaller than its preserved palaeochannels. The megafan is also cut by a smaller river east of the Paraguay, the Nabileque, which follows an ancient meander belt of the Paraguay. In contrast with its other Pantanal counterparts, the Nabileque megafan is in degradational mode; i.e., despite many shared features such as a SW-directed network of meandering distributary palaeochannel patterns (Kuerten and Assine 2011; Kuerten et al. 2013) it also exhibits tributary networks of spring channels (*corixos, vazantes*) eroding into the surface. Because this is the hydrological exit zone of the entire Pantanal Basin, the area also experiences pronounced flooding when seasonal channel-spill occurs. The associated sheetflow erodes and reworks older deposits and landforms, and palaeochannels are commonly reactivated. Fluvial forms comprise well-preserved scrolls, multiple channel cutoffs, abundant

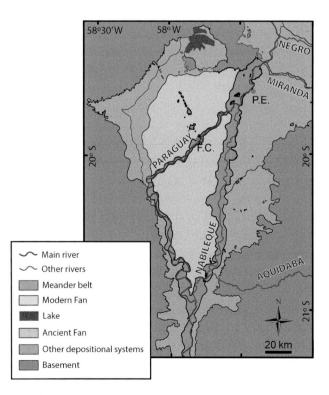

Figure 6.5 **Land system map of the megafan derived from the Pantanal alluvial plains: the Nabileque Megafan**. A colour version of this figure is available in the SOD for Chapter 6.

oxbow lakes, vegetated longitudinal bars, and rare to absent exposures of sand surfaces (Table 6.1).

The fan was formed during the Pleistocene by a then mostly meandering Paraguay River (Assine 2003; Assine and Soares 2004). The preserved palaeochannels reveal a meander belt much larger than the present one, but it was abandoned because of avulsion of the Paraguay River around 4.5–3.9 ka (Kuerten et al. 2013). The river at that time was displaced to its current meander belt, which is incised in Pleistocene deposits to the west of the area. The current degradational setting means that there are no active distributive lobes as in the other megafans of the basin (Kuerten and Assine 2011). The area also, however, presents outstanding examples of palaeochannels with varying sinuosities and dimensions, many of which preserve a population of much larger relict fluvial forms including scrollbars and oxbow lakes. A pronounced feature is the N–S ancient meander belt, which is > 5 km wide and through which the Nabileque River flows for approximately 150 km from its source to the mouth. This smaller river is an underfit stream cutting through a larger, antecedent meander belt. The megafan's abandoned lobe serves as its floodplain, particularly flood during periods which are followed by flooding of the entire meander belt and spill into surrounding areas of the megafan. Sheetflow is the dominant process in those circumstances.

6.4 Discussion

Megafan dimensions are strongly related to the geology, geomorphology (e.g., Assine et al. 2015a), and size of contributing drainage basins (e.g., Davidson and Hartley 2014) (Fig. 6.6). Although catchment area is an important control of megafan dimension, in the Pantanal it is clear that hinterlands in sedimentary rocks produce the largest megafans, the Taquari and São Lourenço being prime examples. This probably results from more easily erodible rocks, which in the case of shale and sandstone provide texturally mature debris that is readily remobilised both as bedload and suspended load.

The Pantanal megafans collectively display a number of fluvial styles which record both allogenic and autogenic processes. Braided planforms are recorded in Pleistocene deposits preserved on abandoned lobes, with an indication of distributary channels which potentially terminated as splays in playa lakes (Fig. 6.7). The conditions promoting a sharp change in current depositional settings are related to climatic changes and tectonic subsidence (as recorded by recurrent seismic activity), resulting in an increase in accommodation space. Modern meander belts in the basin are characterised by common point-bar development and preserved scrollbars, oxbow lakes, channel cut-offs and rare crevassesplay development (Fig. 6.8). Mid-channel and bank-attached sand bars are commonly developed, some of which are stabilised by vegetation. Preserved crevasses are observable along the meander-belt margins, showing that increased discharges were present in the recent past.

The meander belts of the Taquari and São Lourenço rivers present abundant, multiple-style barforms, with comparatively large areas of exposed sand compared with other rivers in the basin (Fig. 6.9). Uptake of these readily available sediments during flood events can directly influence the dynamics of the rivers, and particularly bring them closer to equilibrium (e.g., Nanson and Huang 2008), here evidenced by an abundance and diversity of barform types. Other rivers in the basin do not have such readily available stocks of sediment, with vegetation buffers additionally impeding any attainment of fluvial equilibrium. For example, the Aquidauana and Cuiabá megafans present very limited bar development and a denser vegetation cover, with very small and scattered areas of exposed sand.

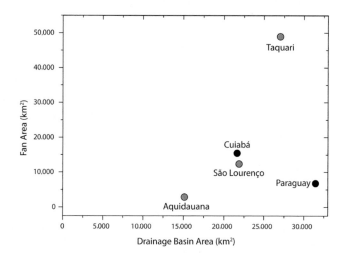

Figure 6.6 **Relationship between megafan catchment area and megafan area.**

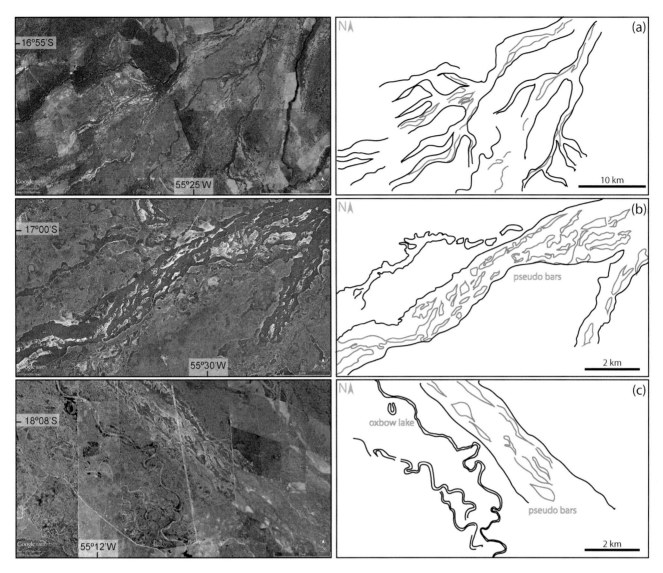

Figure 6.7 **Examples of ancient depositional processes in the Pantanal Basin**. Images on left show preserved palaeochannels on abandoned lobes; interpretation on right. (a), (b) São Lourenço Megafan. (c) Taquari Megafan. Image sources: CNES/Spot 2016. A colour version of this figure is available in the SOD for Chapter 6.

The Paraguay River features a wide range of barforms within its confined meander belt on the megafan, with many vegetation-stabilised bars of numerous types but also commonly occurring areas of bare sand exposure. In contrast, the distal sector of this river exhibits rare barforms and very limited areas of exposed sand. There is thus a trend of rivers developing numerous types of barforms in their proximal areas and displaying few to no bars in their distal zones. This situation is potentially related to the fact that most bedload sediment is being deposited on the proximal floodplain zones of the megafan. In the distal zones, channels multiply, are narrower, and commonly develop anabranching and distributary planforms. Division of river flow into multiple channels probably explains this observed loss of higher-order fluvial forms, and may have resulted in shallower and narrower sandbodies (channels) flanked by finer-grained floodplain deposits. In contrast, proximal areas are composed of well-defined and thicker channel bodies of relatively homogeneous sands, recording multiple stages of the development of the meander belts, with barforms commonly preserved. The confinement of the meander belts likely results in the amalgamation of sandbodies. In contrast to the previously described bar development, the Nabileque

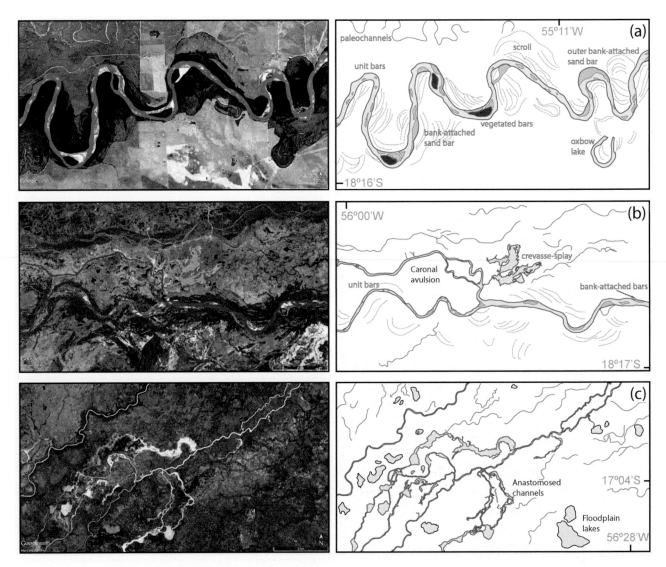

Figure 6.8 **Examples of ancient and modern depositional processes in the Pantanal Basin**. (a) Modern meander belt of the Taquari Megafan (source: Digital Globe 2016). (b) Detail of the Caronal avulsion (source: Landsat 2016). (c) Distal reach of the São Lourenço Megafan (source: Digital Globe 2016). A colour version of this figure is available in the SOD for Chapter 6.

megafan displays quite different dynamics, with multiple types of barforms. Most of these display relatively larger dimensions than on other fans elsewhere in the basin, and extraordinarily good scroll-bar preservation.

Channel-bar dimensions diminish and are poorly developed downstream of knickpoints as a result of diminishing discharge caused by flow division into multiple distributary channels and to decreasing sediment bedload resulting from deposition on floodplains through crevassing. Meander belts are incising as deep as 5 m or more into Pleistocene deposits. Incision depth decreases in the direction of the megafan intersection point, downstream of which depositional lobes

prograde onto floodplains of the older megafan surface. This generates ideal conditions for avulsion processes, as recorded by numerous abandoned lobes arranged radially downstream from the intersection point. Channel re-joining is also common on the lobes, with a common development of anastomosing channels. Channel width decreases downstream of the intersection point, and crevassing and crevasse-splay development are much more common and may evolve into a complete avulsion of the parent channel as new channels develop from crevasses (Fig. 6.10).

Floodplain lakes and swamps commonly develop in the more distal regions of the megafans. There are

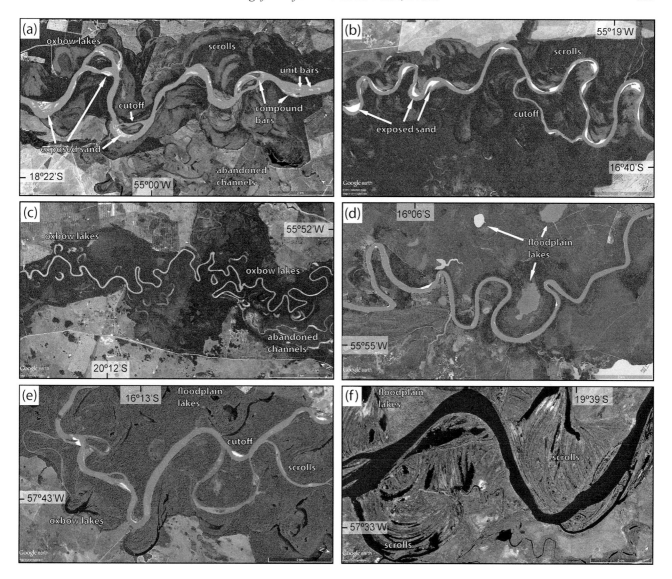

Figure 6.9 **Aspects of the fluvial dynamics occurring among the proximal meander belts of the Pantanal Basin**. Flow from right to left. (a) Taquari Megafan (source: Digital Globe 2016). (b) São Lourenço Megafan (source: CNES/Spot 2016). (c) Aquidauana Megafan (source: Digital Globe 2016). (d) Cuiabá Megafan (source: CNES/Spot 2016). (e) Paraguay Megafan (source: Digital Globe 2016). (f) Nabileque Megafan (source: Digital Globe 2016). A colour version of this figure is available in the SOD for Chapter 6.

many tributary drainage patterns on the Pantanal Basin, most of them being spring channels (*corixos*) which are characteristically degradational and develop ubiquitously on the megafans. They incise and scour older deposits, resulting in an erosional surface quite different from the fluvial system or processes responsible for depositing the underlying sediments. The significance of such erosional surfaces represents a challenge to those interpreting fluvial processes in the rock record. While most of the deposition is accomplished by distributive fluvial systems, such as the Taquari and the São Lourenço, there are tributary

systems commonly operating alongside those systems, on the same alluvial plain and under the same climatic and basin settings – all incised into previously deposited sediments.

6.5 Conclusions

The Pantanal Basin underwent climate- and tectonics-driven fluctuations during the Quaternary, which led to the development and preservation of numerous geomorphic features characteristic of contrasting fluvial styles. Pleistocene braided channel features are

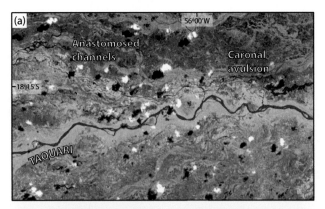

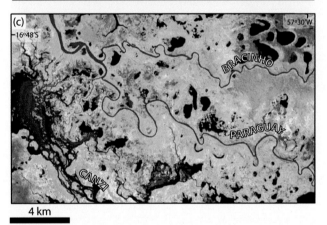

4 km

Figure 6.10 **Avulsion processes occurring on the modern distal depositional lobes**. (a) Taquari Megafan and the Caronal avulsion. (b) São Lourenço Megafan and the Boca Brava avulsion. (c) Paraguay Megafan and the Bracinho, Paraguay and Canzi rivers. Source: Landsat 5 TM.

preserved alongside Holocene meandering channels and modern-day depositional systems. Such contrasting fluvial styles result from external forcing, but also from autogenic controls. The bulk of the modern Pantanal Basin is dominated by the deposition of distributive fluvial systems or megafans, with associated lacustrine environments and a small area dominated by

debris-flow-dominated alluvial fans. Some fluvial systems develop distinct arrays of barforms, and it seems that such features are more varied in systems which present larger non-vegetated areas or no vegetation. Additionally, rivers with more varied barforms are those which develop into larger megafans, and which additionally are sourced from sedimentary rock catchments, such as the Taquari and São Lourenço megafans. The modern morphology of the megafans can be generalised and classifies into four main landform assemblages:

(i) incised meander belts in the proximal areas of the megafans; this hinders avulsion processes near the entrenched fan heads;
(ii) active aggradational lobes featuring anastomosing and distributary channels; the lobes are dominated by avulsion processes and by elevated channel–levee complexes constructed by channel progradation onto coarser-grained floodplain deposits;
(iii) abandoned degradational lobes;
(iv) undifferentiated, mixed-process floodplains.

Sedimentary dynamics and hydrological regimes are influenced by the geomorphology and geology of the drainage basin, and the resulting megafan geomorphology of the Pantanal Basin can be differentiated into distinct depositional tracts as follows: (i) an axial trunk-river system fed by multiple fluvial fans and megafans; (ii) fluvial fans sourced from scarplands in sedimentary rocks; (iii) fluvial fans sourced from lowlands in crystalline Precambrian rocks; and (iv) fluvial interfan systems. The area of each megafan seems to be a function of both catchment size and geology, with megafans fed from sedimentary rock outcrops developing into larger megafans and with more complex barform development than megafans supplied from Precambrian basement catchments.

Acknowledgements

The authors thank the São Paulo Research Foundation (FAPESP 2014/06889-2; 2014/13937-3; 2014/23334-4) for financial support to our research in the Pantanal Basin; the National Council of Technological and Scientific Development (CNPq) for grants to MLA (432985/2018-2); and J. Wilkinson, E. Latrubesse and Y. Gunnell for the invitation to write this chapter and for their editorial assistance.

References

Ab'Sáber, A. N. (1988). O Pantanal Mato-Grossense e a teoria dos refúgios. *Revista Brasileira de Geografia*, 50, 9–57.

Alho, C. J. R. (2005). The Pantanal. In L. H. Fraser and P. A. Keddy, eds., *The World's Largest Wetlands – Ecology and Conservation*. Cambridge University Press, Cambridge, 203–271.

Almeida, F. F. M. (1959). Traços gerais da geomorfologia do Centro-Oeste brasileiro. In F. F. M., Almeida and M. A. Lima, eds., *Planalto Centro-Ocidental e Pantanal Matogrossense*. Guia de Excursão nº 1 do XVIII Congresso Internacional de Geografia. Conselho Nacional de Geografia, Rio de Janeiro, 7–65.

ANA – Agência Nacional de Águas (2004). Implementação de Práticas de Gerenciamento Integrado de Bacia Hidrográfica para o Pantanal e Bacia do Alto Paraguai ANA/GEF/PNUMA/OEA. Programa de Ações Estratégicas para o Gerenciamento Integrado do Pantanal e Bacia do Alto Paraguai: Síntese Executiva. *Brasília*, 64 pp.

Assine, M. L. (2003). *Sedimentação na Bacia do Pantanal Mato-Grossense, Centro-Oeste do Brasil*. Instituto de Geociências e Ciências Exatas, Universidade Estadual Paulista – Unesp, Rio Claro, Brazil, Tese de Livre-Docência, 106 pp.

Assine, M. L. and Soares, P. C. (2004). Quaternary of the Pantanal, west-central Brazil. *Quaternary International*, 114, 23–34.

Assine, M. L. (2005). River avulsions on the Taquari megafan, Pantanal wetland, Brazil. *Geomorphology*, 70, 357–371.

Assine, M. L. and Silva, A. (2009). Contrasting fluvial styles of the Paraguay River in the northwestern border of the Pantanal wetland, Brazil. *Geomorphology*, 113, 189–199.

Assine, M. L., Corradini, F. A., Pupim, F. N., and McGlue, M. M. (2014). Channel arrangements and depositional styles in the São Lourenço fluvial megafan, Brazilian Pantanal wetland. *Sedimentary Geology*, 301, 172–184.

Assine, M. L. (2015). Brazilian Pantanal: A large pristine tropical wetland. In B. C. Vieira, A. A. R., and Salgado, L. J. C. Santos, eds., *Landscapes and Landforms of Brazil*. Springer, Dordrecht, 135–146.

Assine, M. L., Merino, E. R., Pupim, F. N., et al. (2015a). Geology and geomorphology of the Pantanal Basin. In I. Bergier and M. L. Assine, eds., *Dynamics of the Pantanal Wetland in South America*, Springer Nature, Cham, Switzerland, 25–50.

Assine, M. L., Macedo, H., Stevaux, J., et al. (2015b). Avulsive rivers in the hydrology of the Pantanal wetland. In I. Bergier and M. L. Assine, eds., *Dynamics of the Pantanal Wetland in South America*. Springer Nature, Cham, Switzerland, 83–110.

Assine, M. L., Merino, E. R., Pupim, F. N., Macedo, H. A., and Santos, M. G. M. (2015c). The Quaternary alluvial systems tract of the Pantanal Basin, Brazil. *Brazilian Journal of Geology*, 45, 475–489.

Assumpção, M. and Sacek, V. (2013). Intra-plate seismicity and flexural stresses in central Brazil. *Geophysical Research Letters*, 40, 487–491.

Barbiéro, L., Queiróz-Neto, J. P., Ciornei, G., et al. (2002). Geochemistry of water and groundwater in the Nhecolândia, Pantanal of Mato Grosso, Brazil: variability and associated process. *Wetlands*, 22, 528–540.

Braun, E. W. G. (1977). Cone aluvial do Taquari, unidade geomórfica marcante da planície quaternária do Pantanal. *Revista Brasileira Geografia*, 39, 164–180.

Buehler, H. A., Weissmann, G. S., Scuderi, L. A., and Hartley, A. J. (2011). Spatial and temporal evolution of an avulsion on the Taquari river distributive fluvial system from satellite image analysis. *Journal of Sedimentary Research*, 81, 630–640.

Clapperton, C. (1993). *Quaternary Geology and Geomorphology of South America*. Elsevier, Amsterdam, 779 p.

Corradini, F. A. and Assine, M. L. (2012). Compartimentação geomorfológica e processos deposicionais no megaleque fluvial do rio São Lourenço, Pantanal mato-grossense. *Revista Brasileira de Geociências*, 42, 20–33.

Davidson, S. K. and Hartley, A. J. (2014). A quantitative approach to linking drainage area and distributive-fluvial-system area in modern and ancient endorheic basins. *Journal of Sedimentary Research*, 84, 1005–1020.

Facincani, E. M. and Assine, M. L. (2010). Geomorfologia fluvial do rio Aquidauana, borda sudeste do Pantanal Mato-Grossense. In C. Martins Junior and A. F. Oliveira Neto, eds., *Revelando Aquidauana* (Campo Grande, Editora da UFMS, Brazil), (Serie Fronteiras nº 3), 267–284.

Gregório, E. C., Facincani, E. M., and Amorin, G. M. (2016). Mudanças ambientais quaternárias no Megaleque Fluvial do Aquidauana, borda sudeste do Pantanal Matogrossense. *Revista da ANPEGE*, 12, 363–389.

Hartley, A. J., Weissmann, G. S., Nichols, G. J., and Warwick, G. L. (2010). Large distributive fluvial systems: characteristics, distribution, and controls on development. *Journal of Sedimentary Research*, 80, 167–183.

Horton, B. K. and DeCelles, P. G. (1997). The modern foreland basin system adjacent to the Central Andes. *Geology*, 25, 895–898.

Horton, B. K. and DeCelles, P. G. (2001). Modern and ancient fluvial megafans in the foreland basin system of the central Andes, southern Bolivia: implications for drainage network evolution in foldthrust belts. *Basin Research*, 13, 43–63.

Jongman, R. H. G. (2006). *Pantanal-Taquari; Tools for Decision Making in Integrated Water Management. Alterra-rapport 1295*. Wageningen, The Netherlands: Alterra.

Klammer, C. (1982). Die Paläowüste des Pantanal von Mato Grosso und Die Pleistozäne Klimageschichte der Brasilianischen Randtropen. *Zeitschrift für Geomorphologie*, 26, 393–416.

Kuerten, S. and Assine, M. L. (2011). O rio Paraguai no megaleque do Nabileque, sudoeste do Pantanal Mato-Grossense, MS. *Revista Brasileira de Geociências*, 41, 642–653.

Kuerten, S., Parolin, M., Assine, M. L., and McGlue, M. M. (2013). Sponge spicules indicate Holocene environmental changes on the Nabileque River floodplain, southern Pantanal, Brazil. *Journal of Paleolimnology*, 49, 171–183.

Latrubesse, E. M., Stevaux, J. C., Cremon, E. H., et al. (2012). Late Quaternary megafans, fans and fluvio-aeolian interactions in the Bolivian Chaco, tropical South America. *Palaeogeography, Palaeoclimatology, Palaeoecology*, 356, 75–88.

Latrubesse, E. M. (2015). Large rivers, megafans and other Quaternary avulsive fluvial systems: A potential "who's who" in the geological record. *Earth-Science Reviews*, 146, 1–30.

Macedo, H. A., Assine, M. L., Pupim, F. N., et al. (2014). Mudanças paleo-hidrológicas na planície do rio Paraguai, Quaternario do Pantanal. *Revista Brasileira de Geomorfologia*, 15, 75–85.

May, J.-H., Argollo, J., and Veit, H. (2008a). Holocene landscape evolution along the Andean piedmont, Bolivian Chaco. *Palaeogeography, Palaeoclimatology, Palaeoecology*, 260, 505–520.

May J.-H., Zech, R., and Veit, H. (2008b). Late Quaternary paleosol–sediment-sequences and landscape evolution along the Andean piedmont, Bolivian Chaco. *Geomorphology*, 98, 34–54.

McGlue, M. M., Silva, A., Zani, H., et al. (2012). Lacustrine records of Holocene flood pulse dynamics in the Upper Paraguay River watershed (Pantanal wetlands, Brazil). *Quaternary Research*, 78, 285–294.

Meira, F. C., Stevaux, J. C., Torrado, P. V., and Assine, M. L. (2019). Compartimentação e evolução geomorfológica da planície do rio Cuiabá, Pantanal Mato-Grossense. *Revista Brasileira de Geomorfologia*, 20, 159–183.

Merino, E. R. (2017). Evolução geomorfológica e mudanças paleo-geográficas na porção sul do Pantanal: a planície de interleques do Rio Negro e leques fluviais coalescentes. PhD thesis. Universidade Estadual Paulista, Rio Claro, Brazil.

Nanson, G. C. and Huang, H. Q. (2008). Least action principle, equilibrium states, iterative adjustment and the stability of alluvial channels. *Earth Surface Processes and Landforms*, 33, 923–942.

Padovani, C. R. (2010). *Dinâmica das Inundações do Pantanal. Universidade de São Paulo/ESALQ*, Piracicaba – SP, 174 p.

PCBAP, Plano de Conservacão da Bacia do Alto Paraguai – Pantanal (1997). Ministério do meio ambiente, dos recursos hídricos e da amazônia legal. Programa Nacional do Meio Ambiente (PNMA), Brasília.

Pupim, F. N. (2014). *Geomorfologia e paleo-hidrologia dos megaleques dos rios Cuiabá e São Lourenço, Quaternário da Bacia do Pantanal*. Instituto de Geociências e Ciências Exatas – IGCE, Universidade Estadual Paulista – Unesp, Rio Claro – SP, Brazil, 109 pp.

Pupim, F. N., Assine, M. L., Merino, E. R., Macedo, H. A., and Silva, A. (2014). A planície interleques do rio Piquiri, bacia do Pantanal. In *5º Simposio de Geotecnologias no Pantanal*, Campo Grande, MS, Brazil, 848–857.

Shiraiwa, S. (1994). *Flexura da litosfera continental sob os Andes Centrais e a origem da Bacia do Pantanal*. PhD thesis, Universidade de São Paulo, IAG, São Paulo, Brazil.

Silva, A. (2010). *Geomorfologia do megaleque do rio Paraguai, Quaternário do Pantanal Mato-Grossense, Centro-Oeste do Brasil*. PhD thesis, Universidade Estadual Paulista, Rio Claro.

Sinha, R. and Friend, P. F. (1994). River systems and their sediment flux, Indo-Gangetic plains, Northern Bihar, India. *Sedimentology*, 41, 825–845.

Souza, O. C., Araujo, M. R., and Mertes, L. A. K. (2002). Form and process along the Taquari River alluvial fan, Pantanal, *Brazil. Zeitschrift für Geomorphologie*, 129, 73–107.

Stevaux, J. C., Macedo, H. A., Assine, M. L., and Silva, A., (2020). Changing fluvial styles and backwater flooding along the Upper Paraguay River plains in the Brazilian Pantanal wetland. *Geomorphology*, 350, 106906.

Tricart J. (1982). El Pantanal: un ejemplo del impacto geomorfológico sobre el ambiente. *Informaciones Geograficas*, 29, 81–97.

Ussami N., Shiraiwa S., and Dominguez J. M. L. (1999). Basement reactivation in a sub-Andean foreland flexural bulge: The Pantanal wetland, SW Brazil. *Tectonics*, 18, 25–39.

Weissmann, G. S., Hartley, A. J., Nichols, G. J., et al. (2010). Fluvial form in modern continental sedimentary basins: distributive fluvial systems. *Geology*, 38, 39–42.

Whitney, B. S., Mayle, F. E., Punyasena, S. W., et al. (2011). A 45 kyr palaeoclimate record from the lowland interior of tropical South America. *Palaeogeography, Palaeoclimatology, Palaeoecology*, 307, 177–192.

Wilkinson, M. J., Marshall, L. G., and Lundberg, J. G. (2006). River behavior on megafans and potential influences on diversification and distribution of aquatic organisms. *Journal of South American Earth Sciences*, 21, 151–172.

Zani, H. and Assine, M. L. (2011). Paleocanais no megaleque do rio Taquari: mapeamento e significado geomorfológico. *Revista Brasileira de Geociências*, 41, 37–45.

Zani, H., Assine, M. L., and McGlue, M. M. (2012). Remote sensing analysis of depositional landforms in alluvial settings: method development and application to the Taquari megafan, Pantanal (Brazil). *Geomorphology*, 161–162, 82–92.

7

Geomorphic and Chronological Assessment of Aggradation Patterns on the Río Grande (Guapay) Megafan, Eastern Bolivia

JAN-HENDRIK MAY, TIMOTHY COHEN, FRANK PREUSSER, and HEINZ VEIT

Abstract

A fan-wide assessment of modern depositional processes on the Río Grande (Guapay) megafan, coupled with the analysis of its Holocene evolution, reveals that most of the 36,000 km^2 megafan surface has been subject to sedimentation processes and/or reworking in the last 6–7 ka. Today, depositional dynamics as inferred from multi-date satellite imagery are mostly restricted to a $\sim 2{,}000$ km^2 depozone on the distal megafan, and are characterised by avulsive fluvial environments. Combining remote sensing with field observations and geochronology at key locations has allowed to capture links between the dynamics of depositional processes and larger-scale landforms, documenting significant changes in the location of the depozone since ~ 4 ka. On shorter time scales, the human impact on these dynamic sedimentary processes is expressed by artificial levees and re-channelisation following crevasse splays. This is likely to have prevented a number of channel avulsions over the last decades and implies that the use of modern avulsion frequencies and depositional rates alone as analogues for understanding Holocene megafan evolution is of limited use. Instead, our observations emphasise the increasing vulnerability to future avulsion events within an intrinsically unstable fluvial environment that has seen rapid deforestation and population growth in recent time.

At a distance of some 40 miles I could clearly distinguish the white bed of the Río Grande contrasting with the dark forest, while the river itself appeared as a silver thread, reflecting the rays of the rising sun.

—J. B. Minchin (1881)

7.1 Introduction

Megafans and large distributive fluvial systems are key elements in most foreland basins around the Earth (DeCelles and Giles 1996; Hartley et al. 2010; Weissmann et al. 2010). Their morpho-sedimentary complexity is commonly expressed by downstream changes in channel pattern along megafan rivers (Shukla et al. 2001; Davidson et al. 2013), and high spatial heterogeneity is produced by the dynamic and often avulsive depositional processes (Latrubesse 2015). Understanding the spatial and temporal variability of these megafan depositional processes documents the environmental, hydrological, and ecological properties over extended foreland areas (Wilkinson et al. 2006) and thereby improves the assessment and mitigation of severe flooding and sedimentation caused by avulsive dynamics on megafans (Wells and Dorr 1987; Assine 2005; Chakraborty et al. 2010).

In central South America, the evolution of the Andean foreland is closely linked to the uplift and deformation history of the Central Andes, and has developed as the result of crustal thickening, eastwards propagation of the orogenic front, and topographic loading (Isacks 1988; Horton and DeCelles 1997). The foreland basin in eastern Bolivia is bordered by the Subandean fold-and-thrust belt to the west (Baby et al. 1992), and the Alto de Izozog, a topographic high of pre-Mesozoic origin to the east (May 2006; Uba et al. 2006). It roughly coincides with the Gran Chaco plain (Riveros 2004) where large-scale avulsive channel shifts are evident from satellite imagery along a series of extensive megafans, including those of the Río Pilcomayo, the Río Parapetí, and the Río Grande (Iriondo 1993; Horton and DeCelles 2001; May

2006) – the first of these being among the largest megafans of our planet (Fig. 7.1a, b; Horton and DeCelles 2001; Hartley et al. 2010). The large sediment volume stored in these megafans contains valuable records of Andean uplift, denudation, and climate history extending back to the late Oligocene and Miocene (Horton and DeCelles 2001; Uba et al. 2007). Little is known, however, of their more recent, late Quaternary evolution (Latrubesse et al. 2012). Preliminary luminescence- and radiocarbon-based chronologies indicate a link between sedimentary dynamics and palaeoclimate over the last ~30 ka, with

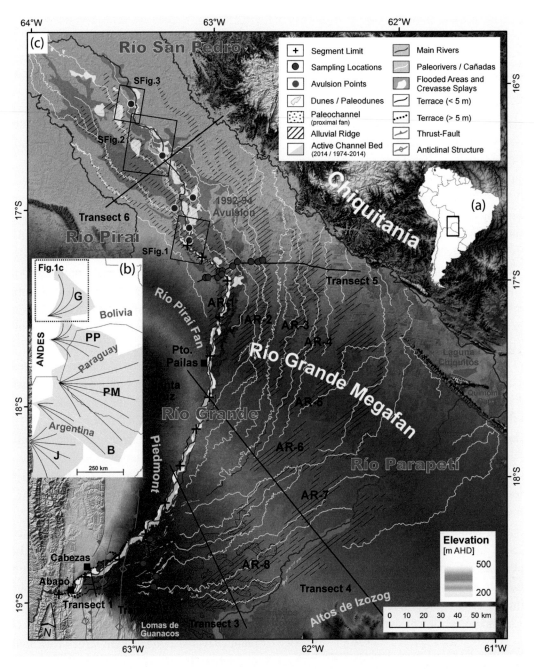

Figure 7.1 **Geographic setting of the Río Grande (Guapay) megafan**. (a) Location in South America. (b) Approximate extent of main Chaco megafans (G– Grande, PP– Parapetí, PM– Pilcomayo, B– Bermejo, J– Juramento). (c) Overview map of main topographic and geomorphic features, and sample sites on the megafan surface; AR1 to AR8 denote alluvial ridges, black lines illustrate topographic profiles shown in Fig. 7.5, and rectangular boxes indicate locations of satellite images in SOD Figs. 7.1–7.3.

the formation of source-bordering dunes and dune-fields possibly corresponding to episodes of changing moisture conditions (Barboza et al. 2000; Kruck et al. 2011; Latrubesse et al. 2012). Large-scale channel shifts, for example, are documented on the Río Parapetí megafan for the mid- to late Holocene, leading to its integration into the Amazon drainage basin (May 2011).

Limited data yet exists for the Río Grande megafan despite its large size, the recent expansion of agriculture (Krüger 2006; Müller et al. 2011), and the ensuing vulnerability to flooding (Wachholtz and Herold-Mergl 2003). The Río Grande drains a $\sim 65{,}000\,km^2$ catchment comprising parts of the Cordillera Oriental, which rises to $> 5{,}000$ m a.s.l. close to Cochabamba and then cuts through the Subandean Zone fold belts before reaching its catchment outlet Abapó. From here, the Río Grande progressively curves around to the northwest before joining the Río Mamoré ~ 500 km downstream of Abapó (Fig. 7.1). The corresponding megafan covers an area of $\sim 37{,}500\,km^2$ (Werding 1977a, 1977b; Barnes and Heins 2009; Latrubesse et al. 2012). Semi-deciduous Chaco dry forest on the megafan surface is interspersed with patches of palm forests and savannah (Navarro and Maldonado 2002; Ibisch et al. 2004). The resulting vegetation pattern reflects local differences in soils and lithology (Agrar- und Hydrotechnik GmbH 1974b; Gerold 2004), and thus mimics the larger-scale geomorphic pattern of relict landforms such as palaeochannels and dune systems. Recent deforestation in many areas of the Río Grande megafan (Steininger et al. 2001; Killeen et al. 2008; Müller et al. 2011) has caused land degradation and soil erosion (Gerold 1985, 1988).

Eastern Bolivia lies at the transition between the tropical humid Amazon basin to the north and the semi-arid Chaco plain to the southeast. This generates a hydroclimatic gradient with annual precipitation of $> 1{,}000$ mm/yr north of Santa Cruz, and < 700 mm/yr in the southeastern part of the megafan and the higher parts of the catchment (Rafiqpoor et al. 2004; Vicente-Serrano et al. 2016). Potential evaporation on the megafan reaches $\sim 1{,}400$ mm/yr with a peak between September and December, whereas most precipitation falls during the wet season from December to March through convective storms related to the South

American Summer Monsoon (SASM) (Zhou and Lau 1998; Vera et al. 2006). Very strong northerly winds related to the South American low-level jet (SALLJ) blow across the Bolivian Chaco throughout the year, with a dry-season peak (July to November; Berri and Inzunza 1993) responsible for significant deflation and the formation of source-bordering dunes along the southern and southeastern margins of the Río Grande (Werding 1977a; May 2006; Latrubesse et al. 2012).

Limited discharge data for the Río Grande shows large seasonal variability, with mean monthly discharges of $> 500\,m^3/s$ during the wet season and mean daily discharge in excess of $10{,}000\,m^3/s$ during the pronounced flood peaks of February–March (Fig. 7.2a; Agrar- und Hydrotechnik GmbH 1974a; Werding 1977a; Guyot et al. 1994). Downstream of the apex, percolation into the sandy fluvial deposits is responsible for transmission losses of up to 25%. These are reflected in the pronounced difference between measured discharge at Abapó and Pto. Pailas (Werding 1977b). In general, suspended sediment yield co-varies with the seasonal discharge pattern but shows an additional peak in the early wet season (Fig. 7.2b). In comparison to the adjacent Río Parapetí to the south, the Río Grande carries much less of its total annual sediment load ($\sim 0.5\%$) as sandy bedload (Werding 1977b). While the suspended load shows a clear peak in the clay and fine-silt fraction, bedload is mainly composed of fine and medium sands with minor percentages of coarser gravel-sized sediments (Fig. 7.2c).

Published values for total annual sediment yield at the megafan apex at Abapó range from 125 to 145×10^6 t/yr (Werding 1977b; Guyot et al. 1994; Latrubesse and Restrepo 2014). Studies on meander migration in the adjacent Amazonian lowlands (Llanos de Moxos, Beni) have suggested that the sediment load of the Río Grande may have played a crucial role since the late Pleistocene in controlling fluvial dynamics further downstream (Lombardo et al. 2012; Plotzki et al. 2013, 2015; Constantine et al. 2014; May et al. 2015; Lombardo 2016). A more detailed assessment of recent fluvial dynamics, however, has recently suggested instead that direct contribution of sediment by the Río Grande may be limited (Lombardo 2016). This is supported by a substantially positive

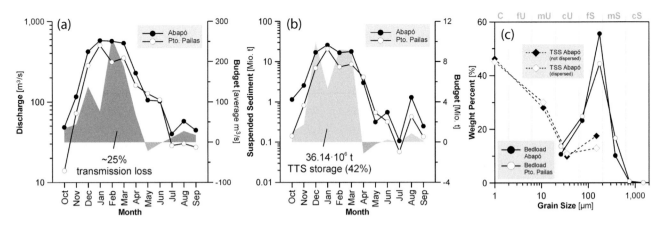

Figure 7.2 **Water and sediment transport regimes**. (a) Average monthly discharge. (b) Average suspended sediment load for the Río Grande at Abapó and Pto. Pailas. (c) Average textural composition of bedload and suspended load (all data from Werding 1977b).

annual sediment budget of ~36×10^6 t on the proximal megafan (Werding 1977b), suggesting that ~42% of the sediments exiting the Andean catchment remains stored on the proximal megafan (Fig. 7.2b). Even though such high sediment loads should have important implications for the dynamic and avulsive processes which characterise megafan evolution, no detailed assessment of depositional processes or resulting landforms on the Río Grande megafan has so far been published. It is nonetheless required as a basis for understanding the controls on megafan evolution over nested geologic time scales.

This study thus aims to unravel the Holocene history of fluvial dynamics along the Río Grande by detailing the active depositional and avulsive processes on the megafan and interpreting its evolution over millennial time scales from geomorphic and sedimentary archives. In combination, these data provide a unique example of megafan evolution via the interplay of geomorphic, hydrologic and sedimentary processes operating on various spatial and temporal scales.

7.2 Methods

Remote sensing and field data are compiled and combined here in order to (i) present a detailed geomorphic assessment of landforms and processes on the megafan; (ii) conduct a multi-temporal analysis of Landsat imagery to assess the history of channel changes since the 1970s (see list of images in SOD Table 7.1); and (iii) interpret and date two sedimentary

transects across the proximal and distal megafan to investigate the timing of major channel shifts (for details, see SOD 7.2).

7.3 Results

7.3.1 Overview

Based on channel pattern, river long profile and downstream slope variations, the Río Grande can be subdivided into five zones, with zones I and II corresponding to the proximal (< ~150 km downstream distance), III to the medial (~150–225 km), and IV–V to the distal megafan (> 200–225 km), respectively (Fig. 7.3). Channel pattern changes covary from sand-braided and anabranching in the proximal (Fig. 7.4a) to sinuous (Fig. 7.4b) in the distal tracts. Fan-wide downstream changes in dominant landforms are reflected in topographic cross-profiles (Fig. 7.5), and diachronous satellite images have allowed a range of active depositional processes to be identified, quantified, and grouped according to their relative distance to the channel.

7.3.2 Modern Fluvial Dynamics

7.3.2.1 Channel Changes

Over the first 300 km downstream of Abapó, channel width and vegetated islands recorded significant variability between 1975 and 2014 (Fig. 7.6a). Increases in average channel width are accompanied by decreases

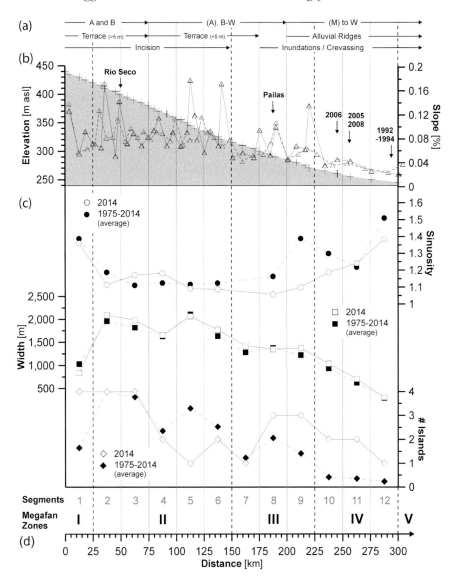

Figure 7.3 **Geomorphic characteristics of the Río Grande channel from the apex to > 300 km downstream, distinguishing five megafan zones (I to V).**
(a) Downstream variation of main geomorphic features (i.e., lateral constraints and channel pattern; A– anabranching, B– braided, W– wandering, M– meandering) and interpreted dominant fluvial processes.
(b) Longitudinal profile averaged slope values (solid line: SRTM data, dashed line: ALOS data) downstream of the apex in relation to river segments (used for GIS analysis).
(c) Downstream variation of GIS-derived average sinuosity, channel width, and number of islands per channel segment (white circles are 2014 measurements, black circles are 1975–2014 averages).
(d) Megafan zones I to V (see text).

in island size. Imagery from 1975 to 1984–1990 documents peak channel widths (~1,350–1,550 m) and a variable but decreasing area of stable and vegetated islands (Fig. 7.6a). Between ~1990 and 1996, average channel width in all segments recorded a marked drop to < 1,000 m, while vegetated island area generally increased. Between 1998 and 2008, morphological changes were smaller but constant and led to channel

widening (average: ~1,500 m in 2008) and island shrinking. This was followed by width and island stability on the proximal and medial megafan belts.

7.3.2.2 Crevasse Splays and Channel Avulsions
Crevasse splays resulting from overbank flow and partial levee breaching have been observed at different

Figure 7.4 **Geomorphological elements of the medial and distal Río Grande megafan.**

locations between ~250 and 350 km downstream of Abapó in 1986, 2005, 2006 and 2008 (e.g., Fig. 7.4d, e). The area of active sedimentation varies for these splays between ~10 and 25 km². None of these events has led to the abandonment of the channel or the permanent establishment of a new channel. The splays are thus interpreted as 'failed random avulsion events' (Stouthamer 2001). A further crevasse splay event in 2007/2008 successfully shifted the channel to a more westerly position, but established a reconnection with the trunk channel only 22 km downstream of the splay two years later. This splay therefore preceded a 'successful but local avulsion event' (Slingerland and Smith 2004). In contrast, at km 300 downstream of Abapó, overbank flow of the Río Grande in 1992 and 1993 led to a significant breach of the channel banks followed by the deposition of a ~5 km² crevasse splay, and catastrophic inundation of the floodplain and backswamp (Wachholtz and Herold-Mergl 2003). This event was the starting point for the evolution of a new channel (or avulsion) belt (Fig. 7.6b, #2), which did not locally recombine with the previously abandoned channel (Fig. 7.6b, #1) and joined it only ~200 km further downstream. Consequently, this event is a 'full regional avulsion' (Slingerland and Smith 2004). As flow was diverted into the flood basin during the 1992–1993 avulsion, the process of abandonment of the previously active channel – i.e., the avulsion duration; Stouthamer and Berendsen (2001) – lasted ~5 years and was accompanied by a reduction in channel width, with in-channel vegetation in 1998 indicating complete cessation of flow (SOD Fig. 7.1). Abandonment of a previously active channel belt by regional avulsion initiated the construction of channel belt #1 before 1966. Insufficient image cover prevents documentation of the avulsion date or duration.

7.3.2.3 Avulsion Effects in the Proximal Floodplain

After the successful avulsion event in 1993, a new trunk channel segment had established itself by 1994 from the distributary channel network on the crevasse splay by exploiting a previously existing road (SOD Fig. 7.1). The ~5-km-long trunk channel runs out in a splay-like zone of narrowing distributaries, and sometimes into terminal channels characterised by sediment deposition (Fig. 7.4f). This zone of deposition is referred to as a 'progradational splay' (Smith et al. 1989; Slingerland and Smith 2004). Together with the newly forming upstream trunk channel segment it successively prograded ~102 km to the northwest over the following 23 years (Fig. 7.6b), establishing a single integrated channel connected to the Yapacaní and Mamoré rivers. Average progradation rates were 4.5 km/yr, but no progradation and/or splay deposition was observed during some years (e.g., 1995, 2000, 2012). Similar progradational splays were mapped downstream of the local avulsion in 2007/2008, and along the previously active channel belt #1 in 1966, 1975, and 1984–1988 (Fig 7.6b). During the 1980s, progradation rates increased to ~10 km/yr (Fig. 7.6b). By 1987, the progradation of the new trunk channel had established a continuous channel up to the confluence with the Yapacaní and Mamoré ~475 km downstream of Abapó. No additional splays were observed between 1988 and channel avulsion in 1992–1994. The depositional areas of all mapped progradational splays since 1966 vary individually

---

Caption for Figure 7.4 (*cont.*)

(a) Aerial photograph of the Río Grande ~35 km from Abapó (date: 07/2007) showing the mixed ('wandering' and anabranching) channel pattern with vegetated islands, a wide sandy channel bed, and multiple channels (channel widths: ~650–3,400 m).

(b) Aerial photograph of the Río Grande ~250 km from Abapó (date: 08/2003) showing the mixed 'wandering' and meandering channel pattern (channel widths: ~270–740 m).

(c) Ground view of abandoned palaeochannel ('cañada') ~110 m wide and ~2 m deep.

(d) Landsat image (date: 02/2008; band combination 7–5–2) of inundation and crevasse splay ~260 km from Abapó showing inundated backswamps on both sides of the Río Grande alluvial ridge, and ~10-km-long splay originating from a breach.

(e) Close-up aerial photograph of crevasse splay ~222 km downstream of Abapó (date: 02/2006), showing distributive channel pattern and extensive sediment cover burying previously used agricultural fields near Puerto Bandegas.

(f) Aerial view of a multi-thread progradational splay two years after deposition ~358 km downstream of Abapó (date: 08/2003).

(g) Geomorphic, hydrological and ecological reorganisation in the flooded backswamp ~415 km downstream of Abapó showing vegetation die-off, transitional anabranching channel pattern, and cross-breaching of previously existing levees (date: 08/2003).

(h) Secondary channel avulsion (date: 08/2003) leading to the formation of a crevasse splay ~417 km downstream of Abapó, prograding into the inundated backswamp

Photographs b, f, g and h courtesy of B. Reynolds.

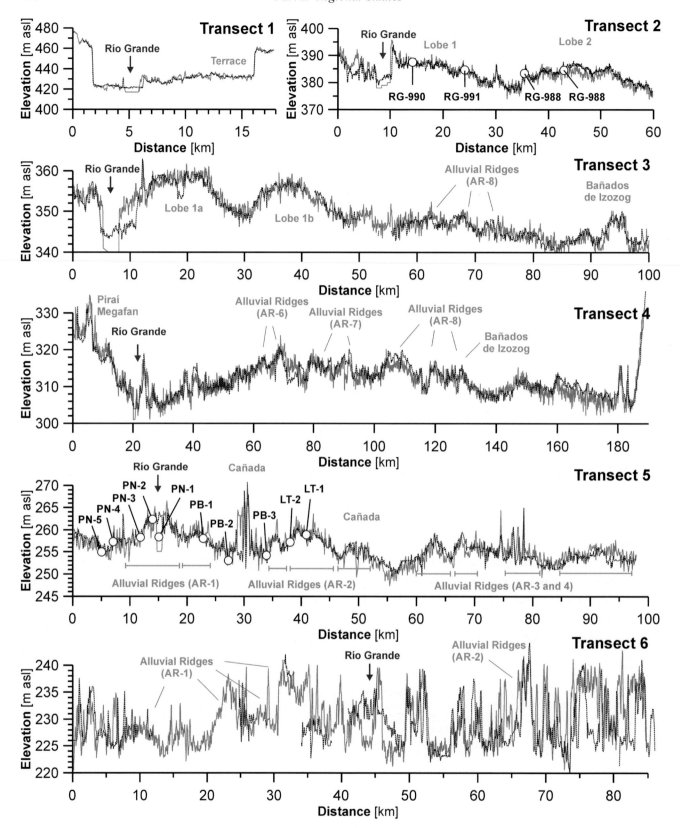

Figure 7.5 **Topographic profiles and main interpreted depositional landforms across the Río Grande megafan**. Derived from 1-arc SRTM (grey line) and ALOS WORLD 3D (black line) elevation data (see Fig. 7.1 for locations). White circles denote the location of field sample points relative to topography.

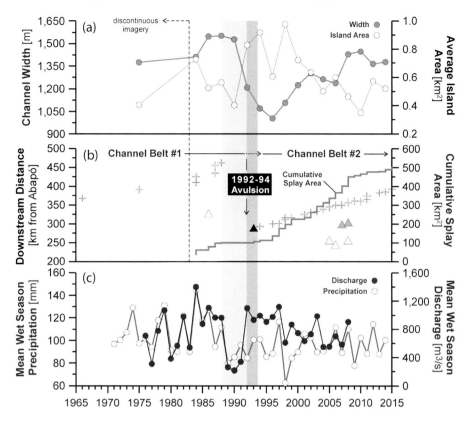

Figure 7.6 **GIS-derived fluvial dynamics of the Río Grande in relation to hydroclimate since the 1960s**.
(a) Channel changes on the proximal megafan.
(b) Depositional dynamics on the distal megafan depozone (crosses: progradational splays; white triangles: crevasse splays associated with random but failed avulsions; grey triangles: minor local avulsions; black triangle: crevasse splay leading to successful regional avulsion ~300 km downstream of Abapó between 1992 and 1994).
(c) Mean wet season discharge at Abapó and mean wet season catchment precipitation based on discharge data from (Diaz 2010) and GPCC v7 precipitation data (Schneider et al. 2016).

between ~2 and more than 50 km², but they are dominantly < 10 km². Together with the crevasse splays associated with overbank flow and levee breaching, the total area of splay deposition amounts to ~598 km², averaging ~15.8 km²/yr.

7.3.2.4 Avulsion Effects in the Backswamp

With the commencing regional avulsion and successive deposition of progadational splays in the proximal floodplain, large quantities of floodwater get diverted into the distal floodplain (Fig. 7.4d). These events cause extensive inundations which can last for several months and lead to the deposition of cm-scale drapes of fine-grained clays and muds (Fig. 7.8a). They have occupied a total area of at least 1,900 km² since 1975, thus covering an area around four times larger than the area affected by splay sedimentation. Over months to years, the effects of the 1992–1994 avulsion in the

affected downstream distal floodplain are more complex. Floodwaters may cover the floodplain either as unchannelised extensive sheets of water (Fig. 7.4d), or start forming wide transitional multi-channel systems where flow is facilitated by topographic constrictions (e.g., palaeo-levees, Fig. 7.4g) or when approaching progradational splays upstream (SOD Figs. 7.2–7.3). Downstream of progradational splays, suspended sediment may be transported further into the floodplain, filling previously existing lakes, swamps, and other depressions over a few years (SOD Figs. 7.2–7.3). Also, floodwaters may overflow and erode pre-existing palaeolevees and contribute to the capture and reactivation of previously inactive palaeochannel segments, as well as their subsequent abandonment by secondary local avulsions and formation of 'secondary crevasse splays' (Fig. 7.4h, SOD Fig. 7.3). Because of the downstream decrease in sediment load in the distal floodplain, these secondary splays are generally smaller and less dynamic than progradational and cre-vasse splays (Fig. 7.4h), and can persist for years to decades. Splay progradation into the backswamp is expressed topographically by creating a new alluvial ridge (Smith et al. 1989). The new ridge subsequently affects regional hydrology and leads to modifications in vegetation cover including die-off (Fig. 7.4g) and regrowth of large forest areas (SOD Fig. 7.2) – likely because of changes in groundwater flow and water-logging, which result from the formation of new lakes and marshland in the backswamp (SOD Figs. 7.1–7.3).

7.3.3 Megafan Landforms

The proximal megafan area is bordered by high scarps eroded into the surrounding piedmont and has incised into its fan head, forming a terrace surface at its apex (Fig. 7.5, transect 1). Further downstream, the fan surface gradually transforms into two convex-up ridges ~20 km wide and ~10 m high, interpreted as composite 'depositional lobes' (Fig. 7.5, transect 2; Chakraborty and Ghosh 2010; Zani et al. 2012). While northern lobe 1 bifurcates into two well-developed lobes ~10 km in width, southern lobe 2 lies closer to the apex and splits into a complex of several smaller ridges a few kilometres wide and generally < 5 m high interpreted as individual alluvial ridges (Fig. 7.5, transect 3). The modern channel is incised by ~10 m into lobe 1. On the proximal megafan sur-face, palaeochannels were mapped from changes in vegetation and soil patterns (Agrar- und Hydrotechnik GmbH 1974b), reflecting east-directed former drainage (Fig. 7.1c). Forest-covered source-bordering dune systems are associated with these palaeochannels and indicate sand deflation and accu-mulation along the southeastern channel margins simi-lar to today (May 2013). Source-bordering dunes are generally larger and higher on lobe 2 and in the vicin-ity to the modern channel, and smaller and more numerous in between (Fig. 7.1c).

With increasing distance from the apex, lobe 1 (visible in transect 3) is absent from the medial fan, where several smaller ridges ~5–10 km wide and < 5 m high occur instead (Fig. 7.5, transect 4). These ridges are interpreted as elevated and laterally exten-sive (tens to thousands of metres) constructional units of fluvial architecture that contrast with the adjacent

flood basin and include smaller-scale channels, levees, and splays (Miall 1996; Bridge 2003). These 'alluvial ridges' (e.g., Bridge 2003) can form interconnected groups (AR-6 to 8; Fig. 7.1c) (Richards et al. 1993). Also, a large number of small, episodic, plains-fed channels ('cañadas') and inactive, sinuous palaeochan-nels were mapped on much of the megafan surface (Figs. 7.1 and 7.4c). Cañadas and palaeochannels become more frequent towards the medial and distal northern megafan, pointing to the increased influence of groundwater with decreasing overall slope away from the apex. The megafan-wide distribution of the cañadas and palaeochannels confirms the existence of topo-graphically elevated alluvial ridges that dictate the pat-tern and orientation of the secondary drainage network, and suggests reoccupation of abandoned alluvial ridges by underfit streams (Brierley 1997; Bristow 1999).

Geomorphology on the distal megafan is dominated by several alluvial ridge groups (Fig. 7.1c, AR-1 to 5), each composed of multiple individual alluvial ridges ~5 km wide and ~3–6 m high (Fig. 7.5, transect 5). Further downstream, the groups – particularly AR-1 and 2 – branch out into narrower ridges but identifica-tion gets increasingly difficult given the effects of vegetation on digital elevation data in a very flat land-scape (Fig. 7.5, transect 6).

7.3.4 Stratigraphy and Chronology

Excavation pits in the source-bordering dunes across the proximal megafan (Fig. 7.5, transect 2) have revealed well-sorted aeolian sands with moderately evolved soils (Cambisols) in the three southern pits associated with depositional lobe 2. OSL ages for these pits range between ~5.8 and 4.1 ka (SOD Table 7.2, Fig. 7.7a). In pit RG-991, a radiocarbon age on char-coal of ~4.4 cal ka BP (SOD Table 7.3) confirms the luminescence-based chronology. The pit in closer proximity to the active channel is characterised by a less mature soil profile lacking a B-horizon. Samples here yielded ages of 2.3 and 1.6 ka. A further pit on the crest of a ~25-m-high palaeodune in the Lomas de Guanacos dunefield bordering the megafan to the south (Fig. 7.1c) yielded ages of 3.7 and 2.5 ka.

Ten cores were extracted from two adjacent alluvial ridge groups on the distal megafan (Fig. 7.1c, 5) across a W–E transect comprising the alluvial ridge of the

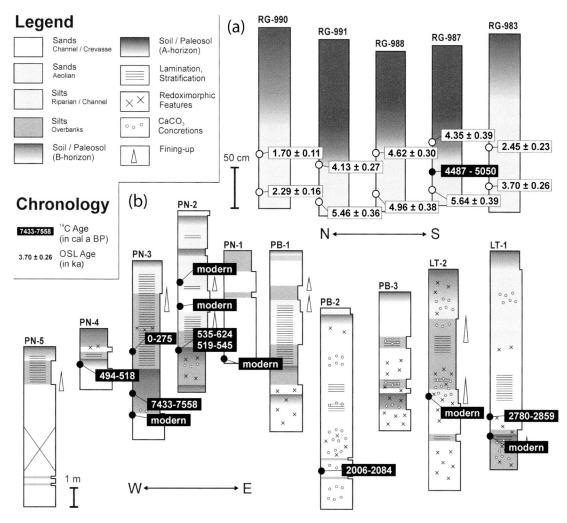

Figure 7.7 **Stratigraphic profiles across the Río Grande megafan**. (a) Transect across the proximal megafan with chronological information from fossil source-bordering dunes. (b) Transect across alluvial ridge groups AR-1 and AR-2 based on sediment cores (see Figs. 7.1 and 7.5 for location). A colour version of this figure is available in the SOD for Chapter 7.

currently active Río Grande (AR-1) and the first inactive alluvial ridge to the east (AR-2). The core locations thus include inter-ridge depressions (PN-5, PB-2 and 3), marginal (PN-4, LT-2) and axial alluvial ridge (PN-2 to PN-3, PB-1, LT-1), and marginal channel positions (PN-1). Sediments were grouped into 7 facies (Röhringer 2006) and interpreted in terms of their dominant depositional environments (SOD Table 7.4 and Fig. 7.7b). In channel settings (e.g., PN-1), sediments were mostly fine (with some medium) sand, generally reflecting cycles of bedload deposition under relatively high flow velocities in the channel bed. Towards the top, channel sands were interrupted by minor layers of silt and clayey silt, indicating a transition to outer channel environments and thus a shift of the main thalweg within its wide

channel, with increasing contribution of suspended sediment. Dominantly sandy sediments are also characteristic for settings in palaeochannels (e.g., 9 m of sand overlying a buried paleosol developed in overbank fines in LT-1). Where very well sorted, these sands may have been reworked by wind. In the distal floodplain between alluvial ridges (e.g., PN-5, PB-2) massive fine sands with some silty interlayers are also the dominant sediments. Soils have formed at the surface at both localities, and thinly bedded overbank fines cap the sands at PN-5. In marginal levee or alluvial ridge settings the stratigraphy is variable, with buried paleosols at the base of a generally coarsening-up sequence (e.g., PN-2, 3, 4, PB-1, 3 and LT-2). In more central alluvial ridge positions, the cores comprise rather sandy to silty sediments (e.g., PN-2, PB-1 and 3), whereas in more distal and lower positions (e.g., PN-3 and 4, LT-2) they include finely bedded overbank and backswamp clays and silts. A clear gradient of increasing amounts of $CaCO_3$ concretions and increasing thickness of surface soils away from the currently active Río Grande is also a prominent feature.

The chronological framework for the cores is based on 13 radiocarbon dates (SOD Table 7.3). The ages of humins and humic acid from a buried paleosol in PN-2 overlap within error, and are thus considered a reliable indication for the commencement of soil burial by increasing sedimentation rates ~1390 CE (i.e., 560 cal yr BP). Independent of depositional environment, depth, or location, six dates on plant remains returned modern radiocarbon dates with pmC (i.e., percent modern carbon, with modern defined as 1950) values of > 100%. These ages are considered to reflect contamination by rootlets related to the densely developed root system of the semi-deciduous natural forest cover and high seasonal groundwater table in the study area (Werding 1977b; Navarro and Maldonado 2002). Due to the absence of significant in-channel vegetation, however, this explanation does not apply to the modern date extracted from the in-channel core PN-1, which instead likely reflects the deposition of 560 cm of sediment over the last 50–60 years. In combination with the paleosol dates from PN-2, the dates available in cores PN-4, PB-2 and LT-1 have yielded older ages in stratigraphically lower positions (e.g., below paleosols, or at the base of the cores) and with increasing distance away from the modern river, and on that basis seem reliable.

7.4 Discussion

7.4.1 Modern Depositional Environments

Two DEMs with 1 arc second (~30 m) ground resolution were used to map and analyse megafan topography. For the detection and interpretation of smaller, landform-scale geomorphic features, particularly relative vertical accuracy (i.e., the point-to-point vertical accuracy of the DEMs) is crucial. The ALOS AW3D DEM is considered the most accurate free DEM (Caglar et al. 2018) with relative vertical accuracy given as ≤5 m (Tadono et al. 2014; Santillan and Makinano-Santillan 2016). SRTM DEMs are reported to have a vertical relative height error of less than 10 m (Farr et al. 2007) which is in the range of average tree height in lowland Bolivia and may thus complicate the identification of landforms at or below this size. Therefore, in combination with use of the global forest cover data (Hansen et al. 2000), the coupled interpretation of both DEMs provides significantly higher confidence in topographic and geomorphological mapping of low relief features on the megafan surface, especially for landforms with horizontal extents of hundreds of metres such as alluvial ridges.

Mapping based on multi-spectral satellite imagery and DEMs has illustrated significant changes in planform and river style of the Río Grande downstream of the megafan apex at Abapó. This common phenomenon for megafans (Shukla et al. 2001; Davidson et al. 2013) likely results from significant transmission losses due to infiltration and evaporation (Werding 1977b). On the proximal megafan, the active channel erodes the piedmont footslopes and appears to be incised, implying the presence of an intersection point around 150 km downstream of the apex in the vicinity of Pto. Pailas. These characteristics therefore reflect a transition from a sediment transfer zone on the proximal megafan, to a depozone on the distal megafan with a variety of depositional processes and recent floodplain inundation dynamics. The documented depositional processes characterise sedimentation patterns typical of highly dynamic and avulsive river landscapes (Perez-Arlucea and Smith 1999; Morozova and Smith 2000) and allow a first estimation of the spatial and temporal dynamics of these depositional processes in the megafan depozone since the 1970s.

Active in-channel deposition on the distal megafan as seen in core PN-1 corresponds to high in-channel sedimentation rates on the order of ~100 mm/yr. Approximately 4 m of channel-bed raising as a result of in-channel aggradation between 1982 and 1997 were also reported from km ~120 downstream of Abapó (Wachholtz and Herold-Mergl 2003), corresponding to sedimentation rates of ~250 mm/yr. These rates probably do not reflect longer-term mean sedimentation because of the frequent in-channel lateral shifts (i.e., channel wandering) and sediment remobilisation during bank-full flood events. However, over a few years in-channel sedimentation and channel-bed raising should significantly contribute to a decrease in the threshold for overbank flow, crevasse splays, and super-elevation with regard to the surrounding floodplain as a set-up condition for channel avulsion (Jones and Schumm 1999; Ashworth et al. 2007). Various processes relating to overbank flow have been observed in satellite imagery and the field. Firstly, in the distal floodplain inundation causes sedimentation of cm-scale layers of fine-grained sediment from suspension (Fig. 7.8a) – a process that has affected a total area of $\sim 2,000\,km^2$ between 1975 and 2014 (Fig. 7.8b). Secondly, crevasse splays lead to the deposition of sandy bedload on the proximal floodplain. A single

Figure 7.8 **Field examples of the hydrological and sedimentary impact of crevasse splays and overbank flow during Río Grande flood events, February 2006, distal region of the megafan.** (a) Thick drape ($\sim 10\,cm$) of dry-cracked overbank muds in the flood basin. (b) Inundation in lowland community following 2006 crevasse splay. (c) Close-up of proximal crevasse splay deposits that have buried buildings to depths > 1 m. Settlement is shown in (d). (d) Aerial view of settlement fallen victim of extensive sedimentation along margin of 2006 crevasse splay. (e) Provisional closure of levee breach (note blue sandbags) after February 2006 crevasse splay event (note thick accumulations of wooden debris along the channel margin, dashed line). (f) example of artificial levee (white arrow) to prevent crevassing and backswamp inundations. Photographs b, d, e and f, date: 02/2006; photographs a and c, date: 08/2006. A colour version of this figure is available in the SOD for Chapter 7.

event may deposit thicknesses of > 1 m (Fig. 7.8c) over areas of > 25 km^2. These events severely affect communities, agricultural land (Fig. 7.8d), and vegetation and ecology in areas of dense forest cover (Lombardo 2017). Since the mid-1980s, more than 30 crevasse splays were observed covering an area of ~ 600 km^2 on the distal megafan between ~ 240 km and 380 km from the apex (Fig. 7.1c, 6).

Only one crevasse splay developed into a full regional avulsion between 1992 and 1994, followed by a dynamic reorganisation of the downstream channel pattern with significant geomorphic and hydromorphic effects extending across floodplain areas into the distant backswamp. The avulsion was not instantaneous; full diversion of flow occurred over the course of a few years and triggered the ongoing construction of a new channel belt downstream of the avulsion point by a series of progradational splays and the formation of a new channel segment, a process documented by Smith et al. (1989). In this context, roads or tracks provide a preferred path for floodwaters, and thus directly influence the development of splays and new channel segments (Jones and Schumm 1999; Wachholtz and Herold-Mergl 2003; CaryGlobal SRL 2008). Progradational splays steadily advance onto the former floodplain, leading to an average area of 16 km^2/yr sediment deposition (Fig. 7.9), and the construction of a topographically elevated alluvial ridge over short time scales of $10–10^2$ years (Smith et al. 1989; Morozova and Smith 2000). There is a significant topographic difference of $\sim 5–7$ m between the SRTM (year 2000) and ALOS DEMs (years 2006–2011) at the position of the Río Grande (Fig. 7.5, transect 6). Despite the uncertainties associated with deriving vertical landscape change from global DEMs (Purinton and Bookhagen 2018), given the excellent agreement of both DEMs along the transect and the high relative vertical accuracy of ≤ 5 m inherent to the ALOS DEM, this observation likely reflects the quick vertical growth characterising initial alluvial ridge formation following the 1992–1994 avulsion. Even though the duration of a complete avulsion cycle is still difficult to estimate from satellite imagery, our data attest to rates that are comparable to the most dynamic depositional systems on the planet (Morozova and Smith 2000; Stouthamer and Berendsen 2001; Bernal et al. 2010, 2013).

The longer term frequency of alluvial ridge construction should be directly linked to the frequency of successful upstream channel avulsions, raising questions regarding their dominant controls, such as alluvial ridge super-elevation, the variability in flood magnitude, and the effect of channel blockage caused by logjams (Jones and Schumm 1999; Slingerland and Smith 2004; Stouthamer and Berendsen 2007). Logjams were reported as key drivers of channel collapse and avulsion along the narrow downstream sections of many rivers in SW Amazonia (Lombardo 2017) and the Chaco (Martín-Vide et al. 2014), and significant quantities of wooden debris are also transported by the Río Grande (Fig. 7.8e). At the locations of the observed crevasse splays, however, logjams are less likely to obstruct the 400–500-m-wide channel and are probably not the main trigger for crevassing. In contrast, the analysis of channel changes across the proximal to distal megafan since 1984, combined with estimates for catchment-wide wet season precipitation and discharge (Fig. 7.6a–c), seem to suggest that the 1992–1994 avulsion event immediately followed three successive years of low wet season discharge that resulted in channel contraction. This implies that

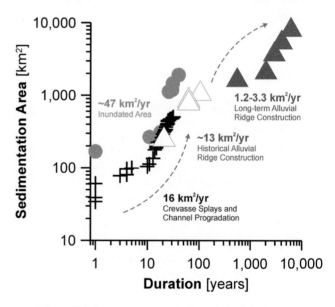

Figure 7.9 **Longer-term evolution of the Río Grande megafan surface, and relationship between spatial and temporal scales for the observed dominant depositional processes and landforms as expressed by cumulative sedimentation area *vs.* cumulative time.** Crosses: crevasse splays. Dots: backswamp inundation. Filled triangles: Holocene alluvial ridge construction. Open triangles: historical alluvial ridge construction.

inter-annual, climatically-induced discharge variations, e.g., in relation to ENSO (Aalto et al. 2003), may trigger major crevasse splays, thus directly influencing avulsion frequency on the distal megafan. Along the Río Grande, overbank flow led to levee breaching and major crevasse splay events covering areas of 10 to $> 25 \text{ km}^2$ at least five times over the last decades. The timing of splays here and elsewhere in SW Amazonia, however, does not exhibit a clear relationship with ENSO (Lombardo 2016). In addition, with the exception of the 1992 event none of these splays was followed by successful regional avulsions. Instead, it seems reasonable to assume that the impact of quite major engineering (Roca Salazar et al. 2003; Wachholtz and Herold-Mergl 2003; CaryGlobal SRL 2008; Fig. 7.8e) has successfully redirected the channel to its abandoned bed and thus prevented the development of full regional avulsions. Consequently, rates and controls on recent splays and avulsions – and thus the overall pace of sediment dispersal on the distal megafan – are difficult to interpret from multi-temporal satellite imagery alone.

Regardless of the exact rates of depositional processes, the active depositional processes imply significant hydrological, biotic, geomorphological, and sedimentary reorganisation of the distal megafan, resulting in an extremely heterogeneous mosaic of sedimentary and ecological environments (Davidson et al. 2013; Weissmann et al. 2013; Lombardo 2017). More importantly, however, the progressive formation of new channel belts and alluvial ridges by progradational splays and channelisation following regional avulsion is probably indicative of disconnection between the megafan and the base-level trunk rivers (e.g., Río Mamoré). During the avulsion period, the downstream transport of bedload and suspended sediment is significantly hindered by the lack of a fully developed channel system. Even though high sediment yield of the Río Grande has been invoked as a key determinant of channel dynamics further downstream along the Río Mamoré (Guyot et al. 1994; Constantine et al. 2014), our observations over the last decades provide strong support for a relatively minor sedimentary contribution of the Río Grande to the Amazon drainage basin as suggested by Lombardo (2016). Thus, under current conditions, significantly more sediment may be stored within the depositional landforms than the $\sim 36 \times 10^6$ t estimated from the limited existing gauging data on the proximal megafan (e.g., Fig. 7.2b), especially when considering that coarse bedload is not measured.

7.4.2 Large-Scale Megafan Geomorphology and Holocene Dynamics

Avulsive channel shifts are common events on megafans. Their frequency and location are therefore crucial aspects for determining the longer-term evolution of the megafan but are difficult to reconstruct because outcrops across the extensive and flat megafan surface are hard to detect. Until recently, the development of detailed maps of palaeo-river courses was limited by the quality and resolution of remote-sensing data – particularly in densely vegetated regions such as eastern Bolivia. As a result, for most megafans information on the chronology of large-scale channel shifts is generally scarce, which limits our understanding of the controlling mechanisms on megafan evolution on time scales of 10^2–10^5 years.

In combination with the first available systematic chronological data from the proximal and distal portions of the megafan, the geomorphological map presented here (Fig. 7.1c) provides a basis for quantifying the spatial and temporal scales of longer-term channel changes and depositional processes on the Río Grande megafan. Surface deposits and landforms across most of the proximal megafan are likely not older than $\sim$ 6–7 ka, whereas alluvial ridges in close proximity to the modern channel are younger than 2.9 cal ka BP. The active branch of the currently active alluvial ridge group AR-1 is younger than $\sim$ 0.6 cal ka BP, i.e., younger than the fourteenth century CE (Fig. 7.10). In combination this suggests that much of the megafan surface has been affected by active transport and sedimentation processes since the mid-Holocene. Consequently, direct inferences on longer-term (i.e., Pleistocene) megafan geomorphic and sedimentary dynamics based on stratigraphic or geomorphic relationships with sites in the adjacent piedmont or downwind dune systems need to be interpreted very carefully (Latrubesse et al. 2012; May 2013) in the absence of field-based sedimentological and chronological evidence from the megafan subsurface (Agrar- und Hydrotechnik GmbH 1973; Werding 1977b; Sinha et al. 2014).

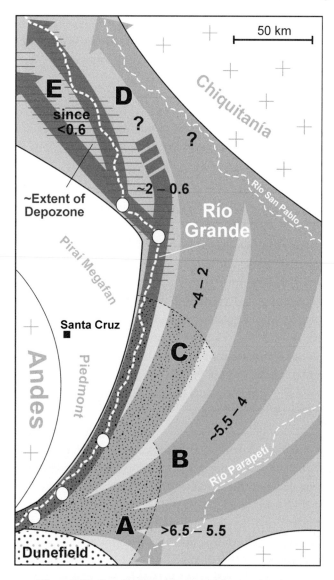

Figure 7.10 **Conceptualised model of the longer-term evolution of the Río Grande megafan since approximately the mid-Holocene (numbers: ages in ka).**

Fossil river courses – and thus the sequential evolution of the megafan – are reflected in the overall organisation of drainage patterns on the megafan surface (May 2006), and in the pattern of palaeochannels associated with source-bordering dunes on the proximal megafan (Werding 1977a; Hanagarth 1993). Consensus seems to exist with regard to an anticlockwise shift from a formerly eastwards-directed drainage – with connection to the Amazonian plains via the Río San Pablo – towards the current position connected to the Río Mamoré in the southern Llanos de Moxos (Figs. 7.1c and 10). The inferred controls on such large-scale channel shifts run from regional

tectonic movements (Hanagarth 1993) to the effect of channel blocking by aeolian sand (Werding 1977a) and climate-driven late Quaternary changes in discharge and sediment supply (May 2006; Latrubesse et al. 2012). The growing quality and availability of DEM data has enabled the identification of subtle, topographically elevated lobes and alluvial ridge groups (Buehler et al. 2011; Syvitski et al. 2012), which together cover at least one third of the megafan surface. Aside from lateral shifts as reflected in the changing orientation of palaeochannels and alluvial ridges on the medial to distal megafan, systematic differences exist between the ridge groups in terms of their distance from the megafan apex (Fig. 7.1c). Ridges in the south (AR-6 to 8) are topographically expressed ~80–130 km away from the apex, whereas AR-1 to 5 commence between 160 and 180 km away and extend > 350 km downstream (Fig. 7.1c). By analogy with the modern channel dynamics, this trend may thus document a downfan progradation of the depozones (Weissmann et al. 2013) and avulsion points (Bridge 2006) superimposed on the lateral and avulsive channel shifts since the early to mid-Holocene.

In combination, progradational and avulsive dynamics provide a conceptual model of the longer-term fluvial dynamics of the Río Grande (Fig. 7.10):

- Before and during the mid-Holocene (phase A; > 6.5 to 5.5 ka), the Río Grande flowed in a west to east direction, and may have even flowed in a southeasterly direction towards the current location of the Bañados de Izozog, as indicated by the presence of several alluvial ridges influencing the zig-zag course of the Río Parapetí (Fig. 7.1c).
- During the mid-Holocene (phase B; 5.5–4 ka), the Río Grande still flowed towards the east as indicated by palaeochannels on the proximal fan, but may have shifted further northeast – then flowing along the margin of the Brazilian Shield towards the Amazonian lowlands. The proximal fan was characterised by aggradation as reflected by a clear depositional lobe (Fig. 7.5, transect 2, 3). Here, source-bordering dunes of sizes comparable to those of today formed along the active channels on the proximal fan. In the upper dunes, ages are younger than lower down by ~0.8–1.2 ka – an anomaly which cannot be ascribed only to error associated with the OSL dates. The anomaly may rather indicate sustained and/or repeated periods of source-bordering dune construction for ~1 ka, either resulting from channel backfilling, multiple channel switches and reactivations from nodal

avulsions near the apex (Bridge 2006; Davidson et al. 2013), or reflecting a distributive system of coexisting active channels similar to rivers along the Subandean ranges further north today (Lombardo 2016). The transition from the proximal to distal fan was located relatively near the apex at Abapó. The combined observations suggest high deposition rates on the proximal megafan – a scenario proposed by Bridge (2006) – possibly through reduced transport capacity and/or relatively high bedload supply during this time. On the proximal tracts of the megafan, coarse gravel deposits (< 1% of the bedload today; Fig. 7.2c) were reported below depths of ~20 m as far as ~100 km away from the apex (Werding 1977b), and provide additional evidence for a significant decrease in stream power and transport capacity some time before the mid-Holocene. More than 600 km further downstream in the Moxos plains, a strong pulse of sedimentation documented by a lobe of fluvial sediment, emplaced around or shortly after ~5 cal ka BP and subsequently abandoned, was generated by increased sediment supply from the Río Grande megafan (Lombardo et al. 2012, 2013b; Plotzki et al. 2015). Given the current low connectivity between the megafan and its foreland and the significant contribution of additional sediment from Subandean rivers (e.g., Río Yapacaní; Lombardo 2016), this hypothesis, however, awaits further testing.

- Around 4 ka, a change in fluvial dynamics was expressed in the progradation of the proximal–distal transition. This is indicated by the formation of alluvial ridges successively further away from the apex (phase C; 4–2 cal ka BP). The avulsion point probably shifted northeast, and drainage lay further west (e.g., AR-2, Fig. 7.10) but likely continued north to connect with the Río San Pablo. On the proximal fan, a well-developed depositional lobe implies continued net aggradation, and source-bordering dunes are more frequent but visibly smaller. This may imply high lateral dynamics and frequent channel shifts on the proximal fan, also expressed by active erosion of the adjacent piedmont ('toe-cutting' *sensu* Leeder and Mack 2001) at some time between 4.3 and 1.5 cal ka BP at Cabezas (May et al. 2008). The palaeohydrological interpretation of these combined observations is not straightforward, but the frequent shifts and sediment aggradation on the proximal megafan, in combination with a downfan progradation of alluvial ridge construction, could imply an increase in flow magnitude and frequency leading to the activation of sediment in the Río Grande catchment. This conclusion concurs with a climatically induced increase in denudation rates since ~4.5 ka inferred from sediment-flux and cosmogenic nuclide data in the Río Grande catchment (Insel et al. 2010).
- Fluvial activity ~2–0.6 cal ka BP (phase D) shifted to the presently still-active alluvial ridge (AR-1). This shift may

have been accompanied by further downstream progradation of the avulsion point (Fig. 7.10), and by the onset of slow but gradual and still ongoing fan-head incision (Wachholtz and Herold-Mergl 2003). Such incision is also detected in exposures of channel facies (~4.3 and 1.6 cal ka BP) in minor terraces along the western channel margin near Cabezas and Río Seco, i.e., on the proximal megafan ~2–3 m above the modern channel (May et al. 2008). The exact timing of this shift is difficult to constrain, but it could have occurred as early as 2 cal ka BP (basal radiocarbon age in PB-2). The sedimentology associated with this core indicates rapid deposition of sandy sediment as a result of repeated crevassing or channelised flow in an inter-ridge setting – possibly as a result of adjacent ridge abandonment (AR-2) and the beginning of alluvial ridge construction (AR-1) following upstream nodal avulsion. Further ages will be required to test this hypothesis. Archaeological excavations of a pre-Columbian indigenous settlement along AR-1 at Pailón yielded radiocarbon ages between ~0.8 and 1.1 cal ka BP (Görsdorf 2002) and suggest that the channel may have been already in close proximity with the existing alluvial ridge developed. A detailed reconstruction of the northwards continuation of the Río Grande channel around 2 cal ka BP – thereby providing an assessment of distant downstream effects in the Amazonian lowlands – is complicated because of the strong recent depositional dynamics and frequent cross-cutting palaeochannels in the very flat landscape. Detailed knowledge of timing, location, and connectivity of the Río Grande, however, could be of crucial importance in evaluating hydrological conditions and geomorphological dynamics during a time that saw the emergence of prehistoric cultures in the Moxos plains further north (Lombardo et al. 2013a). Channel shifts previous to ~1.2 cal ka BP on the adjacent Río Parapetí were generally attributed to increasing transport capacities (May 2011). However, a number of scenarios related to changes in discharge regime and sediment supply could explain the shift from aggradation and rapid channel shifts to the onset of fan-head incision. Such hypotheses require further data on sediment production in the large upstream catchment of the Río Grande (Kober et al. 2015).

- As evidenced by foundations of colonial settlements along the river banks (Köster 1978), the position of the channel and alluvial ridge AR-1 seem to have been spatially stable on the medial belt of the megafan at least since the sixteenth century. In contrast, a further avulsive shift from a northerly to a northwesterly direction likely occurred around or some time after 0.6 cal ka BP on the distal fan (phase E) (radiocarbon dates from below the AR-1 sediments; Fig. 7.5). Avulsive dynamics on the distal megafan during the nineteenth century are also evident from reports of floods and channel shifts in

1825 resulting in the presence of tree trunks in the channel bed (Minchin 1881) and disruption of navigation on the Río Grande (Roca 2016). As a consequence of these shifts, a navigable connection with the Río Piraí existed until at least 1840, but a renewed flood event in the 1930s terminally disrupted navigation on the Río Piraí and likely shifted the Río Grande further north again (Roca 2016). Although the spatial accuracy of these reports is insufficient for assigning course shifts to mapped palaeochannels, it seems reasonable to assume that they correspond to alluvial ridges south of the modern river course (Figs. 7.1 and 7.5 transect 6). Since the mid-Holocene the avulsion point had then prograded ~200 km downfan, implying a significant change in upstream deposition rate and sediment supply (Bridge 2006). This is in agreement with the late Holocene and ongoing tendency for megafan-head incision along the Río Grande (May et al. 2008) and adjacent fluvial systems (Río Piraí; Servant et al. 1981).

- Over the last 50 years, two alluvial ridges were active on the distal Río Grande megafan; the first until 1994, and the second since then. A successful regional avulsion event (1992–1993) diverted the course of the channel and triggered the evolution of a new alluvial ridge, including significant reorganisation in the hydrology and vegetation and sedimentation patterns. Due to progressive capitalisation of the Bolivian lowlands for agriculture, increasing efforts have been directed towards regulation of land use and river channels, and to infrastructure repair to prevent further damage by large-scale avulsive river changes along the lower Río Grande (Fig. 7.8f) (Roca Salazar et al. 2003; CaryGlobal SRL 2008); the efforts, however, appear of limited longer-term effectiveness. Considering the strong fluvial dynamics and large sediment loads conveyed from the Andean catchment of the Río Grande, it also seems doubtful that construction of a large and already approved hydroelectric dam (Forsberg et al. 2017) in the recently declared UNESCO biosphere reserve (Corz 2015), just upstream of the megafan apex, will provide a sustainable and longer-term means for flood control.

The long-term rate of alluvial ridge construction based on mapped areas and the available chronological information averages ~ 1.2–$3.3 \, km^2/yr$. This is significantly slower than the $\sim 13 \, km^2/yr$ estimated for ridges associated with historical channel changes, or $\sim 16 \, km^2/yr$ deduced from the mapped modern splay dynamics (Fig. 7.9). Difficulties in comparing rates across spatial and temporal scales are common (Church and Mark 1980), and may reflect the existence of periods of reduced or absent ridge construction, or show that not all splays effectively contribute to the topographic construction of the ridges. Splays not followed by successful avulsion (e.g., 2005, 2006, 2008) are particularly likely to create local sand bodies without significant topographic expression. The observed differences in rates, however, could also imply a strong and relatively recent increase in fluvial dynamics, thus raising questions regarding climate-driven *vs.* anthropogenic impacts on sediment supply and discharge in the eastern Andes as well as their responsibility in driving megafan dynamics over the last centuries (Assine et al. 2016; Heyvaert and Walstra 2016). This is a key area for future investigation.

7.5 Conclusions

The assessment of temporal and spatial scales of depositional processes on megafans, and thus the understanding of the longer-term evolution of these landforms on Holocene to Pleistocene timescales, still presents a major challenge because of the large size of megafans and their inherently flat topography. Multitemporal satellite imagery combined with fieldwork has provided a first detailed documentation and quantification of recently active depositional processes. These are restricted to a $\sim 2{,}000 \, km^2$ depozone in the distal reaches of the megafan, and they are characterised by processes typical of avulsive and unstable fluvial environments. Fluvial dynamics undergo very fast avulsion processes compared to other fluvial systems. Although these dynamics are overall driven by high sediment loads supplied by the Andes, much of the sediment is currently stored on the megafan without significantly contributing to the fluvial system further downstream in Amazonia. Our data have also documented substantial human interference with the natural sedimentary regime of the megafan, for example river rechannelisation after crevasse splay events and following the construction of artificial levees. These engineering measures have likely prevented a number of channel avulsions over the last decades, suggesting that natural crevasse-splay and avulsion frequencies may in fact be underestimated.

The interpretation of digital elevation data in combination with radiocarbon and luminescence-based dating in two stratigraphic transects has also allowed the various depositional processes to be linked with older landforms over a wide area. Our results imply that most of the 36,000 km^2 megafan surface has been subjected to sedimentation and/or reworking processes within the last 6–7 ka. The Río Grande has undergone significant changes in the location of its channels and depozones since the mid-Holocene (i.e., since 4 ka). Some of these dynamics, for example reflected in the well-developed alluvial ridge system, may result from autogenic processes, but the timing and magnitude of downfan depozone progradation are nonetheless likely to reflect changes in sediment load or calibre and changes in transport capacity driven by discharge and climate. This in turn implies that megafans are not only fascinating and underexplored geomorphic features, but surely also store an integrated record of catchment-wide changes in environmental and hydrological conditions during the Holocene and perhaps even the Pleistocene.

Acknowledgements

The authors thank J. Wilkinson for the invitation to contribute to this book. The research presented here was funded by SNF project 200020-105228/1. We thank Justin Wilkinson, Yanni Gunnell and an anonymous reviewer for detailed and helpful comments on an earlier version of this chapter, and acknowledge permission to use photography taken by Bob Reynolds, Denver.

References

Aalto, R., Maurice-Bourgoin, L., Dunne, T., et al. (2003). Episodic sediment accumulation on Amazonian flood plains influenced by El Niño/Southern Oscillation. *Nature*, 425, 493–497.

Agrar- und Hydrotechnik GmbH (1973). *Proyecto de desarrollo agroindustrial Abapo-Izozog – Perforación de Pozos*. Essen: AHT GROUP AG.

Agrar- und Hydrotechnik GmbH (1974a). *Proyecto de desarrollo agroindustrial Abapo-Izozog – Hidrología*. Essen: AHT GROUP AG.

Agrar- und Hydrotechnik GmbH (1974b). *Proyecto de desarrollo agroindustrial Abapo-Izozog – Pedología. Tomo I - Sector Oeste*. Essen: AHT GROUP AG.

Ashworth, P. J., Best, J. L. and Jones, M. A. (2007). The relationship between channel avulsion, flow occupancy and aggradation in braided rivers: insights from an experimental model. *Sedimentology*, 54, 497–513.

Assine, M. L. (2005). River avulsions on the Taquari megafan, Pantanal wetland, Brazil. *Geomorphology*, 70, 357–371.

Assine, M. L., Macedo, H. A., Stevaux, J. C., et al. (2016). Avulsive rivers in the hydrology of the Pantanal wetland. In I. Bergier and L. M. Assine, eds., *Dynamics of the Pantanal Wetland in South America*. Springer International Publishing, Cham, 83–110.

Baby, P., Herail, G., Salinas, R. and Sempere, T. (1992). Geometric and kinematic evolution of passive roof duplexes deduced from cross section balancing: example from the foreland thrust system of the southern Bolivian subandean zon. *Tectonics*, 11, 523–536.

Barboza, F., Geyh, M. A., Hoffmann, R., et al. (2000). Soil formation and Quaternary geology of the Paraguayan Chaco - Thematic mapping. *Zeitschrift für angewandte Geologie, Sonderheft* 1, 49–53.

Barnes, J. B. and Heins, W. A. (2009). Plio-Quaternary sediment budget between thrust belt erosion and foreland deposition in the central Andes, southern Bolivia. *Basin Research*, 21, 91–109.

Bernal, C., Christophoul, F., Darrozes, J., et al. (2013). Crevassing and capture by floodplain drains as a cause of partial avulsion and anastomosis (lower Rio Pastaza, Peru). *Journal of South American Earth Sciences*, 44, 63–74.

Bernal, C., Christophoul, F., Darrozes, J., et al. (2010). Late Glacial and Holocene avulsions of the Rio Pastaza Megafan (Ecuador–Peru): frequency and controlling factors. *International Journal of Earth Sciences*, 100, 1759–1782.

Berri, G. J. and Inzunza, J. B. (1993). The effect of the low-level jet on the poleward water vapour transport in the central region of South America. *Atmospheric Environment*, 27 A, 335–341.

Bridge, J. (2006). Fluvial facies models: recent developments. In H. W. Posamentier and R. G. Walker, eds., *Facies Models Revisited*. SEPM Society for Sedimentary Geology, Tulsa, 85–170.

Bridge, J. S. (2003). *Rivers and Floodplains: Forms, Processes, and Sedimentary Record*, Blackwell Publishing, Hoboken (NJ).

Brierley, G. J. (1997). What is a fluvial levee? *Sedimentary Geology*, 114, 1–9.

Bristow, C. S. (1999). Gradual avulsion, river metamorphosis and reworking by underfit streams: a modern example from the Brahmaputra River in Bangladesh and a possible ancient example in the Spanish Pyrenees. In N. D. Smith and J. Rogers, eds., *Fluvial Sedimentology IV*. Blackwell, Oxford, 221–230.

Buehler, H. A., Weissmann, G. S., Scuderi, L. A., and Hartley, A. J. (2011). Spatial and temporal evolution of an avulsion on the Taquari River distributive fluvial system from satellite image analysis. *Journal of Sedimentary Research*, 81, 630–640.

Caglar, B., Becek, K., Mekik, C., and Ozendi, M. (2018). On the vertical accuracy of the ALOS World 3D-30 m digital elevation model. *Remote sensing letters*, 9, 607–615.

CaryGlobal SRL (2008). *Estudio Hidrologico-Hidraulico de la Cuenca Baja del Río Grande*. CaryGlobal SRL, Santa Cruz.

Chakraborty, T. and Ghosh, P. (2010). The geomorphology and sedimentology of the Tista megafan, Darjeeling Himalaya: implications for megafan building processes. *Geomorphology*, 115, 252–266.

Chakraborty, T., Kar, R., Ghosh, P., and Basu, S. (2010). Kosi megafan: historical records, geomorphology and the recent avulsion of the Kosi River. *Quaternary International*, 227, 143–160.

Church, M. and Mark, D. M. (1980). On size and scale in geomorphology. *Progress in Physical Geography*, 4, 342–390.

Constantine, J. A., Dunne, T., Ahmed, J., Legleiter, C., and Eli, D. (2014). Sediment supply as a driver of river evolution in the Amazon Basin. *Nature Geoscience*, 7, 899–903.

Corz, C. (2015). La Unesco declara Reserva de la Biosfera al Río Grande. *La Razón*, 10. 07. 2015.

Davidson, S. K., Hartley, A. J., Weissmann, G. S., Nichols, G. J., and Scuderi, L. A. (2013). Geomorphic elements on modern distributive fluvial systems. *Geomorphology*, 180–181, 82–95.

Decelles, P. G. and Giles, K. A. (1996). Foreland basin systems. *Basin Research*, 8, 105–123.

Diaz, J. (2010). *Caracterización de la Cuenca Alta del Río Grande y Sequía en el Chaco Cruceño*. Santa Cruz: Acción Contra el Hambre and Centro Andino Para La Gestión y Uso Del Agua.

Farr, T. G., Rosen, P. A., Caro, E., et al. (2007). The shuttle radar topography mission. *Reviews of Geophysics*, 45, RG2004.

Forsberg, B. R., Melack, J. M., Dunne, T., et al. (2017). The potential impact of new Andean dams on Amazon fluvial ecosystems. *PLoS ONE*, 12, e0182254.

Gerold, G. (1985). Untersuchungen zur Badlandentwicklung in den wechselfeuchten Waldgebieten Südboliviens. *Geoökodynamik*, 6, 35–70.

Gerold, G. (1988). Die Bedeutung von Ariditätswandel und Vegetationsdegradation für die fluviale Morphodynamik in den Äusseren Tropen Boliviens. In J. Hagedorn and H. G. Mensching, eds., *Aktuelle Morphodynamik und Morphogenese in den semiariden Randtropen und Subtropen*. Vandenhoeck and Ruprecht, Göttingen, 277–306.

Görsdorf, J. (2002). Radiocarbon datings from excavations near Pailón, Bolivia. *Beiträge zur Allgemeinen und Vergleichenden Archäologie*, 22, 227–229.

Gerold, G. (2004). Soil: The foundation of biodiversity. In P. L. Ibisch and G. Mérida, eds., *Biodiversity: The Richness of Bolivia*. Santa Cruz: Editorial FAN, 17–31.

Guyot, J. L., Bourges, J., and Cortez, J. (1994). Sediment transport in the Río Grande, an Andean river of the Bolivian Amazon drainage basin. *Variability in Stream Erosion and Sediment Transport*. IAHS Publications, 223–231.

Hanagarth, W. (1993). *Acerca de la geoecología de las sabanas del Beni en el noreste de Bolivia*, Instituto de Ecología, La Paz.

Hansen, M., Defries, R., Townshend, J. R., and Sohlberg, R. (2000). Global land cover classification at 1 km spatial resolution using a classification tree approach. *International Journal of Remote Sensing*, 21, 1331–1364.

Hartley, A. J., Weissmann, G. S., Nichols, G. J., and Warwick, G. L. (2010). Large distributive fluvial systems: Characteristics, distribution, and controls on development. *Journal of Sedimentary Research*, 80, 167–183.

Heyvaert, V. M. A. and Walstra, J. (2016). The role of long-term human impact on avulsion and fan development. *Earth Surface Processes and Landforms*, 41, 2137–2152.

Horton, B. K. and DeCelles, P. G. (1997). The modern foreland basin system adjacent to the Central Andes. *Geology*, 25, 895–898.

Horton, B. K. and DeCelles, P. G. (2001). Modern and ancient fluvial megafans in the foreland basin system of the central Andes, southern Bolivia: implications for drainage network evolution in fold-thrust belts. *Basin Research*, 13, 43–63.

Ibisch, P. L., Beck, S. G., Gerkmann, B., and Carretero, A. (2004). Ecoregions and ecosystems. In P. L. Ibisch and G. Mérida, eds., *Biodiversity: The Richness of Bolivia*. Editorial FAN, Santa Cruz, 47–88.

Insel, N., Ehlers, T. A., Schaller, M., et al. (2010). Spatial and temporal variability in denudation across the Bolivian Andes from multiple geochronometers. *Geomorphology*, 122, 65–77.

Iriondo, M. (1993). Geomorphology and late Quaternary of the Chaco (South America). *Geomorphology*, 7, 289–303.

Isacks, B. L. (1988). Uplift of the Central Andean Plateau and bending of the Bolivian Orocline. *Journal of Geophysical Research*, 93, 3211–3231.

Jones, L. S. and Schumm, S. A. (1999). Causes of avulsion: an overview. In N. D. Smith and J. Rogers, eds., *Fluvial Sedimentology IV*. Blackwell, Oxford, 171–178.

Killeen, T. J., Guerra, A., Calzada, M., et al. (2008). Total historical land-use change in Eastern Bolivia: Who, where, when, and how much? *Ecology and Society*, 13, 36.

Kober, F., Zeilinger, G., Hippe, K., et al. (2015). Tectonic and lithological controls on denudation rates in the central Bolivian Andes. *Tectonophysics*, 657, 230–244.

Köster, G. (1978). *Santa Cruz de la Sierra (Bolivien): Entwicklung, Struktur und Funktion einer tropischen Tieflandstadt*. RWTH, Aachen University, Dept. of Geography.

Kruck, W., Helms, F., Geyh, M. A., et al. (2011). Late Pleistocene-Holocene history of Chaco-Pampa sediments in Argentina and Paraguay. *Eiszeitalter und Gegenwart / Quaternary Science Journal*, 60, 188–202.

Krüger, J.-P. (2006). *Waldkonversion und Bodendegradation im tropischen Tiefland von Ostbolivien. GIS-gestützte*

Analyse zur Regionalisierung der Bodendegradation im Department Santa Cruz. Georg-August-Universität zu Göttingen.

Latrubesse, E. M. (2015). Large rivers, megafans and other Quaternary avulsive fluvial systems: a potential "who's who" in the geological record. *Earth-Science Reviews*, 146, 1–30.

Latrubesse, E. M. and Restrepo, J. D. (2014). Sediment yield along the Andes: continental budget, regional variations, and comparisons with other basins from orogenic mountain belts. *Geomorphology*, 216, 225–233.

Latrubesse, E. M., Stevaux, J. C., Cremon, E. H., et al. (2012). Late Quaternary megafans, fans and fluvio-aeolian interactions in the Bolivian Chaco, Tropical South America. *Palaeogeography, Palaeoclimatology, Palaeoecology*, 356–357, 75–88.

Leeder, M. R. and Mack, G. H. (2001). Lateral erosion ('toe-cutting') of alluvial fans by axial rivers: Implications for basin analysis and architecture. *Journal of the Geological Society, London*, 158, 885–893.

Lombardo, U., May, J.-H., and Veit, H. (2012). Mid- to late-Holocene fluvial activity behind pre-Columbian social complexity in the southwestern Amazon basin. *The Holocene*, 22, 1035–1045.

Lombardo, U., Denier, S., May, J.-H., Rodrigues, L., and Veit, H. (2013a). Human–environment interactions in pre-Columbian Amazonia: The case of the Llanos de Moxos, Bolivia. *Quaternary International*, 312, 109–119.

Lombardo, U., Szabo, K., Capriles, J. M., et al. (2013b). Early and middle holocene hunter-gatherer occupations in western Amazonia: the hidden shell middens. *PLoS ONE*, 8, e72746–e72746.

Lombardo, U. (2016). Alluvial plain dynamics in the southern Amazonian foreland basin. *Earth System Dynamics Discussions*, 7, 453–467.

Lombardo, U. (2017). River logjams cause frequent large-scale forest die-off events in Southwestern Amazonia. *Earth System Dynamics Discussions*, 2017, 1–24.

Martín-Vide, J. P., Amarilla, M., and Zárate, F. J. (2014). Collapse of the Pilcomayo River. *Geomorphology*, 205, 155–163.

May, J.-H. (2006). Geomorphological indicators of large-scale climatic changes in the Eastern Bolivian lowlands. *Geographica Helvetica*, 61, 120–134.

May, J.-H., Zech, R., and Veit, H. (2008). Late Quaternary paleosol-sediment-sequences and landscape evolution along the Andean piedmont, Bolivian Chaco. *Geomorphology*, 98, 34–54.

May, J.-H. (2011). The Río Parapetí – Holocene megafan dynamics and wetland formation in the southernmost Amazon basin. *Geographica Helvetica*, 66, 193–201.

May, J.-H. (2013). Dunes and dunefields in the Bolivian Chaco as potential records of environmental change. *Aeolian Research*, 10, 89–102.

May, J.-H., Plotzki, A., Rodrigues, L., Preusser, F., and Veit, H. (2015). Holocene floodplain soils along the Río

Mamoré, northern Bolivia, and their implications for understanding inundation and depositional patterns in seasonal wetland settings. *Sedimentary Geology*, 330, 74–89.

Miall, A. D. (1996). *The Geology of Fluvial Deposits: Sedimentary Facies, Basin Analysis, and Petroleum Geology.* Springer, Heidelberg, 582 pp.

Minchin, J. B. (1881). Eastern Bolivia and the Gran Chaco. *Proceedings of the Royal Geographical Society and Monthly Record of Geography, New Monthly Series*, 3, 401–420.

Morozova, G. and Smith, N. D. (2000). Holocene avulsion styles and sedimentation patterns of the Saskatchewan River, Cumberland Marshes, Canada. *Sedimentary Geology*, 130, 81–105.

Müller, R., Müller, D., Schierhorn, F., and Gerold, G. (2011). Spatiotemporal modeling of the expansion of mechanized agriculture in the Bolivian lowland forests. *Applied Geography*, 31, 631–640.

Navarro, G. and Maldonado, M. (2002). *Geografía Ecológica de Bolivia: Vegetación y Ambientes Acuaticos, Santa Cruz*: Centro de Ecología Simón I. Patiño.

Perez-Arlucea, M. and Smith, N. D. (1999). Depositional patterns following the 1870s avulsion of the Saskatchewan River (Cumberland Marshes, Saskatchewan, Canada). *Journal of Sedimentary Research*, 69, 62–73.

Plotzki, A., May, J.-H., Preusser, F., and Veit, H. (2013). Geomorphological and sedimentary evidence for late Pleistocene to Holocene hydrological change along the Río Mamoré, Bolivian Amazon. *Journal of South American Earth Sciences*, 47, 230–242.

Plotzki, A., May, J. H., Preusser, F., et al. (2015). Geomorphology and evolution of the late Pleistocene to Holocene fluvial system in the south-eastern Llanos de Moxos, Bolivian Amazon. *Catena*, 127, 102–115.

Purinton, B. and Bookhagen, B. (2018). Measuring decadal vertical land-level changes from SRTM-C (2000) and TanDEM-X (~2015) in the south-central Andes. *Earth Surface Dynamics*, 6, 971.

Rafiqpoor, D., Nowicki, C., Villarpardo, R., et al. (2004). Climate: the abiotic factor that most influences the distribution of biodiversity. In P. L. Ibisch and G. Mérida, eds., *Biodiversity: The Richness of Bolivia*. Santa Cruz: Editorial FAN, 31–46.

Richards, K., Chandra, S., and Friend, P. (1993). Avulsive channel systems: characteristics and examples. In J. L. Best and C. W. Bristow, eds., *Braided Rivers. Geological Society of London, Special Publication*, 75, 195–203.

Riveros, F. (2004). *The Gran Chaco* [Online]. FAO; Agricultural Department, Crop and Grassland Service. Available: www.fao.org/ag/AGP/AGPC/doc/Bulletin/GranChaco.htm [Accessed 15 January 2007].

Roca, O. (2016). Cuatro Ojos - El Histórico Puerto de Ingreso a la Amazonia. Available from: https://ovidioroca.wordpress.com/ [Accessed 21 February 2018].

Roca Salazar, R., Fernádez Ríos, D., and Gutiérrez Guillén, R. (2003). Sistema de Alerta Temprana contra las crecidas del Río Grande (SALTEM-RG). *Gestión del Riesgo Prevención, Mitigación, Preparación*. Bolivia. Federación de Asociaciones Municipales (FAM); Bolivia. Asociación de Municipios de Santa Cruz (AMDECRUZ); Alemania. Cooperación Técnica Alemana (GTZ). Proyecto Gestión Interinstitucional de Riesgos.

Röhringer, I. (2006). *Holozäne Flussdynamik und Auensedimentation des Río Grande (Ostbolivien)*. Unpubl. Diploma Thesis, TU Dresden.

Santillan, J. and Makinano-Santillan, M. (2016). Vertical Accuracy Assessment of 30-M Resolution Alos, Aster, and SRTM Global DEMs Over Northeastern Mindanao, Philippines. *International Archives of the Photogrammetry, Remote Sensing & Spatial Information Sciences*, 41, 149–156.

Schneider, U., Becker, A., Finger, P., et al. (2016). GPCC Full Data Reanalysis Version 7.0: Monthly Land-Surface Precipitation from Rain Gauges built on GTS based and Historic Data. In Research Data Archive at the National Center for Atmospheric Research - Computational and Information Systems Laboratory.

Servant, M., Fontes, J.-C., Rieu, M., and Saliège, J.-F. (1981). Phases climatiques arides holocènes dans le sud-ouest de l'Amazonie (Bolivie). *Comptes Rendus de l'Académie des Sciences, Paris, Série II*, 292, 1295–1297.

Shukla, U. K., Singh, I. B., Sharma, M., and Sharma, S. (2001). A model of alluvial megafan sedimentation: Ganga Megafan. *Sedimentary Geology*, 144, 243–262.

Sinha, R., Ahmad, J., Gaurav, K., and Morin, G. (2014). Shallow subsurface stratigraphy and alluvial architecture of the Kosi and Gandak megafans in the Himalayan foreland basin, India. *Sedimentary Geology*, 301, 133–149.

Slingerland, R. L. and Smith, N. D. (2004). River avulsions and their deposits. *Annual Review of Earth and Planetary Sciences*, 32, 257–285.

Smith, N. D., Cross, T. A., Dufficiy, J. P., and Clough, S. R. (1989). Anatomy of an avulsion. *Sedimentology*, 36, 1–23.

Steininger, M. K., Tucker, C. J., Townshend, J. R. G., et al. (2001). Tropical deforestation in the Bolivian Amazon. *Environmental Conservation*, 28, 127–134.

Stouthamer, E. (2001). Sedimentary products of avulsions in the Holocene Rhine–Meuse Delta, The Netherlands. *Sedimentary Geology*, 145, 73–92.

Stouthamer, E. and Berendsen, H. J. A. (2001). Avulsion frequency, avulsion duration, and interavulsion period of Holocene channel belts in the Rhine-Meuse Delta, The Netherlands. *Journal of Sedimentary Research*, 71, 589–598.

Stouthamer, E. and Berendsen, H. J. A. (2007). Avulsion: the relative roles of autogenic and allogenic processes. *Sedimentary Geology*, 198, 309–325.

Syvitski, J. P. M., Overeem, I., Brakenridge, G. R., and Hannon, M. (2012). Floods, floodplains, delta plains —a satellite imaging approach. *Sedimentary Geology*, 267–268, 1–14.

Tadono, T., Ishida, H., Oda, F., et al. (2014). Precise global DEM generation by ALOS PRISM. *ISPRS Annals of the Photogrammetry, Remote Sensing and Spatial Information Sciences*, 2, 71–76.

Uba, C. E., Strecker, M. R., and Schmitt, A. K. (2007). Increased sediment accumulation rates and climatic forcing in the central Andes during the late Miocene. *Geology*, 35, 979–982.

Uba, E. C., Heubeck, C., and Hulka, C. (2006). Evolution of the late Cenozoic Chaco foreland basin, Southern Bolivia. *Basin Research*, 18, 145–170.

Vera, C., Higgins, W., Amador, J., et al. (2006). Toward a unified view of the American monsoon systems. *Journal of Climate*, 19, 4977–5000.

Vicente-Serrano, S. M., El Kenawy, A., Azorin-Molina, C., et al. (2016). Average monthly and annual climate maps for Bolivia. *Journal of Maps*, 12, 295–310.

Wachholtz, R. and Herold-Mergl, A. (2003). *Amenaza y vulnerabilidad por cambio de cauce e inundación en la cuenca baja del Río Grande*. GTZ, C. T. A., Santa Cruz, La Paz.

Weissmann, G., Hartley, A., Scuderi, L., et al. (2013). Prograding distributive fluvial systems: geomorphic models and ancient examples. In S. G. Driese and L. C. Nordt, eds., *New Frontiers in Paleopedology and Terrestrial Paleoclimatology. SEPM Special Publication*, 104, 131–147.

Weissmann, G. S., Hartley, A. J., Nichols, G. J., et al. (2010). Fluvial form in modern continental sedimentary basins: distributive fluvial systems. *Geology*, 38, 39–42.

Wells, N. A. and Dorr, J. A. J. (1987). Shifting of the Kosi River, northern India. *Geology*, 15, 204–207.

Werding, L. (1977a). Geomorphologie und rezente Sedimentation im Chaco Boreal, Bolivien. *Giessener Geologische Schriften*, 12, 429–446.

Werding, L. (1977b). The Rio Grande ground-water basin, Chaco Boreal, Bolivia. *Geologisches Jahrbuch, C* 17, 19–36.

Wilkinson, M. J., Marshall, L. G., and Lundberg, J. G. (2006). River behavior on megafans and potential influences on diversification and distribution of aquatic organisms. *Journal of South American Earth Sciences*, 21, 151–172.

Zani, H., Assine, M. L., and McGlue, M. M. (2012). Remote sensing analysis of depositional landforms in alluvial settings: Method development and application to the Taquari megafan, Pantanal (Brazil). *Geomorphology*, 161–162, 82–92.

Zhou, J. and Lau, K.-M. (1998). Does a monsoon climate exist over South America? *Journal of Climate*, 11, 1020–1040.

8

Megafans of Southern and Central Europe

ALESSANDRO FONTANA and PAOLO MOZZI

Abstract

Fluvial megafans are uncommon in Europe but a few are recognisable in the Alps and Carpathians foreland zones. Along the southern Alps mountain front, megafans are present from Milan (central Po Plain) to the Venetian–Friulian Plain (Olona, Oglio, Adige, Brenta, Piave and Tagliamento rivers). These systems recorded a strong depositional phase during the Last Glacial Maximum (LGM, 29–19 ka BP), functioning as glacial outwash systems to the larger Alpine glaciers, largely followed by sediment starvation. These Alpine megafans are thus climate-controlled relics of the last glaciation. Megafans also occur on the Little Hungarian Plain (Danube River megafan near Bratislava and Rába River megafan, mainly fed from the Alps), and the Great Hungarian Plain (Maros, Szamos, and Drina megafans). The largest (Maros) consists of two lobes covering an area > 7,000 km². On the Danube, activity was continuous until recent time, whereas on the Maros and Szamos the main aggradational phase occurred during the Lateglacial. Compared to their Alpine counterparts, megafans on the Great Hungarian Plain are fed by larger catchments but these were not severely glaciated during the LGM. They recorded important depositional phases during the humid periods of the Lateglacial.

8.1 Introduction

Fluvial megafans in Europe are generally smaller than those described on other continents, but their limited extent and easier accessibility have enabled intensive research. As a result, for some European megafans very robust chronologies and detailed subsurface information are available.

Modern Quaternary geology began in Europe at the end of the nineteenth century, when pioneers started to investigate the causes and mechanisms of past glaciations (e.g., Agassiz 1840; de Charpentier 1841). However, until a few decades ago research mainly focused on the mountain catchments, with limited attention paid to the geomorphology or stratigraphy of the related alluvial basins. Even in recent years, the terms 'alluvial fan' or 'cone' have been commonly used in the scientific literature and among geological maps of Europe for describing cone-shaped fluvial landforms that extend for thousands of km², and that are characterised by downstream trends in sediment texture and by the presence of a well-defined distal sector dominated by fine-grained sediments (e.g., Ori 1982; Gábris 1994; Castiglioni 1997; Marchetti 2001; Gábris and Nàdor 2007; Kiss et al. 2014; Perşoiu and Rădoane 2017a). Such large depositional landforms only began to be recognised in Europe as alluvial megafans about a decade ago (e.g., Mozzi 2005; Fontana et al. 2008), with scholars beginning to apply concepts and terminology mainly derived from seminal studies carried out in the Himalayan and Andean foreland plains (e.g., Gohain and Parkash 1990; Horton and DeCelles 2001). A further problem for the recognition of large landforms among the alluvial plains of Europe has often been pervasive anthropogenic activity which, since the middle Holocene, has altered the natural landscape and surface topography through ditching and dyking for agriculture, irrigation, land reclamation and the development of artificial waterways, and through modern urban sprawl.

In the alluvial plains of Central Europe, the recognition of large landforms was also partly limited by the political separations of the twentieth century, which in some sectors of the Little and Great Hungarian plains created administrative boundaries unrelated to the physiographic setting. This restricted research cooperation between different countries.

Here we review the late Quaternary alluvial megafans so far recognised and described in Europe, analysing specific literature and interpreting previous information in the light of new geomorphological and stratigraphic data. The areas in Europe where megafans could have formed in late Quaternary times are relatively scarce because of specific tectonic and geomorphic factors that control the formation of these alluvial landforms. In fact, mountain ranges facing vast alluvial plains, where extensive sedimentation can take place, exist only at the foot of the southern Alps (Po Plain and Venetian–Friulian Plain) and along the Carpathian Mountains (Little and Great Hungarian plains, and the Valachian plains of the Getic Basin in southern Romania) (Fig. 8.1).

Unfortunately, it was not possible to include the Getic Basin in this review, as very little specific literature is available on the large fan-shaped landforms that occur along the Southern Carpathians. These landforms apparently consist of different patches of middle and late Pleistocene alluvial and aeolian formations (Howard et al. 2004), deformed by tectonics (Maţenco 2017). Thus, a specific study would be necessary to properly recognise and characterise the potential megafans in the area.

Likewise, megafans that are still detectable in the physical landscape but dissected and thus no longer active, or that are recognisable only from stratigraphic evidence, are not considered here. Such is the case for the Lannemezan megafan in the Aquitaine Basin (northern Pyrenees; Mouchené et al. 2017), the Loire megafan (Gunnell, Ch. 9), and the remnants of the Lower and Middle Pleistocene megafans in the so-called Danube–Tisza interfluve and Nyírség areas of the Hungarian Plain (e.g., Borsy 1990; Gábris and Nádor 2007).

Here we present the late Quaternary megafans and discuss the relationships existing with their catchments, the tectonic setting and the climatic evolution, trying to highlight common factors and local characteristics. The Po and the Venetian–Friulian plains, in particular, are directly connected to the Adriatic Sea and allow the influence exerted on the development of megafans by mountain glaciation and glacio-eustatic base-level variations to be investigated.

8.2 Methods

We use the term megafan to describe alluvial landforms displaying a fan-shaped planform and an areal extent generally much larger than $1,000 \text{ km}^2$. A key factor for their classification is the occurrence of a well-defined distal sector of gentle topographic gradient dominated by fine sediments. These characteristics are satisfied also by several smaller fan-shaped landforms $(200–1,000 \text{ km}^2)$, as documented, for example, along the southern Alps for the Cormor, Chiese, and Mincio rivers, or along the southern Cordillera Cantabrica in Spain for the Órbigo and Carrión rivers. This situation raises problems regarding their classification as fans, however, especially if compared with steep, short (hundreds of metres to a few kilometres in radius), piedmont alluvial fans dominated by hyperconcentrated flow and unconfined runoff processes (Fontana et al. 2008, 2014; Wilkinson, Ch. 17).

In addition to papers dealing with the specific topic of the megafans and large alluvial fans in some regions of the continent (e.g., Borsy 1990; Gabris and Nàdor 2007; Fontana et al. 2008, 2014; Mezősi 2016; Perşoiu and Rădoane 2017a, 2017b) we have included older publications and some grey literature. Important tools in this analysis have been geological, geomorphological, and soil maps from different countries and regions, with highly variable scales and dates of production. It is worth noting that the analyses of reports and maps have been partly limited by difficulties in accessing some local literature and maps produced in different languages.

The landforms defined as fluvial megafans have been recognised on the basis of geomorphological, pedological, and stratigraphic data. Surface topography and morphometry (Table 8.1) are based on Digital Elevation Models (DEMs), analysis of hillshading, and the major variables of topographic slope, aspect, and curvature. A main source for producing the DEMs was the Shuttle Radar Topography

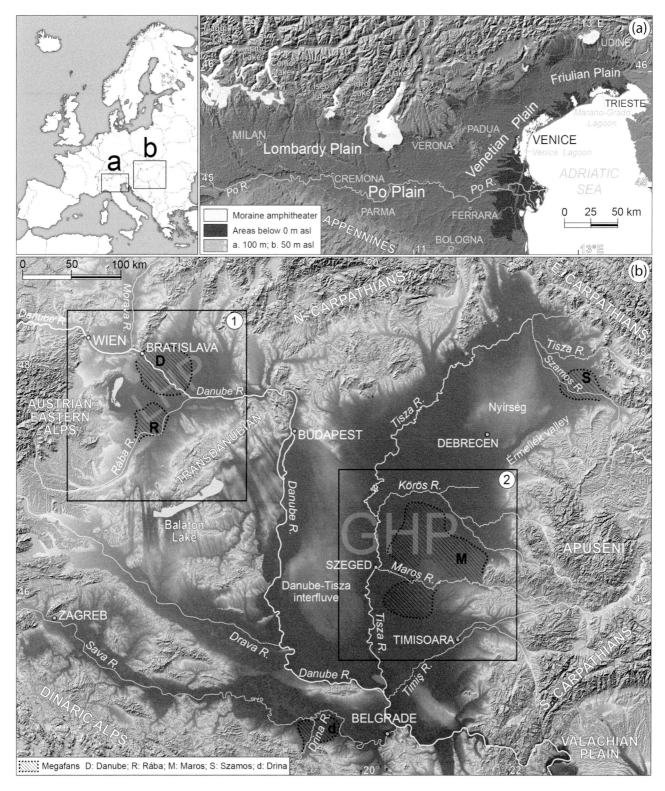

Figure 8.1 **Digital elevation model of northern Italy (a) and of the Carpathian Basin (b)**. GHP: Great Hungaria Plain; LHP: Little Hungarian Plain. Digital elevation source: Shuttle Radar Topographic Mission (SRTM).

Table 8.1 *Morphometric characteristics of the European alluvial megafans and main hydrographic characteristics of their related mountain catchments.*

Alluvial megafan name	Fan area (km²)	Length of longitudinal axis (km)	Length of perpendicular axis (km)	Mountain catchment area (km²)	Mountain range	Length of river (km)	Elevation of present channel at valley mouth (m)	Elevation of higher terrace at valley mouth (m)	Elevation of lower boundary of megafan (m)	Slope of LGM surface in apical sector (‰)	Slope of LGM surface in distal sector (‰)	Mean annual discharge (m³/s)	Flood peak discharge (m³/s)
Tagliamento	1,200	65	35	2,580	South-eastern Alps	172	130	200	0	7.0-4.0	2.0-0.8	109	4,500
Piave	1,050	55	32	3,899	South-eastern Alps	222	77	84	0	5.0-3.0	1.5-0.6	132	4,250
Brenta	2,600	75	50	1,787	South-eastern Alps	160	100	130	0	6.0-4.0	2.0-0.6	71	2,810
Adige	1,600	60	35	11,954	South-eastern Alps	410	80	125	0	5.0-3.0	3.0-0.6	220	4,000
Oglio	1,000	80	20	5,682(1,842)	South-central Alps	280	160	210	40	5.5-4.0	2.0-1.0	59	410
Olona	1,020	70	20	142	South-central Alps	71	257	290	90	6.2-4.0	3.0-1.5	7	n.d.
Danube North	2,400	70	45	131,300	Alps and Sudetes	709/2,259	133	138	110	1.1	0.4	2,047	10,300
Raba	1,000	75	50	14,968	Austrian eastern Alps	398	134	140	110	1.1	0.7	18–80	800
Maros-North	5,500	110	60	30,332	Apuseni/Carpathians	761	115	120	80	0.8	0.3	184	2,450
Maros-South	1,500	45	35	30,332	Apuseni/Carpathians	761	92	98	75	0.4	0.3	184	2,450
Szamos	2,000	50	45	15,282	Carpathians Est	415	135	154	120	0.7	0.3	120	2,360
Drina	1,000	45	40	19,570	Dinaric Alps	375	115	120	80	1.1	0.3	370	4,000

Mission (SRTM; Rabus et al. 2003), which has a ground resolution of 1 arcsec (i.e., ~30 m) in all areas relevant to this study. The topography of northern Italy is also described by other DEMs derived from scanned topographic maps at scales 1:5,000–1:10,000 (e.g., Tarquini et al. 2007; Fontana et al. 2014). In the Venetian–Friulian Plain and in the western sector of the Po Plain, LiDAR-derived information is also available, with a ground resolution of 1 m^2 and a vertical accuracy of ±0.15 m (http://wms.pcn .minambiente.it/).

Remote sensing was an important tool for describing the remnants of channel belts visible on megafan surfaces. This analysis was mainly achieved from colour-composite satellite images available in Google Earth (www.google.com/earth) and Bing Maps (www .bing.com/maps). For Italy, digital colour ortho-imagery was also used.

Compared with other areas in the world, several regions of Europe where megafans have been recognised have many subsurface datasets available, covering depths down to tens of metres below the surface. This situation is common in northern Italy and in some sectors of the Carpathian Basin (also named Pannonian Basin), where stratigraphic data have been collected since the end of the nineteenth century for hydrogeological, geotechnical, and hydrocarbon investigations. These data are particularly abundant in the Venetian–Friulian and Po plains, where they provide constraints on the identification and characterisation of formative phases of the megafans and other fans since the end of the Middle Pleistocene. Moreover, in Europe a large number of geochronological constraints have been acquired for several megafans. By virtue of relatively humid climatic conditions, organic remains and peat layers in some European regions are rather abundant in the stratigraphic record, including from colder periods. This has allowed numerous radiocarbon dates to be recovered, particularly in northern Italy but also in some regions of the Carpathian Basin, thus leading to a robust and high-resolution chronology for the late Quaternary. In the Great Hungarian Plain, most of the chronology is based on luminescence dating. Results obtained are generally accurate, whereas in northern Italy this technique has been affected by certain methodological problems (e.g., Lowick et al. 2010).

Numerical ages are calibrated and expressed in thousands of years before present (cal ka BP). The terms Last Glacial Maximum (LGM) and Late Glacial are used in their global definitions, with the LGM framed here broadly between 29 and 19 ka BP (Clark et al. 2009) and the Late Glacial (19–11.7 cal ka BP) starting at the end of the LGM and ending with the beginning of the Holocene. In a number of publications, the duration of the LGM and of the Late Glacial are sometimes not precisely defined or have variable meanings. In the Hungarian literature describing the Carpathian Basin, for example, the Late Glacial is often described only as the interval from 14.7 to 11.7 cal ka BP, whereas the earlier interval (26.5–14.7 cal ka BP) in that context is referred to as the late Pleniglacial period (cf. Feurdean et al. 2014).

8.3 Results

8.3.1 Megafans along the Southern Alps

In northern Italy a vast alluvial plain extends between the Alps and the Apennines (Fig. 8.1). The central and western portions of this basin, the Po Plain, are crossed by the Po River and its tributaries. The north-eastern sector is the Venetian–Friulian Plain (hereafter VFP), which pertains to a set of Alpine rivers that flow directly into the Adriatic Sea – the largest being the Adige, Brenta, Piave, Tagliamento, and Isonzo (Fig. 8.2). The central sector of the Alpine side of the Po Plain is hereafter named Lombardy Plain (LP).

This geographic setting is strictly related to the geodynamic evolution of the area, driven by the collision between the Eurasian tectonic plate and the Adriatic micro-plate, a northern extension of the African Plate. The Po and Venetian–Friulian plains function as foreland basins to both the Alps and the Apennines in relation to the accretion of the south-verging south-Alpine backthrusts and the north-verging Apennine thrusts (Cuffaro et al. 2010). The sedimentary basin is further delimited to the east by the Dinaric Alps. As a result of crustal flexuring and subduction of the Adriatic plate, a 4–8-km-thick succession of marine to alluvial sediments of Pliocene and Quaternary age was deposited along the Apennines piedmont, progressively thinning northwards to a few hundreds of metres along

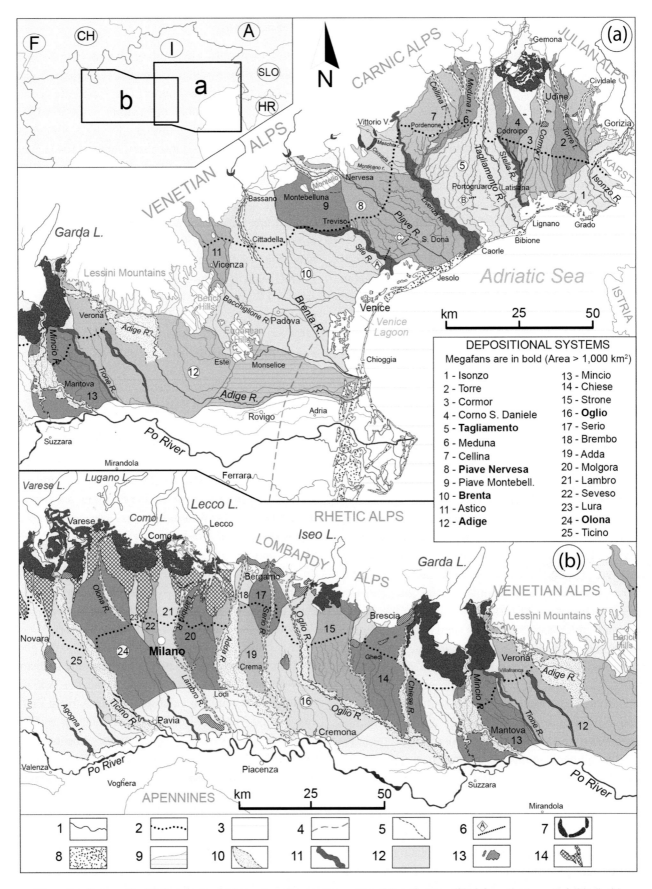

Figure 8.2 **Large alluvial fans and megafans along the southern side of the Alps** (modified from Fontana et al. 2014). (1) river courses, (2) upper limit of the spring belt, (3) sea and lagoons, (4) inner limit of Holocene lagoon deposits, (5) fluvial terrace riser, (6) line of stratigraphic sections in Figure 8.3, (7) terminal moraine, (8) Holocene coastal deposits, (9) alluvial plain of Adige River, (10) post-LGM fluvial incision, (11) alluvial unit related to major groundwater-fed rivers, (12) undifferentiated LGM alluvial deposits, (13) tectonic terrace, (14) middle and early Pleistocene alluvial deposits.

the southern Alps (Muttoni et al. 2003; Ghielmi et al. 2010; Toscani et al. 2016).

In northern Italy, the climate is temperate with hot summers in the alluvial plains, cooler in the hills, and cold in the mountain valleys (Pinna 1982). In the Alps, the average precipitation is generally > 900 mm/yr, with an autumn peak of rainfall. In the central sector of the southern Alps, near Lake Maggiore, the maximum average annual precipitation is about 2,000–2,500 mm/yr, but it reaches 3,500 mm/y in the Julian Alps, at the boundary between Slovenia and Italy.

The existence of large fans along the southern margin of the Alps in the LP and VFP has been reported since the end of the nineteenth century (e.g., Stella 1895; Feruglio 1925; Castiglioni 1940), with reference to predominantly gravelly alluvial bodies developed in the piedmont area (Marchetti 1996, 2001; Castiglioni 1997; Guzzetti et al. 1997). In the last few decades, the continuation of these alluvial distributary systems well downstream of their gravel-dominated proximal sectors has been recognised from evidence provided by improved topographic data and geological surveys. These landforms have been lately classified as fluvial megafans based on the following characteristics (Mozzi et al. 2003; Fontana et al. 2008, 2014): (i) the continuous downstream transition of sedimentary facies from the gravelly apical portions to the fines-dominated distal sectors; (ii) the overall fan-shaped geometry of the alluvial landforms, with slopes gradually decreasing from a maximum of 10–7‰ at the apex to less than 0.5‰ in the distal fringe; (iii) the large dimensions, with areal extent up to 2,600 km^2 and longitudinal axes up to 80 km. In previous papers, the alluvial depositional systems satisfying these conditions were described as megafans, without considering a minimum size; here this term is applied only to those greater than ~ 1,000 km^2.

Along the continuous belt of megafans and fans which extends for about 400 km from east to west (Fontana et al. 2014), the megafans of the LP often merge with one another in the distal sector and merge with the LGM channel belts of the Po River (Ravazzi et al. 2012), whereas those of the VFP extend to the coastal area. Large portions of each megafan were abandoned at the end of the LGM and are now characterised by well-developed soils. The soil profiles are marked by argillic B horizons in the gravelly apical

sectors, while calcic horizons with carbonate concretions and gley features are common in the fines-dominated distal portions, where the groundwater table lies within 2–3 m of the surface.

The apical portions of the megafans generally consist of amalgamated gravel bodies deposited by braided river channels. Stratigraphic sections exposed along fluvial terrace risers of the main Alpine rivers and in gravel pits show a predominance of crudely bedded gravel and secondary sandy gravel, whose stratal geometries suggest deposition in longitudinal bars typical of braided channels (e.g., Paiero and Monegato 2003; Bersezio et al. 2007). Surface gradients (Table 8.1) rapidly drop from 4–3‰ to 2–1‰, corresponding to a marked lithological shift in channel deposits from gravel, to gravelly sand, to sand. This is caused by a decrease in transport capacity. A large, fines-dominated distal sector characterises megafans of the VFP. Here, fluvial channel belts with average widths of 200–800 m are normally present, bordered by a population of natural levees of silty sand that rise 0.5–2 m above the floodplain. Channel bodies consist of fine to medium sand, generally 1–2 m thick and 40–250 m wide. Longitudinal bars, scour-and-fill features and reactivation surfaces are the most common sedimentary structures observed among the channel deposits. These features and the surface traces observed on remote imagery point to the activity of sand-dominated braided channels that shift to a wandering and slightly sinuous style over distal fan sectors.

In addition to the major Alpine river channels, the distal sectors of megafans in northern Italy display dense networks of groundwater-fed streams. The spring-generated run-off channels rise at the downstream boundary of the gravel-rich apical sector of the megafans. Here, part of the groundwater flowing through gravel-dominated sediment is forced to the surface by the occurrence of silt- and clay-rich impermeable beds. In modern times, the number of springs has been reduced by engineering operations and the groundwater-fed rivers have been canalised (Bondesan 2001; Fontana et al. 2014).

In the distal part of the VFP megafans, overbank clays and silts are generally deposited in the inter-ridge areas, and are often interlayered with organic-rich horizons. These organic accumulations mainly formed

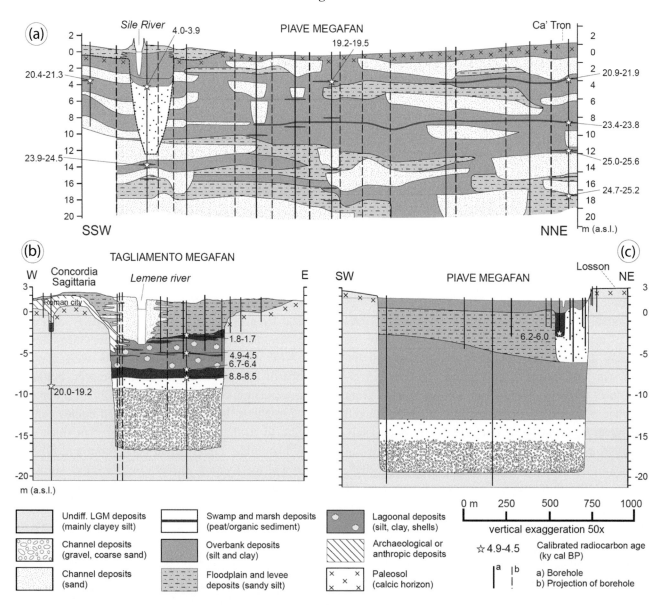

Figure 8.3 **Selected stratigraphic sections from the megafans of the VFP**. Section lines are indicated as (a), (b) and (c) in Figure 8.2a. (a) LGM deposits of the Piave megafan (modified from Miola et al. 2006). (b) Section through the cut and refilled Tagliamento valley near the city of Concordia Sagittaria (modified from Fontana 2006). (c) Section through the cut and refilled valley near Losson, on the Piave megafan (modified from Carton et al. 2009).

during the LGM in areas that were temporarily inactive, when fluvial deposition shifted to more distant megafan sectors (Miola et al. 2006; Fig. 8.3). The abundance of organic deposits has allowed extensive radiocarbon dating (Rossato et al. 2015; Rossato and Mozzi 2016; Hippe et al. 2018) and derived age–depth models show that the main phase of megafan aggradation took place during the LGM, at a time when outlet glaciers had attained the mouths of Alpine valleys (Fig. 8.4). The development of the sedimentary sequences was primarily controlled by changes in the

ratio between sediment supply and water discharge, in response to glacial outwash from the Alps. Basin subsidence was sufficient to allow burial and preservation of successive aggradation events. VFP megafans were able to prograde unrestricted onto the exposed continental shelf of the Adriatic Sea, and thus attained dimensions larger than those of the LP, whose outwards growth was impeded by the axial Po River.

Dates from selected cores from the Tagliamento, Piave and Brenta megafans are plotted with respect to depth in Figure 8.4. The data reveal the strong

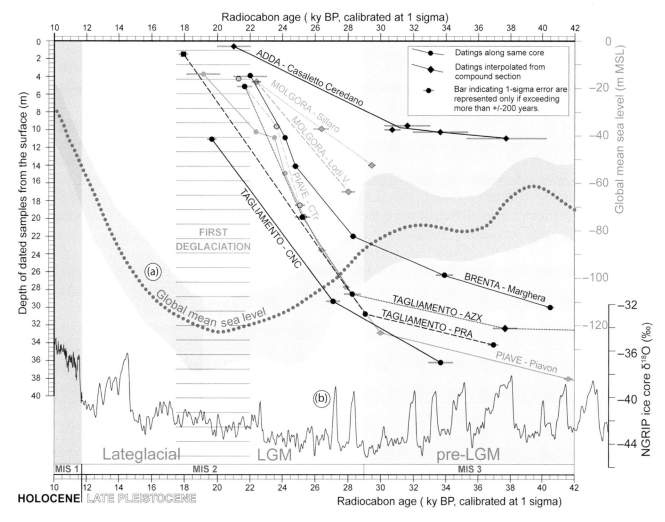

Figure 8.4 **Radiocarbon-dated age–depth profiles for different megafans and fans of northern Italy** (modified from Fontana et al. 2014). Core data sources: CNC, PRA, Piavon, and Marghera from Fontana et al. (2014); AZX from Pini et al. (2009); PIAVE–CTr from Miola et al. (2006). Sections near Lodi Vecchio and Cavo Sillaro after Bersezio et al. (2004) and Baio et al. (2004); Casaletto Ceredano section after Ravazzi et al. (2012). Palaeoclimate of the Northern Hemisphere represented by $\delta^{18}O$ curve (a) from NGRIP ice core (NGRIP Members 2004; Kindler et al. 2014). Curve (b) reports the sea-level changes according to Lambeck et al. (2014).

aggradation of about 20–30 m that occurred in these alluvial systems between ca. 30 and 19–18 cal ka BP, simultaneously with the Alpine glacial expansion (Monegato et al. 2017). The period of maximum aggradation on the plains seems to have generally occurred between 26–22 cal ka BP and was characterised by average depositional rates of up to 3 mm/yr. The Brenta megafan presents marked variations in sedimentation rates (1.8 mm/yr between 40 and 26.7 cal ka BP; 3 mm/yr between 26.7 and 23.8 cal ka BP and 1.4 mm/yr from 23.8 to 17.5 cal ka BP), suggesting a strict relation in phase with fluctuations of the terminal glacial front: highest sedimentation rates

occurred when the terminal glacial front was at its outermost position (Rossato and Mozzi 2016).

The megafans of the LP seem to have experienced the same phases of aggradation displayed as those of the VFP, but the total accumulated thickness during the LGM is significantly lower. On the Chiese fan, near Ghedi (Fig. 8.2), the present surface formed at the end of the LGM and the upper 12 m of subsurface materials consists of gravel overlying an organic horizon, radiocarbon-dated to 28–30 cal ka BP (Garzanti et al. 2011). In the distal sector of the LP near Lodi, several stratigraphic sections obtained through the interpolation of long cores highlight the fact that the uppermost 15 m of sediment were accumulated after 29 cal ka BP, and that the last depositional phase occurred after 21 cal ka BP (Bersezio et al. 2004). At the zone where the Oglio megafan and the large fan of the Adda merge near Crema, the total thickness of the LGM deposits is less than 15 m (Ravazzi et al. 2012).

As described in detail in the discussion, in some cases the streams currently flowing across the mega-fans are supplied by a mountain catchment of much smaller size than their depositional system. This under-fitting anomaly is mainly related to the role of glaciers, which overfilled some catchments overpassing the boundaries of the watersheds and also supplying the adjoining basins. Between 19 and 17.5 cal ka BP, the glaciers had already withdrawn within the valleys, and almost all the Alpine rivers became entrenched in their own megafans (Fontana et al. 2014; Monegato et al. 2017). In the LP, Alpine rivers are still currently confined to these entrenched valleys, and flow from the

fan apex to their respective junctions with the Po River (Figs. 8.2 and 8.4). In the valleys feeding the LP, large lakes such as Garda, Iseo, Como, Lugano, and Maggiore commonly formed in the deep depressions vacated by the ice, just upstream of the terminal moraine amphitheatres. These basins have acted as efficient sediment traps and still prevent transport of coarse material to the plain. In the south-eastern Alps, east of Lake Garda, only some temporary lakes were formed. They were later either completely filled by sediments, or their natural dams were breached.

The megafans of the VFP have thus recorded a complex post-LGM evolution. From the Adige to the Isonzo, all the Alpine rivers presently flow down a single incised channel entrenched in the apical zone of the LGM fan, but each of these rivers generated post-LGM depositional lobes in the distal sector (Fig. 8.5). These lobes have since responded dynamic-ally to Holocene climatic phases and relative sea-level rise. Since about 7.5 ka BP, at the onset of the marine

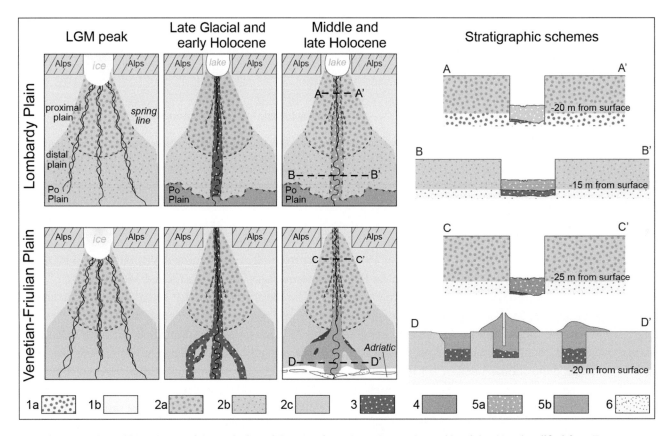

Figure 8.5 **Simplified scheme of the evolution of the megafans along the southern side of the Alps** (modified from Fontana et al. 2014). Key to ornaments: (1a) pre-LGM gravels and sands, (1b) pre-LGM fine deposits, (2a) LGM gravels and sands, (2b) LGM sands, (2c) LGM fine deposits, (3) Late Glacial and early Holocene gravels and sands, (4) Po alluvial plain (sand and silt), (5a) middle and late Holocene gravels, (5b) middle and early Holocene fine deposits, (6) Holocene coastal sands.

highstand when relative sea level was between 10 and 5 m lower than present, the Adriatic coastline was located at a position fairly comparable to the present one north of the Po Delta (Amorosi et al. 2008). This setting triggered the formation of deltas and lagoons, which built a coastal wedge that lapped onto the LGM plain. Megafans that are directly connected to the sea have experienced progressive gradient lowering in their distal sectors because of post-glacial sea-level rise, and consequently the evolution of these sectors is no longer driven exclusively by alluvial processes. Such is the case for the terminal tract of the Adige

megafan, which transitions to an almost flat alluvial surface connected to the deltaic plain of the Po River (Piovan et al. 2012). In contrast, the late Holocene evolution of the distal sectors of the Isonzo, Tagliamento, Piave, and Brenta megafans has been generally driven by avulsions occurring upstream of the coastal plain. During the last millennia, such avulsions have also been influenced by sea-level rise and by the local establishment of lagoon environments (Fig. 8.5).

The Tagliamento megafan provides a good example of a post-LGM lobe (Fig. 8.6). Channel belts show a

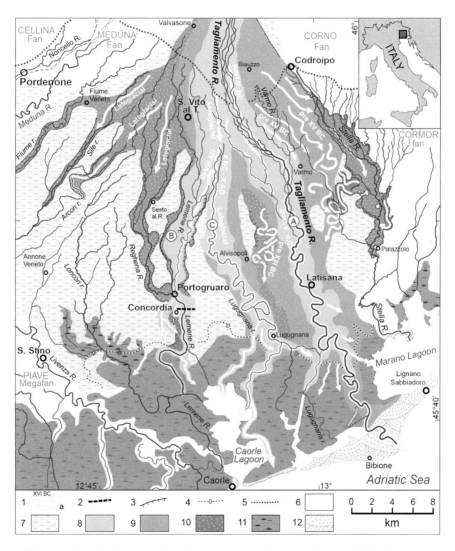

Figure 8.6 **Post-LGM lobe of the Tagliamento alluvial megafan** (modified from Fontana 2006). Key to ornaments: (1) channel belt, with indication of the period of activity (Roman numbers indicate centuries), (1a) buried channel belt, (2) trace of stratigraphic section in Figure 8.3a, (3) terrace riser, (4) 0 m a.s.l. contour, (5) upper limit of the spring belt, (6) deposits of other fluvial systems, (7) LGM alluvial deposits, (8) undifferentiated post-LGM alluvial deposits, (8) Late Glacial incised valleys, subsequently re-occupied by groundwater-fed streams, (9) Late Holocene channel belts with formation of fluvial ridges (A: Latisana unit, B: Concordia unit, C: *Tiliaventum Maius* unit), (10) Holocene lagoon deposits.

divergent pattern from an avulsive node situated at the spring belt, where gradients abruptly change from 3 to 1‰. The oldest channel belts, which formed between the Late Glacial and early Holocene, are deeply incised into the LGM fan surface, whereas the late Holocene channel belts aggraded by a few metres on the plain. In the apical portion of the megafan near the foothills, the Tagliamento still flows in a single valley which has undergone 70 m of post-glacial incision at the fan head, but the incision depth decreases progressively to near zero at the spring belt (Paiero and Monegato 2003; Fontana et al. 2008). Downstream from this area, in the fines-dominated sector, the river has carved several separate valleys, some of which are still morphologically evident (Figs. 8.5 and 8.6); others, however, have been buried by younger depositional units (Fontana et al. 2014). Incised valleys display widths spanning 500–2,000 m, their depths reach 2–4 m near the spring belt, and increase downstream to 15–20 m, over a distance of 20 km, near the lagoon fringes (Fig. 8.3b). Downstream from this zone, valley depths decrease and no evidence of significant fluvial incision is documented seawards of the present coast.

During Late Glacial times, because of the funnelling effect related to the concentration of sediment along the incised fluvial channels, gravels have been transported towards the distal plain. The coarse sediments arrived near the present coast and, thus, about 25 km further downstream than during the LGM. Late Glacial and early Holocene gravels are present near Concordia Sagittaria, south of Portogruaro, where the valley incised in the LGM deposits which formed after 19 cal ka BP was already partly filled by gravels at 14 cal ka BP, and subsequently abandoned by the Tagliamento at 8.5 cal ka BP (Fig. 8.3b; Fontana 2006).

Since about 7.5 cal ka BP the marine transgression reached the present coastline and the abandoned incised valleys were invaded by brackish waters, leading to lagoon deposition of silt and clay (Fontana et al. 2008). Occasionally, the Tagliamento River reoccupied the valley of Concordia Sagittaria and, together with lagoon sediments, progressively brought to completely fill the fluvial incision with silt and sand during late Holocene (Fontana 2006). Since the middle Holocene, following sea-level rise the downstream limit of gravel deposition in the Tagliamento channels

migrated upstream. It currently lies 20 km from the coast near the city of Latisana, where tidal influence stops. In the last 4 ka, the river formed fluvial ridges along its active channels, consisting of population of natural levees (Fig. 8.6). These convex landforms attain widths of 1–2 km and elevations of 2–4 m above the surrounding floodplain.

The Piave and Brenta megafans are also characterised by post-LGM lobes, and their evolution and chronology are similar to the one described for the Tagliamento megafan. Some incised valleys have been detected in the distal sector of the Piave megafan near San Donà (Fig. 8.3c; Carton et al. 2009), while others are conspicuous on the Brenta megafan between Cittadella and Padova (Mozzi et al. 2013).

The dramatic contrast between the strong and widespread aggradation that occurred during the LGM and the limited sedimentation documented during post-LGM times was also recorded during the previous glacial cycles (cf. Muttoni et al. 2003; Fontana et al. 2014). Megafans thus formed in the LP and VFP during the MIS 6 glacial peak, and characterised large sectors of the plains almost until the onset of the LGM because alluvial sedimentation from MIS 5 to MIS 3 was rather limited (Fontana et al. 2010).

8.3.2 Megafans in the Carpathian Basin

The Carpathian, or Pannonian, Basin is surrounded by mountain belts of the Carpathians and Alps, and currently represents the largest intramontane basin of Europe. During the Cenozoic, the area was flooded by an incursion of the Mediterranean Sea, but it later evolved into a freshwater lake (Nagymarosy and Hámor 2012). The palaeo-Danube river entered the basin in the late Miocene and, together with other alluvial systems, completely filled the lake as a result of fan-delta progradation, imposing terrestrial conditions from the Pliocene onwards (Magyar et al. 1999). During Pliocene time, the regional structural setting shifted to a compressional regime, which led to a sharp geomorphic demarcation between the uplifting mountain ranges and the subsiding alluvial basins (Bada and Horvát 2001). Several compressional phases occurred throughout the Pleistocene, increasing the mountain relief and leading to the establishment of the present drainage pattern (Borsy 1990;

Gábris and Nádor 2007). The subsiding basins attracted the major river systems of the Danube and Tisza. The largest basin corresponds to the present Great Hungarian Plain (GHP; Fig. 8.1b), where thicknesses of Quaternary alluvial deposits reach almost 1,000 m. The other important basin is the Little Hungarian Plain (LHP), also called Danube Basin as it is filled by about 700 m of sediment conveyed almost entirely by this river system (Gábris and Nádor 2007). Detailed research on Quaternary deposits has been carried out only in limited parts of the GHP and LHP, mainly through interpretation of satellite imagery, geomorphological mapping, and analysis and dating of shallow boreholes a few tens of metres deep (e.g., Gábris and Nagy 2005; Kasse et al. 2010; Nádor et al. 2011; Kiss et al. 2014).

The present climate of the Carpathian Basin is continental, but some sub-regions can be differentiated on the basis of annual precipitation. A SE–NW trend is defined by values of ~500 mm/yr in the central and south-eastern sectors of the GHP, rising to 600–800 mm/yr in the LHP. Higher precipitation is recorded in the mountain areas (Lóczy 2015; Mezősi 2016). During Pleistocene glacial stages the Carpathian Basin experienced recurrent phases of dominant periglacial, semi-arid environmental conditions, with a marked prevalence of aeolian processes (Sebe et al. 2011; Gábris et al. 2012). Aeolian deposits (loess and blown sands) and related landforms are widespread in the Pannonian alluvial plains (Mezősi 2016).

8.3.2.1 Megafans of the Little Hungarian Plain

8.3.2.1.1 Danube Megafan

The LHP, currently shared between Hungary, Austria and Slovakia, is crossed by the Danube River, which bisects an alluvial megafan with its apex at the Devin Gate gorge, i.e., where the Danube enters the Carpathian Basin (Fig. 8.7a). The Danube catchment area upstream of Bratislava is about 131,330 km^2.

The Danube megafan features two anabranching channels that split from the main Danube course near Bratislava and rejoin it near Győr and Komárno, respectively: the Moson Danube in the west, and Little Danube (Kis in Hungarian, Malý in Slovak), in the east. They roughly mark the divergent planform of the megafan and, with the main course of the Danube,

they isolate two large interfluves: the western Szigetköz, and the eastern Žitný Ostrov (Csallóköz, in Hungarian). These areas are traditionally defined as 'islands' in the region and are characterised by hydromorphic soils with a common occurrence of degraded peat (Pécsi 1996; Papp-Váry 1999).

At the megafan apex, upstream of Bratislava, the Danube River flows at the same altitude as the surrounding plain. Until the beginning of the twentieth century, downstream of the city and almost as far as Győr, the river presented a multichannel, anastomosing pattern (Mezősi 2016). This fluvial style clearly indicates the large sediment load that the river used to transport in this tract, and documents an aggrading or equilibrium state (Carling et al. 2014; Makaske et al. 2017). Today in the LHP, the multichannel pattern of the Danube is strongly regulated by fluvial engineering works carried out in the last 150 years, and especially by the barrage systems of Dunacsúny and Gabčíkovo (Fig. 8.7). In the active channel slightly south of Bratislava, maximum sediment particle size is 15–20 mm, rapidly decreasing downstream to 5 mm (Mezősi 2016).

According to Gábris and Nádor (2007), an older, larger megafan was built by the Danube during the Middle Pleistocene, when the river entered the LHP west of Devin Gate (Pécsi 1996). Remnants of the apical sector of this older landform have been recognised near Parndorf (Fig. 8.7a), as have alluvial terraces in the distal reaches near Győr. The formation of the present Danube megafan probably started after the Middle Pleistocene and, most likely achieved its present form during the LGM and Late Glacial. The Danube megafan also comprises Holocene deposits, which are not clearly recognisable morphologically as they lie at almost the same topographic level as the late Pleistocene alluvial surface.

The convex morphology of the megafan had been already described by Pécsi (1996), who analysed the surface deposits and reconstructed a near-surface stratigraphic cross-section. The megafan has a topographic gradient of about 1‰ near Bratislava, which rapidly decreases to 0.7–0.4‰ and reaches <0.3‰ in the distal sector. The convex geometry of the megafan is evident at least as far as Gabčikovo dam, while it gradually becomes difficult to define the outer boundary of the landform farther downstream. Depending on

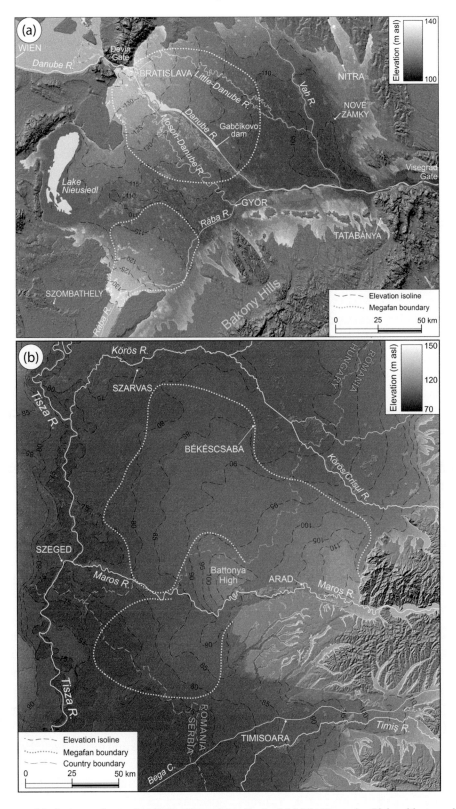

Figure 8.7 **Topographic features of megafans in the Carpathian Basin**. (a) Little Hungarian Plain with general outlines of the Danube and Rába megafans. (b) Maros megafan, with general outlines of the northern and southern lobes. Digital elevation source: Shuttle Radar Topographic Mission (SRTM).

the portion of the distal sector included, the megafan area ranges between 2,300 and 2,700 km^2.

In our reconstruction, the extent of the late Quaternary Danube megafan is significantly smaller than that proposed by previous scholars (e.g., Pécsi 1996; Gábris and Nádor 2007) because we considered only the area where the surface presents a convex, cone-shaped morphology. Previous authors, in contrast, included also all the areas with coarse-grained deposits, even flat areas and patches of the Early to Middle Pleistocene surfaces deformed by tectonic activity. In a geological perspective, it is likely that the subsurface succession of the LHP preserves evidence of several generations of ancestral Danube megafans, some of which could have extended further than that recognised on the current surface.

8.3.2.1.2 Rába Megafan

A megafan was generated by the Rába River (Raab in German), the major right-hand tributary of the Danube in the LHP (Fig. 8.7a). The Rába currently has an average discharge of just 68 m^3/s and transports mostly suspended load (Mezősi 2016). Its catchment extends for about 15,000 km^2 into the Austrian Eastern Alps, where annual precipitation reaches 800 mm/yr.

As shown by the DEM (Fig. 8.7a), the Rába megafan extends into the LHP with a fan-shaped, convex morphology, almost reaching the Danube megafan in its distal sector. The wide, low-lying area between the Rába and Danube megafans hosted swampy environments until modern land reclamation (Pécsi 1996). Aeolian deposits (sand drift and loess) are common on the megafan surface and are attributed to the final phases of the LGM and the Late Glacial (Sávai et al. 2015). This suggests that the Rába megafan grew during the last glaciation and, even though its mountain catchment did not host valley glaciers, periglacial processes were widespread (Van Husen 1997). The Rába River currently flows out onto the megafan surface along its right boundary and the stream is slightly entrenched into the apical sector by a few metres.

The progradation of the Rába megafan forced streams flowing from the western Transdanubian Mountains of Hungary (Figs. 8.1b and 8.7b), which used to reach the LHP directly, to become tributaries of

the Rába River (Sávai et al. 2015). It is thus likely that the Rába megafan prograded partly over its older megafan and partly over smaller alluvial systems emanating from the Transdanubian Mountains.

8.3.2.2 Megafans of the Great Hungarian Plain

The GHP is bounded in the south-west by the Dinaric Alps, and is surrounded in all other directions by the Carpathians (Fig. 8.1b), which feed a number of rivers that mostly belong to the catchment of the Tisza River, i.e., the trunk river of the GHP and major left tributary of the Danube (Gábris and Nádor 2007). The GHP is mainly situated in Hungary, with a large area shared with Romania and Serbia, and a smaller area shared with Ukraine. The GHP lies generally between 200 and 65 m a.s.l., and most of the rivers enter the plain at an elevation below 150 m (Figs. 8.1b and 8.7b).

Literature on the alluvial geomorphology of the GHP (e.g., Borsy 1990; Pécsi 1996; Gábris and Nádor 2007; Kiss et al. 2015; Perşoiu and Rădoane 2017a, b) reports several large fan-like systems built by the main tributaries of the Tisza River, which have their catchments in the Eastern Carpathians and in the Apuseni Mountains. These landforms generally are limited in size, with few exceeding 1,000 km^2. The features that meet our criteria for description here, in order of size, are the depositional fluvial systems of Maros, Szamos, and Drina rivers. The apical portions and related mountain catchments of the Maros and Szamos megafans lie in Romania, where these rivers are named Mureş and Someş, respectively.

The Danube–Tisza interfluve area (Fig. 8.1b), described in previous literature as the largest alluvial fan of the GHP (Gábris and Nádor 2007), is not considered here as a megafan because it is a remnant of an old landscape, strongly disconnected from the present one. Moreover, according to available stratigraphic data and recent geomorphologic investigations, the Danube–Tisza interfluve corresponds to different stacked bodies of various ages and origin, not only alluvial fan-like features. The previous complex and composed topography has been largely reworked and overprinted by regional deformations and mostly by late Pleistocene aeolian processes (Sebe et al. 2011).

8.3.2.2.1 Maros Megafan

The Maros megafan is the largest among the European megafans (Fig. 8.7b), as it consists of a larger northern lobe (~5,500 km^2) and a smaller southern lobe (1,300 km^2). The surface of the Maros megafan is characterised by an extraordinary abundance of abandoned river channels, visible in aerial images and DEMs. These palaeodrainage traces have been mapped in detail by Kiss et al. (2014; 2015), leading to the recognition of 18 different late Quaternary channel belts, now dated using luminescence methods (Fig. 8.8). Sediment cores have been analysed from the upper 10 m (Section line A–A', Fig. 8.9) at the boundary between the northern lobe of the Maros system and the alluvial plain of the Körös River (Crişul in Romanian; Nádor et al. 2011). This area is one of the most actively subsiding zones in the GHP (average rate 3–5 mm/yr), which – together with the sinking basin near Szeged – has influenced the long-term evolution of the Maros megafan (Borsy 1990; Gábris and Nádor 2007). From the valley mouth

to the north-western margin of the depositional system, the thickness of the Quaternary sediments increases from 100 m to 700 m (Borsy 1990). Neotectonic movements also generated the Battonya High, a bedrock outcrop that rises 3–5 m above the level of the alluvial plain (Fig. 8.7b), separating the two lobes of the megafan. It is presently crossed by the Maros River.

As documented by the diverging pattern of the numerous abandoned channel belts, megafan development was mainly guided by river avulsion (Fig. 8.8). The extent and visibility of palaeodrainage traces testify to the fact that overbank sedimentation was limited, as confirmed by the small, discontinuous patches of residual loess and coversands (Kiss et al. 2014). The avulsions that built the southern lobe of the megafan occurred at the mouth of the small valley crossing the Battonya High. In the northern lobe, the main nodal zone was located between the mouth of the mountain valley and the city of Arad, while a secondary nodal

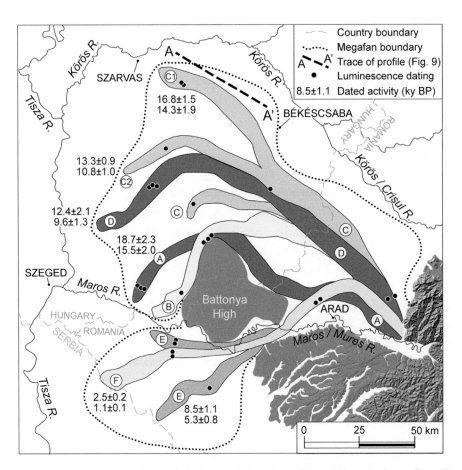

Figure 8.8 **Schematic reconstruction of abandoned channel belts of the Maros River** (redrawn from Kiss et al. 2014). Luminescence dating of sand bars tracks palaeochannel activity.

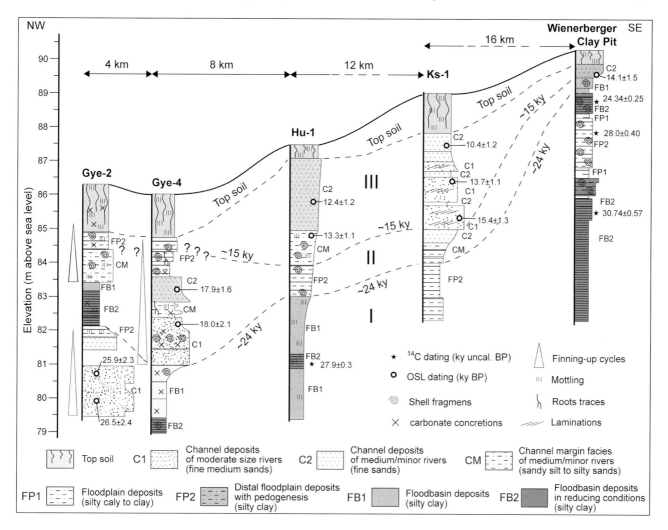

Figure 8.9 **Stratigraphic section of the area at the boundary between the distal sector of the Maros megafan and the Körös basin** (redrawn from Nádor et al. 2011). For location see Figure 8.7b. The subsurface data were derived from cores and stratigraphic sections exposed in clay pits.

point was located near the boundary between Romania and Hungary. This administrative limit roughly corresponds also to the downstream limit of the gravel deposits, while the rest of the megafan consists of sand, silt, and clay (Fig. 8.7b).

According to the geomorphological and geochronological framework reconstructed by Kiss et al. (2014), the northern lobe consists of several channel belts that were active on the megafan surface between 18.7 ± 2.3 and 9.6 ± 1.3 cal ka BP – thus coinciding almost exactly with the span of the Late Glacial period. The oldest palaeodrainage belt (A in Fig. 8.8) shows meandering channels and is dated between 18.7 ± 2.3 and 13.3 ± 1.4 cal ka BP. Part of this interval overlaps with activity in channel belts B and C, which started around 15.2 ± 2.2 cal ka BP. The similar ages

displayed by belts A, B, and C point either to their simultaneous coexistence, or to temporary and rapid avulsion from one to the other (Kiss et al. 2014). Belts C1 and C2 testify to a northwards shift of the Maros River. Palaeodrainage belt D was active soon after for some millennia (12.4 ± 2.1 to 9.6 ± 1.3 cal ka BP), resulting in a stable orientation for a relatively long interval. The final phase of the existence of D probably coincided with activation of the southern lobe and initial stages of belt E activity. This belt, the oldest of the southern lobe, started 8.5 ± 0.9 cal ka BP and was active until 3.5 ± 0.4 cal ka BP. The last recorded palaeochannel belt (F in Fig. 8.8) was active for only a few centuries during the Roman period.

The different palaeodrainage directions were characterised by meandering, anastomosing, or braided patterns and these diverse styles can be seen to coexist along the same channel belt. Generally, the anastomosing pattern characterises the apical portion of the megafan, where braided channels are also documented and associated with slightly steeper slopes and convex topography of the channel (Kiss et al. 2014). The mean bankfull discharge of the meandering channels was qualitatively estimated comparing the planform morphometry of the relict paleochannels with the morphometry of the present meanders of the Tisza River (Sümeghy and Kiss 2011). In the northern lobe, the estimated palaeochannel discharge varied from 2,000 up to 2,650 m^3/s, much larger than the current value for the Maros River (680 m^3/s). The estimated bankfull discharge was apparently still high during the first part of the Holocene, thereafter decreasing to the present amount (Kiss et al. 2014).

Where the Maros megafan meets the alluvial plain of the Körös River, several cores have been retrieved and stratigraphic units have been recognised on the basis of facies associations and depositional cycles. Dates place formation of the uppermost 10 m of sediment in the interval between the LGM and the Late Glacial (Fig. 8.9; Nádor et al. 2011). The geochronological results are in good agreement with the ages obtained by Kiss et al. (2014) in the same sector of the Maros megafan.

The geomorphological and chronological data suggest that the main formative phase of the Maros megafan occurred during the Late Glacial. Palaeoclimatic reconstruction suggests that this interval recorded some much more humid phases than the LGM, and that these probably favoured the greater production and export of sediments from the mountains to the GHP (Gábris and Nádor 2007; Kiss et al. 2015). Moreover, during Late Glacial times the vegetation in the plain and mountain valleys experienced strong and rapid changes (cf. Feurdean et al. 2014) that also modified the availability of sediment flux on the megafans. The mountain catchment of the Maros River was not affected by significant glacial activity in the LGM, but periglacial and aeolian processes were particularly important until the end of the Late Glacial (cf. Perşoiu and Rădoane 2017a).

The larger expanse of the northern lobe of the Maros megafan is probably partly related also to an abrupt diversion which led the Tisza River to abandon the valley of Érmellék and shift well north to its present path, north of the Nyírség area (Fig. 8.1b). This dramatic change was mainly induced by tectonic movements and differential subsidence, and occurred during the Late Glacial, between 18 and 14 cal ka BP (Timár et al. 2005; Nádor et al. 2007). The precise chronology, however, is still debated. Before this important diversion, the Tisza used to flow directly through the area now occupied by the depositional systems of the Körös and Maros rivers. The new configuration made space available for fan progradation, especially for the northern lobe of Maros megafan. The Maros fluvial system mainly expanded over the Tisza River's alluvial plain, which had been prone to strong differential subsidence and probably conditioned megafan progradation.

8.3.2.2.2 Szamos Megafan

The northwards diversion of the Tisza also possibly allowed the Szamos River to form its own megafan (Fig. 8.1b). This has an area of ~2,000 km^2, and with its thick sandy deposits represents an important aquifer for the region (Posea 1997). The Szamos megafan also presents several relict channel belts at the surface. Their diverging pattern has been interpreted as the consequence of several avulsions, the last dating to about 5 cal ka BP (Perşoiu and Rădoane 2017b). The abandoned channel belts are mainly characterised by meandering channels, with some braided patterns documented. The meanders that formed during the Late Glacial or in the early Holocene display significantly larger dimensions than the meanders of mid and late Holocene age suggesting higher discharges (Posea 1997; Perşoiu and Rădoane 2017a).

8.3.2.2.3 Drina Megafan

Slightly west of the GHP, the megafan of the Drina River is located on the south-western margin of the Pannonian Basin, where the river marks the boundary between Bosnia–Herzegovina and Serbia, about 100 km west of the city of Belgrade (Fig. 8.1b). The 1,000 km^2 megafan has been partly eroded in its distal sector by Holocene lateral migration of the Sava River. The Drina megafan mainly consists of Pleistocene sand and fine gravel, but silt and clay deposits are present in the distal sector, where the topographic gradient is < 0.5‰ (Rajčević 1982; Vrhovčić et al. 1984).

Geochronological data are not available but, according to the evolution of the upstream sector of the Sava River valley (Bačani et al. 1999), it is likely that the Drina formed its megafan during LGM and Late Glacial times, and that during the Holocene the stream shifted near to its present position. Currently, the Drina River almost bisects the megafan displaying mostly a sinuous multithread, wandering channel, but traces of Holocene abandoned meanders are abundant (some are modern). The Pleistocene surface lies at the same level as the Holocene surface.

8.4 Discussion

To the exception of the outsize Maros megafan, all the other megafans described in this survey of Middle Europe meet our criteria for areal extent, ranging between 1,000 and 2,700 km² (Table 8.1). However, the mountain catchments feeding the megafans of the VFP are much smaller than those feeding the megafans of the Carpathian Basin. In general, megafan areas are clearly not proportional to the dimensions of the catchments currently feeding the relative fluvial systems. For example, in the Carpathian Basin, the Danube megafan is comparable in areal extent to those built by the Rába and Szamos rivers, although the Danube catchment (131,330 km²) is one order of magnitude larger. In northern Italy, the largest megafan is the Brenta River (2,600 km²), whose present catchment (1,787 km²) is relatively small; in contrast, the Adige River, with a catchment of about 12,000 km², has formed a megafan of only 1,600 km².

These discrepancies are mainly related to the different geomorphic evolutions experienced by the montane catchments during Pleistocene glaciations. The valleys of the southern Alps were largely occupied by ice masses during the LGM, whereas the Carpathians were not populated by valley glaciers and the LGM climate was particularly dry (Kühlemann et al. 2008). Periglacial processes, however, produced large volumes of debris (e.g., van Husen 1997; Onaca et al. 2017). The megafans of the southern Alps are mostly relics of the last glacial peak (26–22 cal ka BP), when glaciers acted as the main agent of highland erosion, producing sediment and conveying it to the mouths of the Alpine valleys. Occurrences of glacial transfluences from larger to smaller catchments allowed the oversupply of some alluvial systems which, in non-glacial conditions, would have retained the characteristics of much smaller catchments. Examples are the Brenta and Olona catchments (Fig. 8.2 and Table 8.1).

From the beginning of deglaciation (i.e., 22–17.5 cal ka BP), a major period of sediment starvation began for the Italian megafans, leading the rivers to cut into the LGM surface and form incised valleys. A different behaviour is evident in the Carpathian Basin, where for some systems (e.g., Maros and Szamos megafans) the major phase of sedimentation coincided with the Late Glacial (19–11.7 cal ka BP). It is likely that where LGM glaciers remained small and climate was dry, the occurrence of Late Glacial humid phases mobilised and transported hillslope materials out onto the plain (cf Gábris and Nádor 2007; Kiss et al. 2015).

A peculiar situation characterises the Danube River. Upstream of Bratislava, this river is fed almost solely by Alpine tributaries. During the LGM, the main valley was free of ice, although glaciers were present in the highest portions of the major Alpine catchments (Ehlers and Gibbard 2004). Moreover, also upstream of Bratislava, large intermediate depositional basins (e.g., Wien, Stockerau, and Machlande) probably buffered onwards transport of the sediment load supplied by the glaciers. The particular situation of the Danube is also reflected in the continuation of sedimentation on the megafan surface until modern times. All the other megafans of the LHP and GHP recorded minor activity during the last millennia or have been almost entirely inactive since the beginning of the Holocene.

The formation and evolution of the Danube megafan was probably driven by the presence of a structural threshold at Devin Gate, the narrow passage between the Vienna Basin and the subsiding LHP (Fig. 8.7a). Active tectonics exerted evident forcing on the GHP where, for example, Late Pleistocene and Holocene deformation strongly conditioned the evolution of the Szamos megafan, and depocentres located in the Körös Plain and near Szeged attracted some channel belts of the Maros megafan. In contrast, active tectonics apparently had a limited influence in northern Italy during the phase of megafan accumulation. Long-term megafan evolution was influenced instead by structural settings that created major differences between the LP and the VFP. An example is the occurrence of large valley lakes in the central sector of the southern Alps, interpreted as a legacy of valley incision in this segment of the mountain range during the Messinian salinity crisis (Finckh et al. 1984).

Unlike the continental Carpathian Basin, the rivers at the foot of the southern Alps drain directly into the Adriatic Sea. They thus allow the relationships between major global sea-level variations and fluvial sedimentary system to be discussed, as described for Tagliamento megafan in the Holocene. Notably, the megafans of northern Italy experienced their maximum aggradation and widening during the LGM marine lowstand, when the coast retreated 250–350 km basinward, so that the influence of sea level was negligible. Sea level has only affected fluvial sedimentation since about 8.5 ka BP, i.e., when the coastal environments almost reached their present position. During the marine highstand, progradation started again in the distal sector of the megafans and triggered the formation of Holocene depositional lobes (Figs. 8.5 and 8.6).

In northern Italy, the channel belts forming the LGM lobes of the megafans are all gravel-textured and characterised by braided river patterns in their apical sectors, whereas downstream they become sandy while remaining braided, and wandering to sinuous in the distal sector. In contrast, megafans of the LHP and GHP display anastomosing and meandering channel patterns also in their proximal sectors. This characteristic is mainly explained by the low declivity (1.2–0.6‰) of the apical portion of the Danube, Rába, Maros, Szamos and Drina megafans, whereas at the foot of the southern Alps the proximal sector of the LGM megafans is significantly steeper (i.e., 7–3‰) and does not allow the formation of meandering or anastomosing channel. In the VFP, meanders are typical of distal megafan sectors during the Holocene, probably because sea-level rise strongly reduced the slope of the most distal river tracts.

When compared to the typical megafans existing along the Andean and Himalayan fronts, the European ones have much smaller dimensions and are limited in number. They are more comparable to those along the foothills of the Sierra Nevada in California (Weissmann et al. 2002, 2005) and the Southern Alps in New Zealand (Leckie 1994, 2003), where megafans also experienced a formative phase

during glacial periods, whereas minor or very limited activity occurred in the Holocene.

As proposed by Fontana et al. (2014), these similarities and differences mainly relate to the 'size' of the orogenic belts and the climate of the continent. Most of the mountain ranges in Europe have limited tectonic activity and low average elevation (i.e., < 2,000 m). Moreover, excluding a few exceptions, the main valleys debouching from active mountain fronts onto the piedmont and subsiding receiver basins are supplied by rather small catchments (from 1,000 to 20,000 km^2). Concerning climate, Europe does not have prolonged and strong seasonal rainfall (such as the monsoon), which would favour fluvial instability and induces frequent river avulsions and alluvial sedimentation.

Thus, few areas in Europe have a regional setting with suitable conditions for megafan formation and, generally, only the glacial and periglacial conditions, which occurred during the coldest phases of the Late Pleistocene, supported enough sedimentary production in the mountain catchments for generating significant deposition in the related piedmont plains.

8.5 Conclusions

Several megafans are present in Europe in the foreland basin of the southern Alps (Po and Venetian–Friulian plains) and in the subsiding Carpathian foreland region of the Pannonian Basin. Their surface areas generally range between 1,000 and 2,700 km^2. The Maros River in the GHP has formed the largest megafan in Europe (~7,000 km^2).

The megafans of northern Italy can be mainly considered as relict landforms produced during the LGM by deposition of sediments entrained by Alpine valley glaciers down to the piedmont domain. The thickness of LGM deposits spans 15–35 m, much of it aggraded during the peak of the glaciation between 26 and 22 cal ka BP. The apical portions of these LGM lobes present gravelly, braided channels, and can be differentiated from distal sectors where rivers were sandy, braided, and wandering. The activation of glacial transfluences promoted an oversupply of sediment to some megafans.

Since the beginning of deglaciation (22–17.5 cal ka BP), the sediment flux from the Alpine catchments to the plain has decreased dramatically. Between the Late

Glacial and early Holocene, a major phase of entrenchment led the Alpine rivers to incise the LGM surface of the megafans. Another phase of limited deposition occurred in the distal tract of the megafans after about 8.5 ka BP, when relative sea level rose to about – 10 m and led the alluvial systems to fill those incised valleys almost completely.

During the LGM, the Carpathians were devoid of valley glaciers and the area was characterised by relatively dry conditions, with evidence for aeolian processes in the alluvial plains. Some of the megafans in the GHP and LHP, such as those formed by the Maros and Szamos rivers, attained their maximum extent during the Late Glacial (i.e., 19–11.5 cal ka BP), when moister climatic conditions probably allowed more efficient sediment transfer from mountain catchments. In the Carpathian Basin, Holocene fluvial activity on the megafans was generally limited, and in several cases large sectors were abandoned by river incision (e.g., Rába and Drina megafans).

A distinctive evolutionary style applies to the Danube in the LHP, where anastomosing and anabranching channels were still delivering sediment to wide portions of the megafan until the river became heavily regulated in the nineteenth and twentieth centuries. It is likely that this behaviour of the Danube has been influenced by the vastness and characteristics of its catchment, which is much larger than others described in the study area, and where a succession of sedimentary basins has tended to regulate the stock and flow pattern of sediment.

However, in general there is no direct relationship between the areal extent of the megafans and the size of their present catchments. In the Alps, this is largely ascribable to glacial transfluence processes. In the unglaciated Carpathians, several other factors may have concurred in controlling the different sediment yield from specific catchments during cold periods, such as contrasts in bedrock erodibility and regional palaeoclimatic gradients. Their roles have not yet been investigated in detail.

The European megafans are one order of magnitude smaller than those described in the piedmont plains of the Himalaya and Andes. They are comparable in size to the megafans fed by other mountain ranges of the temperate zone with topography similar to the Alps and Carpathians. The European megafans are mainly interpreted as the result of abundant sediment production by glacial and periglacial processes during the coldest periods of the late Quaternary.

Acknowledgements

The authors thank Edgardo Latrubesse and Justin Wilkinson for their invitation to write this chapter. This paper benefited from comments and suggestions by J. Wilkinson and two anonymous reviewers. Elisabetta Starnini, Janos Makkay, Roger Langhor, and Erzsébet Horváth are thanked for introducing the authors to the vastness of the Nagy Alföld.

References

Agassiz, L. (1840). *Etudes sur les glaciers*. Jent et Gassmann, Soleure.

Amorosi, A., Fontana, A., Antonioli, F., Primon, S., and Bondesan, A. (2008). Post-LGM sedimentation and Holocene shoreline evolution in the NW Adriatic coastal area. *GeoActa*, 7, 41–67.

Bada, G. and Horváth, F. (2001). On the structure and tectonic evolution of the Pannonian basin and surrounding orogens. *Acta Geologica Hungarica*, 44, 301–327.

Bačani, A., Šparica, M. and Velić, J. (1999). Quaternary deposits as hydrogeological system of Eastern Slavonia. *Geologia Croatica*, 52, 141–152.

Baio, M., Bersezio, R. and Bini, A. (2004). Assetto geologico nel sottosuolo tra Melegnano e Piacenza. *Il Quaternario – Italian Journal of Quaternary Sciences*, 17, 355–359.

Bersezio, R., Pavia, F., Baio, M., et al. (2004). Aquifer architecture of the quaternary alluvial succession of the southern Lambro basin (Lombardy, Italy). *Il Quaternario – Italian Journal of Quaternary Sciences*, 17, 361–378.

Bersezio, R., Giudici, M., and Mele, M. (2007). Combining sedimentological and geophysical data for high-resolution 3-D mapping of fluvial architectural elements in the Quaternary Po plain (Italy). *Sedimentary Geology*, 202, 230–248.

Borsy, Z. (1990). Evolution of the alluvial fans of the Alföld. In A. H. Rachocki and M. Church, eds., *Alluvial Fans: A Field Approach*. Wiley, Chichester, 229–247.

Bondesan, M. (2001). Hydrography. In G. B. Castiglioni and G. B. Pellegrini, eds., *Illustrative Notes of the Geomorphological Map of Po Plain (Italy)*, Suppl. 4. Geografia Fisica Dinamica Quaternaria, 33–44.

Carton, A., Bondesan, A., Fontana, A., et al. (2009). Geomorphological evolution and sediment transfer in the Piave River watershed (north-eastern Italy) since the LGM. *Géomorphologie: relief, processus, environnement*, 3, 37–58.

Castiglioni, B. (1940). L'Italia nell'età quaternaria. In G. Dainelli, ed., *Atlante fisico economico d'Italia*, Milano, Consociazione Turistica Italiana, tav. 3.

Castiglioni, G. B. (1997). *Geomorphological Map of Po Plain*. Firenze, MURST–S.El.Ca, 3 sheets, scale 1:250,000.

Clark, P., Dyke, A., Shakun, J., et al. (2009). The Last Glacial Maximum. *Science*, 325, 710–714.

Cuffaro, M., Riguzzi, F., Scrocca, D., et al. (2010). On the geodynamics of the northern Adriatic plate. *Rendiconti Lincei*, 21 (Suppl. 1), 253–279.

Carling, P., Jansen, J., and Meshkova, L. (2014). Multichannel rivers: their definition and classification. *Earth Surface Processes and Landforms*, 39, 26–37.

de Charpentier, J. (1841). *Essai sur les glaciers et sur le terrain erratique du bassin du Rhône*. Lausanne, Ducloux.

Ehlers, J. and Gibbard, P., eds. (2004). *Quaternary Glaciations Extent and Chronology*, Part 1: *Europe*. Elsevier, Amsterdam.

Feruglio, E. (1925). *La zona delle risorgive del basso Friuli tra Tagliamento e Torre*. Annali Stazione Chimica Agraria Sperimentale serie III 1, Udine.

Feurdean, A., Perşoiu, A., Tanţău, I., et al. (2014). Climate variability and associated vegetation response throughout Central and Eastern Europe (CEE) between 60 and 8 ka. *Quaternary Science Reviews*, 106, 206–224.

Finckh, P., Kelts, K., and Lambert, A. (1984). Seismic stratigraphy and bedrock forms in perialpine lakes. *Geological Society of America Bulletin*, 95, 1118–1128.

Fontana, A. (2006). *Evoluzione geomorfologica della bassa pianura friulana e sue relazioni con le dinamiche insediative antiche*. Monografie Museo Friulano Storia Naturale, 47, Udine. Enclosed Geomorphological Map of the Low Friulian Plain scale 1:50,000.

Fontana, A., Mozzi, P., and Bondesan, A. (2008). Alluvial megafans in the Venetian–Friulian Plain (north-eastern Italy): evidence of sedimentary and erosive phases during Late Pleistocene and Holocene. *Quaternary International*, 189, 71–90.

Fontana, A., Mozzi, P., and Bondesan, A. (2010). Late Pleistocene evolution of the Venetian-Friulian Plain. *Rendiconti Lincei*, 21 (Suppl. 1), 181–196.

Fontana, A., Mozzi, P., and Marchetti, M. (2014). Alluvial fans and megafans along thesouthern side of the Alps. *Sedimentary Geology*, 301, 150–171.

Gábris, G. (1994). Pleistocene evolution of the Danube in the Carpathian Basin. *Terra Nova*, 6, 495–501.

Gábris, G. and Nagy, B. (2005). Climate and tectonically controlled river style changes on the Sajó-Hernád alluvial fan (Hungary). In A. M. Harvey, A. E. Mather, and M. Stokes, eds., *Alluvial Fans: Geomorphology, Sedimentology, Dynamics*. Geological Society of London, Special Publication, 251, 61–67.

Gábris, G. and Nádor, A. (2007). Long-term fluvial archives in Hungary: response of the Danube and Tisza rivers to tectonic movements and climatic changes during the Quaternary: a review and new synthesis. *Quaternary Science Reviews*, 26, 2758–2782.

Gábris, G., Horváth, E., Novothny, Á., and Ruszkiczay-Rüdiger, Z. (2012). Fluvial and aeolian landscape evolution in Hungary. The results of the last 20 years research. *Netherlands Journal of Geosciences*, 91, 111–128.

Garzanti, E., Vezzoli, G., and Andò, S. (2011). Paleogeographic and paleodrainage changes during Pleistocene glaciations (Po Plain, Northern Italy). *Earth-Science Reviews*, 105, 25–48.

Ghielmi, M., Minervini, M., Nini, C., et al. (2010). Sedimentary and tectonic evolution in the eastern Po–Plain and northern Adriatic Sea area from Messinian to Middle Pleistocene (Italy). *Rendiconti Lincei*, 21 (Suppl. 1), 131–166.

Gohain, K. and Parkash, B. (1990). Morphology of Kosi megafan. In A. H. Rachocki and E. M. Church, eds., *Alluvial Fans: A Field Approach*, Wiley, Chichester, 151–178.

Guzzetti, F., Marchetti, M., and Reichenbach, P. (1997). Large alluvial fans in the north–central Po Plain (Northern Italy). *Geomorphology*, 18, 119–136.

Hippe, K., Fontana A., Hajdas, I., and Ivy-Ochs, S. (2018). A high-resolution ^{14}C chronology tracks pulses of aggradation of glaciofluvial sediment on the Cormor megafan between 45 and 20 ka BP. *Radiocarbon*, 60, 857–874.

Horton, B. K. and DeCelles, P. G. (2001). Modern and ancient fluvial megafans in the central Andean foreland basin system, southern Bolivia. *Basin Research*, 13, 43–63.

Howard, A. J., Macklin, M. G. Bailey, D. W., and Andreescu, A. (2004). Late-glacial and Holocene river development in the Teleorman Valley on the southern Romanian Plain. *Journal of Quaternary Science*, 19, 271–280.

Kasse, C., Bohncke, S., Vandenberghe, J., and Gábris, G. (2010). Fluvial style changes during the last glacial-interglacial transition in the middle Tisza valley (Hungary). *Proceedings of the Geologists' Association*, 121, 180–194.

Kindler, P., Guillevic, M., Baumgartner, M., et al. (2014). Temperature reconstruction from 10 to 120 kyr b2k from the NGRIP ice core. *Climate of the Past*, 10, 887–902.

Kiss, T., Sümeghy B., and Sipos G. (2014). Late Quaternary paleodrainage reconstruction of the Maros River alluvial fan. *Geomorphology*, 204, 49–60.

Kiss, T., Hernesz, P., Sümeghy, B., Györgyövics, K., and Sipos, G. (2015). The evolution of the Great Hungarian Plain fluvial system – fluvial processes in a subsiding area from the beginning of the Weichselian. *Quaternary International*, 388, 142–155.

Kühlemann, J., Rohling, E. J., Krumrei, I., et al. (2008). Regional synthesis of Mediterranean atmospheric circulation during the last glacial maximum. *Science*, 321, 1338–1340.

Lambeck, K., Roubya, H., Purcell, A., Sun, Y., and Malcolm, S. (2014). Sea level and global ice volumes from the Last Glacial Maximum to the Holocene. *Proceedings of the National Academy of Sciences*, 111, 15,296–15,303.

Leckie, D. A. (1994). Canterbury Plains, New Zealand – implications for sequence stratigraphic models. *American Association of Petroleum Geologists Bulletin*, 78, 1240–1256.

Leckie, D. A. (2003). Modern environments of the Canterbury Plains and adjacent offshore areas, New Zealand – an analog for ancient conglomeratic depositional systems in nonmarine and coastal zone settings. *Bulletin of Canadian Petroleum Geology*, 51, 389–425.

Lóczy D., ed. (2015). *Landscapes and Landforms of Hungary*. Springer, Berlin.

Lowick, S. E., Preusser, F., Pini, R., and Ravazzi, C. (2010). Underestimation of fine grain quartz OSL dating towards the Eemian: comparison with palynostratigraphy from Azzano Decimo, northeastern Italy. *Quaternary Geochronology*, 5, 583–590.

Magyar, I., Geary, D. H., and Müller, P. (1999). Paleogeographic evolution of the Late Miocene Lake Pannon in Central Europe. *Palaeogeography, Palaeoclimatology, Palaeoecology*, 147, 151–167.

Makaske, B., Lavoii, E., De Haas, T., Kleinhans, M. G., and Smith D. G. (2017). Upstream control of river anastomosis by sediment overloading, upper Columbia River, British Columbia, Canada. *Sedimentology*, 64, 1488–1510.

Marchetti, M. (1996). Variazioni idrodinamiche dei corsi d'acqua della Pianura Padana centrale connesse con la deglaciazione. *Il Quaternario – Italian Journal of Quaternary Sciences*, 9, 465–472.

Marchetti, M. (2001). Fluvial, fluvioglacial and lacustrine forms and deposits. In G. B. Castiglioni and G. B. Pellegrini, eds., *Illustrative Notes of the Geomorphological Map of the Po Plain*, Geografia Fisica Dinamica Quaternaria (Suppl. 4), 73–104.

Maţenco, L. (2017). Tectonics and exhumation of Romanian Carpathians: inferences from kinematic and thermochronological studies. In M. Rădoane and A. Vespremeanu-Stroe, eds., *Landform Dynamics and Evolution in Romania*, Springer, Berlin, 15–56.

Mezősi, G. (2016). Physical geography of the Great Hungarian Plain. In G. Mezősi, ed., *The Physical Geography of Hungary, Geography of the Physical Environment*. Springer, Berlin, 195–229.

Miola, A., Bondesan, A., Corain, L., et al. (2006). Wetlands in the Venetian Po Plain (north–eastern Italy) during the Last Glacial Maximum: vegetation, hydrology, sedimentary environments. *Review of Palaeobotany and Palynology*, 141, 53–81.

Monegato, G., Scardia, G., Hajdas, I., Rizzini, F., and Piccin, A. (2017). The Alpine LGM in the boreal ice-sheets game. *Scientific Reports*, 7, 1–8.

Mouchené, M., van der Beek, P., Mouthereau, F., and Carcaillet, J. (2017). Controls on Quaternary incision of the Northern Pyrenean foreland: chronological and geomorphological constraints from the Lannemezan megafan, SW France. *Geomorphology*, 281, 78–93.

Mozzi, P., Bini, C., Zilocchi, L., Becattini, R., and Mariotti Lippi, M. (2003). Stratigraphy, palaeopedology and palynology of Late Pleistocene and Holocene deposits in the landward sector of the Lagoon of Venice (Italy), in relation to the 'caranto' level. *Il Quaternario – Italian Journal Quaternary Science*, 16, 193–210.

Mozzi, P. (2005). Alluvial plain formation during the Late Quaternary between the southern Alpine margin and the Lagoon of Venice (northern Italy). *Geografia Fisica e Dinamica Quaternaria, Suppl.* 7, 219–230.

Mozzi, P., Ferrarese, F., and Fontana, A. (2013). Integrating digital elevation models and stratigraphic data for the reconstruction of the post-LGM unconformity in the Brenta alluvial megafan (North-Eastern Italy). *Alpine and Mediterranean Quaternary*, 26, 41–54.

Muttoni, G., Carcano, C., Garzanti, E., et al. (2003). Onset of major Pleistocene glaciations in the Alps. *Geology*, 31, 989–992.

Nádor, A., Thamó-Bozsó, E., Magyari, Á., and Babinszki, E. (2007). Fluvial responses to tectonics and climate change during the Late Weichselian in the eastern part of the Pannonian Basin (Hungary). *Sedimentary Geology*, 202, 174–192.

Nádor, A., Sinha, R., Magyari, Á., et al. (2011). Late Quaternary (Weichselian) alluvial history and neotectonic control on fluvial landscape development in the southern Körös plain, Hungary. *Palaeogeography, Palaeoclimatology, Palaeoecology*, 299, 1–14.

Nagymarosy, A. and Hámor, G. (2012). Genesis and evolution of the Pannonian Basin. In J. Haas, ed., *Geology of Hungary*, Springer, Berlin, 149–198.

NGRIP (North Greenland Ice Core Project) Members (2004). High-resolution record of Northern Hemisphere climate extending into the last interglacial period. *Nature*, 431, 147–151.

Onaca, A., Urdea, P., Ardelean, A. C., Şerban, R., and Ardelean, F. (2017). Present-day periglacial processes in the Alpine zone. In M. Rădoane and A. Vespremeanu-Stroe, eds., *Landform Dynamics and Evolution in Romania*. Springer, Berlin, 147–176.

Ori, G. G. (1982). Braided to meandering channel patterns in humid-region alluvial fan deposits, River Reno, Po Plain (northern Italy). *Sedimentary Geology*, 31, 231–248.

Papp-Váry, A., ed. (1999). *Magyarország Atlas*. Cartographia, Budapest.

Pécsi, M. (1996). *Geomorphological Regions of Hungary*. Geographical Research Institute, Hungarian Academy of Sciences, Budapest, 121 pp.

Perşoiu, I. and Rădoane, M. (2017a). River behavior during Pleniglacial-Late Glacial. In M. Rădoane and A. Vespremeanu-Stroe, eds., *Landform Dynamics and Evolution in Romania*. Springer, Berlin, 443–468.

Perşoiu, I. and Rădoane, M. (2017b). Fluvial activity during the Holocene. Landform dynamics and evolution in Romania. In M. Rădoane and A. Vespremeanu-Stroe, eds., *Landform Dynamics and Evolution in Romania*. Springer, Berlin, 469–488.

Pini, R., Ravazzi, C., and Donegana, M. (2009). Pollen stratigraphy, vegetation and climate history of the last 215 ka in the Azzano Decimo core (plain of Friuli, north-eastern Italy). *Quaternary Science Reviews*, 28, 1268–1290.

Pinna, M. (1982). *Climatologia*. UTET, Torino, 442 pp.

Piovan, S., Mozzi, P., and Zecchin, M. (2012). The interplay between adjacent Adige and Po alluvial systems and deltas in the late Holocene (Northern Italy). *Géomorphologie*, 4, 427–440.

Posea, G. (1997). *Câmpia de vest a României (The Western Romanian plain)*. Editura Fundaţiei România de Mâine, Bucharest, 430 pp.

Rabus, B., Eineder, M., Roth, A., and Bamler, R. (2003). The shuttle radar topography mission – a new class of digital elevation models acquired by spaceborne radar. *Photogrammetry and Remote Sensing*, 57, 241–262.

Rajčević, D. (1982). *Osnovna geološka karta SFRJ 1:100,000. Tumač za list Šabac L34–112*. Geol. Inst. Beograd. Savezni geološki zavod, Beograd, 56 pp.

Ravazzi, C., Deaddis, M., De Amicis, M., et al. (2012). The last 40 ka evolution of the Central Po Plain between the Adda and Serio rivers. *Géomorphologie: Relief, Processus, Environnement*, 2, 131–154.

Paiero, G. and Monegato, G. (2003). The Pleistocene evolution of Arzino alluvial fan and western part of Tagliamento morainic amphitheatre (Friuli, NE Italy). *Il Quaternario – Italian Journal of Quaternary Sciences*, 16, 185–193.

Rossato, S., Fontana, A. and Mozzi, P. (2015). Meta-analysis of a Holocene ^{14}C database for the detection of palaeohydrological crisis in the Venetian-Friulian Plain (NE Italy). *Catena*, 130, 34–45.

Rossato, S. and Mozzi, P. (2016). Inferring LGM sedimentary and climatic changes in the southern eastern Alps foreland through the analysis of a ^{14}C ages database (Brenta megafan, Italy). *Quaternary Science Reviews*, 148, 115–127.

Sávai, S., Molnár, D., and Sümegi, P. (2015). Late glacial river-bed changes on the Little Hungarian Plain, based on preliminary chronological, geological and paleontological data. *Open Geoscience*, 7, 572–579.

Sebe, K., Csillag, G., Ruszkiczay-Rüdiger, Z., et al. (2011). Wind erosion under cold climate: a Pleistocene periglacial mega-yardang system in Central Europe (Western Pannonian Basin, Hungary). *Geomorphology*, 134, 470–482.

Stella, A. (1895). Sui terreni quaternari della valle del Po in rapporto alla carta geologica italiana. *Bollettino Regio Comitato Geologico Italiano*, 51–108.

Sümeghy, B. and Kiss, T. (2011). Discharge calculation of paleochannels on the alluvial fan of the Maros River, Hungary. *Journal of Environmental Geography*, 4, 11–17.

Tarquini, S., Isola, I., Favalli, M., et al. (2007). TINITALY/01: a new Triangular Irregular Network of Italy. *Annals of Geophysics*, 50, 407–425.

Timár, G., Sümegi, P., and Horváth, F. (2005). Late Quaternary dynamics of the Tisza River: evidence of climatic and tectonic controls. *Tectonophysics*, 410, 97–110.

Toscani, G., Marchesini, A., Barbieri, C., et al. (2016). The Friulian-Venetian Basin I: architecture and sediment flux into a shared foreland basin. *Italian Journal of Geosciences*, 135, 444–459.

van Husen, D. (1997). LGM and Late-glacial fluctuations in the Eastern Alps. *Quaternary International*, 38–39, 109–118.

Vrhovčić, J., Mojićević, M., Andelković, J., et al. (1984). *Osnovna geološka karta SFRJ 1:100,000. Tumač za list Bjeljina L34–111*. Geological Institute of Belgrade, Belgrade, 56 pp.

Weissmann, G. S., Mount, J. F., and Fogg, G. E. (2002). Glacially driven cycles in accumulation space and sequence stratigraphy of a stream-dominated alluvial fan, San Joaquin Valley, California, U.S.A. *Journal of Sedimentary Research*, 72, 270–281.

Weissmann, G. S., Bennett, G. L., and Lansdale, A. L. (2005). Factors controlling sequence development on Quaternary fluvial fans, San Joaquin Basin, California, USA. In A. M. Harvey, A. E. Mather, and M. Stokes, eds., *Alluvial Fans: Geomorphology, Sedimentology, Dynamics*. Geological Society of London, Special Publication, 251, 169–186.

9

The Loire Megafan, Central France

YANNI GUNNELL

Abstract

Centrally situated in France, hosting famous chateaus and vineyards yet oddly secluded, the extensive, diamond-shaped region known as Sologne is
a geological elephant in the room. Nationally one
of the last regions to be fully surveyed, this
deceptively uniform yellow patch on small-scale
geological maps is presented here and portrayed
for the first time as a fluvial megafan. Its attributes
are compiled and reviewed from a collection of
reports, maps, handbooks and articles. Evidence
shows that this ~ 100-m-thick Neogene accumulation of quartz-rich sand and clay was primarily
generated by the Loire River and remained intermittently functional until early Quaternary time,
whereupon dissection by the Loire River itself
and by its fan-fed tributary streams prevailed.
The ~ 120-km-long megafan was fed from the
south by the rising volcanic swell of the Massif
Central in a succession of pulses, relatively well-
dated by different generations of accessory volcanic minerals. Sediment aggradation conforms
to a pattern for which modern analogues may
only exist in Africa, a continent hosting many
dozens of swells of a similar nature to the
Massif Central, with adjacent basins populated
by very large megafans fed by swell-flank rivers.
As such, the Loire megafan is a unique occurrence in Europe.

9.1 Introduction

Our current understanding of megafans is based
on imagery from the air or from space, is rarely
field-based (Latrubesse 2015), and internal sedimentary architecture is often a complete unknown despite
very recent progress in understanding the lithostratigraphy of some megafans in the context of groundwater and oil exploration (Miller et al., Ch. 15). This
explains much of the debate currently surrounding the
significance and present-day geographical distribution of distributive fluvial systems (DFS), the uniqueness of their diagnostic facies assemblages (if any),
the vocabulary that should be used to characterise
them (i.e., establishing the true distinction beween
wide axial rivers, avulsive fluvial systems, and megafans), and their relative preponderance in the
Cenozoic sedimentary record (Fielding et al. 2012;
Latrubesse 2015; Weissmann et al. 2015; Wilkinson,
Ch. 17).

This study provides an overview of an under-
reported intracratonic megafan, remote from any convergent continental margin or Cenozoic collisional
orogen in western Europe: the Loire megafan, situated
in a region known as Sologne, central France
(Fig. 9.1). Because it occupies a major bend in the
Loire River near the city of Orléans, Sologne has been
described geologically as an 'interior delta' by implicit
analogy with the interior delta of the Niger in Mali
(Rasplus 1978, 1982). Situated just 150 km south of
Paris, the greatest claim to fame of Sologne are its
Renaissance chateaus such as Chambord, Blois, and
Cheverny.

The Sologne region has until now been addressed as
a geological region, but has never previously been
described as a megafan, or indeed even as a landform.
Its size, ~ 7,500 km^2, is almost three times that of the
most conspicuous of the Pyrenean retroforeland fans,
the Lannemezan megafan (Fig. 9.1), whose late

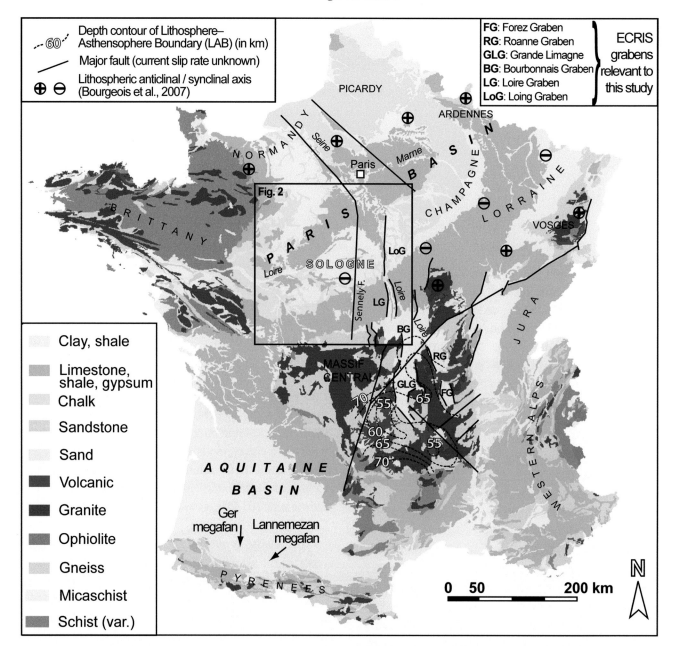

Figure 9.1 **Lithological map of France, showing tectonic and geophysical features of the Massif Central relevant to the Loire megafan**. Depths to the LAB after Sobolev et al. (1997). Yellow and pale green outcrops in the Loire and Allier catchments, including Sologne, are all areas of Miocene and younger fluvial outcrops. Cenozoic volcanic centres coincide with sites where the asthenosphere is shallowest (~55 km) and where late Neogene crustal uplift has been greatest. Other volcanic areas in the east of the Massif Central are of Paleozoic age and irrelevant to Cenozoic geodynamics. Base map: Bureau des Ressources Géologiques et Minières, http://infoterre.brgm.fr/viewer/MainTileForward.do.

Miocene to Quaternary wedge-top sequence is about 2,700 km² (Mouchené et al. 2017). Piecemeal descriptions of the stratigraphic, sedimentological and palaeogeographical features of Sologne are provided in the companion handbooks to the 1:50,000 scale geological maps produced by the Bureau des Recherches Géologiques et Minières (BRGM) – whose national headquarters are situated on the megafan itself, south of Orléans. Given the level of detail acquired through mapping (e.g., Fleury et al. 1991, 1992, 1997; Cruz Mermy et al. 2007; Debrand-Passard et al. 2010), the region is suited to gaining insight into the characteristics of other continental megafans in regions where such detailed databases are lacking.

9.2 Regional Characteristics of the Loire Megafan

9.2.1 Sologne, a Large Sand Body in Central France

Sologne is a distinctive geographical region of central France (Fig. 9.2), a sandy 'wasteland' (Sutton 1971) where rye cultivation prevails over wheat and which was probably named after the Latin for rye (genus 'secale', hence 'Secalonia'; Denizet 1900). While still a mosaic of heathland, marshland and lakes, the region was actively afforested between 1850 and 1879, with intense marling and draining efforts to improve soils for agriculture. The area presents abrupt boundaries with the adjacent wheatbaskets of central France, namely the

loess-covered limestone Beauce plateau to the north, and the fertile limestone plateaus of Berry to the south. It is flanked by the wine-growing regions of Touraine to the west and Sancerre to the east. Situated just north of the Massif Central, Sologne also constitutes the youngest onshore depocentre of the Paris Basin.

Sologne is thus a conspicuously coherent land unit, definable today by:

(i) its diamond shape (Fig. 9.2);
(ii) the land use – a relatively uniform mosaic of forest and heathland (Fig. 9.2) on poor soils (often podzols), either too dry (sand) or too waterlogged (clay) for agriculture;

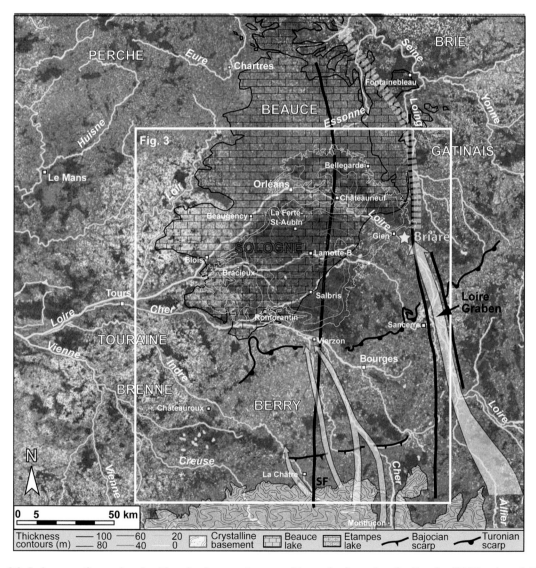

Figure 9.2 **Loire megafan regional setting: land use patterns and isopachs**. Isopachs after Rasplus (1982), adapted. Yellow arrows: clastic influx routes from the Massif Central, primarily by an ancestral Loire River, secondarily via an ancestral Cher River. Yellow star: proposed megafan apex and point of avulsion. Outcrops and arrows after Larue and Etienne (1998), Fleury et al. (1991), Debrand-Passard et al. (2010), corroborated by other sources.

(iii) the entirely allocthonous nature of its mostly unconsolidated sand or sandy clay deposits (Fig. 9.3)
(Lacquement et al. 2010);

(iv) the fine-grained texture of its entirely alluvial deposits,
which among other criteria is detectable from airborne
gamma spectrometry maps that show excesses of
potassium associated with the widespread presence of
clay and sandy clay (SOD Fig. 9.1; Lacquement et al.
2015; Tissoux et al. 2017). These radiometric facies

produce similar spectral signatures to those of the
basement rocks of the Massif Central, which contain
comparatively large amounts of U, Th and K, thus
providing a clue to this as the source region.

Despite further reclamation efforts after World War II,
Sologne has remained a region of stagnant agriculture,
given instead to forestry, wild game hunting and fish-
ing – much as in the Renaissance period. Groundwater

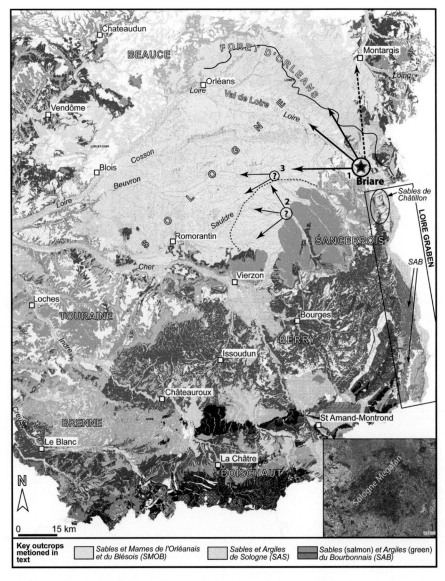

Figure 9.3 **Geological resources of central France: a lithological map**. Note uniformity of materials making up the Loire
megafan irrespective of age or depositional origin. SMOB, SAS and SAB outcrops of direct relevance to the Loire megafan are
specified in key below map, and the 'Sables de Châtillon' formation in the apex region is also labelled. Stipple texture over the
SMOB and SAS polygons indicates Quaternary overburden. General colour coding is as follows: pale orange, yellow, and grey
all depict fine clastic materials (unconsolidated and weakly consolidated) mapped as suitable for aggregate, ballast, and
industrial-grade silicates. Deeper green: clays. Red: Paleogene laterite or red beds. Blue to purple: mostly limestone, tufa, etc.
(outcrops suitable for lime, cement, and building-stone industries). Solid black line: present-day drainage divide between the
Seine and Loire watersheds. Briare (star): megafan apex 1. Hypothetical apex 2 of possible (Sauldre River) smaller alluvial fan
(dashed line). Subapex 3 reminiscent of a downfan subapex of the Loire River (see text). More detailed legend freely available in
the public report by Thauvin et al. (2011, http://infoterre.brgm.fr/rapports/RP-59248-FR.pdf).

pH is acid, often iron-rich, and the potential for irrigation is low given the high technical costs of harnessing groundwater from sand. As a result, traditional wells are shallow, and the low topographic gradients and intraformational geological dips explain the very low aquifer productivity (maxima of 10 m^3/h; Cruz Mermy et al. 2007).

9.2.2 The Megafan in Its Continental Environment

The first-order features of the Loire megafan are its remoteness from marine shores for most of its geological history, the steady position of its depocentre, and its connection to a Cenozoic topographic swell about five times the diameter of the megafan itself: the Massif Central, a roughly triangular upland region with each side approximately 350 km in length (Fig. 9.1, and SOD 9.2).

9.2.2.1 The Sediment Sink: A Large Clastic Depocentre

About 20% of the modern Loire River catchment, which is situated half in the Massif Central and half in the Paris Basin, consists of outcrops of alluvial deposits ranging in age from Miocene to modern (Fig. 9.1). Components of the European Continental Rift System (ECRiS), forming extensional grabens within the Massif Central, represent about half of those alluvial deposits, whereas Sologne, situated outside the Massif Central in an intracontinental sag basin, consists roughly of the other half. In the Massif Central we are dealing with axial fluvial belts guided by rift tectonics, and in Sologne with a megafan, where tectonics have played a subsidiary role.

The Sologne depocentre is located at the southwestern extremity of a NE-striking lithospheric furrow linking central Germany to the southern portion of the Paris Basin (Guillocheau et al. 2000). It is tentatively known as the Sologne–Franconian Basin Syncline (SFBS: Bourgeois et al. 2007; Bessin et al. 2017). Two anticlines parallel to the SFBS have been advocated to explain late Cenozoic uplift of the Swabian Jura, Black Forest, Vosges and Burgundy to the south, and the Vogelsberg, Hunsrück, Taunus, Eifel, Ardennes and Normandy to the north of the SFBS (Fig. 9.1). The relative relief of these lithospheric undulations attains maxima of 1 km for wavelengths of ~200 km. The Sologne synclinal depocentre, therefore, is approximately 100 km in diameter, which is exactly the size of the extant megafan (Fig. 9.2).

The region situated between the current Massif Central upland and the Paris Basin was a land of large, shallow lakes from as far back as mid-Eocene time, functioning as a sediment sink for much of the Cenozoic. The various lakes filled with lacustrine limestone in the Brenne, Brie, and Beauce regions (Figs. 9.2 and 9.3), with marine (nummulitic limestone) conditions occurring only farther north in connection with the North Sea and the Atlantic. As the sea retreated northwards from the Paris Basin (early Oligocene), of special interest is the vast (Rupelian) Etampes Lake, which established itself across the Sologne, Beauce, and Ile-de-France areas. Thereafter the lake fluctuated in size, and younger depositional limestone sequences indicate that the depocentre migrated south towards Sologne, where the Aquitanian Pithiviers limestone ('Calcaires de Pithiviers') succeeded the Etampes limestone (Fig. 9.4). The Beauce limestone is thus an umbrella term encompassing the early Oligocene to early Miocene lacustrine limestone sequence (43–25 Ma) that accumulated in the large, shallow bowl of the Beauce–Sologne depocentre (Figs. 9.1 and 9.2). Its history relates to a low-gradient, low-energy environment with limited clastic inputs (note very low sedimentation rates in Fig. 9.4) prevailing for about 14 Myr under semi-arid conditions (Turland et al. 1990) and internal drainage.

These last depositional stages in the intracratonic Paris Basin, before it underwent tectonic inversion, could perhaps be compared to an environment similar to present-day Lake Chad in sub-Saharan Africa. The Beauce limestone has accordingly been studied for its palustrine stromatolites and calcrete duricrusts (Freytet 1965; Freytet and Plaziat 1982). Importantly, this relatively thin sequence of lacustrine limestones forms the geological plinth upon which the Loire megafan deposits began to aggrade during the middle Miocene – a first indication of steeper slope systems and river channels driven by uplift in the Massif Central. Sologne continued thereafter to function as a regional depocentre where aggradation consistently involved clastic sequences entirely devoid of limestone beds (Fig. 9.4).

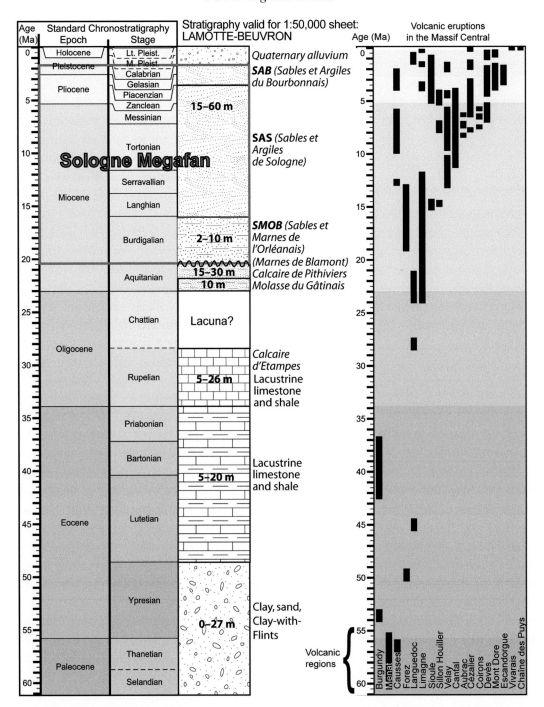

Figure 9.4 **Synthetic stratigraphy of Sologne** (from Lamotte-Beuvron 1:50,000 scale geological sheet: Debrand-Passard et al. 2010), with chronology of volcanic eruptions in the Massif Central (modified after Wilson and Patterson 2001). A colour version of this figure is available in the SOD for Chapter 9.

9.2.2.2 The Sediment Source: A Rising Volcanic Swell

The sediment fill was sourced from the Massif Central (Fig. 9.1), a topographic swell capped by volcanoes. In this respect it is similar to many volcanic swells of the African continent, around which a number of intracratonic megafans have also been reported (Burke and Gunnell 2008; Wilkinson et al., Ch. 3). The Massif Central is a Hercynian highland consisting of high-grade metamorphic nappes and plutonic intrusions, with only vestiges of post-orogenic cover rocks. The northern edge of the crystalline outcrops occurs barely 80 km south of the southern edge of the Loire megafan (Fig. 9.2). ECRiS-related rifting started during mid-Eocene time, when lacustrine environments were already prevailing in the Paris Basin.

Late Cenozoic uplift of the Massif Central was dynamic, with topographic buoyancy ensured by a shallow, 200 km wide 'hot finger' of upwelling asthenosphere (Olivetti et al. 2020). Lithospheric thinning above the corresponding mantle diapir ensued (see SOD 9.2 for further details). The timing of uplift in the Massif Central is directly relevant to the Loire megafan, its age being indirectly a function of the geodynamic process responsible for generating this highland swell of central France. The Loire megafan clearly records the sedimentary history of post-14 Ma updoming, and not that of the ECRiS grabens internal to the Massif Central itself. Prior to 14 Ma, palaeontological evidence supports the idea that marine transgressions in the Grande Limagne, Roanne, and Forez grabens (Fig. 9.1) occurred intermittently, indicating that before Burdigalian time (~20 Ma), most of the Massif Central still lay close to sea level and was, like the Paris Basin at the time, a low-elevation patchwork of shallow lakes. For example, during the early Oligocene, the Alpine molasse sea made incursions from the east. Lacustrine conditions prevailed in the ECRiS grabens during Chattian and Aquitanian times (Sissingh 2001, 2006; Ziegler and Dèzes 2007). ECRiS graben overfilling by middle Miocene time, when extensional fault activity had abated, together with regional uplift (the topographic floors of the Grande Limagne, Roanne, and Forez graben currently lie at elevations of 350 m) tipped the regional balance in favour of sediment transfer out of the internal ECRiS depocentres to external depocentres such as Sologne. Layers of river alluvium sandwiched between dated Pliocene and Pleistocene lava flows have shown that deep valley incision into the Massif Central highlands (with some valleys up to 400 m deep) was rapid. It occurred as recently as the Pliocene and early Pleistocene (Etienne 1984; Goër de Herve and Etienne 1991), thus emphasizing the youthfulness of relief on the volcanic swell.

9.3 First-Order Anatomy of the Loire Megafan

Figure 9.1 clearly suggests that the Massif Central source area and the Paris Basin were connected by large rivers, with an ancestral Loire floodway guided by active faults (Loire Graben) and reaching a fork, or avulsion point, where the transfer of sand and clay could either continue northwards to the Seine catchment via the Loing Graben, or westwards to Sologne. Both scenarios in fact occurred intermittently during the late Cenozoic (see Section 9.5.2), but Sologne remained by far the dominant aggradational depocentre.

The outcrop of Loire megafan sand-and-clay deposits is 120 km long and 80 km wide. Mean elevations range between 100 and 110 m, dropping to 80 m on the westernmost (distal) fringe but rising to nearly 190 m in the east (above the towns of Léré, Châtillon-sur-Loire, Briare) near the inferred apex; and to 177 m in the Forêt d'Orléans, which is a continuation of the megafan north of the Loire (Figs. 9.2 and 9.3). The resulting slope is 0.08% to the west. Modern rivers on the fan currently follow that slope, e.g., the Cosson, Beuvron, and Sauldre (SOD Fig. 9.1). Maximum megafan deposit thicknesses occur around Salbris (90 m) and La Ferté-Saint-Aubin (80 m), with two secondary maxima near Soings (60 m) and Châteauneuf-sur-Loire (60 m, north of the modern Loire valley) (Fig. 9.5). The Sologne depocentre also accumulated greater thicknesses of detrital material west of the N–S-trending Sennely Fault (Figs. 9.1 and 9.2), which was moderately active throughout Cenozoic time and generated excess accommodation space in the downthrown block west of the fault. Although it remains difficult to reconstruct regional-scale geometries within the body of the megafan, hydrogeological data indicate water-table slopes beneath interfluves of ~0.001% (geological sheet: Aubigny-sur-Nère; Fleury 1990). Water-table gradients steepen to 0.005% on valley slopes, where rivers have incised the megafan deposits since the Quaternary.

The Beauce limestone platform, with dips $< 1°$ to the NW, underlies the Loire megafan deposits (Fig. 9.2). The Beauce palaeolake, last avatar of the larger Oligocene lake (Etampes Lake), was a residual water body lacking uniform depth. As the lake dried up, the depositional limestone surface formed a gentle ramp sloping northwest, i.e., away from the Massif

Central and its north-flowing river systems. It also underwent minor karst-generated undulations, although never exceeding amplitudes of ~20 m. Peaks and troughs in Loire megafan deposit thicknesses correlate well with highs and lows in the buried topography of this limestone plinth – except along the modern Loire valley itself, where Quaternary incision has removed most of the Neogene material (a residual thickness of 10 m is preserved beneath the Loire alluvial belt at Orléans).

The Mesozoic strata resting on the basement rocks of the Massif Central also dip by 1° to the north and have been sculpted into low-relief, scarp-and-vale topography (south-facing scarps in Bajocian and Turonian limestone in Berry, Fig. 9.2). In the eastern third of the area, the megafan deposits cover a 40 m-thick layer of Clay-with-Flints (Fig. 9.3). This residual deposit results from chemical weathering of the underlying Cretaceous chalk under hothouse climatic conditions during the Paleogene. The Cretaceous chalk dips gently northwest towards the Beauce paleolake and plunges beneath it.

9.4 The Loire Megafan Sedimentary Sequence

9.4.1 Data Sources

The 1940 geological map of central France depicted Sologne as a uniformly pale pink mass, a reflection of the superficial monotony of its geology first studied by Denizot (1927). Detailed investigations began in 1986, complete with numerous boreholes. Synthesis of the data was achieved between 1998 and 2010 by the BRGM, with field evidence progressively completing, but also altering, earlier interpretations of the stratigraphic sequence such as those of Rasplus (1982). Rasplus (1978) provided the first comprehensive mineralogical provenance analysis of the deposits. The present overview is based on an additional 22 geological maps (1:50,000 scale) displaying outcrops of Neogene sand and clay deposits, with accompanying handbooks succinctly describing outcrop and subcrop geology, hydrogeology, economic geology, and palaeogeography. Because of dense afforestation and private land ownership, geological exploration is difficult, and some maps were among the last in France to be completed. Quarrying in this geological formation

has been confined historically to small-scale clay workings within the sand bodies for brick and tile manufacturing. No quarries in the Neogene formations are still active. They have been either overgrown by forest, converted to lakes, or turned into landfill sites. The deepest vertical cuts are ~5 m. Overall, conditions for field observation are poor, and the entire stratigraphy has been mostly reconstructed from a fairly dense population of borehole logs.

9.4.2 The Basal Sequence: 'Sables et Marnes de l'Orléanais et du Blésois' (SMOB, 20–14 Ma)

The 'Sables et marnes de l'Orléanais et du Blésois' (SMOB; chiefly sand and marl) constitute the lower sequence of the megafan, but in detail the basal strata consist of a thin layer of poorly sorted clastic debris of fluvial origin ('Molasse du Gâtinais': sand, mudstone and marl) and of marl ('Marnes de Blamont') with a lacustrine limestone interlayer ('Calcaire de Pithiviers') (Fig. 9.4). Typically, within the first metre immediately above the Beauce limestone, the deposits contain a large number of Aquitanian (24–20 Ma) to early Burdigalian fossil remains. The stratigraphic age has thus been determined on the basis of fossil megafauna (biozone MN 3) on the north side of the Loire River: *Rhinoceros turonensis*, *Rhinoceros brachypus*, *Anchitherium aurelianense*, *Listriodon lockharti*, *Amphycion giganteus*, *Mastodon angustidens*, *Dinotherium cuvieri*, as well as crocodiles and tortoises, beaver- and otter-related species (*Steneofiber depereti carnutense*, and *Potamotherium miocenicum*), and palm treetrunks. These biota suggest freshwater environments.

The SMOB marks the termination of carbonate sedimentation in the Paris Basin and the steady onset of clastic influx from the south, at a time of transition towards wetter climatic conditions and of topographic change in the Massif Central. The SMOB is currently considered as a package comprising the 'Marnes et Calcaires de l'Orléanais' and 'Marnes et Sables de l'Orléanais', the latter being equivalent to the 'Marnes et Sables du Blésois' in the older literature. Maximum thickness does not exceed 40 m. The marl and limestone near the base of the SMOB occur as discrete islands ~10 m thick or more, but the overlying formation consisting of marl and sand likewise forms islands or lenses of irregular thickness. It reaches thickness maxima of 35 m near Romorantin but thins to a few metres in other places. The causes for such variations are unknown but can be ascribed to exogenic factors, such as differentiated subsidence and fault movements producing localised, secondary depocentres; or to autogenic, fluvially-driven patterns of floodplain or levee aggradation and erosion through space and time. Lateral variations in sediment texture range from silt to rudite, upwards to sub-angular grains of coarse sand ('gros sel' in the literature, i.e., cooking salt texture), with occasional small rounded pebbles that seem to concentrate in palaeochannels.

Overall, the pattern suggests multiple floodways with relatively stable islands. Figure 9.5 shows that the sediment 'islands' standing at +80 or +90 m are separated by channels lying at +60 to +75 m, which are filled with the younger 'Sables et Argiles de Sologne' (SAS) unit (see below). None of the borehole logs report layers of organic material, suggesting that the clastic floodplain accretion has preserved little evidence of past vegetation on these islands. Pollen counts from the SMOB nonetheless indicate a mix of conifers (*Pinus, Sequoia, Tsuga, Cathaya*), moisture-tolerant hardwoods (*Quercus, Betula, Corylus, Alnus, Ulmus*), and a range of herbaceous plants. The hardwood pollen may have been transported from the Massif Central, but its presence is more likely indicative of stable, elevated islands hosting terrestrial hardwood instead of riparian softwood species.

9.4.3 The Middle and Top Sequences: 'Sables et Argiles de Sologne' (SAS, 13–3.4 Ma), and 'Sables et Argiles du Bourbonnais' (SAB, 3.1–1.9 Ma)

Unlike the SMOB, the SAS and SAB units are entirely devoid of fossil remains. The SAS formation ranges in thickness between 15 and 60 m. It was once argued by Rasplus (1978, 1982) that the Cher River was the main conveyor of sediment because heavy mineral suites in Sologne lacked volcanic minerals from the upper reaches of the Loire and Allier river catchments (the Allier is a major tributary of the Loire; Fig. 9.2). However, it has since been shown by Gigout and Desprez (1977), Fleury (1990), and Fleury et al. (1991) that the SAS are instead a continuation of the middle Miocene ('Vindobonian') 'Sables de

Châtillon', with outcrops mapped at elevations of up to 190 m on left-bank plateau surfaces of the Loire over a distance of 25 km upstream of Briare (Fig. 9.3). On that basis, it would seem reasonable to infer that the fan head/avulsion point of the Loire megafan was situated near Briare, and that the Loire Graben (terminating near the fan head, Figs. 9.1 and 9.2) was a transfer zone for most of the Sologne deposits (Debrand-Passard et al. 1992; Debrand-Passard 1995). In support of this interpretation of an avulsion point near Briare, the 'Sables de Châtillon' formation is described as exhibiting "stratification tourmentée de type torrentiel" (poorly sorted and poorly stratified debris-flood deposits: Gigout and Desprez 1977, p. 7). It features a much coarser-textured load than sediments at locations farther north and west within the SAS unit, and consists of quartz and flint pebbles (the latter remobilised from Clay-with-Flints outcrops in the local catchments of the Sancerre region, Figs. 9.2 and 9.3) embedded in a clay-rich matrix. The maximum known thickness of the 'Sables de Châtillon' is 30 m.

The SAB formation is younger than the SAS, although defining it on maps as Pliocene or Quaternary has also fluctuated as a result of the Pliocene–Pleistocene boundary being pushed back to 2.588 Ma by the International Commission on Stratigraphy in 2010 (geological maps of the region are mostly pre-2010). Radiometric dating is thus of the essence. Specific mineralogical signatures relating to eruptions of the Mont-Dore stratovolcano in the Massif Central have helped confirm the SAS as stratigraphically older than the SAB. On the basis of dated mineral suites diagnostic of Mont-Dore volcanic ejecta and of their cross-cutting relations with the SAS, the SAB are Upper Pliocene to Lower Pleistocene in age (3.1–1.9 Ma), whereas SAS sedimentation ceased during the Zanclean (i.e., prior to ~3.4 Ma, Tourenq 1989). The main index minerals used for establishing this distinction are pumice-derived quartz crystals and volcanic zircons present in small amounts in the SAB but not in the SAS. Despite debate over the source and transport routes of the large masses of detrital sediment, it seems clear that ancestors to the Loire and Allier rivers have been the main conveyors of the SAS and SAB via the Bourbonnais and Loire grabens (see location of SAB outcrops in Fig. 9.3). The formation contains pebble-filled channels in which volcanic and nonvolcanic clasts up to 5 cm in diameter decrease in mean size with distance downstream (Thauvin et al. 2011). Vestiges of the SAB sequence also occur in Berry but are < 10 m thick. In contrast, the Loire–Allier interfluve (Fig. 9.2) still displays SAB thicknesses of ~50 m. The steeper gradient of Pliocene deposits compared to that of inset Quaternary alluvial terraces in the southern Paris Basin (Larue and Etienne 1998; Larue 2003a) testifies to the continuing northwards crustal tilt of the landsurface away from the rising Massif Central towards the Paris Basin.

The ancestor to the Cher River was a fan-margin river, and thus not a major feeder to the SAS. Larue and Etienne (1998) argued nonetheless that at least the SW portion of the SAS was supplied by the Cher catchment, and thus by northern exposures of the Massif Central basement. Evidence for this comes from the presence of grossular in the SAS deposits around Romorantin – grossular being an index mineral typical of metamorphic outcrops in the Cher watershed and encountered in its alluvial sequences. The presence of barytine in a few SAS samples also suggests that nearby silcrete outcrops (silicified Eocene lacustrine limestone) in Berry were minor contributors to the megafan. In other, more central parts of the megafan such as around Salbris (Fig. 9.2, SOD Fig. 9.1), almandine, more typical of the Loire catchment, is predominant, whereas grossular is absent from those areas. This mixed provenance is also confirmed by the andalusite/staurotide ratios mapped by Rasplus (1982), who reported high ratios in the east and centre of the SAS, consistent with inputs from the Loire catchment – and low ratios in the southwestern tracts, consistent with conrtibutions from the Cher headwaters.

The extant channel orientations that arguably explain the aforementioned distributions are the following. The present-day smaller rivers (Fig. 9.3, SOD Fig. 9.1), especially the Beuvron and Cosson, nominally qualify as fan-fed rivers (as opposed to the Loire itself: the mountain-fed river; Sinha and Fiend 1994) and likely occupy beheaded palaeochannels of the Loire River; such palaeochannels are typically oriented radially from a primary avulsion point, in this case near Briare (point 1, radiating heavy arrows, Fig. 9.3). A subapex appears to be situated near point 3 (radiating arrows, Fig. 9.3), with prior courses of the Loire related to the modern Beuvron and lower Sauldre courses.

9.4.4 Sedimentological Features and Palaeohydrological Inferences

SMOB sediments are predominantly quartz and feldspar sands with clay lenses, marl, and local occurrences of limestone. Mineralogical composition is poorly distinguishable from the overlying SAS. The dominant sand masses lack grain sorting, and feldspar grains (mostly orthoclase) can reach proportions of up to 30–35% at any given location, particularly in the 0.315–0.500 mm fraction. Such high proportions of feldspar suggest low-intensity weathering of crystalline rocks in the source area, reminiscent of semi-arid tropical granitic basement environments. The feldspar grains are mostly poorly rounded, with an admixture of likewise poorly rounded quartz clasts (< 2 cm). Clay occurs more often as the finer matrix of the sand and gravel mass than as uniform, monomineralic bodies, although clay pits and quarries also document the existence of economic grade clay lenses. The presence of rare limestone lenses and stringers is the main feature that distinguishes the SMOB from the overlying, carbonate-free SAS. Heavy-mineral content of the SMOB is dominated by zircon (average: 26%) and epidote (maximum: 20%), with lower proportions of garnet (< 10%), staurotide (13%) and andalusite (9%). Among clay minerals, smectite dominates with 50 to 60% of the total, suggesting (by analogy with smectite-dominated environments in Sub-Saharan Africa: e.g., Ruxton and Berry 1978) semi-arid weathering conditions in the source area. The sediment does not result from post-depositional granular disintegration of larger, polymineralic pebbles. The absence of pebbles is thus less an instrinsic indication of limited stream power than of hillslopes in the source areas (i.e., the rising Massif Central) being covered by a deep weathering mantle and thus delivering a predominance of fine-textured debris.

Grain sizes and textural percentages in the SAS are similar to those of the older SMOB unit, i.e., the entire grain size spectrum is present but dominated by medium sand fractions. Sorting is expressed in the stratigraphy by an occurrence of equigranular sand sheets. According to Rasplus (1982), median sand grain size in the SAS ranges between 0.08 and 0.99 mm, with a Krumbein quartile deviation $((d_{75}-d_{25})/2)$ of 1.00–1.82. As in the SMOB sequence, quartz in the SAS is dominant but feldspar (80–90% K-feldspar) represents up to 35% of the fraction. Pumice-derived quartz from Mont-Dore pyroclastic eruptions is conspicuously absent. Smectite appears to be slightly less abundant (44–91%) than in the SMOB, with greater relative proportions of kaolinite and somewhat less illite.

Stratification features are commonly reported as poorly identifiable although clay beds, when present, tend to highlight stratification within the sand bodies. From quarry exposures, Rasplus (1982) reported cross-bedding with clearly distinguishable laminae a few millimetres thick. From boreholes, Rasplus (1982) reported metre-thick lenses of sand or clay. Alternating beds of sand and clay, or coarse sand and finer sand, are also reported in certain areas within the SMOB and SAS, including graded bedding within individual sand sheets. The size or frequency of cross-bedding sets is not, however, an intrinsically reliable indicator of whether the sand deposits were produced by a small to medium-sized river with a mean discharge of a few hundred m³/s (as the modern Loire, for example), or by a large river with tens of thousands of m³/s of mean discharge (Latrubesse 2015).

Given that extremely poor grain sorting is the rule compared with that in most mid-latitude fluvial environments, Denizot (1927) even interpreted feeble quartz rounding and the rich clay and silt matrix surrounding sand-sized frame-silicate grains (frequent texture of sandy mud) as the work of 'mudflow'-type (e.g., molasse) dynamics. Rasplus (1982), however, has expressed doubts over the inference of high turbidity levels because of a presence of fossil faunas such as pike and other Cyprinids, which require clearwater habitats. He emphasised that Denizot's conclusions had been limited to the observation of surface exposures, and argued that the clay matrix and clay coatings, where present, resulted from post-depositional pedogenesis. Here I suggest that the angularity and poor sorting of sand particles, despite the considerable transport distances from the Massif Central, could be a feature of deposits fairly typical of some large tropical rivers, where peak flow in a seasonal climate involving

high discharges transports sand grains mostly as suspended load. Abrasion of sand particles thus only occurs at best during intermittent conditions of bedload transport.

9.5 Palaeogeography of the Loire Megafan

The palaeogeography of sediment routing between the Massif Central and the Paris Basin have long been a challenging puzzle. Olmalius d'Halloy (1828), for example, speculated that the large bend in the Loire hinted at a history of drainage capture and prior connections between the Loire and the Seine drainage basins. Two constraints have been key in establishing the palaeodrainage of the streams feeding or bypassing the megafan:

(i) the presence of shelly calcarenite (or crag) deposits ('falun', in French) as far east as the longitude of Blois (Fig. 9.5), which records the last and highest sea-level in northwestern France ca. 15 Ma (Fig. 9.6);
(ii) the presence, absence, or relative ratios of metamorphic index minerals and volcanic minerals, particularly in reference to the sudden rise in activity of the rhyolitic Mont-Dore volcano between 3.5 and 0.25 Ma (e.g., Féraud et al. 1990). The heavy mineral suites supplied by its ejecta to river loads have been carefully finger-printed by tephrostratigraphic analysis of the stratovolcano. These have helped authors such as (Pastre 1992) to reconstruct the course of an ancestral Loire floodway exiting the mid-Cenozoic Grande Limagne and Bourbonnais grabens (Fig. 9.1) and flowing periodically either north to the English Channel or west to Touraine and the Atlantic Ocean (Figs. 9.2 and 9.6).

Evidence for both these sets of clues is examined below.

9.5.1 Neogene Sea Levels

Onset of regional uplift in the Massif Central during the middle Miocene promoted the formation of large, poorly incised floodways which began by depositing the SMOB. At the time of this Miocene eustatic highstand, the shelf sea in western France (the 'Mer des Faluns', Fig. 9.6) formed deep coastal indentations in the form of drowned valleys. Fossil marine invertebrates and mammals indicate that the Mer des Faluns was a relatively warm, subtropical sea, consistent with the Mid-Miocene Climatic Optimum prevalent

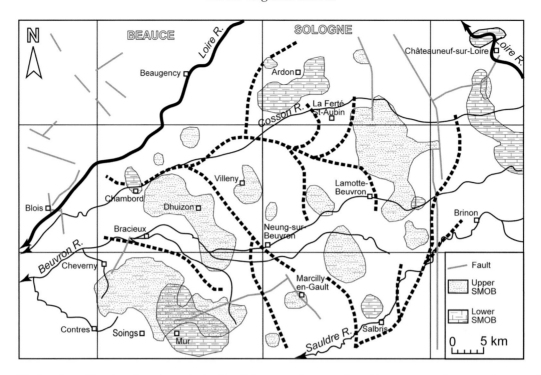

Figure 9.5 **Syn-depositional palaeochannels in the Loire megafan stratigraphic sequence** (lower and upper SMOB palaeogeography).

ca. 17–14 Ma. By cutting off insular western Brittany from the mainland, this shelf sea formed a broad, shallow strait linking the Atlantic Ocean to the English Channel seaway between 16 and 11 Ma (Fig. 9.6). One estuarine finger of the Mer des Faluns extended eastwards up the present-day lower Loire valley, reaching the SMOB depositional area from the west and generating an outcrop of shelly calcarenite as far east as 1°32'E (Soings-en-Sologne) (Fig. 9.5).

Towards the end of the Miocene, the Armorican Massif in the west underwent uplift (Dugué et al. 2012; Bessin et al. 2015, 2017) and the sea accordingly retreated to the Atlantic Ocean and Western Approaches of the English Channel. Overall, the basal SMOB sequence of the Loire megafan was thus deposited in the vicinity of a marine ria, but the much thicker SAS and SAB sequences (i.e., most of the Loire megafan stratigraphy: 13–1.9 Ma) were deposited in a continental environment after the retreat of the Mer des Faluns. Note that heavy minerals in the Mer des Faluns have Armorican signatures and contain none from the Massif Central whatsoever (Tourenq et al. 1971), thereby confirming that the Loire megafan was at no time a megafan-delta *sensu* (Lane et al., Ch. 12).

9.5.2 Loire–Seine Connections, and Histories of Avulsion at the Fan Head

For almost 200 years, competing palaeogeographic scenarios have stumbled over whether the Loire River used to join up with the English Channel via the Seine, or whether the Seine was a river originating in the Massif Central. Efforts to clarify these issues have relied on regolith compilation maps (e.g., Lacquement et al. 2010; SOD Fig. 9.2), and on provenance analyses of a 60-km-wide and 200-km-long ribbon of discontinuous alluvial outcrops across a SE–NW swathe of the Paris Basin. These outcrops extend from Gien, at a point where the modern Loire is only a few kilometres from breaching the current Seine–Loire drainage divide, to the Seine estuary in the chalklands of coastal Normandy. Across the Paris Basin, intensely studied outcrops of fluvial deposits known as the *Sables de Lozère* (type area: Lozère, a disused sand quarry in the southern suburbs of Paris; e.g., Denizot 1927; Pomerol 1951; Tourencq 1989; Debrand-Passard 1995; Tourenq and Pomerol 1995; Larue and Etienne 1998; Larue 2003a, b) contain minerals with Massif Central signatures and the 'cooking salt' texture of the quartz grains, but lack indicator

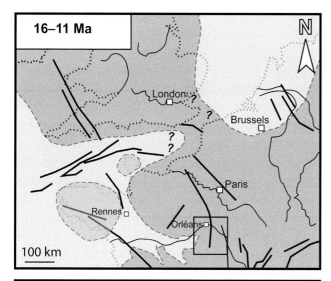

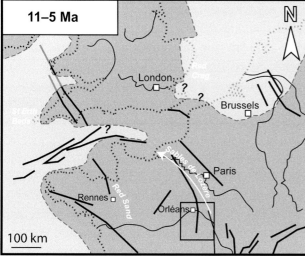

Figure 9.6 **The Loire megafan in relation to Neogene marine base levels**. Named geological formations in England and Brittany share similar facies and are considered coeval. Palaeogeographic outlines after Dugué et al. (2012).

fossils (Fig. 9.6). Many other occurrences along that swathe of terrain between Sologne and Normandy show up on 1:50,000 scale geological maps as a succession of confetti outcrops situated on plateau summits (altitudes of 180 m or less), but often partly covered by Quaternary loess or reworked as colluvium on valley slopes as a result of postdepositional plateau incision (Larue and Etienne 2000).

The sedimentological study of these residual pockets of alluvium has progressively improved our understanding of the Neogene floodways that alternately operated as feeders and bypasses of the Loire megafan. The floodways were wide and shallow,

presumably more similar to large fluvial floodways in cratonic regions of the Tropics than to any modern river valley of the mid latitudes (the floodzone of the interior Niger megafan, for example, is approximately 120 km wide; e.g., Brunet-Moret et al. 1986). Etienne and Larue (2000, 2002, 2011) have reviewed the controversy around the connections between Loire and Seine during the Neogene. Two basic theories currently exist: the first (minority) view is that ancestral floodways from the Loire catchment never reached the English Channel area directly (Freytet et al. 1989). Instead, minerals from the Massif Central present in alluvium deposits in the Seine catchment are considered to have been redistributed by tributaries of the Seine through a process of headward erosion, capturing and transporting deposits stored in the northern part of the Loire megafan, close to the drainage divide (Forêt d'Orléans upland area, in particular; Fig. 9.3). The second theory advocates that an ancestor to the Loire reached the English Channel area directly but intermittently during the last 20 Ma. Debate focuses on whether this happened once or more than once, and this has given rise to three palaeogeographic scenarios.

(i) Scenario 1 advocates a single episode of connection between the ancestral Loire and Seine catchments, and this places the *Sables de Lozère* firmly in the Burdigalian (20–16 Ma) on the basis of their superficial resemblance with the SMOB or SAS. Paris Basin inversion in the Beauce, Brie, and Normandy, and subsidence in the lower Loire valley, promoted the establishment and stabilisation of the modern Loire alluvial belt, definitively ending the connection to the English Channel area (e.g., Denizot 1927; Larue and Etienne 2002).

(ii) Scenario 2 adds to the Burdigalian ancestor a second Loire floodway episode during the late Pliocene, before its extinction and restoration of a westerly flow through Sologne to the Atlantic during the early Pleistocene. This fostered the notion of two palaeofloodways through the Paris Basin, with distinct trails of deposits and distinct mineral signatures distinguishable on the basis of their volcanic mineral content.

(iii) Scenario 3 supports the notion of three intermittent floodways: Burdigalian, Pliocene, and also early Pleistocene. Flow northwards to the Seine catchment would have continued at least into the early Pleistocene as a result of local tectonic adjustments in the Loire Graben and Loing Graben (Freytet et al. 1989; Debrand-Passard et al. 1992, 1998; Tourenq and Pomerol 1995; Pasquiou 1995; Larue 2003b).

This indication is supported by an augite-rich sand deposit (Larue 1999), which is traceable from Briare northwards but is absent from alluvium occurring to the west of the Loire bend at Gien (Pastre 1986; Larue 2003a), i.e., in Sologne. The augite has been traced to a 1.75 Ma Mont-Dore eruption. The proportion of augite crystals (green clinopyroxene) in the deposits distributed along this 'augite floodway' ('Fleuve à augite') can attain 37% of the total mineral content, mixed with zircons probably re-eroded out of the SAB formation. The fluvial origin of the augite grains, which lie up to 500 km from the volcanic source, has been advocated on the basis of grain size (0.08–0.4 mm) (research on Massif Central volcanism and other sites around the world shows that grain sizes of this calibre are never projected more than ~110 km from a volcano, and thus cannot be airborne in the present case; Tourenq and Pomerol 1995). The 1.75 Ma 'augite floodway' is the oldest and most elevated among the Loire's Pleistocene terraces in the Briare area.

All three scenarios are implicitly based on allogenic causes driving the regional patterns of epeirogeny, and thus on changes in sediment transport rates through the transfer zone between the rising Massif Central and the subsiding Sologne depocentre. A fourth possibility would be to consider the Loire megafan as a 'divide megafan' (Wilkinson, Ch. 17), with autogenic Loire avulsions occurring in the vicinity of Briare and feeding the west and north alternately (but unequally) over a long period of the Neogene. This autogenic scenario was nonetheless terminated by Quaternary valley incision in the Paris Basin, made evident by southwards recession of the Loire–Seine drainage divide now encroaching on the elevated ridge (max. altitude: 177 m) of SMOB and SAS megafan outcrops north of Orléans (Fig. 9.3). Prior to this, the sandbed floodways of the Loire were able to convey their loads unobstructed by steep topographic relief (e.g., valleys) all the way to Normandy.

9.6 Quaternary Incision of the Loire Megafan

The Miocene and Pliocene history of the Loire megafan was one of two alternating floodways inferred mostly from provenance studies and the presence or absence of diagnostic minerals produced in the Massif Central by well-dated eruptions. These floodways supplied the megafan to the west or, subsidiarily, the English Channel to the north. By Pleistocene time, the ancestral Loire was definitively diverted westwards across Sologne, generating a flight of Quaternary fluvial terraces in response to continuing tectonic inversion of the Paris Basin. As a result, the late Neogene Loire megafan is no longer evolving under an aggradational regime. Continual subsidence of the Touraine region (Macaire 1981) could explain likewise the bend in the Cher River near Vierzon, and more generally the convergence of the Cher, Indre, Vienne, and several other smaller rivers (Cosson, Beuvron, Sauldre) that flow parallel to the Loire – also westwards across Sologne (SOD Fig. 9.1).

Maximum Quaternary incision of the megafan units by the modern Loire River at Orléans is 60 m, which is only about half the depth of incision accomplished by the Seine River at Paris (> 120 m) during roughly the same period despite the two cities being located at identical along-channel distances from marine base level (373 km for the Loire, and 365 km for the Seine). Based on available data, little evidence exists of valley incision in the south of the Paris Basin before at least the late early Pleistocene. This is supported by the age sequences of alluvial terraces along the Creuse, Cher, and Loire rivers (Fig. 9.2), which have yielded predominantly Middle Pleistocene ESR (electron spin resonance) depositional ages, with an age of 1.7 Ma on one terrace of the Creuse River (Voinchet et al. 2007). The 'augite floodway' in the Loing, mid-Seine, and Eure catchments was argued by Tourenq and Pomerol (1995) to be inset well below extant levels of the *Sables de Lozère*, confirming that valley incision was under way during the early Pleistocene.

Finding a consensus on Quaternary deposit nomenclature and age is still an ongoing process. Signatures obtained from land classifications based on airborne gamma-ray spectrometric imagery have nonetheless helped to clarify the distribution of Quaternary terraces (SOD Fig. 9.1; Lacquement et al. 2015; Tissoux et al. 2017). Debrand-Passard et al. (2010) and Tissoux et al. (2013) inventoried up to nine terraces among the fan-fed rivers of Sologne, and at least six along the Loire alluvial belt itself. Tissoux et al. (2013) and Liard et al. (2017) dated a selection of Quaternary terraces of the Loire in the Val d'Orléans and in the upper catchments of the Sauldre, Cosson and Beuvron rivers. The ESR depositional ages range between ~600 ka and 150 ka, with stages broadly consistent with the orbitally driven climatic cycles of the Middle Pleistocene. Importantly, they reported no ESR ages older that 600 ka among the rivers of Sologne and Sancerrois (highest sampled terrace: 30–40 m above the modern channel), inferring thereby two possible alternatives: either the 'augite floodway' joining the Seine catchment persisted until as late as 0.5 or 0.4 Ma (Tourenq and Pomerol 1995, who ventured its persistence as late as 200 ka); or the uppermost Quaternary fluvial terraces of Sologne (including deposits potentially containing augite) have been stripped away. This chronology suggests at least that the emergence of the Loire and Seine as separate drainage basins is a geologically recent feature, probably occurring mostly during the Middle and Late Pleistocene. More research is nonetheless needed on the regional geochronology and through standardisation of the Pliocene and Pleistocene deposits nomenclature (see Larue and Belizal 2016).

Quaternary fluvial incision has removed substantial swathes of the original SAB, SAS and SMOB stratigraphy, which are thus very extensively buried beneath veneers of younger regolith (Fig. 9.3, SOD Figs. 9.1 and 9.2). Whereas the Quaternary terrace sequence of the mountain-fed Loire contains a rich variety of volcanic minerals from the Massif Central, the fan-fed rivers since early Pleistocene time have instead been mostly cannibalising the megafan's Neogene sediments on very low channel gradients of 0.0002–0.0005 (Debrand-Passard et al. 1998). Some distinguishing features of the Quaternary terrace deposits are that they are texturally better sorted than older megafan deposits, overwhelmingly displaying quartz within the sand fraction; and they display less intense postdepositional weathering, contain more abundant and more diverse heavy mineral assemblages as a consequence of remixing, and exhibit intraformational overprints ascribable to climatic oscillations alternating between tundra permafrost (frost wedges and a variety of thermokarst features) and temperate forest environments.

9.7 Overview, Discussion, and Conclusions

Sologne can be classed as a megafan laid down by the Loire River, comprising three fluvially-deposited units. The landform is ~120 km long, 80 km wide, and its deposits are 0.1 km thick. The low slope (0.08%) to the west is broadly consistent with an extremely low-gradient, fan-like continental deposit (Stanistreet and McCarthy 1993). Modern rivers that have cut sequences of Quaternary alluvial terraces into the fan currently follow this slope, e.g., the Cosson, Beuvron, and Sauldre (SOD Fig. 9.1). The megafan displays a classic apex in the Briare region, with evidence of up to three major avulsions distributing successive ribbons of fluvial sediment across the Paris Basin to Normandy, and with behaviour tentatively suggestive of a 'divide megafan'. The Cher, and the Loire itself below Orléans, fit a general pattern of megafan landscapes that would define them as fan-margin rivers (see Wilkinson, Ch. 17), i.e., they are larger than the on-fan misfit streams, which themselves are possibly beheaded remnants of ancestral palaeochannels of the Loire. Such beheaded and smaller misfit streams (Cosson, Beuvron) occupy radial courses that probably followed radially oriented ancient courses of the Loire. The ancestral Loire, for example, may, on repeated occasions, have occupied the course of the modern Sauldre in the Salbris–Romorantin–Selles-sur-Cher sector (Fig. 9.3). On the basis of drainage pattern, the upper Sauldre River area is likely (i) the trunk stream of an alluvial fan (ii) modifying / indenting the Loire megafan diamond shape in the Salbris region. Streams from the south apparently explain the unique mineralogical mix in the SW portion of the fan because they feed the fan-margin depression (Sauldre–Cher).

The Loire megafan was formed in a low-relief landscape over a period of ~20 Ma at the receiving end of a source-to-sink sedimentary continuum, with no tectonic, structural or erosional landforms across its path (e.g., valleys) likely to impede sediment transfer. The ancestral Loire River sustained discharges large enough to compensate for topographic gradients low enough to generate a long reach into the receiving Beauce sag basin. Quartz and (subsidiarily) feldspar are the most abundant minerals, and crystal subangularity – despite prior travel distances of 100–300 km – confirms the notion of great discharges capable of transporting sand-sized particles most of the time as suspended load. The megafan deposit has experienced a long succession (i) of climatic changes, from tropical swamps in the Miocene to periglacial frost-wedging and aeolian coversands in the Pleistocene; (ii) of perceptible but not radical changes in its hydrological and sedimentological regime; (iii) of relative changes in its sediment sources, which are traceable from heavy mineral suites – some relating to volcanic eruptions in the Massif Central; and (iv) of epeirogenic events in its avulsive behaviour.

Today, Sologne is a fossil megafan, i.e., it is no longer actively aggrading or prograding. Quaternary valley incision, chiefly in response to regional uplift of the Paris Basin, has reached local maxima of ~60 m for the Loire and ~40 m for its Sologne tributaries. Dissection of the megafan is nonetheless geologically recent, which explains why it has retained the surface features and patterns summarised above. It is also remarkable that an anabranching fluvial pattern can be reconstructed from the interpolation of isopachs and from the topographic pattern of each respective formation of the Sologne stratigraphic sequence (Fig. 9.5). Anabranching rivers are defined as multiple channel systems with islands (usually vegetated) that may persist for millennia, with relatively stable banks (Latrubesse 2008). The differentiated (undulating) topography of the intraformational stratigraphic boundaries, including the basal unconformity with the 'Calcaire de Beauce' (Cruz Mermy et al. 2007), is consistent with the diagnostic presence of internal and basal erosional surfaces previously reported from the record of very large alluvial rivers (Latrubesse 2015). In Figure 9.5, the distributions of the older islands (marly limestone, relative variation in thickness and surface topography 0–15 m, lower SMOB) and the younger islands (sandy marl, upper SMOB) generally do not overlap. This suggests that the younger deposits filled palaeochannels between the pre-existing islands before eventually aggrading over them, thus documenting two distinct aggradational phases with an intervening discontinuity. The pattern is also quite consistent with a fan-forming trunk river generating broad, shallow floodways into the prior poorly consolidated megafan surface, i.e., in the familiar cut-and-fill style of rivers but applied here to broad megafan surfaces (see examples in Wilkinson et al., Ch. 4) with successions of aggradation (lower SMOB), erosion, aggradation (upper SMOB), erosion, then SAS aggradation, followed by the SAB and 'augite floodway' (Figs. 9.3 and 9.4). All of these events are probably allogenically controlled (tectonics, climate), and difficult to untwine in the geological record from the

strictly autogenic factors which are nonetheless endemic to megafan behaviour.

In summary, Sologne relates more closely to the fluvial megafan end-member rather than to the major avulsing systems (MAS) of Latrubesse (2015) or the distributary fluvial systems (DFS) of Weissmann et al. (2015). MAS have topographic apices so low as to be non-existent. DFS are viewed as upscaled versions of smaller alluvial fans, with a radial pattern of channels from a single and well-defined apex, and a pattern of decreasing grain size and channel width in a downstream direction. There is so far no clear evidence on the Loire megafan of consistent changes in grain size (or increase in clay content) with distance from the apex. This aspect could perhaps benefit in future from a reexamination, for example by revisiting the databank of borehole logs and sediment samples collected by Rasplus (1978) and subsequent workers, and perhaps by adopting Passega's *CM* imaging technique as a tool for classifying alluvial suites and floodplain environments (overview, e.g,. in Peiry and Bravard 1999). Given that secondary avulsions and apices are not unusual on classic megafans (e.g., the Pilcomayo; see Wilkinson, Ch. 17), the presence here of two closely-spaced radial drainage patterns (i.e., Briare, and downstream at the toeslope of the putative Sauldre alluvial fan; Fig. 9.3) is strongly suggestive of a fluvial megafan formed by the Loire River. As also illustrated by Wilkinson (Ch. 17), megafans do not only display the normal radial pattern near the apex, but medial and distal parts can often be the opposite of radial, with parallel drainages tending in some cases to a somewhat contributary planform distally – as observed at least in the modern drainage net of Sologne.

Lastly, the Loire megafan is probably the only existing analogue in Europe of the volcanic swell–megafan systems of Africa (see Wilkinson et al., Ch. 3), in the present case of the diamond-shaped (34% of the African inventory) and relict (30% of the African inventory) variety. The Sologne region has been subsiding continuously since mid-Eocene time, but it became the receptacle of a megafan only from early Miocene time, in response to the onset of uplift in the Massif Central to the south (Fig. 9.1) and to a change in climate to wetter conditions during the Burdigalian, thereby increasing river discharge and sediment flux towards the Paris Basin. Two main pulses of crustal deformation occurred in the uplifting sediment source area: one during the late Miocene, and another during the Pliocene and Quaternary (Fig. 9.4). Accordingly, the Loire megafan is diamond-shaped, was produced in a context where a neighbouring crustal swell was activating consequent drainage, and formed over a period of millions of years under conditions where the topographic and stream profiles lay higher than the floor of the neighbouring basin, thereby allowing sediment accumulation in the form of a megafan.

Acknowledgements

This manuscript owes much to enthusiastic guidance and crisp commentary from Justin Wilkinson, who also introduced me to the widespread occurrence of swell-flank megafans in Africa.

References

Bessin, P., Guillocheau, F., Robin, C., Schroëtter, J. M., and Bauer, H. (2015). Planation surfaces of the Armorican Massif (western France): denudation chronology of a Mesozoic land surface twice exhumed in response to relative crustal movements between Iberia and Eurasia. *Geomorphology*, 233, 75–91.

Bessin, P., Guillocheau, F., Robin, C., et al. (2017). Quantification of vertical movement of low elevation topography combining a new compilation of sea-level charts and scattered marine deposits (Armorican Massif, Western France). *Earth and Planetary Science Letters,* 470, 25–36.

Bourgeois, O., Ford, M., Diraison, M., et al. (2007). Separation of rifting and lithospheric folding signatures in the NW-Alpine foreland. *International Journal of Earth Sciences*, 96, 1003–1031.

Brunet-Moret, Y., Chaperon, P., Lamagat, J. P., and Molinier, M. (1986). *Monographie hydrologique du fleuve Niger*. Vol. II, *Cuvette lacustre et Niger moyen*. Monographies hydrologiques de l'IRD (ORSTOM), 8, 521 pp.

Burke, K. and Gunnell, Y. (2008). *The African Erosion Surface: A Continental-scale Synthesis of Geomorphology, Tectonics, and Environmental Change over the Past 180 Million Years*. Geological Society of America Memoir, 201, 66 pp.

Cruz Mermy, D., Giot, D., Maget, P., et al. (2007). *Notice explicative, carte géologique de la France (1:50,000 scale), feuille Bracieux*. Bureau des Recherches Géologiques et Minières, Orléans, 171 pp.

Debrand-Passard, S., Gros, Y., Lablanche, G., et al. (1992). Âge, genèse et évolution du fossé de la Loire : nouvelle approche stratigraphique, morphologique et structurale. *Bulletin d'Information des Géologues du Bassin de Paris*, 29, 63–74.

Debrand-Passard, S. (1995). Histoire géologique résumée du sud du Bassin parisien. *Bulletin d'Information Géologique du Bassin de Paris*, 32, 15–25.

Debrand-Passard, S., Macaire, J.-J., Clozier, L., and Fleury, R. (1998). Particularités de l'évolution du système fluviatile solognot dans le bassin de la Loire. Corrélations possibles. *Géologie de la France*, 2, 55–68.

Debrand-Passard, S., Giot, D., Cruz Mermy, D., et al. (2010). *Notice explicative, carte géologique de la France (1:50,000 scale), feuille Lamotte-Beuvron*. Bureau des Recherches Géologiques et Minières, Orléans, 143 pp.

Denizet, H. (1900). *La Sologne*. Herluison et Michau, Orléans, 224 pp.

Denizot, G. (1927). *Les formations continentales de la région orléanaise*. Imprimerie Launay et fils, Vendôme, 582 pp.

Dugué, O., Bourdillon, C., Quesnel, F., and Lautridou, J.-P. (2012). The Neogene and Lower Pleistocene crags of Upper Normandy: biostratigraphic revision and paleogeographic implications. *Compte Rendus Geoscience*, 344, 415–422.

Etienne, R. (1984). Mouvements tectoniques différentiels et soulèvement d'ensemble du Massif Central à partir de la limite mio-pliocène. *Bulletin du Laboratoire Rhodanien de Géomorphologie*, 15–16, 3–14.

Etienne, R. and Larue, J.-P. (2011). Contribution à l'étude des liaisons Loire–Seine : mise en évidence par l'étude des minéraux lourds de l'antécédence de la Loire en Sologne (Bassin Parisien, France). *Physio-Géo*, 5, 269–291.

Féraud, G., Lo Bello, P., Hall, C., et al. (1990). Direct dating of Plio-Quaternary pumices by $^{40}Ar/^{39}Ar$ step-heating and single-grain laser fusion methods: the example of the Mont-Dore massif (Massif Central, France). *Journal of Volcanology and Geothermal Research*, 40, 39–53.

Fielding, C. R., Ashworth, P. J., Best, J. L., Prokocki, E. W., and Sambrook Smith, G. H. (2012). Tributary, distributary and other fluvial patterns: what really represents the norm in the continental rock record? *Sedimentary Geology,* 261–262, 15–32.

Fleury, R. (1990). *Notice explicative, carte géologique de la France (1:50,000 scale), feuille Aubigny-sur-Nère*. Bureau des Recherches Géologiques et Minières, Orléans, 44 p.

Fleury, R., Debrand-Passard, S., Gros, Y., et al. (1991). *Notice explicative, carte géologique de la France (1:50,000 scale), feuille Argent-sur-Sauldre*. Bureau des Recherches Géologiques et Minières, Orléans, 62 pp.

Fleury, R., Charnet, F., Debrand-Passard, S., et al.. (1992). *Notice explicative, carte géologique de la France (1:50,000 scale), feuille Salbris*. Bureau des Recherches Géologiques et Minières, Orléans, 50 pp.

Fleury, R., Charnet, F., Corpel, J., et al. (1997). *Notice explicative, carte géologique de la France (1:50,000 scale), feuille Romorantin.* Bureau des Recherches Géologiques et Minières, Orléans, 93 p.

Freytet, P. (1965). Sédimentation microcyclothématique avec croûte zonaire à algues dans le Calcaire de Beauce de Chauffour-Etrechy (Seine et Oise). *Bulletin de la Société Géologique de France*, 7, 309–313.

Freytet, P. and Plaziat, J. C. (1982). Continental carbonate sedimentation and pedogenesis in Late Cretaceous and early Tertiary in Southern France. In B.H. Purser, ed., *Contribution to Sedimentology*, vol. 12, E. Schweitzerbart, Stuttgart, 213 pp.

Freytet, P., Dewolf, Y., Joly, F., and Plet, A. (1989). L'évolution de la section Loire-Loing-Seine à la fin du Tertiaire. Réinterprétation géomorphologique des relations entre les Sables de Sologne et ceux de Lozère. Signification du complexe alluvial de la Montagne de Trin. *Bulletin d'Information des Géologues du Bassin de Paris*, 26, 49–57.

Gigout, M. and Desprez, N. (1977). *Notice explicative, carte géologique de la France (1:50 000 scale), feuille Gien.* Bureau des Recherches Géologiques et Minières, Orléans, 26 pp.

Goër de Herve, A. de and Etienne, R. (1991). Le contact Margeride–Cézalier–Cantal, les incidences de la tectonique et du volcanisme sur la sédimentation et l'hydrographie. *Bulletin du Laboratoire Rhodanien de Géomorphologie*, 27–28, 3–21.

Guillocheau, F., Robin, C., Allemand, P., et al. (2000). Meso-Cenozoic geodynamic evolution of the Paris Basin: 3D stratigraphic constraints. *Geodinamica Acta*, 13, 189–245.

Lacquement, F., Prognon, F., Prognon, C., et al. (2010). *État des lieux de la connaissance cartographique du régolithe de la France métropolitaine à 1/1 000 000, Rapport final.* BRGM/RP-57932-FR, Bureau des Recherches Géologiques et Minières, Orléans, 55 pp.

Lacquement, F., Prognon, F., Tourlière, B., et al. (2015). *Méthodologie de cartographie du régolithe à partir de données radiométriques aéroportées en région Centre—établissement de cartes lithologiques, Rapport final.* BRGM/RP-64932-FR, Bureau des Recherches Géologiques et Minières, Orléans, 133 pp.

Larue, J.-P. and Etienne, R. (1998). Les formations détritiques miocènes, pliocènes et quaternaires entre le Massif Central et la Sologne. Nouveaux éléments d'interprétation. *Géologie de la France*, 1, 39–56.

Larue, J.-P. (1999). Le fleuve à augite dans le Bassin parisien: nouveaux éléments d'interprétation. *Géologie de la France*, 3, 1–16.

Larue, J.-P. and Etienne, R. (2000). Les Sables de Lozère dans le Bassin parisien: nouvelles interprétations. *Géologie de la France*, 2, 81–94.

Larue, J.-P. and Etienne, R. (2002). Les Sables de Lozère et les Sables de Sologne: nouvelles interprétations de deux décharges détritiques du Miocène inférieur, issues de la paléo-Loire (Bassin parisien, France).

Bulletin de la Société Géologique de France, 173, 185–192.

Larue, J.-P. (2003a). L'encaissement inégal de la Seine et de la Loire dans le Bassin parisien (France). *Géographie Physique et Quaternaire*, 57, 21–36.

Larue, J.-P. (2003b). L'encaissement de l'Allier et de la Loire supérieure et moyenne (France) au Pliocène et au Pléistocène. *Géomorphologie: Relief, Processus, Environnement*, 9, 135–149.

Larue, J.-P. and Bélizal, E. de (2016). Les formations superficielles entre la Loire et le Loir (sud-ouest du Bassin parisien): les enseignements de l'analyse sédimentologique. *Norois*, 240, 43–57.

Latrubesse, E. (2008). Patterns of anabranching channels: the ultimate end-member adjustments of mega-rivers. *Geomorphology,* 101, 130–145.

Latrubesse, E. M. (2015). Large rivers, megafans and other Quaternary avulsive fluvial systems: a potential "who's who" in the geological record. *Earth-Science Reviews*, 146, 1–30.

Liard, M., Tissoux, H., and Deschamps, S. (2017). Les alluvions anciennes de la Loire en Orléanais (France, Loiret), une relecture à l'aune de travaux d'archéologie préventive et d'un programme de datation ESR. *Quaternaire*, 28, 105–128.

Macaire J.-J. (1981). *Contribution à l'étude géologique et paléopédologique du Quaternaire dans le SW du Bassin de Paris (Touraine et abords). Thèse d'Etat,* Tours, 450 pp.

Mouchené, M., van der Beek, P., Carretier, S., and Mouthereau, F. (2017). Autogenic versus allogenic controls on the evolution of a coupled fluvial megafan–mountainous catchment system: numerical modelling and comparison with the Lannemezan megafan system (northern Pyrenees, France). *Earth Surface Dynamics*, 5, 125–143.

Olivetti, V., Balestrieri, M. L., Godard, V., et al. (2020). Cretaceous and late Cenozoic uplift of a Variscan Massif: the case of the French Massif Central studied through low-temperature thermochronometry. *Lithosphere*, 12, 133–149.

Olmalius d'Halloy, J. B. J. (1828). *Mémoires pour servir à la description géologique des pays bas de la France et de quelques contrées voisines.* Imprimerie D. Gerard, Namur, 307 pp.

Pasquiou, X. (1995). Étude des flux détritiques sableux tertiaires et quaternaires dans la région de Gien. *Bulletin du Laboratoire Rhodanien de Géomorphologie*, 33–34, 35–49.

Pastre, J.-F. (1986). Altération et paléo-altération des minéraux lourds des alluvions pliocènes et pléistocènes du bassin de l'Allier (France). *Bulletin de l'Association française pour l'Étude du Quaternaire*, 3–4, 257–269.

Pastre, J.-F. (1992). Les pyroclastites du Mont-Dore (Massif Central français): place dans l'évolution du massif et dispersion périphérique. In Y. Lageat and J.-C. Thouret, eds., *Rythmes morphogéniques en domaine volcanisé.* Actes du Colloque de l'Association des Géographes

Français, 5 December 1992, CERAMAC, Université Blaise-Pascal, Clermont-Ferrand, 115–136.

Peiry, J.-L. and Bravard, J.-P. (1999). The CM pattern as a tool for the classification of alluvial floodplains along the river continuum. In S. B. Marriott and J. Alexander, eds., *Floodplains: Interdisciplinary Approaches.* Geological Society of London, Special Publication, 163, 259–268.

Pomerol, C. (1951). Origine et mode de dépôt des sables granitiques miocènes entre Paris et la Manche. *Bulletin de la Société Géologique de France*, 6, 251–263.

Rasplus, L. (1978). *Contribution à l'étude géologique des formations continentales détritiques tertiaires de la Touraine, de la Brenne et de la Sologne.* Thèse d'État (3 vols.), Univ. of Orléans, 454 pp.

Rasplus, L. (1982). *Contribution à l'étude des formations continentales détritiques tertiaires du Sud-Ouest du bassin de Paris.* Sciences géologiques, Mémoire 66, Université Louis Pasteur, Strasbourg, 227 pp.

Ruxton, B. P. and Berry, L. (1978). Clay plains and geomorphic history of the central Sudan: a review. *Catena*, 5, 251–283.

Sinha, R. and Friend, P. F. (1994). River systems and their sediment flux, Indo-Gangetic plains, Northern Bihar, India. *Sedimentology*, 41, 825–845.

Sissingh, W. (2001). Tectonostratigraphy of the West Alpine foreland: correlation of Tertiary sedimentary sequences, changes in eustatic sea-level and stress regime. *Tectonophysics*, 233, 361–400.

Sissingh, W. (2006). Syn-kinematic palaeogeographic evolution of the West European Platform: correlation with Alpine plate collision and foreland deformation. *Netherlands Journal of Geoscience*, 85, 131–180.

Sobolev, S. V., Zeyen, H., Granet, M., et al. (1997). Upper mantle temperatures and lithosphere–asthenosphere system beneath the French Massif Central constrained by seismic, gravity, petrologic and thermal observations. *Tectonophysics*, 275, 143–164.

Stanistreet, I. G. and McCarthy, T. S. (1993). The Okavango fan and the classification of subaerial fan systems. *Sedimentary Geology*, 85, 115–133.

Sutton, K. (1971). The reduction of wasteland in the Sologne: nineteenth-century French regional improvement. *Transactions of the Institute of British Geographers*, 52, 129–144.

Thauvin, M., Colin, S., and Saint Martin, S. (2011). *Carte des ressources en matériaux de la Région Centre, élaborée dans le cadre de la révision des Schémas Départementaux des Carrières, Rapport final.* BRGM/RP-59248-FR, Bureau des Recherches Géologiques et Minières, Orléans, 134 pp.

Tissoux, H., Prognon, F., Voinchet, P., et al. (2013). Apport des datations ESR à la connaissance des dépôts sableux plio-pléistocènes de Sologne, premiers résultats. *Quaternaire*, 24, 141–153.

Tissoux, H., Prognon, F., Martelet, G., et al. (2017). Interprétation d'un levé de spectrométrie gamma pour la connaissance des dépôts silico-clastiques fluviatiles en centre France (Loire et Sologne). *Quaternaire*, 28, 87–103.

Tourenq, J., Decaillot, P., and Pomerol, C. (1971). Origine armoricaine des minéraux lourds de la mer des faluns. Mise en doute doute de la capture de la pré-Loire à l'Helvétien inférieur. *Compte Rendus Sommaires de la Société Géologique de France*, 65–67.

Tourenq, J. (1989). Les sables et argiles du Bourbonnais: une formation fluvio-lacustre d'âge pliocène supérieur, étude minéralogique, sédimentologique et stratigraphique. *Documents du BRGM*, 174, 333 pp.

Tourenq, J. and Pomerol, C. (1995). Mise en évidence, par la présence d'augite du Massif central, de l'existence d'une pré-Loire–pré-Seine coulant vers la Manche au Pléistocène. *Comptes Rendus de l'Académie des Sciences, Série 2, Sciences de la Terre et des Planètes*, 320, 1163–1169.

Turland, M., Hottin, A. M., Cojean, R., et al. (1990). *Notice explicative, carte géologique de la France (1:50,000 scale), feuille Hérisson.* Bureau des Recherches Géologiques et Minières, Orléans, 118 pp.

Voinchet P., Despriée J., Gageonnet R., et al. (2007). Datation par ESR de quartz fluviatiles dans le bassin de la Loire moyenne en région Centre: mise en évidence de la tectonique quaternaire et de son influence sur la géométrie des systèmes de terrasses. *Quaternaire*, 18, 335–347.

Weissmann, G. S., Hartley, A. J., Scuderi, L. A., et al. (2015). Fluvial geomorphic elements in modern sedimentary basins and their potential preservation in the rock record: a review. *Geomorphology*, 250, 187–219.

Wilson, M. and Patterson, R. (2001). Intraplate magmatism related to short wavelength convective instabilities in the upper mantle: evidence from the Tertiary–Quaternary volcanic province of western and central Europe. In R. E. Ernst and K. L. Buchan, eds., *Mantle Plumes: Their Identification Through Time.* Geological Society of America Special Paper, 352, 37–58.

Ziegler, P. A. and Dèzes, P. (2007). Cenozoic uplift of Variscan massifs in the Alpine foreland: timing and controlling mechanisms. *Global and Planetary Change*, 58, 237–269.

10

Megafans of the Gangetic Plains, India

RAJIV SINHA, KANCHAN MISHRA and SAMPAT K. TANDON

Abstract

The Eastern Gangetic Plains (EGP) hosts three actively aggradational megafans, the Teesta, Kosi, and Gandak, which are directly fed by rivers draining the Himalayan Mountains in the north; and a fourth, the Sone, which is fed from the Indian shield in the south. Topography in the Western Gangetic Plains (WGP) consists instead of a 900-km-long and 100-km-wide raised interfluve between the incised valleys of the Yamuna and Ganga rivers. None of the commonly-advocated megafan criteria, such as mappable fluvial sediment entities with an apex, distributary drainage, convex-up transverse topographic profiles, or a distinct fan boundary with a break in slope, seem to apply. There are thus no active megafans in the WGP. While not ruling out the possible occurrence of relict megafans, evidence suggests that the WGP landscape is essentially a coalescing floodplain in a valley–interfluve setting. Contrasts between the EGP and WGP are controlled by (i) differential late Quaternary and Holocene basin subsidence, which governs regional-scale variations in accommodation space; (ii) along-strike tectonic and climatic variability, primarily reflected in differential uplift rates, rainfall gradients, mountain-front tectonics, and river exit-point spacing; and (iii) hydrological characteristics of the feeder channel manifested as variability in stream power and sediment flux.

10.1 Introduction

The Indo-Gangetic plains are bordered by the > 2500-km-long Himalayan mountain front to their north and by the Bundelkhand craton to their south. Although some large Himalayan rivers enter the plains, only a few of them form megafans (Geddes 1960; Gole and Chitale 1966; Wells and Dorr 1987; Gohain and Parkash 1990; Mohindra et al. 1992; Singh 1996). Figure 10.1 shows the location and boundaries of these megafans along with their respective hinterland and drainage pattern. When rivers exit the Himalaya and reach the flat and unconfined foreland, they progressively build cone-shaped structures through sediment deposition and lateral migration of the channel. Plains-fed radial systems that develop on the surface of these megafans rework the fan deposits (Geddes 1960; Sinha and Friend 1994). Sedimentologically, the Himalayan fans are mainly composed of thick, multilayered sandy sequences in the northern proximal part, and of fine sands and silts in the southern distal part (Singh et al. 1993; Sinha and Friend 1994; Jain and Sinha 2003; Tandon and Sinha 2007).

Several large rivers such as the Sarda, Ghaghara, Ganga, and Yamuna exit the mountain front west of the Gandak drainage basin. Previous workers have mapped the Ganga–Yamuna as a single megafan (Fig. 10.1) along with the Sarda (Singh 1996; Shukla et al. 2001; Srivastava et al. 2003). Detailed drainage maps of the rivers such as the Teesta, Kosi, Gandak, and Sone draining the eastern Gangetic plain (EGP) show typical radial drainage patterns, whereas the others such as the Rapti and Ghaghra do not (Fig. 10.2). Furthermore, none of the rivers draining the western Gangetic plains (WGP), such as the modern Ganga and Yamuna (in the Ganga drainage), display radial drainage (Fig. 10.2, nor is convex-up topography typical of megafans preserved in the WGP. Based on subsurface data from nine localities

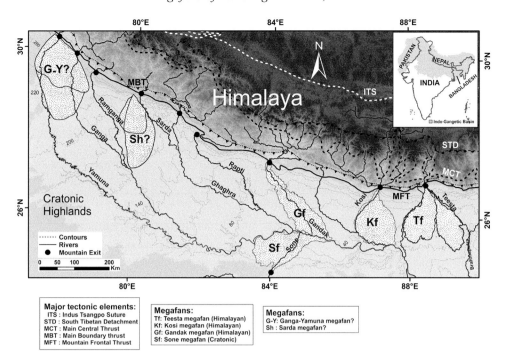

Figure 10.1 **Location and boundaries of the megafans in the Gangetic plains, with major tectonic elements**. All four active megafans (Teesta, Kosi, Gandak, Sone) have formed in the eastern Gangetic plains.

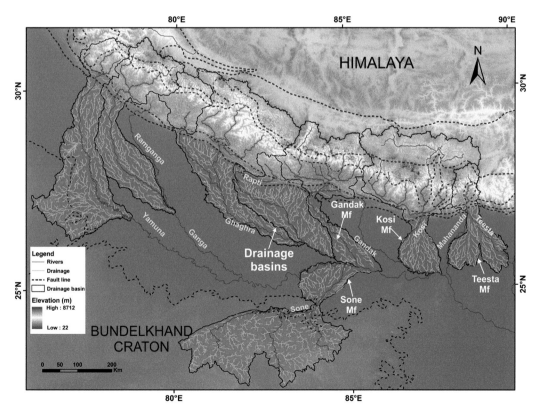

Figure 10.2 **Detailed drainage maps for the rivers draining the eastern Gangetic plains**. All four rivers in the EGP, the Teesta, Kosi, Gandak, and Sone, form megafans and display typical radial drainage. Rivers draining the western Gangetic plains e.g., Ganga and Yamuna, do not display any radial drainage typical of megafans.

along the modern Ganga river, Shukla et al. (2001) proposed the idea of a relict Ganga megafan; however, no geomorphic signatures exist, nor does any detailed information on subsurface facies architecture or palaeocurrents (Tandon et al. 2008). Further west, more recent work in the Sutlej and Yamuna plains by Van Dijk et al. (2016) has demonstrated the possibility of relict megafans in this region. The Sutlej lies in the Indus watershed and thus falls outside the Ganga plain, to which we restrict the present study. It is nonetheless important to differentiate between (i) the active mega-fans, with clear geomorphic signatures in the modern landscape of the EGP and the incised plains and (ii) relict megafans of the WGP. This rarely emphasised regional contrast and its significance on the distribution of the megafans across the Indo-Gangetic plains will be the focus of this chapter. We first discuss the geomorphic processes controlling this variability based on the hydrological and sediment flux data of the feeder channels, and then elaborate the geological controls on the long-term evolution of the megafans across the Indo-Gangetic plains.

10.2 Methods

We used two remote-sensing products for drainage network and megafan mapping and analysis: the SRTM 1-degree digital elevation model (DEM) with ~90 m ground resolution, and different sets of cloud-free Landsat TM satellite images made available by the U.S. Geological Survey (http://earthexplorer.usgs.gov/). Geo-rectification was applied to all the data to remove some acquisition errors, a low-pass convolution (3 × 3) filter was applied to the DEM data for noise elimination and to avoid a minimum loss of information, and the DEM was subjected to a sink-filling routine in ArcGIS 10.2 in order to suppress 'no data' and negative altitude values and obtain a depressionless topography of the study area. Thereafter, we followed the extraction procedures for drainage and topographic attributes described in Miliaresis and Argialas (2000), Miliaresis (2001), Argialas and Tzotsos (2004), and Norini et al. (2016).

The methodology for delineating the shape and extent of the alluvial fans is based on a set of geometric criteria that depend on the DEM-defined topography of the area. These criteria are as follows:

(i) The required topographic setting is a contributing river channel that flows from a source catchment area and drains into a lowland.
(ii) A conical shape and associated drainage pattern that radiates from the fan apex to the surrounding lowland areas.
(iii) The fan apex is recognised by a topographic break in the slope at, or a little downstream of, the mountain exit point.
(iv) The longitudinal slope of the fan, measured from topographic profiles generated from the DEM, is very low.

A combination of spatial analyses based on GIS algorithms was used to delineate the alluvial fan and the upstream drainage contributing area. A slope map was derived from the DEM using the ArcGIS slope tool, which is an extension of the Spatial Analyst module that calculates slope. The slope map was processed by using the aspect tool to generate the aspect map, which defines the maximum rate of change in value from each cell to its neighbours in the downslope direction. The aspect map gives the direction of the slope of the alluvial fan. The combination of slope and aspect maps defines the spatial arrangement of the alluvial fan at the mountain front. It also helps in identifying the apex of the alluvial fans in addition to the visual interpretation. The drainage network was defined through a hydrological analysis of the DEM. After delineation of the megafan and its hinterland, various morphometric parameters of the megafans were measured as defined in Figure 10.3 (see Table 10.1 for definitions).

The processed maps are not error-free. The SRTM DEM, for example, reports vertical errors of less than ± 7 m, but on low-gradient and low-relief fans this is quite a large uncertainty interval. The apex of the alluvial fan, and the radial and transverse profiles to define the edges of the fan, were thus also drawn on the basis of interpreter's knowledge of the terrain. Further, the delineation of the distal boundary of the alluvial fan is based on the combination of the above analysis and the visual interpretation. Many of these operations could potentially propagate locational errors to the map. However, no GIS technique currently in use can provide the confidence limits associated with the results of the analysis, and therefore the uncertainties associated with mapping are difficult to quantify.

Table 10.1 *Morphometric attributes of megafans in the Gangetic plains*

Parameters	Definition and measurements	Teesta	Kosi	Gandak	Sone
Contributing area, A (km^2)	Total upland area	8370	58,219	36,800	65,898
Fan Area, F_A (km^2)	Total area from apex to toe	17,128	11,486	9932	11,882
Upland/plains ratio (u/p)	Ratio of upland area to alluvial area	0.46	4.95	3.65	4.84
Highest basin elevation, U_h (m)	Highest elevation in the upland area	8421	8712	8072	1217
Basin relief (m/m)	Difference between highest and lowest elevation in the upland basin	123	133	65	66
Fan slope (deg.) Average	Average slope between fan apex and toe	0.057	0.038	0.015	0.031
Upper/middle/lower fan	Average slope for the upper, middle and lower fan	0.23/0.03/0.01	0.13/0.25/0.01	0.04/0.02/0.01	0.04/0.03/0.02
Angle of divergence (deg.)	Angle of divergence of the fan margins at its apex	63°–66°	42°–45°	31°–34°	37°–39°
Fan axial length, F_L (km)	Maximum length between fan apex and toe	176	159	254	144
Fan axial width, F_W (km)	Maximum across-fan width	155	116	60	207
Fan length/width ratio, F_L/F_W	Ratio of maximum length and width	1.182	1.105	4.231	0.990
Maximum channel length in fan region, L_{max} (km)	Length of main feeder channel on the megafan	301	252	306	172
Sinuosity, P	Average sinuosity of main feeder channel on the megafan	1.16	1.96	1.31	1.25
Braid-channel ratio	Ratio of sum of lengths of all primary and secondary channels to length of primary channel	3.38	4.33	3.03	3.31
Fan drainage density (m/m^2)	Sum of channel lengths (active and abandoned) on fan surface divided by fan area	0.167	0.196	0.162	0.174
Highest fan elevation, H_{max} (m)	Highest elevation on the fan surface	192	162	113	118
Lowest fan elevation, H_{min} (m)	Lowest elevation on the fan surface	16	32	44	46
Relief ratio of the fan (m/m)	Elevation difference ($H_{max} - H_{min}$) divided by horizontal distance along the longest dimension of the fan parallel to the main stream	0.012	0.001	0.0003	0.005
Lobe system	Present and past depositional areas of the megafan	Double/Multiple	Single	Single	Single

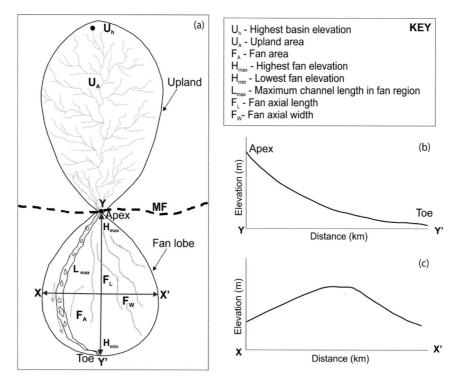

Figure 10.3 **Morphometric variables, and generic longitudinal and cross-profiles of the megafans.** (a) Morphological and morphometric attributes of the megafans (see Table 10.1 for definitions) illustrating the essential characteristics of megafans – a river exiting an upland or highland, fan apex, radial drainage. (b) Downfan profile illustrates flattening of the fan surface from apex to toe. (c) Profile across the megafan surface illustrates convex topography.

10.3 Systematic Description of the Major Fans in the Gangetic Plain

The Gangetic Plain is one of the largest alluvial plains in the world, with a drainage area of 980,000 km² and a depositional area of 575,430 km². It is marked by several large axial river systems such as the Ganga and the Yamuna, and transverse drainages such as the Kosi and Gandak flowing from the north and the Sone flowing from the south (Fig. 10.1). Several of the transverse drainages in the eastern part of the plains have formed megafans, some are typical fan-shaped such as the Kosi, Teesta, and Sone; and others, such as the Gandak, are elongate. Of these, the Sone megafan is the only one formed by drainage flowing from the Deccan shield.

Figure 10.1 shows the locations of the megafans across the Gangetic plains and Table 10.1 lists their major morphological attributes. The Teesta is the largest megafan in the Gangetic Plain, with two major lobes. The next largest is the Sone, followed by the Kosi and Gandak – two megafans known to be the most active systems in terms of water and sediment delivery from the Himalaya. The Gandak has the longest axial channel (306 km), followed by the Teesta (301 km), the Kosi (252 km) and the Sone (172 km). The axial length and width describe the shape of the megafan from elongate (Gandak) to lobate (Kosi, Teesta). Table 10.1 also highlights the upland/plains (u/p) ratio, which shows little relationship to fan area. The Kosi has the highest u/p ratio but is closely followed by the Sone and Gandak. The Teesta is the largest megafan but displays a very low u/p ratio.

10.3.1 Teesta Megafan

The Teesta is the easternmost megafan in the Gangetic Plain and covers parts of West Bengal and Bangladesh (Fig. 10.4a). It occurs 830 km west of the eastern Himalayan syntaxis. The western edge of the megafan is bounded by the foothills-fed Mahananda River and its eastern edge by the Teesta River itself. A large part of the megafan (total area: 17,128 km²) therefore lies

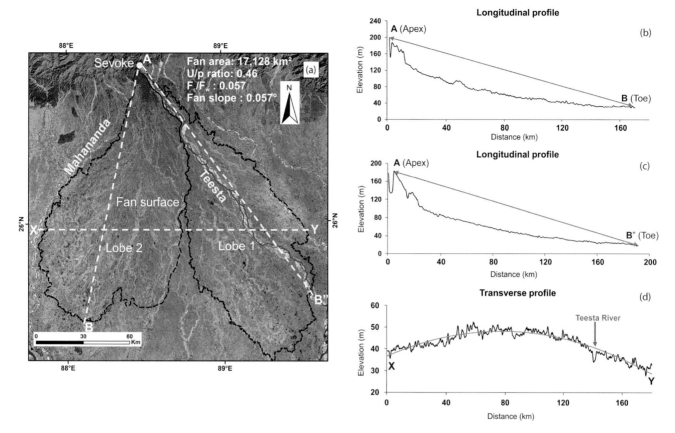

Figure 10.4 **Aspects of the Teesta megafan**. (a) False-colour composite (FCC) of the Teesta megafan prepared from Landsat TM satellite imagery using a band combination of 5, 4, 2 on RGB. Note the two distinct lobes of the Teesta fan and radial drainage. (b, c) Longitudinal profiles along AB and AB" based on the SRTM DEM data; both lobes of the megafan show a typical concave-up downfan profile. (d) Typical convex-up profile across the megafan surface drawn from the SRTM DEM also reflects the multi-lobe morphology. A colour version of this figure is available in the SOD for Chapter 10.

west of the modern Teesta channel. A typical radial drainage pattern characterises the megafan surface, which is drained by several plains-fed rivers and their palaeochannels (Fig. 10.4a). The highest elevation in the Teesta watershed is 8,421 m, and the lowest 16 m. Maximum elevation on the fan surface is 192 m. Average fan slope angle is $0.057°$, but the fan is much steeper $(0.23°)$ in its proximal than its distal part $(0.01°)$. The longitudinal profile of the Teesta River from fan apex to fan toe is close to a graded concave-up profile (Fig. 10.4b). One of the most characteristic features of the Teesta fan is its two major lobes. Chakraborty and Ghosh (2010) considered the eastern lobe to be older, based on drainage discordance; these authors also mapped a third, much smaller lobe between lobes 1 and 2. A cross-profile of the megafan clearly shows a convex-up profile and also illustrates a multi-lobe morphology with a slight depression between two convexities (Fig. 10.4c). The multi-lobe

morphology of the megafan and several palaeochannels on the megafan surface suggest random and repeated lobe switching through nodal avulsions (Chakraborty and Ghosh 2010). The plains-fed channels are at places entrenched deeply into the megafan surface and highly sinuous, suggesting incision events during periods of high discharge.

Based on the limited sedimentological data, Chakraborty and Ghosh (2010) suggested significant proximal to distal variation in facies association of the Teesta megafan deposits. The proximal parts have been documented as 50–90 cm thick units of stratified sand deposits formed by unconfined, high-energy braided rivers with flashy flood discharges and rapidly decelerating flow. These units have a characteristic basal erosion surface strewn with coarse to pebbly sand and mud clasts. The medial parts of the megafan are characterised by trough and planar cross-stratified pebbly coarse-textured sand and horizontally stratified medium-grained sand units. The distal parts of the megafan feature a varied facies assemblage ranging from stratified sand alternating with silty clay layers, to thick muddy layers with interlayers of clay and silt. These are interpreted as mixed-load, low-energy sinuous channel deposits typical of floodplain marsh environments.

The present-day Teesta River is incised up to 35 m into the megafan deposits (Chakraborty and Ghosh 2010), and three alluvial terrace levels have been mapped (Meetei et al. 2007). Terraces are generally present between the tributary confluences with the main Teesta River. Based on the geomorphic setting and facies associations, which comprise hyperconcentrated-flow, debris-flow, channel-flow, and sheet-flow variants, Meetei et al. (2007) suggested episodic erosion and deposition driven by an interplay of tectonics and climate during the late Quaternary. Detachment of the main feeder channel from the Himalayan catchment, and reworking of the megafan surface by plains-fed rivers including incision, is suggested to result from hydroclimatic changes (Chakraborty and Ghosh 2010). New cosmogenic [10]Be and IRSL ages of terraces suggest that the proximal lobe of the Teesta megafan could be as old as ~135 ka, and that the incision started ~3.7 ka (Abrahami et al. 2018). Two distal lobes developed successively from this proximal lobe; the western lobe was abandoned in early Holocene time (~10–11 ka) whereas deposition across the eastern lobe continued until ~1 ka (Abrahami et al. 2018). While the main Teesta megafan is no longer active and the major fan-building processes have ceased to operate since the early Holocene, it has been suggested that a faintly incised lobe was built after the main Teesta River avulsed to the east. This lobe may still be active (Abrahami et al. 2018). The form of the main Teesta megafan is still very well preserved, and the fan surface is being reworked by modern plains-fed systems through seasonal flooding and sediment redistribution.

10.3.2 Kosi Megafan

The Kosi megafan has a maximum length of 159 km from its apex to the confluence with the Ganga, and a maximum width of 116 km in its mid-fan region, with a total fan area of 11,486 km^2. The Kosi megafan mostly lies in the plains of northern Bihar, except for its apical part, which falls in Nepal. The megafan's principal watershed, the Kosi River, originates at an elevation of > 5500 m in Tibet and has a very large hinterland mostly lying in the Himalaya of Tibet and Nepal (Fig. 10.2). The Kosi River enters the plains at Chatra (immediately upstream of Barahkhetra, Fig. 10.5a) in Nepal, and then flows for ~80 km in Nepal before entering the north Bihar plains in India. The river then flows for another ~160 km in north Bihar before meeting the Ganga River near Kursela.

The natural flow of the Kosi River is regulated through a barrage at Bhimnagar (Birpur, constructed in 1963), and the river is regulated by embankments on both sides (completed in 1956) throughout its alluvial part. Despite water flow regulation, the hydrology of the Kosi River reflects a typical monsoonal climatic regime with a peak of monthly discharge in August/ September mimicking the rainfall pattern. The current average annual discharge (Q_{av}) of the Kosi at the most downstream station, Baltara, is 2256 m^3/s; the average monsoon discharge computed from 1985 to 2002 is 5156 m^3/s, which is approximately five times higher than the non-monsoon discharge of 1175 m^3/s (Sinha et al. 2008, 2019). Such a large difference between monsoonal and non-monsoonal discharges makes the river vulnerable to flooding as the shallower river sections (average channel width: ~50 m) cannot accommodate the excess flow. The average annual suspended sediment flux for the Kosi is ~80 Mt at Birpur (close to the apex of the megafan) and ~43 Mt at the most downstream station (Baltara) (Sinha and Friend 1994; Sinha et al. 2019). It suggests large-scale aggradation in the alluvial part of the river even though a large part of total sediment flux from the Himalayan source areas is trapped at the barrage. A first-order sediment budgeting calculation suggests that the total mass of sediments accumulated in the channel belt between Birpur and Baltara during the last 54 years (post-embankment period) could be ~2862 Mt (1080 M m^3), and this may have accumulated at a rate of 3.94 cm/yr (Sinha et al. 2019).

Palaeochannels on the Kosi megafan surface are testimony to the frequent lateral migration of the Kosi river during historical and pre-historic periods (Fig. 10.5a). Frequent avulsive movement of the Kosi River has been described as autocyclic and stochastic,

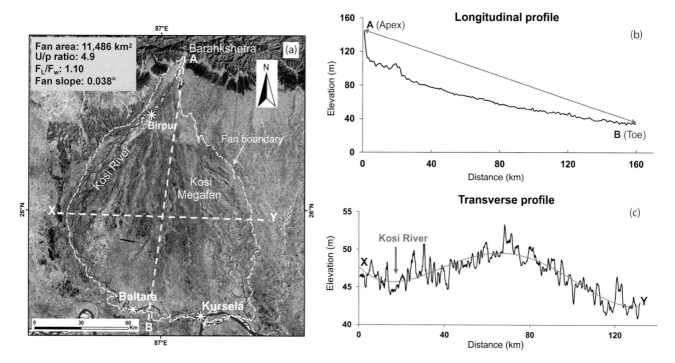

Figure 10.5 **Aspects of the Kosi megafan**. (a) False-colour composite (FCC) of the Kosi megafan prepared from Landsat TM satellite imagery using a band combination of 5, 4, 2 on RGB. Typical cone shape, with numerous radial palaeochannels seen on the satellite image formed by repeated avulsions, makes the Kosi a classic example of megafan. (b) Typical concave-up downfan profile along AB based on the SRTM DEM data; (c)Typical convex-up profile across the megafan surface drawn from the SRTM DEM. A colour version of this figure is available in the SOD for Chapter 10.

which is typical of most alluvial fans across the world. However, the average avulsion frequency for the Kosi is known to be well below once in 24 years, and is thus among the highest in the world compared to, for example, up to 1400 years for the lower Mississippi River (Slingerland and Smith 2004). More recent work has also confirmed that the dynamics of the Kosi River is primarily controlled by local slope changes influenced by excessive sedimentation in the channel belt, and that the situation has become worse after the construction of embankments on both sides of the Kosi River (Sinha 2009; Chakraborty et al. 2010; Sinha et al. 2014a). Figures 10.5b and c illustrate the typical down-fan slope profile and convex-up cross-fan morphology.

Sinha et al. (2014b) documented the subsurface stratigraphy and alluvial architecture for the upper ~100 m of the Kosi megafan based on electrical resistivity soundings, borehole data, and drill cores. The proximal to medial parts of the megafan show significant variability in sediment grain size and layer thickness. The proximal part of the fan displays a gravel unit at shallow levels (< 15 m), which is under- and overlain

by medium- to coarse-grained sand units. The medial part of the megafan displays ~20–30 m thicknesses of medium- to coarse-grained sand units, with significant lateral variability. A detailed description of the subsurface stratigraphy and controls on proximal to distal spatial variability are presented by Sinha et al. in Ch. 11.

10.3.3 Gandak Megafan

The Gandak megafan was formed by the Gandak River, which rises in the Higher Himalaya of Tibet. The Gandak is braided throughout in its alluvial reach, and its hydrology displays typical monsoonal characteristics. Water discharge starts rising in early June, with a peak discharge in August. The mean annual discharges of the Gandak at both upstream (Triveni) and downstream (Dumariaghat) stations, about 200 km apart, are quite similar (~1500 m^3/s), and the mean annual floods at both these stations are 6600 and 8450 m^3/s, respectively (Sinha and Friend 1994). Such a large difference in the average and mean annual discharge of the Gandak make this river extremely flood-prone, particularly at the downstream station (Sinha and Jain 1998). Annual suspended sediment load of the Gandak at the upstream (Triveni) and downstream (Dumariaghat) stations are 79 Mt and 82 Mt, respectively (Sinha and Friend 1994). A sediment yield of 1.73 Mt /km²/yr at Dumariaghat for the Gandak is considerably higher than that of the Kosi at Baltara. The high sediment flux of these rivers reflects exceptionally high topographic relief and intense rainfall in parts of their catchments as well as on the plains.

The Gandak megafan displays the unusual morphology of an elongate fan, but with a typical radial drainage pattern (Fig. 10.6a). The Gandak has a very large hinterland (36,800 km²) and a relatively smaller fan area (9932 km²) compared to other megafans; this results in a high u/p ratio of 3.65. The average fan slope angle is 0.015°, but the upper part of the fan is steeper (0.04°) than the lower part (0.02°), giving rise to the typical contrasting slope pattern already highlighted for the Kosi (Fig. 10.6b). Unlike the Kosi, however, a transverse profile across the widest portion of the Gandak megafan does not produce a typical convex up pattern (Fig. 10.6c), although some convexity in the eastern and western parts is observed. At the apex of the Gandak megafan, close to the mountain exit point, a very sharp convexity in the main channel of the Gandak is noticed. This may point to a strong tectonic control but needs further investigation.

The megafan surface features several palaeochannels, ponds, and lakes on both sides of the main channel, pointing to a very dynamic regime of the river. A dominantly west-to-east movement of the Gandak River has been suggested by previous workers and attributed to slow eastward neotectonic tilting of the megafan (Mohindra et al. 1992; Mohindra and Parkash 1994). Detailed geomorphic mapping and soil chrono-association studies have suggested that the oldest soils on the Gandak megafan surface might date back to ~ 10 ka whereas the youngest could be < 500 years old (Mohindra et al. 1992).

No detailed sub-surface information exists for this megafan, but a recent study by Sinha et al. (2014b) based on resistivity soundings, borehole logs and a few drill cores has suggested that the Gandak megafan is made up of two major lithological units. The upper fan

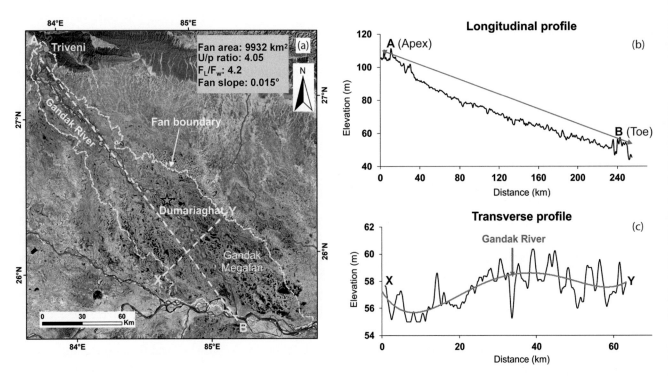

Figure 10.6 **Aspects of the Gandak megafan.** (a) False-colour composite (FCC) of the Gandak megafan prepared from Landsat TM satellite imagery using a band combination of 5, 4, 2 on RGB. The Gandak is an elongate megafan with numerous radial palaeochannels formed as a result of repeated avulsions. (b) Typical concave-up downfan profile along AB based on the SRTM DEM data. (c) Typical convex-up profile across the megafan surface drawn from the SRTM DEM. A colour version of this figure is available in the SOD for Chapter 10.

succession has a higher stacking density of smaller sand bodies, perhaps reflecting the migratory behaviour of the river; whereas the lower succession shows narrow but thick sand fills reflecting incised channels. The western part of the Gandak megafan has more abundant sand bodies compared to the eastern side of the river. There are no significant differences between proximal and medial transects across the Gandak megafan. Unlike the adjacent Kosi megafan, the hinterland of the Gandak megafan has a prominent intermontane valley that has acted as a sediment buffer, thereby trapping most of the coarser fraction (Sinha et al. 2014b; Densmore et al. 2016). It explains the lack of coarser facies in the deposits of the Gandak megafan both in the proximal as well as in the distal parts.

10.3.4 Sone Megafan

The Sone (also known as Son) is the only non-Himalayan megafan in the Gangetic plains and was produced by an upland river originating in the Amarkantak Hills at an elevation of ~1050 m in peninsular India. This oval-shaped megafan covers an area of around 11,882 km^2. The axial length of the megafan is ~144 km and its width is ~207 km. The mean slope of the megafan surface is ~0.031°. A large segment of the Sone river in the upland terrain is tectonically controlled and flows for more than 500 km along the Narmada–Sone lineament (Sahu et al. 2015).

The Sone river currently flows in a position fairly central to the fan and meets the Ganga at Babua; but several palaeochannels have been mapped on the megafan surface (Fig. 10.7a), suggesting that the confluence point with the Ganga has shifted considerably over time (Sahu et al. 2015). Based on satellite imagery, as many as nine avulsion events in discrete steps towards the west were mapped for the Sone River by Sahu et al. (2010). They all occurred in the recent geological past although no absolute chronology is available for them. These avulsions were suggested to have been triggered by tectonic uplift along the East Patna Fault and associated lateral tilting of the hangingwall block (Sahu et al. 2010). The main channel of the modern Sone River is fairly straight and moderately braided, and the palaeochannels on both sides of the modern river also show a similar morphology. Only a few palaeochannels show moderate

sinuosity, and unlike the Himalayan rivers the Sone does not possess a regular floodplain for most parts of its course. However, significant longitudinal variation in sinuosity, braiding, width-depth ratio and convexity in longitudinal profile (Fig. 10.7b) are observed, and this is interpreted as a response to tectonic tilting and valley slope modifications (Sahu et al. 2010). Like the other megafans built by the Himalayan rivers in the north, the Sone in the south also shows a typical convex-up morphology across the fan (Fig. 10.7c).

The average annual discharge of the Sone River is estimated as ~1050 m^3/s, but the minimum flow goes down to 500 m^3/s (Sahu et al. 2015). The maximum observed flood discharge in the Sone was 33,980 m^3/s, in 1975. It has a fairly high annual sediment load of ~37 Mt at Koilwar, a few kilometres upstream of its confluence with the Ganga. The sediment flux of the Sone River is one of the highest among the north-flowing tributaries of the Ganga, and this has been attributed to various climatic and lithological controls in the catchment (Sahu et al. 2015). Several episodes of prolonged aggradation have been reported in the Sone megafan during the late Quaternary period, typically associated with weaker monsoon conditions (Williams and Clarke 1984; Kale and Rajguru 1987; Williams et al. 2006). Stratigraphic data available from the Sone indicates interfingering of Himalayan and the cratonic sediments in the subsurface, particularly in the western distal parts of the megafan. This has been related to relative migration of the Ganga and Sone rivers through time.

Borehole data suggest that the Sone megafan sedimentary sequence in the western part, and in most of its central and eastern part, is more than 200 m thick, consisting of oxidised, brown-yellow sand (called Sone sand) with gravels and calcrete (Sahu et al. 2018). In the distal part of the megafan, unoxidised, grey to dark grey fine sand with mica flakes and minor iron nodules (called Ganga sand) onlap the brown-yellow sand in the upper ~50 m of the sedimentary sequence (Sahu et al. 2015). The occurrence of brown-yellow Sone sand in the subsurface well to the north of the present position of the axial Ganga river seems to suggest that, at some time in the past, the Sone megafan extended at least 15–20 km north of the modern Ganga River channel (Sahu et al. 2015). However, the modern geomorphic extent of the active Sone megafan

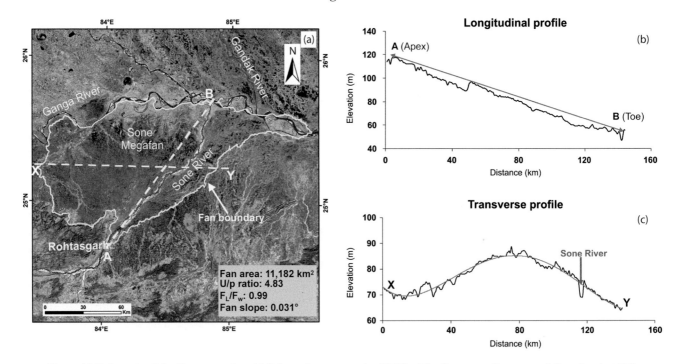

Figure 10.7 **Aspects of the Sone megafan**. (a) False-colour composite (FCC) of the Sone megafan prepared from Landsat TM satellite imagery using a band combination of 5, 4, 2 on RGB. The oval-shaped Sone is the only megafan in the Gangetic plains fed by a non-Himalayan river, and strong tectonic controls have been suggested to explain the upland river course as well as its westward avulsions. (b) Typical concave-up downfan profile along AB based on the SRTM DEM data. (c) Typical convex-up profile across the megafan surface drawn from the SRTM DEM. A colour version of this figure is available in the SOD for Chapter 10.

is currently limited by the Ganga River, as shown in Figure 10.7a. Such basement-derived sediment wedges situated well to the north of the present-day Ganga river have also been documented in the Ganga–Yamuna interfluve, suggesting that rivers from the Indian shield might have contributed a significantly increased amount of sediments to the Himalayan foreland in the geological past (Sinha et al. 2009).

10.4 Geomorphic Processes Controlling the Basin-Scale Distribution of Megafans

A general point that emerges from our analysis is that all major megafans in the Gangetic Plains have formed in the eastern part. Significant differences in geomorphic processes in the EGP and WGP have likewise been highlighted by previous workers (Sinha et al. 2005, 2007; Tandon et al. 2006, 2008; Roy and Sinha 2017, 2018; Swarnkar and Sinha 2020; Kaushal et al. 2020). The rivers draining the EGP are characterised by aggrading systems with high w/d ratios and frequent avulsions, whereas those in the WGP have developed incised valleys with low w/d ratios and are

laterally stable. Such spatial heterogeneity is attributed to the modern hydrological characteristics of the rivers expressed by stream power, sediment transport characteristics and the marked variation in precipitation patterns, both in the Himalayan upstream tracts and in the alluvial parts (Sinha et al. 2005; Roy and Sinha 2017, 2018; Swarnkar and Sinha 2020; Kaushal et al. 2020). The EGP is characterised by high rainfall, low stream power ($10–15$ W/m^2), and high sediment flux, whereas the WGP is marked by relatively lower rainfall, much lower sediment flux, and higher erosive stream power ($40–45$ W/m^2) as a result of the lower sediment flux. The variability in sediment flux, in turn, is governed by higher uplift rates (15 mm/yr; Lave and Avouac 2000) in the eastern Himalaya covering the upper catchments of the Teesta, Kosi, and Gandak compared with lower uplift rates in the western part (6.9 ± 1.8 mm/yr; Wesnousky et al. 1999) covering the hinterlands of the Ganga and Yamuna rivers. Sinha et al. (2005) argued that such hydrological differences between the rivers draining the EGP and WGP, which are controlled by climatic and tectonic variability, may have existed for a long time. We suggest that higher rainfall,

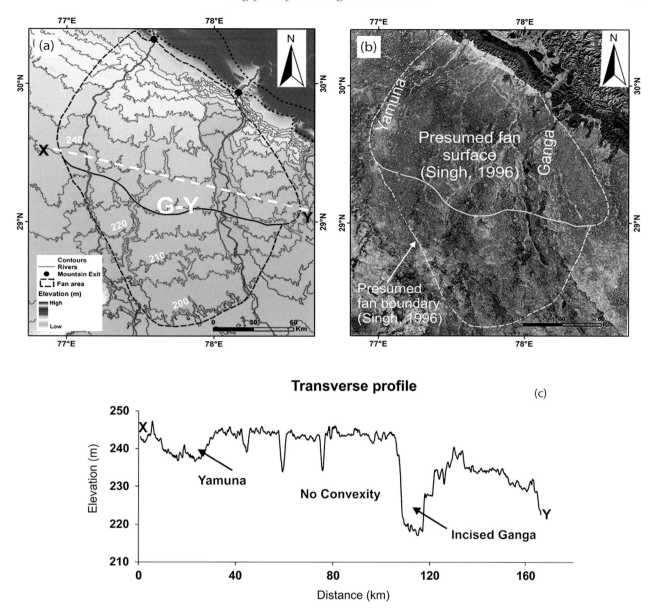

Figure 10.8 **Aspects of the Ganga–Yamuna megafan-like landform**. (a) Topographic contours. (b) Satellite images along with presumed fan boundaries by Singh (1996) and Shukla et al. (2001). There is no defined fan apex, no radial drainage, nor any other morphological characteristics of an active megafan such as seen on the Gandak, Kosi, Teesta, and Sone. (c) A profile along X–Y shows incised valleys instead of any convexity across the surface.

higher sediment flux and low stream power in the EGP have resulted in the formation of megafans. In contrast, lower rainfall, lower sediment flux and higher stream power have resulted in incised valleys in the WGP in the modern landscape.

Several workers have mapped relict fans in the WGP drained by the modern Ganga and Yamuna rivers (Fig. 10.8a), but without any distinguishable geomorphic boundary with the Sarda megafan (Singh 1996; Shukla et al. 2001). More importantly, these relict megafans were categorised by later workers as large

alluvial megafans on the same basis as the modern Kosi and Gandak megafans (Goodbred 2003; Blum 2007). Previous maps by Geddes (1960), as well as our analysis do not provide enough evidence to describe them as relict fans; furthermore, there is no convexity in transverse topographic profiles, there are no radial drainage systems on the presumed fan surfaces (Fig. 10.8b, c), and there are no systematic sub-surface data to substantiate the existence of buried fans in this region.

We suggest instead that the WGP represents a distinctive land system that may be best described as a large, inactive *coalescing floodplain* in a valley–interfluve setting formed by the combined fluxes from closely spaced rivers such as the Ganga, Ramganga and Yamuna, and which has emerged as a high surface following incision events. During early stages of its geomorphic evolution, i.e., before incision or when incision was still shallow, floods were able to inundate parts of the surface. These incised surfaces in the Ganga and Yamuna alluvial plains have been described as 'detached floodplains' in our earlier work (Gibling et al. 2005), i.e., they resulted from long-term monsoonal fluctuations during late Quaternary period. These are regionally extensive surfaces that extend way beyond the valley margin of the rivers along the southern banks, and therefore are disconnected from modern fluvial dynamics but nonetheless do not qualify, and thus cannot be mapped, as terraces – unlike what has been suggested by earlier workers (Singh et al. 1990; Srivastava et al. 2003; Singh 2004).

Late Quaternary stratigraphic records from the Ganga–Yamuna interfluve have suggested that they consist of thick muddy deposits interrupted by thin sand bodies representing minor channels (Gibling et al. 2005; Sinha et al. 2007; Tandon et al. 2008). In contrast, the valley fill deposits of the Ganga and Yamuna valleys consist of thick (20–30 m) channel sands (Sinha et al. 2007; Roy et al. 2012) with little or no pedogenesis (Srivastava et al. 2018), and often occurring as fining-upward sequences (see Summary presented in Tandon et al. 2008).

While emphasising that topographic analysis in the modern Ganga and Yamuna plains quite clearly features incised fluvial valleys as opposed to the fan topography of EGP rivers, we suggest that the WGP in general represents a unique land system of coalescing floodplains in the interfluves of the major large rivers over a wide lateral scale. The practice of describing megafans from the modern landscape of the WGP, therefore, should be viewed with caution.

10.5 Long-Term Evolution and Geological Controls: A Discussion

Megafans have mainly formed along the northern fringe of the Gangetic plains bordering the advancing Himalayan tectonic front (Fig. 10.2), the sole exception being the Sone fan. In contrast to the western Ganga plain where the rivers are incised, resulting in a valley–interfluve landscape (Sinha. et al. 2005; Tandon et al. 2006), river channels draining the eastern Ganga plain are characteristically raised above the floodplain.

Analysing the megafan morphologies of the Gangetic plain has revealed the following features, which call for a better understanding of their controlling factors:

(i)   A distinct fan morphology in which the fan apex, its radiating/distributary drainage pattern and its boundary are readily mappable.
(ii)   The geological characteristics such as lithology and fault- and fold-generated geometries of the mountain front through which the feeder channel debouches into the plains, which may control megafan location and the sediment grade supplied to the plains.
(iii) Distance between the mountain exit points of the Himalayan rivers, which may control accommodation space and the lateral mobility of channels.
(iv)   Attributes of the source area of the fan system, with a potential influence of catchment size, annual precipitation, uplift rate, and rock erodibility.
(v)   Variations in the foreland subsidence pattern at regional basin scale over the late Quaternary period.

Some of the issues listed above have been already addressed, while others, including the role of mountain exit points and basin subsidence, are discussed below.

Exit-point spacing in the development of megafans has been examined by previous workers. Gupta (1997) suggested that the megafan feeder rivers break through the mountain front at well-defined, structurally controlled locations. In several instances, these are linked to the tectonically forced deflection of drainages in the course of mountain building. The subsequent aggregation of deflected drainages produces augmented discharges that then flow behind growing fault-related anticlines along the Himalayan mountain front. These major rivers then debouch into the plains through exit points that are mostly linked to transverse faults occurring in fault segmentation zones of the tectonically active mountain front. The location and number of such exit points, therefore, influences the total water and sediment fluxes conveyed onto the plains and the resulting formation of megafans.

Spacing between the mountain exit points is also critical as it controls the accommodation space available in the foreland for lateral migration of the rivers –

one of the essential conditions for megafan formation through avulsion (see Wilkinson, Ch. 17). Figure 10.9 shows that exit-point distances in the EGP between the Teesta and Kosi, and the Kosi and Gandak are greater than those in the WGP, e.g., between Ganga–Yamuna, Ganga–Ramganga, Ramganga–Ghaghra, and Ghaghra–Rapti.

Table 10.1 also highlights another feature of these megafans regarding the upland/plains (u/p) ratio, which shows little relationship to fan size (as seen elsewhere) but appears to relate more to available lateral variation in the accommodation space.

Figure 10.10 shows a conceptual model to explain the role of the geological and geomorphological controls of megafan formation in the Gangetic plains and also highlights the contrasts between the EGP and WGP. There are significant variations in the precipitation and uplift rates along the strike of the orogen (Wesnousky et al. 1999; Lave and Avouc 2000); together, these have resulted in differences in catchment erosion and stream power among Himalayan rivers (Fig. 10.10a). Sinha et al. (2005) advanced this as a cause for the occurrence of high sediment supply leading to superelevated channels (Bryant et al. 1995; Mohrig et al. 2000) along with the development of the fan morphologies in the EGP, as opposed to the

development of incised channels in a valley–interfluve landscape in the WGP (Fig. 10.10b, c).

Dingle et al. (2016) analysed the possible controls on this geomorphic diversity of the Himalayan foreland fan–interfan landscape in the EGP *vs.* the valley–interfluve landscape in the WGP through a combined assessment of basin subsidence rates, sediment grain-size data, and sediment supply rate. These studies were carried out on the Yamuna, Ganga, Karnali, Gandak, and Kosi rivers. These researchers approximated the subsidence rate by combining a seismically derived basement profile with known India–Tibet convergence rates. From these data, they recognised slower subsidence in the western part of the Himalayan foreland, corresponding to the WGP, compared to the EGP. Dingle et al. (2016) further proposed that higher subsidence rates were responsible for a deeper basin in the east with perched, low-gradient river systems, and thus more accommodation space for sediment aggradation. In contrast, lower subsidence rates in the west were associated with a more elevated basin topography, and entrenched river systems recording climatically induced river base-level lowering during the Holocene. The factor of along-strike regional variation in basin subsidence, as advanced by Dingle et al. (2016), in conjunction with along-strike regional

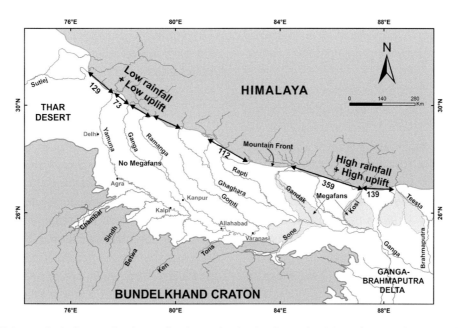

Figure 10.9 **Major geological controls of megafan formation in the Gangetic plains**. These can be summarised as: along-strike tectono-climatic variability manifested as mountain uplift rates and rainfall variation; spacing between mountain exit points (values in km); distance between the mountain front and axial river; and basin subsidence rate. The last three variables control accommodation space.

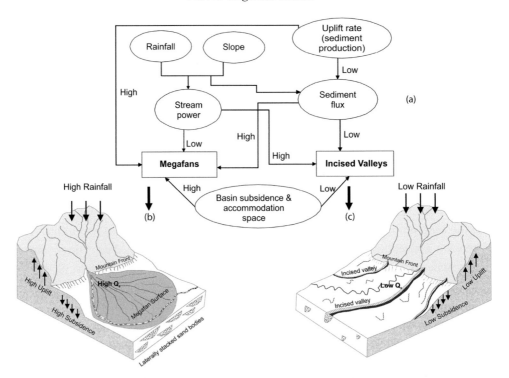

Figure 10.10 **Schematic process–response model illustrating the factors controlling the formation of megafans in the Gangetic plains**. (a) The process-response system; (b) cartoon illustrating the formation of megafans in the EGP; (c) cartoon illustrating the formation of incised valleys in the WGP.

variation in the stream power and sediment supply suggested by previous workers (Sinha et al. 2005; Tandon et al. 2008; Roy and Sinha 2017, 2018; Swarnkar and Sinha 2020; Kaushal et al. 2020), offers a reasonable explanation for the development histories of the fan morphologies to the north of the axial Ganga River and the resulting basin-scale geomorphic contrast between the EGP and WGP.

Another important question concerns the antiquity and stratigraphic evolution of the megafans in the Gangetic plains. Very limited data are available on the sub-surface stratigraphy and sediment chronology of the megafan deposits. However, as discussed earlier, some published records from the Teesta megafan suggest that fan deposition might have started at ~135 ka and then ceased at ~10–11 ka on the megafan surface, although some minor deposition on one of the smaller lobes continued until < 1 ka (Abrahami et al. 2018). Similarly, sedimentary cores from the Kosi megafan down to a depth of ~50 m have yielded the oldest OSL dates of ~65 ka (Sinha et al. 2015). Even though the Kosi river is now embanked on both sides, frequent breaches such as that in August 2008 (Sinha 2009) have enabled opportunities for deposition on the fan

surface. While more chronostratigraphic data may be required to constrain the antiquity of these megafans, it seems likely that those in the Gangetic plains have persisted since the late Pleistocene (MIS 3). The megafans continue to dominate the modern landscape of the EGP, while the modern landscape of the WGP may only have buried/relict fans whose surfaces are being incised into by the modern rivers.

10.6 Conclusions

The term 'fan/megafan' has been generically applied to most of the sediment bodies of river systems that flow from the Himalayan source area to the Ganga plains. In the present study, we apply this term 'megafan' restrictively to those landscape features that conform to the defining characteristics of the megafans, which are: mappable fluvial sediment entities with an apex, distributary drainage, convex-up transverse topographic profiles, and a distinct fan boundary with a break in slope. Following the application of these criteria, there are only four active megafans in this region, all of them occurring in the Eastern Gangetic plains (EGP). Three of them, namely the Teesta, Kosi, and Gandak, are fed by the rivers draining from the Himalaya whereas one of them, the Sone is fed from the uplands of a part of the Indian shield. There are no active megafans in the WGP, although their occurrence as relict megafans is quite possible. We suggest that the WGP landscape can be best described as that of a coalescing floodplain in a valley–interfluve setting. These observations do not preclude the occurrence of smaller fans and fan-like features in the WGP, particularly in the piedmont zone (Goswami 2009, 2010).

We suggest that the formation of megafans and their limited occurrence in the EGP are controlled by several parameters: (i) mountain-front tectonics manifested as spacing between river exit points along the strike of the range; (ii) differential late Quaternary and Holocene basin subsidence between the EGP and WGP, which governs regional-scale variations in the creation of accommodation space; (iii) along-strike tectonic and climatic variability, primarily reflected in differential uplift rates and rainfall gradients; and finally (iv) hydrological characteristics of the feeder channel manifested as stream power and sediment flux variability.

Acknowledgements

We thank Edgardo Latrubesse and Justin Wilkinson for inviting us to write this chapter, and this has been a learning experience for ourselves. Some parts of the work presented in this chapter were supported by various projects at different times funded by the Ministry of Earth Sciences (MOES), New Delhi; International Center for Integrated Mountain Development (ICIMOD), Kathmandu; and Indo-French Centre for Advanced Research (IFCPAR), new Delhi, and we sincerely acknowledge their support. Several students at IIT Kanpur worked on these projects and generated valuable data – particularly Surya Gupta and Surjodoy Ghoshal, who also helped at various stages of the project and in the preparation of this chapter.

References

Abrahami, R., Huyghe, P., van der Beek, P., et al. (2018) Late Pleistocene – Holocene development of the Tista megafan (West Bengal, India): [10]Be cosmogenic and IRSL age constraints. *Quaternary Science Reviews*, 185, 69–90.

Argialas, D. P. and Tzotsos, A. (2004). Automatic Extraction of Alluvial Fans from ASTER L1 Satellite Data and a Digital Elevation Model using Object-oriented Image Analysis. XXth ISPRS Congress, Commission 7, Istanbul, Turkey.

Blum, M. D. (2007). Large river systems and climate change. In A. Gupta, ed., *Large Rivers: Geomorphology and Management*. Wiley, Chichester, 627–659.

Bryant, M., Falk, P., and Paola, C. (1995). Experimental study of avulsion frequency and rate of deposition. *Geology*, 23, 365–368.

Chakraborty, T. and Ghosh, P. (2010). The geomorphology and sedimentology of the Tista megafan, Darjeeling Himalaya: implications for megafan building processes. *Geomorphology*, 115, 252–266.

Chakraborty, T., Kar, R., Ghosh, P., and Basu, S. (2010). Kosi megafan: historical records, geomorphology and the recent avulsion of the Kosi River. *Quaternary International*, 227, 143–160.

Densmore, A. L., Sinha, R., Sinha, S., Tandon, S. K., and Jain, V. (2016). Sediment storage and release from Himalayan piggyback basins and implications for downstream river morphology and evolution. *Basin Research*, 28, 446–461.

Dingle, E. H., Sinclair, H. D., Attal, M, Mildodowski, D.T., and Singh, V. (2016). Subsidence control on river morphology and grain size in the Ganga plain. *American Journal of Science*, 316, 778–812.

Geddes, A. (1960). The alluvial morphology of the Indo-Gangetic Plain: its mapping and geographical significance. *Institute of British Geographers, Transactions and Papers*, 28, 253–276.

Gibling, M. R., Tandon, S. K., Sinha, R., and Jain, M. (2005). Discontinuity-bounded alluvial sequences of the southern Gangetic plains, India: aggradation and degradation in response to monsoonal strength. *Journal of Sedimentary Research*, 75, 373–389.

Gohain, K. and Parkash, B. (1990). Morphology of the Kosi Megafan. In A. H. Rachocki and M. Church, eds., *Alluvial Fans: A Field Approach*. Wiley, Chichester, 151–178.

Gole, C. V. and Chitale, S. V. (1966). Inland delta building activity of Kosi River. *Journal of Hydraulic Division Proceedings ASCE*, 92, 111–126.

Goodbred, S. L. (2003). Response of the Ganges dispersal system to climate change: a source-to sink view since the last interstade. *Sedimentary Geology*, 162, 83–104.

Goswami, P., Pant, C. C., and Shefali, P. (2009). Tectonic controls on the geomorphic evolution of alluvial fans in the Piedmont Zone of Ganga Plain, Uttarakhand, India. *Journal of Earth System Science*, 118, 245–259.

Goswami, P. and Yhokha, Y. (2010). Geomorphic evolution of the Piedmont Zone of the Ganga Plain, India: a study based on remote sensing, GIS and field investigation. *International Journal of Remote Sensing*, 21, 5349–5364.

Gupta, S. (1997). Himalayan drainage patterns and the origin of fluvial megafans in the Ganga foreland basin. *Geology*, 25, 11–14.

Jain, V. and Sinha, R. (2003). River systems in the Gangetic plains and their comparison with the Siwaliks: A review. *Current Science*, 84, 1025–1033.

Kale, V. S. and Rajaguru, S. N. (1987). Late Quaternary alluvial history of northwest Deccan Upland region. *Nature*, 325, 612–614.

Kaushal, R. K. Sarkar, A., Mishra, K., et al. (2020). Spatio-temporal variability in stream power distribution in the Upper Kosi River basin, Central Himalaya: controls and geomorphic implications. *Geomorphology*, https://doi.org/10.1016/j.geomorph.2019.106888.

Lave, J. and Avouac, J.-P. (2000). Active folding of fluvial terraces across the Siwalik Hills, Himalayas of central Nepal. *Journal of Geophysical Research*, 105, 5735–5770.

Meetei, L. I., Pattanayak, S. K., Bhaskar, A., Pandit, M. K., and Tandon, S. K. (2007). Climatic imprints in Quaternary valley fill deposits of the middle Teesta valley, Sikkim Himalaya. *Quaternary International*, 159, 32–46.

Miliaresis, G. and Argialas, D. P. (2000). Extraction and delineation of alluvial fans from digital elevation models and Landsat Thematic Mapper images. *Photogrammetric Engineering and Remote Sensing*, 66, 1093–1101.

Miliaresis, G. (2001). Extraction of bajadas from digital elevation models and satellite imagery. *Computers & Geosciences*, 27, 1157–1167.

Mohindra, P. S. and Parkash, B. (1994). Geomorphology and neotectonic activity of the Gandak megafan and adjoining areas, middle Gangetic plains. *Journal of the Geological Society of India*, 43, 149–157.

Mohindra, R., Parkash, B., and Prasad, J. (1992). Historical geomorphology and pedology of the Gandak megafan, Middle Gangetic plains, India. *Earth Surface Processes and Landforms*, 17, 643–662.

Mohrig, D., Heller, P. L., Paola, C., and Lyons, W. J. (2000). Interpreting avulsion process from ancient alluvial sequences: Guadalope–Matarranya system (northern Spain) and Wasatch Formation (western Colorado). *Geological Society of America Bulletin*, 112, 1787–1803.

Norini, G., Zuluaga, M. C., Ortiz, I. J., Aquino, D. T., and Lagmay, A. M. F. (2016). Delineation of alluvial fans from Digital Elevation Models with a GIS algorithm for the geomorphological mapping of the Earth and Mars, *Geomorphology*, 273, 134–149.

Roy, N. G., Sinha, R., and Gibling, M. R. (2012). Aggradation, incision and interfluve flooding in the Ganga Valley over the past 100,000 years: Testing the influence of monsoonal precipitation. *Palaeogeography, Palaeoclimatology, Palaeoecology*, 356–357, 38–53.

Roy, N. G. and Sinha, R. (2017). Linking hydrology and sediment dynamics of large alluvial rivers to landscape diversity in the Ganga dispersal system, *India*. *Earth Surface Processes and Landforms*, 42, 1078–1091.

Roy, N. G. and Sinha, R. (2018). Integrating channel form and processes in the Ganga River: implications for geomorphic diversity. *Geomorphology*, 302, 46–61.

Sahu, S., Raju, N. J., and Saha, D. (2010). Active tectonics and geomorphology in the Sone-Ganga alluvial tract in mid-Ganga Basin, India. *Quaternary International*, 227, 116–126.

Sahu, S, Saha, D., and Dayal, S. (2015). Sone megafan: A non-Himalayan megafan of craton origin on the southern margin of the middle Ganga Basin, India. *Geomorphology*, 250, 349–369.

Sahu, S., Saha, D., and Shukla, R. R. (2018). Sone megafan: a non-Himalayan megafan of craton origin, forming a potential groundwater reservoir in marginal parts of Ganga Basin, India. *Hydrogeology Journal*, 26, 2891–2917.

Shukla, U. K., Singh, I. B., Sharma, M., and Sharma, S. (2001). A model of alluvial megafan sedimentation: Ganga Megafan. *Sedimentary Geology*, 144, 243–262.

Singh, I. B., Bajpai, V. N., Kumar, V. N., and Singh, M. (1990). Changes in the channel characteristics of Ganga River during Late-Pleistocene-Holocene. *Journal of Geological Society of India*, 36, 67–73.

Singh, I. B. (1996). Geological evolution of Ganga plain – an overview. *Journal of the Paleontological Society of India*, 41, 99–137.

Singh, I .B. (2004). Late Quaternary history of the Ganga Plain. *Journal of the Geological Society of India*, 64, 431–454.

Singh, H., Parkash, B., and Gohain, K. (1993). Facies analysis of the Kosi megafan deposits. *Sedimentary Geology*, 85, 87–113.

Sinha, R. (2009). The great avulsion of Kosi on 18 August 2008. *Current Science*, 97, 429–433.

Sinha, R., Priyanka, S., Jain, V., and Mukul, M. (2014a). Avulsion threshold and planform dynamics of the Kosi river in north Bihar (India) and Nepal: a GIS framework. *Geomorphology*, 216, 157–170.

Sinha, R., Ahmad, J., Gaurav, K., and Morin, G. (2014b). Shallow subsurface stratigraphy and alluvial architecture of the Kosi and Gandak megafans in the Himalayan foreland basin, India. *Sedimentary Geology*, 301, 133–149.

Sinha, R. and Friend, P. F. (1994). River systems and their sediment flux, Indo-Gangetic plains, Northern Bihar, India. *Sedimentology*, 41, 825–845.

Sinha, R. and Jain, V. (1998). Flood hazards of north Bihar rivers, Indo-Gangetic Plains. In V. S. Kale, ed., *Flood Studies in India*. Geological Society of India Memoir, 41, 27–52.

Sinha, R., Jain, V., Prasad Babu, G., and Ghosh, S. (2005). Geomorphic characterisation and diversity of the fluvial systems of the Gangetic plains. *Gemorphology*, 70, 207–225.

Sinha, R., Bhattacharjee, P., Sangode, S. J., et al. (2007). Valley and interfluve sediments in the southern Ganga plains, India: exploring facies and magnetic signatures. *Sedimentary Geology*, 201, 386–411.

Sinha, R., Bapalu, G. V. Singh, L. K., and Rath, B. (2008). Flood risk analysis in the Kosi river basin, north Bihar using multi-parametric approach of Analytical Hierarchy Process (AHP). *Journal of Indian Society of Remote Sensing*, 36, 293–307.

Sinha, R., Kettanah, Y., Gibling, M. R., et al. (2009). Craton-derived alluvium as a major sediment source in the Himalayan Foreland Basin of India. *Geological Society of America Bulletin*, 121, 1596–1610.

Sinha, R., Gupta, A., Kanchan M., et al. (2019). Basin scale hydrology and sediment dynamics of the Koshi River in the Himalayan foreland. *Journal of Hydrology*, 570, 156–166.

Srivastava, P., Singh, I. B., Sharma, M., and Singhvi, A. K. (2003). Luminescence chronometry and Late Quaternary geomorphic history of the Ganga Plain, India. *Palaeogeography, Palaeoclimatology, Palaeoecology*, 197, 15–41.

Srivastava, P., Sinha, R. V., Deep, S. A. K., and Upreti, N. (2018). Micromorphology and sequence stratigraphy of the interfluve paleosols from Ganga plains of alluvial cyclicity and paleoclimate during the Late Quaternary. *Journal of Sedimentary Research*, 88, 105–128

Swrankar, S., Sinha, R., and Tripathi, S. (2020). Morphometric diversity of supply-limited and transport-limited systems in the Himalaya foreland. *Geomorphology*, 348. https://doi.org/10.1016/j.geomorph.2019.106882

Tandon, S. K., Gibling M. R., Sinha R., et al. (2006). Alluvial valleys of the Gangetic Plains, India: causes and timing of incision. In R. W. Dalrymple, D. A. Leckie, and R. W. Tillman, eds., *Incised Valleys in Time and Space*. SEPM Special Publication, 85, 15–35.

Tandon, S. K. and Sinha, R. (2007). Geology of large river systems. In A. Gupta, ed., *Large Rivers: Geomorphology and Management*. Wiley, Chichester, 7–28.

Tandon, S. K., Sinha, R., Gibling, M. R., Dasgupta, A. S., and Ghazanfari, P. (2008). Late Quaternary evolution of the Ganga Plains: myths and misconceptions, recent developments and future directions. *Journal of the Geological Society of India Memoir*, 66, 259–299.

Van Dijk, W. M., Densmore, A. L., Singh, A., et al. (2016). Linking the morphology of fluvial fan systems to aquifer stratigraphy in the Sutlej-Yamuna plain of northwest India. *Journal of Geophysical Research: Earth Surface*, 121, 201–222.

Wells, N. A. and Dorr Jr., J. A. (1987). Shifting of the Kosi River, northern India. *Geology*, 15, 204–207.

Wesnousky, S. G., Kumar, S., Mohindra, R., and Thakur, V. C. (1999). Uplift and convergence along the Himalayan Frontal Thrust of India. *Tectonics*, 18, 967–976.

Williams, M. A. J., and Clarke, M. F. (1984). Late Quaternary environments in North-Central India. *Nature*, 308, 633–635.

Williams, M. A. J., Pal, J.N., Jaiswal, M., and Singhvi, A. K. (2006). River response to Quaternary climatic fluctuations: evidence from the Son and Belan valleys, north-central India. *Quaternary Science Reviews*, 25, 2619–2631.

11

The Kosi Megafan, India
Morphology, Dynamics, and Sedimentology

RAJIV SINHA, KUMAR GAURAV, KANCHAN MISHRA, and SURYA GUPTA

Abstract

This overview of the well-documented, ~11,200 km^2 Kosi megafan updates many aspects of its geomorphology, and maps the detail of its modern trunk channel. The axis of the Kosi megafan is orthogonal to the Himalaya and Ganga trunk river, but with a mean annual flow of 52×10^9 m^3 it has constructed a relatively small megafan (~150 km long) constricted between neighbouring megafans and the Ganga floodplain. However, while the coarser sediment load of other megafans farther to the west, such as the Gandak, has been trapped upstream in piggyback basins of the Terai belt, this upstream filter does not exist in the case of the Kosi River. Consequences for the Kosi channel are thought to be a more continuous supply of sediment, a higher proportion of coarse debris, higher rates of bed aggradation, and a more avulsive style of river behaviour on the megafan. Extensive construction of artificial levees, initially designed to mitigate the hazards arising from excessive flooding, has accelerated natural rates of channel aggradation – thereby raising the channel bed in several reaches and resulting in more frequent levee breaching and flood-related damage than on other Himalayan megafans.

11.1 Introduction

The Sapt Kosi ('seven channels of the Kosi River') is the main feeder channel of the Kosi megafan. It has seven major tributaries, the Arun, Tamor, Sun Kosi, Tama Kosi, Dudh Kosi, Bhote Kosi, and the Indrawati, all of which originate in the Higher Himalaya in Tibet and Nepal (Fig. 11.1a). Among these, the Arun, Sun Kosi, and Tama Kosi are the major tributaries and these meet near Tribeni, from which point a much-enlarged Kosi River flows through the Barahkshetra gorge for about 15 km before emerging onto the plains at Chatara (Fig. 11.1b). The river then flows for about 250 km before it joins the Ganga River near Kursela (Fig. 11.1b).

The hinterland of the Kosi River, mostly lying in Nepal and Tibet, consists of six geological and climatic belts: the Tibetan plateau, the Higher Himalaya, the Midland hills, the Lesser Himalaya (Mahabharat), the Siwaliks and the Terai (foothills). The elevation of this region varies from > 8,000 m to 95 m above mean sea level. It has been estimated that there are around 36 glaciers and 296 glacial lakes in the Kosi catchment (Bajracharya et al. 2007). The Sun Kosi, Arun and Tamor are considered to be antecedent drainage systems (Wagner 1937; Gansser 1964; Oberlander 1985) that flowed into the Tethys Sea before the Himalaya began to rise about 70 million years ago. With the rise of the Himalaya, these rivers have since cut deep gorges for a large part of their courses in the Higher Himalaya. The Mahabharat mountains acted as a barrier to the south-flowing rivers and forced the Sun Kosi and Tamor rivers to flow parallel to the range as far as Tribeni, where they join the Arun flowing in from the north.

The climatic setting of the Kosi catchment is monsoonal and the rainfall occurs from June to September. The entire catchment is marked by large regional and temporal variations because of considerable orographic contrasts. The average rainfall in the western

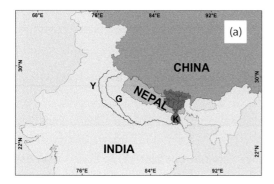
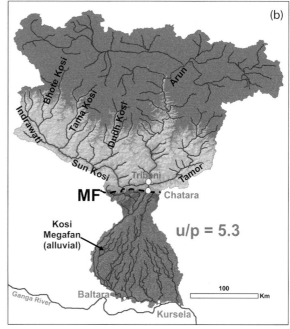

Figure 11.1 **The Kosi megafan, north Bihar plains, and its Himalayan catchment**. (a) Regional location map. (b) Digital elevation map of the Kosi megafan, its hinterland in Nepal and Tibet, and seven major tributaries feeding it. A colour version of this figure is available in the SOD for Chapter 11.

hinterland of the Kosi basin in ∼2,810 mm, whereas the eastern part receives only 1,238 mm. The Tibetan part experiences very low rainfall as it lies on the leeward side of the Himalaya and a gradual decrease in rainfall from the foothills in Nepal towards the alluvial part in Bihar (∼2,100 mm) has been documented (Sinha et al. 2019). Sudden cloudbursts in the upper catchment can generate up to 500 mm of rainfall in a single day. Such extreme climatic variations in the Kosi catchment are manifested clearly in the hydrology of the Kosi River, particularly in the alluvial reaches where the megafan has formed.

The Kosi megafan has attracted widespread attention mainly because of its conspicuous outline, large size, morphology, active Himalayan hinterland, and high sediment influx, as well as being one of the first megafans to be described scientifically. The first systematic morphologic description of the Kosi megafan was provided by Geddes (1960) based on the analysis of contour spacing. Geddes (1960) used the term *cone* to describe the Kosi fan and its radial drainage. He further suggested that high discharge and silt load were the main causal factors for generating this large landform. Even before any detailed work on the Kosi was started, Geddes (1960, p. 263) documented the

fact that 'even in a normal season, the Kosi sands may raise the belt of the flood plain by a foot'. Gole and Chitale (1966) described the Kosi megafan as an 'inland delta' built by the large sediment flux from the Himalayan orogeny, displaying a conical delta with the centre situated in the vicinity of Belka Hill, west of Chatara. The mobile behaviour of the Kosi River was attributed to the cone-building process. The term 'megafan' was coined by Gohain and Parkash (1990) using the Kosi as an example. They distinguished four topographically distinct geomorphic units, namely the active channel belt, active floodplain surface, sparsely vegetated surface, and an older floodplain unit for the Kosi megafan and the feeder channel.

Frequent avulsive shifting of the course of the Kosi River has been recognised as one of the manifestations of the fan building processes, and available records suggest a net westward migration of about 150 km in the past 200 years (Gole and Chitale 1966; Wells and Dorr 1987). Geddes (1960) recognised that water and sediment discharge were two of the major factors controlling the size of the fan, and attributed the rate of migration of the Kosi River and fan building to the rate of sediment accumulation on the fan and the very low channel gradients. Gole and Chitale (1966) also conceived that fan building, and therefore channel migration, would be slower in the central part compared to the eastern and western edges of the fan; this was attributed to the maximum radial distance of the river apex to the toe of the fan in its central part. It was further suggested that fan-building activity by the Kosi was still incomplete, which might explain the unstable behaviour of the Kosi River. More recent work has also suggested that geological controls such as subsidence rates (Dingle et al. 2016), spacing between the mountain exits, and basin dimension (Sinha et al., Ch. 10) also influence the megafan formation.

This chapter presents critical information about the Kosi megafan and summarises the latest research on its features and behaviour. Much of it has been acquired from satellite remote sensing methods, digital elevation models (DEMs), and a compilation of hydrological and subsurface geological data.

11.2 Methods

The SRTM (1 arcsecond) DEM was used for deriving topographic characteristics of the megafan and its watersheds. Landsat-8 images were processed for mapping various morphological elements of the Kosi megafan and its feeder channel. Infrared, red and green wavelength bands were used to map various morphological elements on the fan surface such as minor active channels, abandoned channels, ponds, and large waterlogged patches. In-channel features were also mapped from Landsat imagery for multiple years to reconstruct planform dynamics of the feeder channel. Hydrological data (monthly discharge and sediment load) from different stations along the main feeder channel were analysed to assess the spatial and temporal variability in hydrology and sediment transport characteristics. Rating curves were generated between discharge and sediment load to establish their interrelationships. We also utilised available lithologs from groundwater wells maintained by the Central Groundwater Board (CGWB) to map the subsurface stratigraphy of the Kosi megafan. Lithologs were plotted along predefined transects to understand the stratigraphic variability along and across the megafan. Statistical analysis of sand body thicknesses in different lithologs provided insights into sedimentological characteristics and depositional environments.

11.3 Geomorphology of the Kosi Megafan

The Kosi megafan has an aerial extent of 10,351 km^2 and a radius of 115–150 km (Wells and Dorr 1987; Gupta 1997; Chakraborty et al. 2010). It exhibits a partial-cone morphology (Fig. 11.2a) with associated radiating channels, as typically observed across many other fans in the world. The longitudinal slope varies from 0.0008 at the apex to 0.00006 near the toe (Gole and Chitale 1966; Wells and Dorr 1987; Chakraborty et al. 2010). The radial profile of the Kosi megafan is close to linear, with a slight upward concavity arising from being steeper in the north (gradients: 0.00055–0.00075) and flatter in the south (0.00006) (Fig. 11.2b). The transverse profile is broadly convex-up, but highly irregular (Fig. 11.2c) because of numerous paleochannel belts as each one of these has its own channel belt (depressions). The eastern part is relatively lower than the western part (Chakraborty et al. 2010).

Based on the multiple groups of radiating palaeochannels, Chakraborty et al. (2010) identified three accretionary lobes (1a, 1b, and 2) on the Kosi megafan. This multi-lobe pattern is reflected in the transverse profile (Fig. 11.2c). Although no absolute chronology is available, it has been suggested that the central lobe (2) is younger than its western (1a) and eastern (1b) counterparts (Chakraborty et al. 2010). This relative chronology of the lobes is supported by the geomorphic classification of the Kosi megafan by Gohain and Parkash (1990), who recognised that lobe 1a displays patches of the remnant Old Alluvial Plain (OAP) featuring ancient meander belts, channel cutoffs, oxbow lakes, swamps, and dense vegetation. In contrast, the rest of the megafan surface is characterised by geomorphic features of the Young Alluvial Plain (YAP) such as abandoned channels, interchannel areas, a low-lying plain, distal fan plain, a proximal triangular alluvial plain, all covered by grey unconsolidated sediments (Gohain and Parkash 1990). Based on a more detailed analysis of historic maps, Chakraborty et al. (2010) suggested that the Kosi River had been migrating over the fan surface in a random and oscillating manner, except for a prolonged period in the

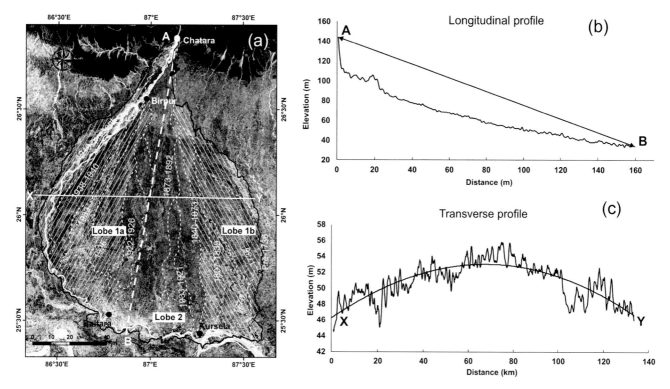

Figure 11.2 **Some morphological attributes of the Kosi megafan**. (a) Landsat image of the Kosi megafan showing its apex and toe along with several palaeochannels mapped by Gole and Chitale (1966). (b) Radial (N–S). (c) Transverse (E–W) topographic profiles based on SRTM data.

central part of the fan, thereby revising the idea of progressive westward migration of the Kosi River hypothesised by earlier workers (Gole and Chitale 1966; Wells and Dorr 1987).

Figure 11.3 shows a detailed geomorphological map of the Kosi megafan and the feeder channel prepared from Landsat 8 (OLI) of 2013. The wider megafan surface is marked by the numerous active and abandoned channels, locally called 'dhars', formed by frequent migration of the main channel during historic times (pre-embankment period). These palaeochannels are typically fed by rainwater and groundwater (plains-fed channels of Sinha and Friend 1994). They are activated during the monsoon season and cause large-scale inundation during floods (Sinha et al. 2014a). The morphology of these palaeochannels is highly sinuous (sinuosity > 1.5), much different from the modern Kosi channel. Lateral migration of these paleo-channels on the megafan surface has left behind numerous meander scars and oxbow lakes. Numerous sand bodies on the megafan surface represent channel fills and avulsion deposits, including those from the most recent event in August 2008. At present, many of

the palaeochannels have been disrupted as a result of construction of roads, railway lines and embankments (levees) for flood protection (Fig. 11.3), and this has intensified waterlogging on the low-lying, flat megafan surface (Kumar et al. 2014). Permanently waterlogged areas are locally called 'chaurs' and are generally fed by groundwater and/or monsoon flows through the minor channels. Adjacent to the eastern embankment, widespread waterlogged areas probably represent water seepage through the embankment. Over the years, these seepages combine to form well-defined channels, and one of them can be seen along the eastern embankment in the mid-fan area (Fig. 11.3).

The feeder channel of the megafan shows a braided pattern (Gaurav et al. 2015) in the proximal part, with numerous mid-channel bars. In the medial part, the Kosi channel is clearly anabranching, with large stable alluvial islands. Further downstream, the multi-channel braided pattern picks up again but channel sinuosity increases significantly. Since the Kosi River in the megafan sector is now confined within embankments, most of the area between the embankments and the active channel belt has been classified as active

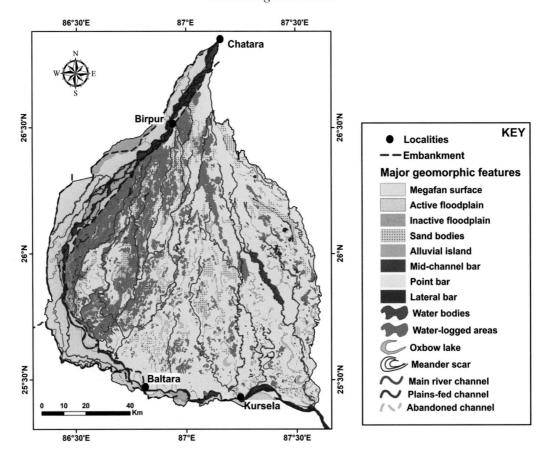

Figure 11.3 **Geomorphological map of the Kosi megafan from Landsat imagery.** Note characteristic morphological features such as abandoned channels, meander scars, waterlogged areas, and sand patches as well as the engineered embankments on both sides of the channel belt.

floodplain, marked by little or no vegetation and high moisture content, presumably caused by annual flooding. There are patches of inactive/older floodplain along the edges of the embankments that are generally vegetated and beyond the reach of annual flooding (Fig. 11.3).

11.4 Feeding the Megafan: Hydrology and Sediment Transport Characteristics of the Kosi River

11.4.1 Hydrology of the Kosi River

Seven major tributaries of the Kosi River – the feeder channel of the megafan – meet at Chatara (Fig. 11.1), and the combined flow and sediments are transported downstream to build the megafan. The catchment area of the Kosi basin at Chatara (i.e., approximately at the apex of the megafan) is 52,730 km^2, which is 62% of the total area of the basin, i.e., 84,739 km^2. The mean annual discharge at Chatara is 1,545 m^3/s, decreasing to 1,452 m^3/s at Birpur (immediately upstream of the barrage), but increases significantly to 2,256 m^3/s at Baltara, the farthest downstream station (Table 11.1). Among several other factors responsible for the megafan formation, hydrological characteristics – particularly the sediment flux – of the feeder channel play an important role in the generation of megafans, as suggested by theoretical considerations and demonstrated from laboratory experiments (Parker 1999; Reitz and Jerolmack 2012; Delorme et al. 2017).

Previous hydrological analysis of the Kosi basin has suggested that ~56% of the total discharge at Chatara is contributed by the tributaries flowing from the west (e.g., Indrawati, Bhote Kosi, Tama Kosi and Sun Kosi), whereas the tributaries draining from the central part (e.g., Arun) and eastern part (e.g., Tamor) contribute 38% and 16%, respectively (Sinha et al. 2019). This variability in discharge is a direct manifestation of the significant decrease in rainfall from west to east.

Table 11.1 *Hydrology and sediment transport characteristics of the Kosi River*

Parameters	Chatara	Birpur	Baltara
Catchment area (A) (km^2)	52,730	54,089	84,739
Average annual discharge (Q) (m^3/s)	1,545	1,452	2,256
Average monsoonal discharge (m^3/s)	3,313	3,252	4,812
Average non-monsoonal discharge (m^3/s)	663	681	804
Mean annual flood discharge (m^3/s)	6,695	7,711	8,153
Average annual sediment load (10^6 t/yr)	101	81	43
Specific sediment load (t/km^2/yr)	1,915	1,497	507
Unit discharge (Q/A) (m^3/s/km^2)	0.03	0.03	0.03

Source: Sinha et al. (2019)

Figure 11.4a shows the hydrographs of the Kosi River at three stations: Chatara, Birpur and Baltara. Peak discharge occurs during the monsoon season from June to October. Monsoon rains provide about 74% of total river discharge, whereas the rest originates from snow melt (Gohain and Parkash 1990). The average monsoonal discharge is 4–6 times higher than the non-monsoonal discharge at all stations (Table 11.1) on the feeder channel, clearly emphasising the monsoon's influence (Fig. 11.2a). Both non-monsoonal and monsoonal discharges are considerably higher at the downstream Baltara station compared with that at Birpur, opposing the general trend of downstream decline in discharge observed in many megafans (attributed to channel bifurcation, infiltration, and evapotranspiration by Hartley et al. 2010). Detailed hydrological analysis (Sinha et al. 2019) has shown that two major interfan rivers, the Baghmati and Kamla-Balan, meet the Kosi (through the inlets provided in the western embankment) upstream of Baltara, and contribute 189 m^3/s and 68 m^3/s of discharge annually. Furthermore, numerous paleochannels on the fan surface are groundwater-fed (Gohain and Parkash 1990; Sinha and Friend 1994) and they also add significantly to the monsoonal flows of the Kosi at Baltara.

11.4.2 Sediment Dynamics and Fluxes

One of the most characteristic features of the Kosi River is its high sediment flux. The Kosi has a very large upland catchment area and a number of tributaries from west and east bring large quantities of sediments from the Higher and Lesser Himalaya (Fig. 11.1). The relative contributions of from these tributaries are significantly different. At Chatara, close to the fan apex, the average annual sediment load is 101 Mt, of which 40% is contributed from the western tributaries (Indrawati/Bhote Kosi, and Tama Kosi) through the axial Sun Kosi, in contrast to only 16% from the eastern tributaries (mainly Tamor). The remaining 44% of sediment load at Chatara is contributed by the central tributaries (Arun and Dudh Kosi).

Figure 11.4b shows the discharge–sediment relationships for Chatara, Birpur and Baltara, all situated on the feeder channel of the megafan. The rating curves for Chatara and Birpur are similar but that for Baltara is strikingly different (Fig. 11.4b), suggesting contrasting processes operating at the upstream and downstream stations. Both Chatara and Birpur stations show a steady increase in sediment load at low discharges but a rapid increase at high discharges, suggesting significant contribution from mass wasting processes in the Himalaya and high erosive power of the river to transport sand and silt-sized sediments. In contrast, Baltara shows a much higher dispersion of loads at both low and high discharges, indicating the dominance of easily erodible sediments contributing to wash load (Morgan 1995) at this location. Given such spatial variability in sediment transport characteristics, it has been suggested that the Kosi River is supply-limited at Chatara and Birpur, but transport-limited at Baltara (Sinha et al. 2019). This contrasts with other rivers draining the western Ganga plains where rivers are mostly supply-limited (Roy and Sinha 2017) and have not developed megafans in the modern landscape (Sinha et al., Ch. 10).

Sediment load decreases steadily from upstream to downstream stations: 101 Mt at Chatara, 81 Mt at

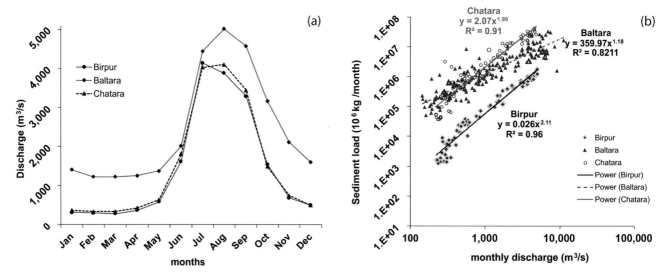

Figure 11.4 **Hydrology and sediment transport characteristics of the Kosi River.** (a) Hydrographs for three major stations on the feeder channel of the Kosi megafan: Chatara, Birpur and Baltara. (b) Rating curves showing discharge–sediment relationships for these stations. A colour version of this figure is available in the SOD for Chapter 11.

Birpur, and 43 Mt at Baltara. This happens despite an increase in discharge at Baltara and suggests a highly aggradational regime throughout the feeder channel. A first-order sediment budgeting suggests that ~20 Mt of sediments may be accumulating annually between Chatara and Birpur, which translates to $408 \times 10^6 \, \text{m}^3$ of sediments accumulated in the last 54 years since the embankments were constructed (Sinha et al. 2019). Budgeting between Birpur and Baltara – including the contributions from the interfan rivers, Baghmati (7 Mt/yr) and Kamla-Balan (8 Mt/yr) – suggests 53 Mt of annual and $1080 \times 10^6 \, \text{m}^3$ of total sediment accumulation (Sinha at al. 2019). Using the active channel belt of the feeder channel as the main accumulation zone, Sinha et al. (2019) computed the average sedimentation rates as 5.31 cm/yr and 3.94 cm/yr for the Chatara–Birpur and Birpur–Baltara sectors, respectively. These are exceptionally high sedimentation rates, primarily induced by channel confinement between embankments.

11.5 Planform Dynamics, Avulsive Processes, and Flood Pathways

11.5.1 Planform Dynamics Reconstructed Through Temporal Remote Sensing Data

Lateral migration of the Kosi River has now been restricted by construction of embankments on both sides of the river since 1955–1956, and by a barrage

at Birpur operating since 1963. The lengths of the eastern and western embankments are currently 144 km and 123 km, respectively. The distance between these two levees varies from 5 to 16 km (Desai 1982; Sinha et al. 2013). As a consequence, the Kosi River has been disconnected from its fan, and is forced to deposit large volumes of sediment in the zone enclosed by the embankments, as discussed above. This has resulted in rapid aggradation of the river bed, which is now 'superelevated' with respect to the adjacent fan surface by ~4 m (Sinha et al. 2013, 2014a). Such high rates of channel bed aggradation have significant implications for morphodynamics of the river and floods, particularly in the downstream reaches.

Figure 11.5a shows the planform maps of the Kosi channel downstream of the Kosi barrage between 1972 and 2016. The channel was divided into 33 reaches at 5 km interval and all in-channel bars were mapped to understand the aggradation pattern. It is also noted that the Kosi River has been shifting its position between the western and eastern embankments, primarily guided by the in-channel aggradation. Temporal variability of the reach-wise bar area (BA) and the channel area (CA) shows that a major aggradation zone starts immediately downstream of the barrage (reach 8) and continues down to reach 21, below which the channel belt becomes quite narrow and aggradation is much lower (Fig. 11.5b). The barrage acts as a significant barrier to flow, thereby inducing excessive channel aggradation.

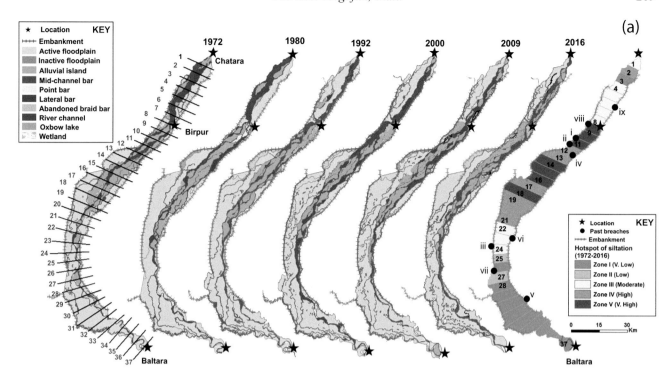

Figure 11.5 **Changes in floodplain morphology and distribution of siltation hotspots along the Kosi River, 1972–2006.**
(a) Planform dynamics of the feeder channel of the Kosi megafan between 1972 and 2016 reconstructed from satellite images. Note the major aggradation zones upstream and downstream of the barrage at Birpur. The map far right shows hotspots of siltation based on BA/CA ratios. Solid circles show the breaching points in recent years (i- Dalwa 1963, ii- Kunauli 1967, iii- Jamalpur 1968, iv- Bhatania 1972, v- Bahurawa 1980, vi- Navhatta 1984, vii- Ghoghepur 1987, viii- Joginia 1991, ix- Kusaha 2008).
(b) Reach-wise plot of channel area and bar area for different years during the period 1972–2016. Note the sharp rise in bar area downstream of the barrage.
(c) Reach-wise plot of BA/CA ratios for different years highlighting the hotpots of siltation (rectangles). The recent breaches in the embankment have also been plotted; several match the moderate to high siltation hotspots.

We therefore use the BA/CA ratio as a good indicator of hotspots of siltation (Fig. 11.5c). We have mapped several such hotspots (high BA/CA values) both upstream and downstream of the barrage, and have qualitatively classified the reaches into high, moderate and low zones of siltation. Several reaches downstream of the barrage area classify as very high and high zones of siltation, and several reaches upstream of the barrage register as moderate zones. Most of the lowest reaches are classified as low to very low zones of siltation, and planform changes through time here have been minimal. This approach has also provided insights into potential sites of future breaching and avulsion.

11.5.2 Avulsion Through Breaching of the Embankment

Since construction of the embankments, the natural migration of the Kosi River across the megafan surface through nodal avulsions (cf. Slingerland and Smith 2004) as documented by various workers (Gole and Chitale 1966; Wells and Dorr 1987; Chakraborty et al. 2010), has been constrained. However, frequent breaches of the embankment have been occurring – as many as nine since 1963 in both the eastern and western embankments (Fig. 11.5a), resulting in severe flooding (Mishra 2008; Sinha et al. 2014a). Although most of these breaches did not result in complete avulsions, the latest breach – on 18 August 2008 at Kusaha (close to the fan apex) 12 km upstream of the Birpur barrage – moved the main channel eastward by ~60 km in a single avulsive shift (Sinha 2009; Chakraborty et al. 2010; Sinha et al. 2014a). The maximum shift to the east was recorded as ~120 km (Sinha 2009).

Several breaching points fall in the reaches marked as high and very high zones of siltation, and some in the moderate zone including Kusaha resulted in a major avulsion in 2008. Theoretical and experimental

(b)

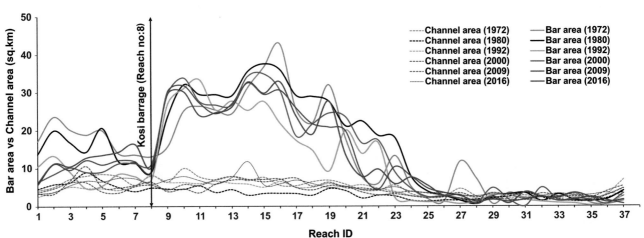

(c)

Figure 11.5 (*cont.*)

work has established that a channel tends to avulse when the cross-valley slope (S_{cv}) exceeds the down-valley slope (S_{dv}) at any given point (Jones and Schumm 2009); the S_{cv}/S_{dv} ratio defines the avulsion threshold and emphasises the superelevated condition of a channel (Mackey and Bridge 1995). When the avulsion threshold is exceeded, the river tries to establish a new course for its channel. The hotspots mapped in Figure 11.5a therefore represent the points where the river is close to the avulsion threshold; this is manifested by frequent breaching, and by at least one major avulsion in recent times. Importantly, if the river is close to the avulsion threshold it may not require a high discharge for breaching or avulsion: a small

trigger may be sufficient to dislodge the river from its present channel to a new course. Detailed analysis of the 2008 avulsion of the Kosi River at Kusaha by Sinha et al. (2014a) revealed that this breach occurred at relatively low flow, i.e., when discharge was only 4,078 m³/s, compared to the design discharge (26,901 m³/s) of the embankment (Sinha et al. 2013). It has been argued, therefore, that this avulsion occurred primarily because of large-scale aggradation that resulted in a 'superelevated' channel, thereby pushing the river close to the avulsion threshold at that point (Sinha 2009; Sinha et al. 2014a).

Further, it is noted that Kusaha is located in a reach classified as a moderate BA/CA hotspot, despite which the 2008 breach led to a major disaster. Several reaches downstream of the barrage classified as high and very high hotspots are extremely vulnerable to breaching, but whether or not they convert to points of complete avulsion depends on the fan slope. The Kusaha breach was converted into a complete avulsion because it occurred in the proximal part of the megafan, where slope is steeper than in the distal part. Such proximal to distal variability in slope of the megafan is governed by the ratio of sediment to water discharges (Q_s/Q_w) and segregation of coarser and finer sediments as demonstrated in experimental studies (Parker et al. 1998; Reitz and Jerolmack 2012; Delorme et al. 2017).

11.5.3 Flood Pathways

The August 2008 avulsion resulted in unprecedented floods in parts of Nepal and Bihar. The flood followed multiple paths as mapped by the Dartmouth Flood Observatory (Fig. 11.6a, b). The main flow followed one of the palaeochannels of the Kosi in the central part of the fan (Fig. 11.6). The floodwave inundated an area of 2,722 km^2 at a velocity of ~1 m/s in the avulsion belt (Sinha et al. 2013) and affected more than 3 million people in India and Nepal. The breach in the embankment was plugged on 26 January 2009, and the

river was relocated to its pre-avulsion position within the embankments.

The August 2008 breach provided an opportunity to understand the influence of avulsive processes on flood pathways primarily governed by the connectivity structure of the megafan surface. Connectivity is defined as the potential for energy and material flux (water, sediments, nutrients, heat, etc.) to navigate within or between the landscape compartments (Jain and Tandon 2010). Sinha et al. (2013) used a topography-driven connectivity model to simulate the avulsion pathway on the megafan surface and explored the way the hydrological flux was connected in the avulsion belt downstream. This study concluded that there was adequate connectivity embedded in the fan topography to forecast both the avulsion/flood pathways and their extent. Analysis of the connectivity structure before and after the avulsion event demonstrated that connectivity of the fan surface increased after the hydrological flux was connected to the avulsion belt through reactivation of paleochannels and creating tens of kilometres of new channels. At the same time, several existing channels were filled by a very large quantity of sediment that was eventually deposited over the fan surface (Sinha et al. 2014a).

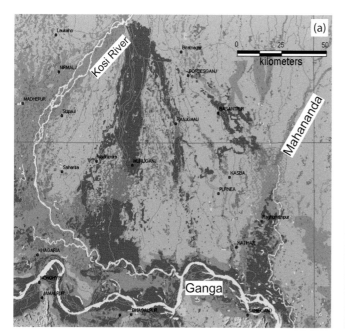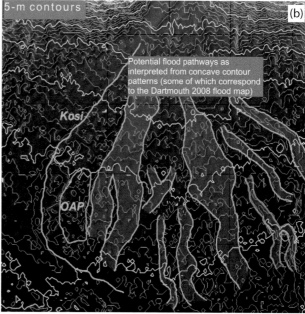

Figure 11.6 **Bar and channel systems**. (a) Flood inundation map of 2 September 2008 event. (b) Topographic map of the Kosi megafan showing 5 m contours and potential flood pathways as interpreted from the concave contour patterns, some of which were taken by the 2008 flood shown in (a).

Further analysis of the flood pathways and associated sedimentation of the August 2008 breach also suggested that the floodwave propagated mostly as unconfined sheet flow in the initial stage, and later as a broad zone of associated shallow, braided channels (Majumdar and Ghosh 2017). Significant flow widening and branching of flow occurred in the initial stage primarily as a consequence of disconnectivity generated by artificial barriers (Kumar et al. 2014), which delay the establishment of the main flowline connection to the local base level of the Ganga. Most of the aggradation occurred on the proximal and medial parts along a radially oriented sector of the fan, providing an example of 'creation of secondary, topographically positive lobes superimposed on profiles transverse to the megafan' (Majumdar and Ghosh 2017:334).

As shown earlier, the megafan surface consists of multiple lobes (Chakraborty et al. 2010) and various geomorphic surfaces (Gohain and Parkash 1990; this work). Figure 11.6b suggests that the avulsion channel and flood wave followed the depression between the lobes and was also guided by the older and higher topographic surfaces such as the Older Alluvial Plain (OAP) mapped by Gohain and Parkash (1990). This confirms the findings of Sinha et al. (2013) that the megafan surface retains adequate topography to define the pathways of the avulsion channels and floods. It also emphasises the contribution from geomorphologists to managing river hazards such as avulsion and floods in this region.

11.6 Understanding the Evolution of the Kosi Megafan: Evidence from the Subsurface and from the Past

11.6.1 Sedimentary Facies and Depositional Environments

The subsurface stratigraphy of the Kosi megafan has intrigued workers in the past, and most of the early work speculated that it should primarily consist of large, laterally stacked sand sheets, as found in the ancient sedimentary records of large braided rivers. Initial description of sedimentary facies of the Kosi megafan has been based on river cut-bank sections (0.8–4 m thick) and artificial trenches, along with borehole data (15 m deep) from several locations (Singh et al. 1993). Two depositional units were identified: (i) the upper part (8–10 m) containing medium sand associated with historical shifting of the river, and (ii) the lower part (40–80 m) containing coarse sand with gravel associated with late Holocene glacier melt-out events. There are no radiometric dates available, and the chronological and climatic interpretations are therefore rather speculative. Chakraborty et al. (2010) described the facies of the upper 2–5 m succession exposed in bank sections of the north-central part of the megafan, and reported an overwhelming dominance of meandering river deposits comprised of fine- to medium-grained sand with coarse sand at the base. These deposits were interpreted to have formed by lateral accretion, and therefore stand in opposition to depositional models suggesting the formation of the uppermost megafan succession by lateral sweeping of a modern Kosi-like braided river (cf. Singh et al. 1993). Chakraborty et al. (2010) also argued that deposition of 8–10-m-thick units during historical times, as suggested by Singh et al. (1993), required improbable sedimentation rates in excess of 50 mm/yr. Assuming a more probable rate of 1–2 mm/yr, these deposits would take 2.5–5 ka to accumulate and are hence unlikely to be related to the frequency and magnitude of Kosi River avulsions in historical times (Chakraborty et al. 2010).

11.6.2 Spatial Variability in Megafan Stratigraphy Inferred from Borehole Data

Detailed stratigraphic data from the Kosi megafan remains limited to the information available from lithologs provided by the Central Groundwater Board (CGWB). Sinha et al. (2014b) attempted to integrate the borehole data with resistivity soundings and limited drilling for unravelling the subsurface stratigraphy. The borehole data used by Sinha et al. (2014b) is reanalysed here with two objectives: (i) to understand the spatial variability in the subsurface stratigraphy, and (ii) to compare the sand-body distribution with that recorded in the relict megafans of northwest India (Sinha et al., Ch. 10). We used CGWB lithologs from 54 sites on the Kosi megafan extending down to about 100 m (Fig. 11.7a). Figure 11.7b–d shows the representative borehole data along three transects to

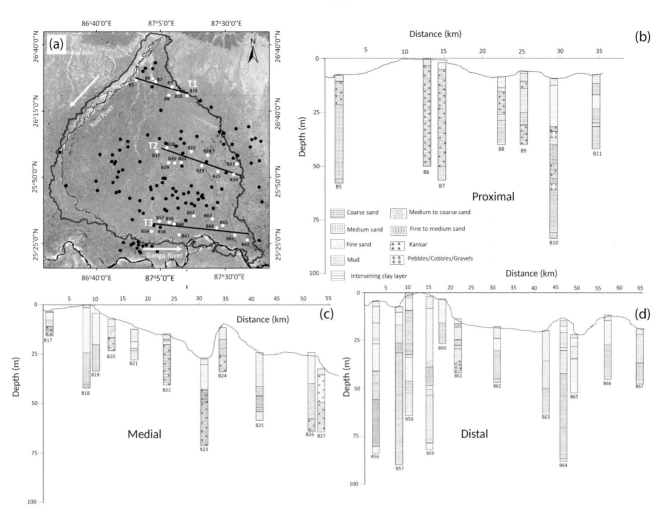

Figure 11.7 **Shallow subsurface lithology of the Kosi megafan**. (a) Location of boreholes. (b–d) Transects along T1, T2, and T3 in proximal, medial, and distal parts.

illustrate the stratigraphy in the proximal, medial, and distal parts. Distinct differences are noted in the proximal and medial parts in terms of grain-size distribution, sand-body thickness, and interpreted widths. Whereas the modern Kosi River does not transport gravel downstream of Chatara, drill cores and resistivity surveys (Sinha et al. 2014b) in the proximal part of the megafan (~40 km downstream of Chatara) show the presence of a gravel layer at shallow levels (< 15 m below ground level), above which onlapping coarse- to medium-textured sand layers (10–20 m thick) are recorded. In contrast, the medial part of the megafan consists of a lower unit (below 50 m depth) made up of fine sand (about 30 m thick) overlain by a 20–30-m-thick upper unit of medium to coarse sand with pockets of clay and silt. Resistivity surveys by Sinha et al. (2014b) also interpreted 8–10-km-wide and

20–30-m-thick, multi-storied medium- to coarse-textured sand sheets in the subsurface of the mid-fan area. As observed in litarea. As observed in litarea. As observed in litologs to depths of ~100 m, the distal part of the megafan shows a variable lithology but primarily consists of fining upward sequences, starting with medium to coarse sand and ending with fine sand or muddy layers. Such sequences are typical of distributary channel systems (meandering type) in distal parts of megafans (Gohain and Parkash 1990; Singh et al. 1993; Assine 2005; Gibling 2006).

For further analysis, we merged the borehole data into two groups, sand and mud, and then computed the probability of finding a sand body of at least a given thickness for the entire megafan (exceedance probability curves). This allowed quantitative comparison of sand-body thickness patterns for varying distances from the fan apices at different depth intervals.

A similar set of data was generated by van Dijk et al. (2016) for boreholes on the Sutlej and Yamuna plains, a region where relict megafans have been inferred at depth (Sinha et al., Ch. 10). Following the methodology outlined by van Dijk et al. (2016), we compared the Kosi megafan data with the Sutlej and Yamuna plains.

Figure 11.8 compares the thickness and exceedance probability curves of all Kosi sand bodies for all boreholes to depths of about 100 m with those of the Sutlej and Yamuna regions (van Dijk et al. 2016), i.e., at both ends of the Indo-Gangetic plains (Table 11.2). A heavy-tailed distribution for the Sutlej and Yamuna plains suggests that sand-body thickness (mean ~9 m) does not change appreciably with distance from the apex. However, a general downslope decrease in sand-body percentage was reported in both the Yamuna and Sutlej regions, suggesting that sand bodies are less

common in distal parts. However, where they do occur, they areof comparable thickness to those in the proximal parts (van Dijk et al. 2016). In contrast, data for the Kosi megafan shows a different trend. The mean thickness of sand bodies (68 m) is higher by an order of magnitude, and the 50th percentile thickness is even higher (81.4 m). This once again reflects a very dynamic behaviour of the Kosi River in the past, with rapid sedimentation resulting in both lateral and vertical stacking of sand bodies. In contrast, both the Sutlej and Yamuna (relict) megafans probably evolved through infrequent avulsions, allowing a higher residence time of channels and a more dominant component of vertical accretion through time.

Limited stratigraphic data available for megafans in general, and the Kosi megafan in particular, highlights the fact that subsurface sedimentary architecture can be extremely variable depending on the geological setting

Table 11.2 *Spatial variability in sand-body thickness distribution (to 100 m depth)*

Megafan	No. of boreholes	Percentile thickness (m)			Mean thickness (m)	No. of sand bodies	Total fraction	
		25th	50th	75th			Sand	Mud
Sutlej	107	4	7	12	9.3	459	0.36	0.64
Yamuna	105	4.5	6.5	11	9.3	489	0.35	0.65
Kosi	104	42	81.4	97	68.6	72	0.92	0.08

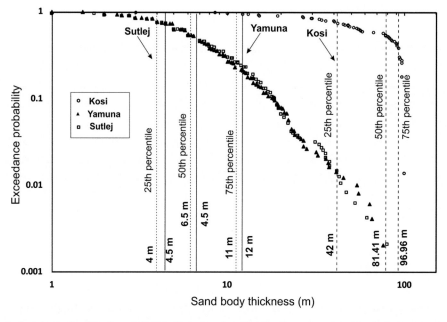

Figure 11.8 **Statistical analysis of the borehole lithologs showing distribution of sand bodies in the subsurface**. Also plotted are data for the buried Sutlej and Yamuna megafans in the western Indo-Gangetic plains, published by van Dijk et al. (2016).

and fan building processes, including sediment fluxes. Apart from the differences with the relict megafans in the western Indo-Gangetic plains, it has also been shown that the Kosi River may have built a distinctive subsurface stratigraphy that is different even from the one built by the adjacent Gandak megafan, which is located ~300 km to the west of the Kosi. For example, Sinha et al. (2014b) documented the lack of gravel deposits in the lithologs from the proximal part of the Gandak megafan. Instead, ~10–30-m-thick medium to fine sand bodies down to 100 m were recorded, and this was attributed to the presence of a prominent intermontane basin (or 'dun') in the hinterland of the Gandak megafan, which acted to trap or filter out coarse sediments (Sinha et al. 2014b).

Figure 11.9a illustrates the conceptual model to explain the variability in subsurface stratigraphy of the Kosi in its proximal to mid-fan areas. The presence of gravels at shallow levels in the proximal part of the Kosi megafan is attributed to a mountain-front setting of the Kosi characterised by the absence of an inter-montane basin in the hinterland (Fig. 11.9a), which

allowed gravels to be transported much farther down-stream in the past. This gravel front has now receded to Chatara, >40 km from the location of the lithology data. This points to a much more dynamic and higher-energy regime of the feeder channel in the past (Sinha et al. 2014b). With reduction in gravel flux, the coarse-to medium-textured sand units onlapped the gravel layers (Fig. 11.9b), thereby generating distinctive stratigraphies in the proximal (Fig. 11.9c) and medial (Fig. 11.9d) parts. The spatial variability in the hinter-land setting can thus play a key role in the develop-ment of megafan stratigraphy.

11.6.3 Ancient Analogues

Another reason why the Kosi megafan has attracted international attention is that it has often been cited as an important modern analogue for river systems that have generated thick, multistoried sand bodies – such as the sandstones of the Siwaliks in the NW Himalaya, which reach widths of 300–1,000 m (Kumar 1993; Willis 1993a, b; Khan et al. 1997; Zaleha 1997). Two

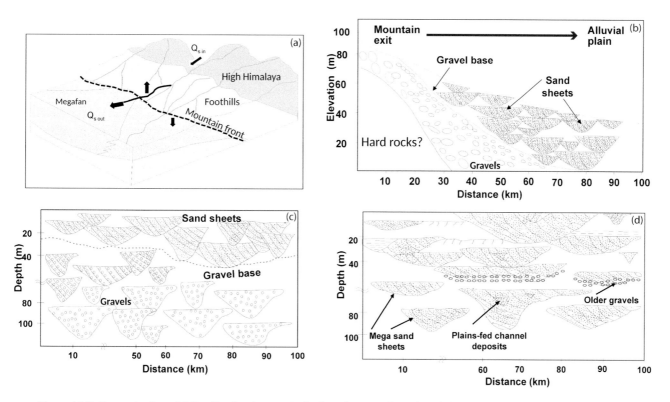

Figure 11.9 **Conceptual model for the development of subsurface stratigraphy of the Kosi megafan.** (a) Mountain-front setting of the Kosi megafan without an intermontane basin, allowing efficient transfer of sediments to the megafan. (b) Proximal to medial section showing the onlap of sand sheets over the gravels. (c) Proximal fan stratigraphy showing gravels at shallow depth overlain by sand sheets. (d) Mid-fan stratigraphy showing the dominance of sand-sheet stacks in the upper 100 m.

classical papers by Friend et al. (1979) and Friend (1983) defined major types of sand bodies in the ancient sequences on the basis of width-to-thickness ratios (W/T) as (i) sheets (W/T > 15) and (ii) ribbons (W/T < 15).

Sedimentary successions that have been considered to be ancient analogues of mefagans include those from the Siwaliks. These successions consist of more than 5,000-m-thick fluvial sediments deposited by rivers that once flowed transverse to the strike of the Himalaya. Detailed studies in the Lower and Middle Siwaliks of the Potwar Plateau (Willis 1993a, b; Zaleha 1997) have shown that major sand bodies are tens of metres thick and several kilometres wide. These were interpreted to be sand sheets (high W/T ratio) deposited by megafans analogous to the modern Kosi. Behrensmeyer and Tauxe (1982) described the 'blue-grey' fluvial system from the Dhok Pathan region of the Middle Siwaliks in Pakistan as widespread sand sheets with low sand/mud ratios. These were interpreted to have been formed by mountain-fed rivers in megafan landscapes. Similar sequences have also been recorded in the Middle Siwaliks in the Garhwal Himalaya, India (Kumar and Nanda 1989), and in the Nepal Himalaya (Nakayama and Ulak 1999).

The understanding of megafan building processes from the Kosi megafan has allowed explanations of the variations in the alluvial architecture of the Siwaliks in terms of intra-basin and inter-basin controls (Willis 1993b; Khan et al. 1997; Zaleha 1997). For example, small-scale variations in sand-body thickness (< 10 m) have been linked to local avulsion and flooding as seen on the Kosi megafan surface. Medium-scale variations (~ 100 m) can be explained by the migration of the Kosi River across the megafan surface. Large-scale changes (> 1,000 m) have been related to hinterland tectonics or climate change causing a change in flows, sediment flux, and grain size, as in the shift of the gravel front in the proximal stratigraphy of the Kosi.

In the foreland basin of the Pyrenees in Spain, Hirst (1991) documented narrow sand sheets and ribbons in the Cenozoic Huesca fan systems and noted a downstream trend in channel-body proportions. Thicker, amalgamated sand bodies were relatively rare. In the modern Kosi system, such near-surface sand sheets (< 40 m thick) are generally formed by the lateral sweep of small channels (Singh et al. 1993). Marzo

et al. (1988) documented sand ribbons (up to 8 m thick) in the Castissent Formation of Spain that are either marginal to, or underlie, large sand sheets. Gibling (2006) considered that such sequences might be generated by processes similar to the lateral sweep of the Kosi, where periodic avulsion of the main channel can divert flows to smaller channels on the megafan surface; these smaller channels in turn may be filled rapidly by coarser sediments and then overlain by sand sheets. The August 2008 avulsion of the Kosi River may be a good analogue: the initial avulsion started as narrow channels, and then widened to a sheet flow 15–20 km wide (Sinha 2009) and generated a very wide zone of avulsion deposits (Sinha et al. 2013).

11.7 Summary and Conclusions

The Kosi is one of the most studied megafans in the world and is characterised by a typical conical, transverse convex morphology. The main channel of the Kosi megafan is highly dynamic, and frequent avulsions on historical time scales have been recorded, although the avulsion frequency has decreased since the construction of flanking embankments in 1955 and 1956. The high sediment flux of the Kosi River is one of the most important characteristics of the feeder channel and most of this flux is delivered by tributaries draining from Nepal. Such high sediment fluxes coupled with anthropogenic interventions such as embankment construction have accelerated channel aggradation, leading to 'superelevation' in several reaches and resulting in frequent breaching of the embankments and flooding. A major avulsion in August 2008 resulted in reoccupation of palaeochannels on the megafan surface during this large flooding event. The flood reconnected several palaeochannels, and the floodwave followed a predictable pathway determined by subtle topographic differences on the megafan surface. Stratigraphic investigations of the subsurface sedimentary architecture based on cut-bank sections, artificial trenches, and shallow borehole data suggest a dominance of large, vertically and laterally stacked sand sheets in the midfan region. This is a manifestation of the rapid lateral shifting of the feeder channel. However, significant proximal-to-distal variability in stratigraphy is noted in terms of grain-size distribution, sand-body thickness, and interpreted widths; this is attributed to variation in sediment supply and fan building processes through space and time. Apart from frequent avulsions and high sediment delivery, a peculiar geotectonic setting makes the Kosi megafan a distinctive landform on this planet that has often been used as a modern analogue to explain the processes and controls that generated large, multistoried sand bodies in ancient fluvial sequences in a foreland setting.

References

Assine, M. L. (2005). River avulsions on the Taquari megafan, Pantanal wetland, Brazil. *Geomorphology*, 70, 357–371.

Bajracharya, B., Shrestha, A. B., and Rajbhandari, L. (2007). Glacial lake outburst floods in the Sagarmatha region: hazard assessment using GIS and hydrodynamic modeling. *Mountain Research and Development*, 27, 336–344.

Behrensmeyer, A. K. and Tauxe, L. (1982). Isochronous fluvial systems in Miocene deposits of Northern Pakistan. *Sedimentology*, 29, 331–335.

Chakraborty, T., Kar, R., Ghosh, P., and Basu, S. (2010). Kosi megafan: historical records, geomorphology and the recent avulsion of the Kosi River. *Quaternary International*, 227, 143–160.

Delorme, P., Voller, V., Paola, C. et al. (2017). Self-similar growth of a bimodal laboratory fan. *Earth Surface Dynamics*, 5, 239–252.

Desai, C. (1982). *The Kosi River: Its Morphology and Mechanics in Retrospect and Prospect*. Central Water Commission, Ministry of Irrigation, Government of India, New Delhi.

Dingle, E. H., Sinclair, H. D., Attal, M., Milodowski, D. T., and Singh, V. (2016). Subsidence control on river morphology and grain size in the Ganga plain. *American Journal of Science*, 316, 778–812.

Friend, P. F. (1983). Towards the field classification of alluvial architecture or sequence. In J. D. Collinson and J. Lewin, eds., *Modern and Ancient Fluvial Systems*. International Association of Sedimentologists, Special Publication, 6, 345–354.

Friend, P. F., Slater, M. J., and Williams, R. C. (1979). Vertical and lateral building of river sandstone bodies, Ebro Basin, Spain. *Journal of the Geological Society, London*, 136, 39–46.

Gansser, A. (1964). *Geology of the Himalaya*. Wiley-Interscience, New York.

Gaurav, K., Métivier, F., Devauchelle, O., et al. (2015). Morphology of the Kosi megafan channels. *Earth Surface Dynamics*, 3, 321–331.

Geddes, A. (1960). The alluvial morphology of the Indo-Gangetic Plain: its mapping and geographical significance. *Institute of British Geographers, Transactions and Papers*, 28, 253–276.

Gibling, M. R. (2006). Width and thickness of fluvial channel bodies and valley fills in the geological record: a literature compilation and classification. *Journal of Sedimentary Research*, 76, 731–770.

Gohain, K. and Parkash, B. (1990). Morphology of the Kosi Megafan. In Rachocki, A. H. and Church, M., eds., *Alluvial Fans: A Field Approach*. Wiley, Chichester, 151–178.

Gole, C. B. and Chitale, S. V. (1966). Inland delta building activity of Kosi River. *Journal of the Hydraulics Division, American Society of Civil Engineers*, HY2, 111–126.

Gupta, S. (1997). Himalayan drainage patterns and the origin of fluvial megafans in the Ganges foreland basin. *Geology*, 25, 11–14.

Hartley, A. J., Weissmann, G. S., Nichols, G. J., and Warwick, G. L. (2010). Large distributive fluvial systems: characteristics, distribution, and controls on development. *Journal of Sedimentary Research*, 80, 167–183.

Hirst, J. P. P. (1991). Variations in alluvial architecture across the Oligo-Miocene Huesca fluvial system, Ebro Basin, Spain. In A. D. Miall and N. Tyler, eds., *The Three-Dimensional Facies Architecture of Terrigenous Clastic Sediments and Its Implications for Hydrocarbon Discovery and Recovery*. SEPM Concepts in Sedimentology and Paleontology, vol. 3, Tulsa, Oklahoma, 111–121.

Jain, V. and Tandon, S. K. (2010). Conceptual assessment of (dis)connectivity and its application to the Ganga River dispersal system. *Geomorphology*, 118, 349–358.

Jones, L. S. and Schumm, S. A. (2009). Causes of avulsion: an overview. In Smith, N. D. and Rogers, J., eds., *Fluvial Sedimentology VI*. Blackwell, Oxford, 171–178.

Khan, I. A., Bridge, J. S., Kappelman, J., and Wilson, R. (1997). Evolution of Miocene fluvial environments, eastern Potwar plateau, northern Pakistan. *Sedimentology*, 44, 221–251.

Kumar, R., (1993). Coalescence mega fan: multistorey sandstone complex of late orogenic (Mio-Pliocene) Sub-Himalayan belt, Dehra Dun, India. *Sedimentary Geology*, 85, 327–337.

Kumar, R., Jain, V., Prasad Babu, G., and Sinha, R. (2014). Connectivity structure of the Kosi Megafan and role of rail-road transport network. *Geomorphology*, 227, 73–86.

Kumar, R. and Nanda, A. C. (1989). Sedimentology of the Middle Siwalik sub-Group of Mohand area, Dehra Dun valley, India. *Journal of the Geological Society of India*, 34, 597–616.

Mackey, S. D. and Bridge, J. S. (1995). Three-dimensional model of alluvial stratigraphy: theory and application. *Journal of Sedimentary Research*, 65, 7–31.

Majumdar, D. and Ghosh, P. (2017). Characteristics of the drainage network of the Kosi megafan, India and its interaction with the August 2008 flood flow. In D. Ventra and L. E. Clarke, eds., *Geology and*

Geomorphology of Alluvial and Fluvial Fans: Terrestrial and Planetary Perspectives. Geological Society of London, Special Publication, 440, 307–326.

Marzo, M., Nijman, W., and Puigdefábregas, C. (1988). Architecture of the Castissent fluvial sheet sandstones, Eocene, South Pyrenees, Spain. *Sedimentology*, 35, 719–738.

Mishra, D. K. (2008). The Kosi and the embankment story. *Economic and Political Weekly*, 43, 47–52.

Morgan, R. P. C. (1995). *Soil Erosion and Conservation*, 2nd edn, Longman, London.

Nakayama, K. and Ulak, P. D. (1999). Evolution of fluvial style in the Siwalik Group in the foothills of the Nepal Himalaya. *Sedimentary Geology*, 125, 205–224.

Oberlander, T. M. (1985). Origin of drainage transverse to structures in orogens. In M. Morisawa and J. T. Hack, eds., *Tectonic Geomorphology*. Allen and Unwin, Boston, 155–182.

Parker, G. (1999). Progress in the modeling of alluvial fans. *Journal of Hydraulic Research*, 37, 805–825.

Parker, G., Paola, C., Whipple, K. X., and Mohrig, D. (1998). Alluvial fans formed by channelized fluvial and sheet flow. I: Theory. *Journal of Hydraulic Engineering*, 124, 985–995.

Reitz, M. D. and Jerolmack, D. J. (2012). Experimental alluvial fan evolution: channel dynamics, slope controls, and shoreline growth. *Journal of Geophysical Research: Earth Surface*, 117, F02021, doi:10.1029/2011JF002261

Roy, N. G. and Sinha, R. (2017). Linking hydrology and sediment dynamics of large alluvial rivers to landscape diversity in the Ganga dispersal system, India. *Earth Surface Processes and Landforms*, 42, 1078–1091.

Singh, H., Parkash, B., and Gohain, K. (1993). Facies analysis of the Kosi megafan deposits. *Sedimentary Geology*, 85, 87–113.

Sinha, R. (2009). The great avulsion of Kosi on 18 August 2008. *Current Science*, 97, 429–433.

Sinha, R. and Friend, P. F. (1994). River systems and their sediment flux, Indo-Gangetic plains, Northern Bihar, India. *Sedimentology*, 41, 825–845.

Sinha, R., Gaurav, K., Chandra, S., and Tandon, S. (2013). Exploring the channel connectivity structure of the August 2008 avulsion belt of the Kosi River, India: Application to food risk assessment. *Geology*, 41, 1099–1102.

Sinha, R., Sripriyanka, K., Jain, V., and Mukul, M. (2014a). Avulsion threshold and planform dynamics of the Kosi River in north Bihar (India) and Nepal: A GIS framework. *Geomorphology*, 216, 157–170.

Sinha, R., Ahmad, J., Gaurav, K., and Morin, G. (2014b). Shallow subsurface stratigraphy and alluvial architecture of the Kosi and Gandak megafans in the Himalayan foreland basin, India. *Sedimentary Geology*, 301, 133–149.

Sinha, R., Gupta, A., Mishra, K., et al. (2019). Basin scale hydrology and sediment dynamics of the Koshi River in the Himalayan foreland. *Journal of Hydrology*, 570, 156–166.

Slingerland, R. and Smith, N. D. (2004). River avulsions and their deposits. *Annual Review of Earth and Planetary Sciences*, 32, 257–285.

Van Dijk, W. M., Densmore, A. L., Singh, et al. (2016). Linking the morphology of fluvial fan systems to aquifer stratigraphy in the Sutlej–Yamuna plain of northwest India. *Journal of Geophysical Research: Earth Surface*, 121, 201–222.

Wells, N. A. and Dorr, J. A. (1987). Shifting of the Kosi River, Northern India. *Geology*, 15, 204–207.

Wagner, L. R. (1937). The Arun River drainage pattern and the rise of the Himalaya. *Geographical Journal*, 89, 239–250.

Willis, B. J. (1993a). Ancient river systems in the Himalayan foredeep, Chinji Village area, northern Pakistan. *Sedimentary Geology*, 88, 1–76.

Willis, B. J. (1993b). Evolution of Miocene fluvial systems in the Himalayan foredeep through a two-kilometer-thick succession in northern Pakistan. *Sedimentary Geology*, 88, 77–121.

Zaleha, M. J. (1997). Fluvial and lacustrine palaeoenvironments of the Miocene Siwalik Group, Khaur area, northern Pakistan. *Sedimentology*, 44, 349–368.

12

The Holocene Mitchell Megafan, Gulf of Carpentaria, Australia

TESSA I. LANE, RACHEL A. NANSON, R. BRUCE AINSWORTH,
and BOYAN K. VAKARELOV

Abstract

As the world's lowest and flattest continent, Australia exhibits numerous gently inclined topographic basins that have generated a large number of megafans. One of the most extensive contiguous arrays occurs on the western flank of Cape York Peninsula. These megafans are derived from a complex network of modern and ancient distributary channels that have debouched into the epicontinental seaway of the Gulf of Carpentaria since at least Pliocene times. The Mitchell megafan-delta represents the largest of these fans. In this case study we examine the modern processes and resulting sedimentary facies and architecture of a type example. Mapping, field exploration and chronologic investigations were used to characterise these channel-belt facies and to link them to specific forcing conditions. These Holocene surficial deposits are associated with a network of modern streams and recent palaeochannels overlying, and sometimes incised into, older, fine-grained, partially indurated Quaternary alluvium. Where present, the partially indurated alluvium restricts channel migration, thereby promoting anabranching as a means of channel adjustment. Additionally, channel-belt distribution, morphology and sediment infilling are strongly affected by marine backwater dynamics operating in the downstream reaches. This has had an overriding influence on the position of avulsion nodes and channel-belt distribution in this zone.

12.1 Introduction

12.1.1 Megafans of Australia

Australia supports an array of megafans that are located in a variety of environmental settings, from large, inland, sub-humid, semi-arid, and arid rivers to those on the northern and northwestern monsoon-tropical coastal plains (Fig. 12.1). They are best developed on the western slopes of the central and southern Great Dividing Range, where they have created an extensive coalescing alluvial surface in the form of the Riverine Plains of the Murray–Darling basin. A similar but smaller surface called the Carpentaria Plains has formed on the western slopes of Cape York along the eastern shoreline of the Gulf of Carpentaria. Several megafans emerge from low ranges in arid central Australia, and extensive axial accretionary fluvial deposits characterise the large, low-gradient valleys draining into Lake Eyre from the northeast. Australia's studied megafans are briefly outlined below, followed by a detailed description of the Carpentaria Plains and Mitchell River case study.

12.1.1.1 Riverine Plains of Southeastern Australia

The Riverine Plains of southeastern Australia take the form of large bajadas flanking the Murray-Darling hydrological basin, which is named after the two major rivers that drain it. Subsidence of the geological Murray Basin has created a ~1 million km^2 saucer-shaped intracratonic depression flanked by low

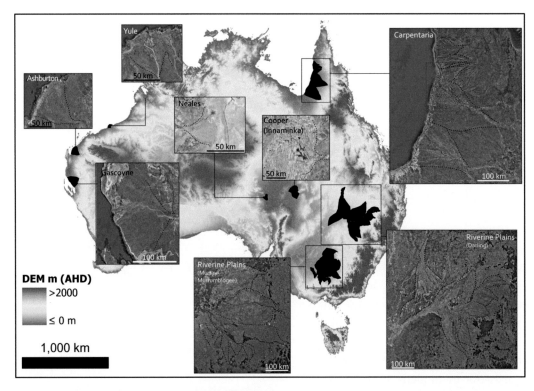

Figure 12.1 **Digital elevation model of Australia showing location and Google Earth images locating known Australian megafans** (elevation data source: Geoscience Australia and CSIRO Land & Water).

mountain ranges to the east and southeast. Its hydrology can be divided into a northern Darling watershed and a southern Murray watershed. The late Quaternary deposits of the Murray and Murrumbidgee megafans have been thoroughly investigated, the Darling less so.

The Murray and Murrumbidgee rivers currently consist of mostly single-thread meandering channels, but the uppermost stratigraphy of their megafans exhibits abundant anabranching and distributary bedload-dominated late Quaternary palaeochannels flanked by related floodplains and source-bordering aeolian dunes (Page and Nanson 1996; Page et al. 2001). A detailed analysis of the chronostratigraphy of late Quaternary fluvial sedimentation has shown that the distributary 'prior stream' and 'ancestral stream' systems carried much larger-than-current peak discharges over extensive areas of the plain, up to and including the Last Glacial Maximum (LGM; e.g., Page et al. 1996, 2009; Forbes et al. 2020). Since then, conditions returned to those much like those of the present, and both the Murray and Murrumbidgee reverted to single channel meandering rivers that have incised into their respective

megafans (e.g., Page and Nanson 1996; Hesse et al. 2018). Kapteinis et al. (Ch. 13) focus on the southern part of this large region.

The Lachlan megafan also feeds into the Murray River and has a complex fluvial-laucustrine relationship with the Willandra Lakes (Kemp et al. 2017). Kemp and Rhodes (2010) have shown that, over the 220 km of the middle and upper reaches of this fan, the mean and peak flows, channel size, and bed particle size all diminish downstream. Dimensions of the unconfined single channel contract with distance downvalley, and anabranches are occupied at high flow stages (Kemp et al. 2017).

The Gwydir megafan lies in the Darling portion of the Murray–Darling basin. In a study of the late Quaternary changes of flow-regimes on the Gwydir distributive fluvial system, Pietsch et al. (2013) found a pattern of cyclical climate-driven anabranching and distributive channel formation, very similar to phases recorded on the Murray and Murrumbidgee megafans. Although Hesse et al. (2018) noted some significant differences, it appears that the Quaternary climatic drivers and associated chronostratigraphy for the northern and southern parts of the Murray–Darling basin were broadly similar.

12.1.1.2 Lake Eyre Basin

Callen and Bradford (1992), Cohen et al. (2010), and Wakelin-King and Amos (2016) describe the fluvial and aeolian interaction that is characteristic of the 8,500 km^2 Cooper megafan, which has developed where Cooper Creek emerges from lateral confinement of the Innamincka Dome in South Australia. Source-bordering transverse dunes and superimposed linear dunes have impeded the course of Cooper Creek and provided a repository of evidence for Quaternary climate change as well as the interactive processes between transverse and linear dune formation on this megafan system.

The megafans in arid central Australia have been less intensively researched. The understanding of these systems is represented primarily in the work of Croke et al. (1998) and Lang et al. (2004) on the Neales River system to the west of Lake Eyre. The present Neales megafan delta extends across 1,200 km^2 and has preserved several phases of activity, which are revealed by 9 m of incision by the Neales River through Quaternary strata and into underlying Cenozoic deposits (Croke et al. 1998).

12.1.1.3 Western Australia

There has been minimal work published on megafan distribution or geomorphology in Western Australia, which has relatively few megafans because of a scarcity of Cenozoic basins (or crustal swells; see Wilkinson et al., Ch. 3) for their accumulation. The largest known Western Australian megafan is the Gascoyne, spanning 11,500 km^2; it has been described by Leonard et al. (2013) and Whitney and Hengesh (2015). At its downstream extent, it is an ephemeral, low-sinuosity, single-thread channel that debouches into the Indian Ocean where it is constructing a delta (Johnson 1982). Further north, the Yule and De Grey megafan-deltas drain the Archean Pilbara Craton, but to date these are unstudied.

12.1.1.4 Carpentaria Plains

One of the most extensive suites of megafans in Australia is located on the western side of the Cape York Peninsula. The search for natural resources initiated early geological and stratigraphic work in the region (Whitehouse 1941; Perry et al. 1964; Twidale 1966; Grant 1968; Isbell et al. 1968; Doutch et al. 1970; Galloway et al. 1970; Smart and Grimes 1971; Smart et al. 1975, 1980; Doutch 1976; Simpson and Doutch 1977; Doutch and Nicholas 1978; Grimes and Doutch 1978). More recently, detailed investigations have provided insights into their unique morphology, controls on their architecture, their late Quaternary chronology, and the rapid induration of the Gilbert (Nanson et al. 1991, 2005; Porritt et al. 2020) and Mitchell deposits (Lane 2016; Lane et al. 2017). Investigations have been undertaken on the largest of these systems to determine the hydrology of the Mitchell River and the effects of land use on sedimentation and gullying (Brooks et al. 2008, 2009; Rustomji 2010; Rustomji et al. 2010; Shellberg 2011; Shellberg et al. 2012, 2013, 2016). Rhodes (1980, 1982) and Rhodes et al. (1980) pioneered investigations into the delta plains of the eastern Gulf, and Nanson et al. (2012, 2013a, b), Massey et al. (2014), and Lane et al. (2017) have most recently investigated the geometry and form–process links on the Mitchell River Delta, the subject of this chapter.

12.1.2 The Mitchell River Megafan

The Mitchell River catchment extends across the Cape York Peninsula (Fig. 12.1) and has formed Australia's second largest externally draining (ocean-connected) megafan, adjacent to the largest, the Gilbert River megafan. The Gulf of Carpentaria receiver basin provides an ideal case study for investigating the morphological response of a megafan-delta's extrinsic and intrinsic drivers. The modern and palaeochannel belts of the Mitchell River megafan can be distinguished easily from the surrounding floodplain by their denser vegetation communities, which have aided investigation of planform geometry from satellite imagery. The coalescing fans of the western Cape York Peninsula have been building into the Gulf of Carpentaria (Fig. 12.1) since at least the Pliocene (Grimes and Doutch 1978), a basin that has repeatedly fluctuated through open shallow marine, land-locked freshwater lacustrine, and complete subaerial conditions (Reeves et al. 2007, 2008). The Mitchell River megafan has thus

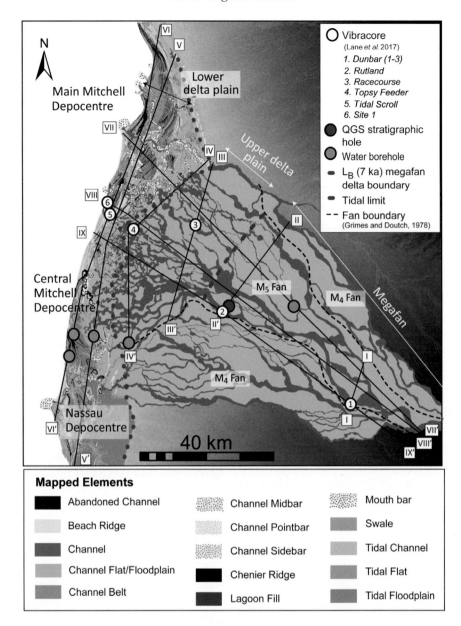

Figure 12.2 **Mitchell River M$_5$ megafan and delta**. Channel-belt-scale mapping and element scale mapping (*sensu* Vakarelov and Ainsworth 2013) indicating depositional zones. Transect lines as shown in Figure 12.9. Modified from Lane et al. (2017).

been subjected to the varying sea-level positions and corresponding climatic changes throughout its development. Grimes and Doutch (1978) proposed a five-stage model for the coalescing fans of the Carpentaria Plains, of which the Mitchell megafan is a part. Their model is based on channel cross-cutting relationships, and ages are thus approximate. Stages 1 and 2 (M$_1$ and M$_2$) of fan development occurred in response to tectonics, eustasy, and climate change prior to ~150 ka; Stage 3 units correspond to the 125 ka Pleistocene highstand; Stage 4 and 5 (M$_4$ and M$_5$) deposits are separated from earlier

stages by the Holroyd Surface, which developed during the LGM either by subsurface solution and subsidence (Grimes 1974), or by aeolian deflation (Doutch 1976). This chapter focuses on Stage 5 (M$_5$, Holocene) fan development, which is inset within, and flanked by, older Stage-4 fan deposits (M$_4$, Fig. 12.2).

12.2 Controls on the Mitchell Megafan

The Holocene Mitchell River megafan spans 31,000 km^2 (Shellberg et al. 2016) and has a drainage basin of

71,630 km^2 (Brooks et al. 2009) that extends across the Cape York Peninsula to the escarpment immediately west of Cairns. This Holocene megafan lobe, termed Stage 5 of the Mitchell megafan (M$_5$, Grimes and Doutch 1978), has been active since the Holocene maximum marine transgression ca. 7 ka, and developed throughout the subsequent forced regression (regression induced by sea-level fall). The Mitchell megafan feeds an extensive delta system comprised of the major Main, Central and Nassau depocentres (Fig. 12.2). Their morphologies are affected by a mix of tidal, wave-related and fluvial processes (for detailed sedimentology and stratigraphy of these deltas see also: Jones et al. 1993, 2003 concerning the Gilbert River delta; and Nanson et al. 2012, 2013a, 2013b; Massey et al. 2014; Lane et al. 2017 for the Mitchell River delta).

Two dominant extrinsic drivers have exerted strong and dynamic influences on the processes and morphology of the Mitchell megafan: climate, and relative sea level. Shulmeister (1999) linked Holocene climate in northern Australia to changes in Holocene Effective Precipitation (EP) caused by the precessional control of Walker Circulation in the Pacific Ocean and the intensification of the Northern Australian Monsoon, and proposed three distinct climatic periods: a gradual increase in EP and temperatures until 5 ka, increased EP from 5 to 3.7 ka, and a sharp reduction in EP and increased climate variability from 3.7 ka to Recent. EP is used in the following sections as a proxy for the rate of sediment supply.

A critical control on the morphologies of the Holocene Mitchell megafan is the inland extent of the backwater effect resulting from marine and palaeo-Lake Carpentaria base-levels (Reeves et al. 2007, 2008). The reaches of channel belts farthest downstream are affected by the receiving body of water (Lane et al. 2017), as channel flow decelerates as it nears the river mouth in the backwater zone (Lamb et al. 2012). This sector, in which the channel dynamics are affected by base level, is termed the backwater length (L$_B$), a zone that is dynamic over short time scales (such as a flood event) because it is sensitive to flow regime, river plume dynamics, tides, and storm surges (Chatanantavet et al. 2012; Lamb et al. 2012). More significantly, backwater length shifts dramatically in response to changes in sea level and lake level,

and is the topic of discussion below. Significant differences in channel hydraulics thus subdivide the depositional zones of the Mitchell River megafan into three depositional zones: the megafan proper, the upper delta, and the lower delta. The effect of base level is evident in the downstream portion of the modern Holocene megafan, and the movement of the upstream limit of the backwater length (up- and down-dip migration of the backwater zone) in response to changing base level is a key focus.

Today the Mitchell River catchment and megafan experience a tropical savanna climate, with a long dry season (April–November) associated with trade winds that travel from the east to the northwest, and a shorter wet season (December–March) associated with a reversal of wind direction (Suppiah 1992). Tropical cyclones also affect this coast, and average two per year (BOM 2018). These climatic variations overlap with mid- to late-Holocene relative sea-level fall in the Gulf of Carpentaria, the timing and pattern of which has been investigated by several research groups. Chappell (1983) determined that isostatic rebound caused the southeastern Gulf of Carpentaria coastal systems to undergo an essentially constant rate of forced regression of up to 1.5 m since the Holocene highstand, which is now generally accepted as ~7 ka BP (Sloss et al. 2018). Jones et al. (1993, 2003), Nanson et al. (2013b) and Lane et al. (2017) similarly provided geomorphic evidence for a ~6–7 ka highstand in these coastal systems, followed by downstepping at approximately 2 ka. (Fig. 12.3).

More recently, Sloss et al. (2018) collated previously published datasets with new data to produce a revised sea-level curve for the region. They extended the sea-level highstand to ~4 ka, and proposed a rapid fall to 3.5 ka, since which time they suggest sea-level has been within 0.5 m of its present level. Global modelling and satellite altimetry suggest modern sea-level transgression and sea-level rise of 18 cm near the Mitchell River coast since the satellite record started for this area in 1993 (CSIRO 2019).

12.3 The Modern Hydrological System

The modern Mitchell River catchment discharges 80,000 ML of water annually, predominantly during the wet season, and causes monsoonal flooding (see

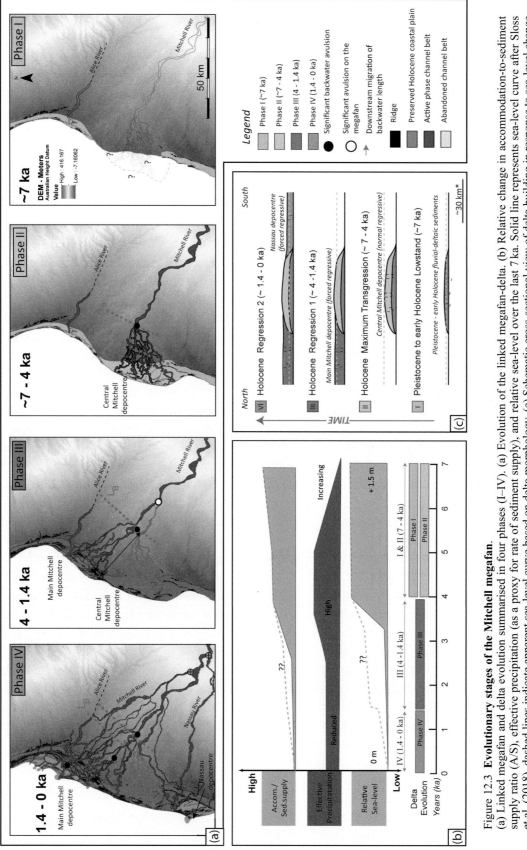

Figure 12.3 **Evolutionary stages of the Mitchell megafan.**

(a) Linked megafan and delta evolution summarised in four phases (I–IV). (a) Evolution of the linked megafan-delta. (b) Relative change in accommodation-to-sediment supply ratio (A/S), effective precipitation (as a proxy for rate of sediment supply), and relative sea-level over the last 7 ka. Solid line represents sea-level curve after Sloss et al. (2018), dashed lines indicate apparent sea-level curve based on delta morphology. (c) Schematic cross-sectional view of delta building in response to sea-level change and shift in depositional loci: I- maximum Holocene marine transgression; II- Central Mitchell depocentre active as a series of lobate units during a highstand (much of this section is fluvially dominated); III– a major avulsion on the megafan shifts the sediment supply to the north (Central Mitchell depocentre abandoned; decline in effective precipitation after 3.7 ka causes a shift from fluvial to tide dominance; the Main Mitchell depocentre builds a series of predominantly tide dominated units); IV- Nassau depocentre initiated at 1.4 ka; both Nassau and Main Mitchell depocentres are currently prograding. Partial avulsion of the Mitchell River near the apex of the megafan causes an influx of sediment to the Nassau depocentre. Backwater-mediated avulsions on the lower delta plain cause localised shifts in deposition. Modified from Lane et al. (2017). A colour version of this figure is available in the SOD for Chapter 12.

Brooks et al. 2009). The modern channels have incised 5–15 m below the surrounding alluvial surface (Rhodes 1980; Rustomji 2010), and are characterised by large sediment slugs (e.g., Fig. 12.4a), and by extensive sand sheets and dunes with reactivation surfaces indicating fluctuating flow conditions (Fig. 12.4b; Nanson et al. 2013a). Large individual flows are capable of reorganising the distribution of pools and bars, and individual events frequently scour to depths of 3–10 m (Brooks et al. 2008). Ferricrete nodules, which develop in weathered alluvium, are transported by the channels and armour bar surfaces (Fig. 12.4c). Sednet modelling (Prosser et al. 2001) estimates that approximately 2.9×10^6 t of fine-grained suspended sediment and washload currently reaches the Mitchell River delta and is deposited in the Gulf of Carpentaria each year (Rustomji et al. 2010). This value is probably higher than the natural level as gully erosion following European settlement is a major source for fine-grained sediment in the Mitchell River channel system. Shellberg et al. (2016), for example, show that gully erosion has increased in the Mitchell megafan by 1.2–10 times since 1949, and they postulate that the upper megafan entered a widespread phase of gullying somewhere between 1880 and 1950, corresponding with European settlement.

During the dry season, low flow in the main Mitchell River is characterised by multiple sinuous and often multi-thread channels separated by vegetated

Figure 12.4 **Sand slugs, comprising dunes and ripples, as often observed in the ephemeral channels during the dry season**. (a) Aerial view, with flow from right to left. (b) Sand slug, flow from left to right (Inset – internal reactivation surfaces, likely the result of fluctuating flow conditions). (c) Ferricrete nodules develop in weathered alluvium in the mid to upper catchment and are transported downstream, often armouring large bars. (d) Mitchell River low-flow channel. The alluvial sediment has been partly indurated (Nanson et al. 2005). Modified from Nanson et al. (2013a). A colour version of this figure is available in the SOD for Chapter 12.

islands and bars (Brooks et al. 2008). The courses of the low-flow channels can adjust significantly during individual floods (Brooks et al. 2008). During the wet season, these channels coalesce to form a low-sinuosity, single-thread channel. Satellite observations indicate that flood flows from this main channel also spill into distributary channels across the megafan (Rustomji 2010), where they follow former palaeochannel networks and incise new channels into megafan deposits, forming an anabranching pattern.

Much of the megafan surface has become pedogenically altered (Grimes and Doutch 1978), as the monsoonal wet-dry climate cycle causes much of the megafan surface to become indurated with calcrete and Fe/Mn oxyhydroxides and oxides. These indurated horizons provide resistance to flow that is analogous to that of soft bedrock (Fig. 12.4d; Nanson et al. 1991). On the Gilbert megafan immediately south of the Mitchell River system, Nanson et al. (2005) described the way this resistant surface has restricted lateral channel activity, thereby promoting in-channel aggradation and avulsion. Furthermore, indurated sediment resists upstream knickpoint migration and thereby inhibits tidal incursions upstream of them (Nanson et al. 2005). Indurated sediment also frequently crops out on the Mitchell megafan delta zones and has the same effect.

In the Gulf of Carpentaria region, low megafan interfluves between active and ancient channels are sporadically inundated by freshwater floods during the monsoon (Simpson and Doutch 1977) and deposit fine-grained material on older alluvial surfaces (Rhodes 1980). Large termite mounds occupy less frequently inundated interfluve surfaces (Nanson et al. 2013a). These surfaces are generally less densely vegetated with grasses and sparse trees and shrubs, and contrast with the palaeochannel belts; as a result, the two can be distinguished in satellite images (Nanson et al. 2013a). After surveying the Carpentaria Plains in 1974 during a severe wet season, Simpson and Doutch (1977) noted that runoff in the interfluve areas was generally clear, with silt-laden flows only near active channels. Rhodes (1980) reported sheets of fine to medium sand migrating across interfluve areas during the summer monsoon. Parts of the modern megafan surface are in a phase of minor erosion, commonly in the form of gullying (Simpson and Doutch 1977;

Shellberg et al. 2012, 2016). Depressions in the interfluve areas form seasonal wetlands and support a variety of wildlife (Nanson et al. 2013a).

12.4 Methods

We present both new and published mapping and field analyses, which reveal the modern and palaeogeography of the Mitchell megafan. These data support our revised analysis of the relative roles of autogenic (river avulsion) and allogenic (climate and base-level) controls. New mapping and stratigraphic data are synthesised to determine the primary processes influencing the Holocene megafan, and their impacts on (i) megafan evolution, (ii) the modern hydrology, (iii) channel-belt facies, and (iv) channel-belt geometry and distribution.

12.4.1 Mapping

Mapping of the Mitchell megafan and delta was undertaken in ArcMap using Google Earth imagery and 30-m Shuttle Radar Topography Mission Digital Elevation Model (STRM DEM; Geoscience Australia and CSIRO Land & Water 2011). Mapping has been updated from previously published datasets (Nanson et al. 2012, 2013a, 2013b; Massey et al. 2014; Lane et al. 2017). The lower delta plain was mapped to the element scale (*sensu* Vakarelov and Ainsworth 2013), then organised into hierarchical units representing pulses of shoreline progradation (Nanson et al. 2013b, Lane et al. 2017). Mapping of the megafan was undertaken at the channel-belt scale, where the channel belt and the interfluve boundary were clearly marked by changes in vegetation and abrupt changes in elevations (often as levees that were also observed in field inspection). Relative timing of channel-belt activity could be inferred in places due to cross-cutting relationships and their association with lower delta plain deposits. Radiocarbon, thermal luminescence (TL), and optically stimulated luminescence (OSL) dates were used to constrain the timing of events. The details of stratigraphic dating across the Mitchell megafan and delta are detailed in Nanson et al. (2012) and Lane et al. (2017). All recognisable avulsion points were recorded.

12.4.2 Sedimentary Facies

Sedimentary facies data for channel belts was obtained through vibracoring (Fig. 12.6) and supported by trenching and hand augering. Resin peels were taken of sand-rich vibracore and mud-rich vibracore, and were X-Rayed on a Toshiba Sedecal Reference DX to resolve subtle sedimentary and biogenic structures. Lane (2016) contains detailed stratigraphic logs and photographs of all samples.

12.4.3 Transects

Topographic transects were generated with the publicly available 30 arcsec SRTM DEM (Geoscience Australia and CSIRO Land & Water 2011), and amended, near the vibracore sites, with higher resolution Topcon Real Time Kinematic (RTK) GPS transects giving an estimated error of ± 10 cm relative to the Australian Height Datum (AHD) (Lane 2016). Channel-belt spacing and geometries were analysed along transverse transects set at 5 km intervals for the 115 km distance between the Mitchell megafan apex (M_5 Holocene apex) and the shoreline. The cross-sections are true to the transect angle (i.e., channel belts have not been rotated to a palaeoflow-normal orientation).

12.5 Evolution of the Megafan-Delta

Using new geomorphic mapping and dated sediment, in conjunction with previously published datasets (Nanson et al. 2013b; Massey et al. 2014), Lane et al. (2017) defined four key phases of evolution of the Mitchell River megafan-delta (Fig. 12.3a). These phases are critical to understanding the distribution of the up-dip megafan facies, and are briefly summarised and slightly revised below.

12.5.1 Phase I (7 ka)

Phase I of M_5 megafan-delta accumulation commenced at the time of the Holocene maximum marine transgression, ~7 ka (Nanson et al. 2013b) when sea level was up to 1.5 m higher than present (Murray-Wallace and Woodroffe 2014; Fig. 12.3b–c, Phase I). Evidence for the extent of fan-delta building during this phase is scant, though nearshore sediment surveys (Rhodes 1980) demonstrate unusually high concentrations of sand (80%) in the modern subtidal zone of the Central Mitchell depocentre. The Holocene transgression drowned the downstream portion of the megafan, and the maximum marine transgression palaeoshoreline is evidenced from sparsely preserved chenier ridges occurring 10 km inland from the modern coastline (Fig. 12.3a Phase I). Onlapping relationships evidenced by preserved beach and chenier ridges suggest that Phase I deposits were typically older than Phase II deposits, but there was significant temporal overlap (Fig. 12.3a).

12.5.2 Phase II (7–4 ka)

Accumulation during this phase was concentrated in the Central Mitchell depocentre (Figs. 12.3a, c, Phase II). Limited OSL dating, and the recent sea-level data compilation of Sloss et al. (2018), suggest that the central delta was actively building until ~4 ka, and may have been operating since maximum transgression. This period coincides with the maximum Holocene EP (~5–3.7 ka: Shulmeister 1999), which we use as a proxy for rates of sediment supply, during which extensive delta progradation occurred. These delta units are largely overprinted by fluvial deposits.

12.5.3 Phase III (4–1.4 ka)

A major avulsion on the megafan caused the abandonment of the Central Mitchell depocentre and shifted deposition to the new (and current) Main Mitchell depocentre (Fig. 12.3a, Phase III). The subtle downstepping of the delta at this time is evidenced by low-lying fluvial and high-tidal floodplain, such that Phase III represents the first pulse of forced regression in the system. This observation contrasts slightly with previous interpretations of the rate of sea-level fall (steady rate of fall since 7 ka: Chappell 1983) and Mitchell delta evolution (Nanson et al. 2013; Lane et al. 2017), but it better explains the downstepping that occurred between depositional Phases II and III. This timing also matches the recently revised sea-level curve for the Gulf of Carpentaria (Sloss et al. 2018; relative sea-level curve Figure 12.3b), whereby the majority of the

Holocene forced regression occurred between 3.5–4 ka, during Phase III development.

12.5.4 Phase IV (1.4 ka to Present)

After a hiatus around 2 ka, evidenced by pronounced and well-dated 2 ka shorelines (Nanson et al. 2013a, b, Lane et al. 2017), and likely driven in part by a marked reduction in EP at that time (Shulmeister 1999), delta building recommenced and a partial avulsion on the megafan caused the division of deposition into two depocentres: the northern Main Mitchell and the southern Nassau (Fig. 12.3a, Phase IV). These Phase IV portions of the delta generally lie at a lower elevation, a characteristic that is particularly evidenced by extensive, seasonally inundated tidal flats. They suggest system downstepping resulting from continued forced regression from about ~1.4 ka. A relative sea-level fall has also been recorded on the eastern side of Cape York Peninsula at Cleveland Bay after 2.3 ka (Woodroffe 2009). While the Mitchell record is consistent with an extended highstand until ~4 ka (Sloss et al. 2018), followed by a sea-level fall until at least 3.5 ka, the subsequent downstepping of the Mitchell and Nassau deltas at approximately ~1.4 ka also is likely the combined result of a hiatus in sediment supply in response to reduced EP, combined with more subdued but continued sea-level fall since that time (Lane et al. 2017).

12.6 Palaeochannel Belts

12.6.1 Classification, Facies, and Architecture

Active channels created floodplains that were ultimately abandoned and preserved on the megafan surface as palaeochannel belts. Lane et al. (2017) defined and characterised four types of channel belts that occur on the Mitchell megafan (Table 12.1 and Fig. 12.5). The first are purely fluvial channels (F), situated on the megafan proper, and the second are located downstream and affected by backwater dynamics (F_{BW}), and occupy the upper delta plain. The third type are fluvial-dominated but tide-influenced (Ft), and the fourth are tide-dominated but fluvial-influenced (Tf; see also Ainsworth et al. 2011). Their relative distribution and facies are summarised below and illustrated in Figures 12.5 and 12.6.

12.6.1.1 Megafan Channel Belts (F)

Major palaeochannel belts on the megafan typically display low sinuosity. They are generally 500–1,200 m wide and can span to up to 2,500 m at their maximum extent (Fig. 12.5). Smaller, narrower (200 m), ana-branching channels (unmapped) also traverse the megafan surface. Vibracoring suggests that major channel-belt thickness decreases downstream, e.g., from > 11 m in Dunbar Core to < 9 m in Rutland Core, Fig. 12.6). Channel belt fills fine upward from gravel to coarse sand, and demonstrate a slight decrease in calibre downstream, from B-axis (D_B) = 50 mm at Dunbar Core to D_B = 40 mm at Rutland Core (Fig. 12.6), a distance of 55 km downstream. Cross-lamination was observed in the basal 3–5 m of channel fills, and current ripples were present in the top 3 m (Fig. 12.6). Bioturbation and carbonaceous material are common in the top 3 m of the sediment. Carbonaceous material comprises mostly roots, and most burrows were constructed by reptiles, insects and worms.

12.6.1.2 Upper Delta Plain Channel Belts (F_{BW})

As fluvial channels reach the delta plain they are affected by backwater dynamics (F_{BW}), whereby channel energy

Table 12.1 *Mitchell River M_5 megafan data sets by channel belt type*

Site Name and Number	Channel belt Type	Data set
1. Dunbar	F (Apex)	3 vibracore, 2 OSL, RTK GPS
2. Rutland	F	1 vibracore, 2 OSL, RTK GPS
3. Racecourse	F_{BW}	1 vibracore, ~40 auger hole, 3 OSL, RTK GPS
4. Topsy Feeder	F_{BW}	1 vibracore 2 OSL, RTK GPS
5. Scroll Site	Ft	1 vibracore, 1 auger, 1 OSL, RTK GPS
6. Site 1	Tf	2 vibracore, RTK GPS

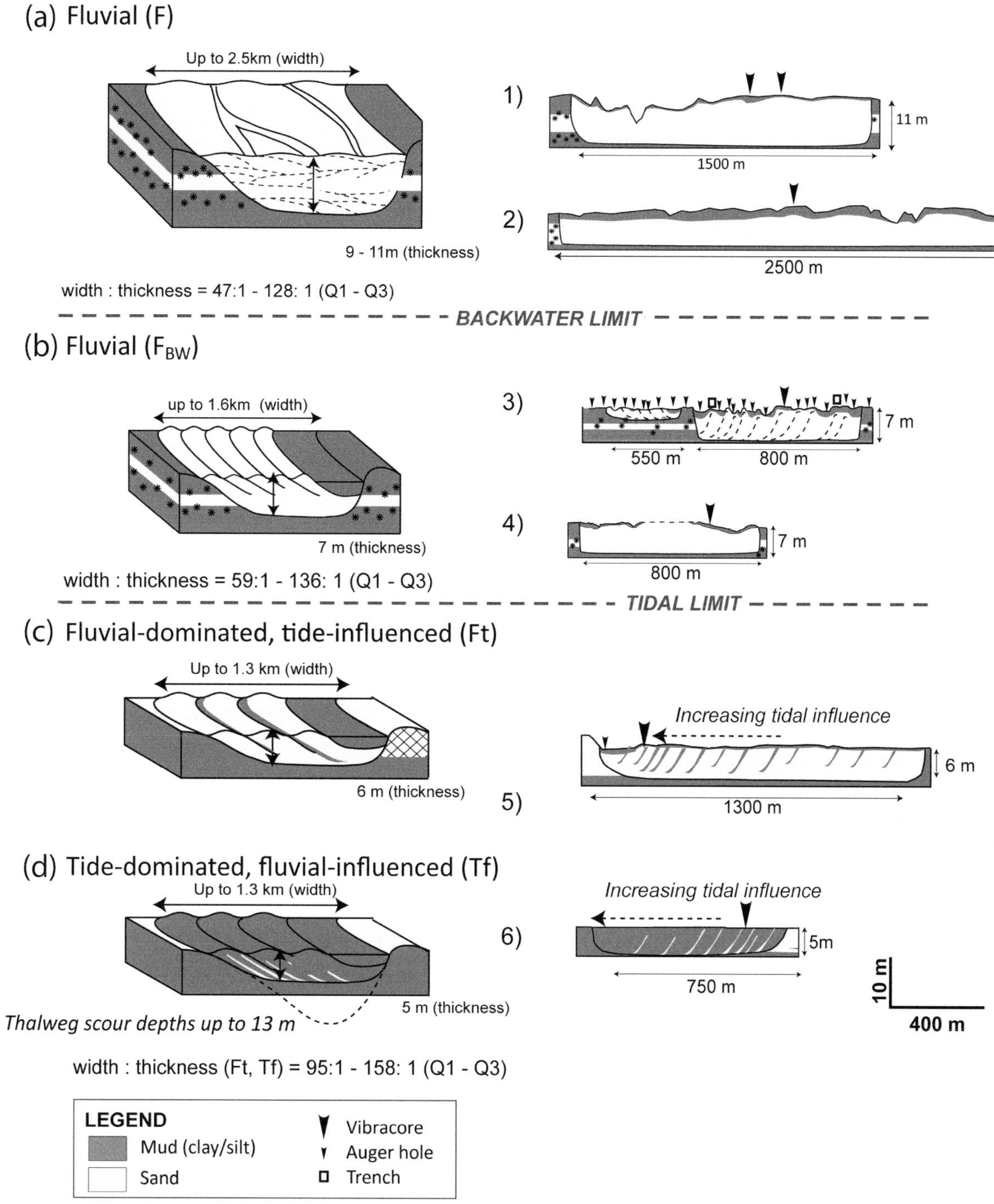

(a) Fluvial (F)

Up to 2.5km (width)

9 - 11m (thickness)

width : thickness = 47:1 - 128: 1 (Q1 - Q3)

1) 11 m

1500 m

2) 9 m

2500 m

– – – – – – – – – – – – – – – – – – *BACKWATER LIMIT* – – – – – – – – – – – – – –

(b) Fluvial (F$_{BW}$)

up to 1.6km (width)

7 m (thickness)

width : thickness = 59:1 - 136: 1 (Q1 - Q3)

3) 7 m

550 m   800 m

4) 7 m

800 m

– – – – – – – – – – – – – – – – – – – *TIDAL LIMIT* – – – – – – – –

(c) Fluvial-dominated, tide-influenced (Ft)

Up to 1.3 km (width)

6 m (thickness)

Increasing tidal influence

6 m

1300 m

5)

(d) Tide-dominated, fluvial-influenced (Tf)

Up to 1.3 km (width)

Thalweg scour depths up to 13 m

5 m (thickness)

width : thickness (Ft, Tf) = 95:1 - 158: 1 (Q1 - Q3)

6)

Increasing tidal influence

5m

750 m

10 m

400 m

LEGEND

- Mud (clay/silt)
- Sand

- ▼ Vibracore
- ▾ Auger hole
- ▫ Trench

Figure 12.5 **Successive belts of the Mitchell megafan, from fluvial to tide-dominated**. (a) Fluvial channel belts (F) upstream of the backwater length display thick, coarse-grained fill and single-thread morphology. (b) Fluvial channel belts on the megafan affected by base level (F$_{BW}$) display a scrolled morphology and a reduction in their cross-sectional geometry. (c) Ft and (d) Tf channel belts are influenced by tides, which introduce significant heterogeneity into their deposits. Modified from Lane et al. (2017). A colour version of this figure is available in the SOD for Chapter 12.

exceeds that required for sediment transport (Sv > Sfmin – i.e., valley slope is greater than the minimum energy slope required to transport the imposed water discharge and sediment load: Nanson and Huang 2008). To expend their excess energy in order to achieve dynamic equilibrium, the channels lengthen their paths by migrating laterally, resulting in high-sinuosity channels (see also Jones et al. 1993), often with scrolled

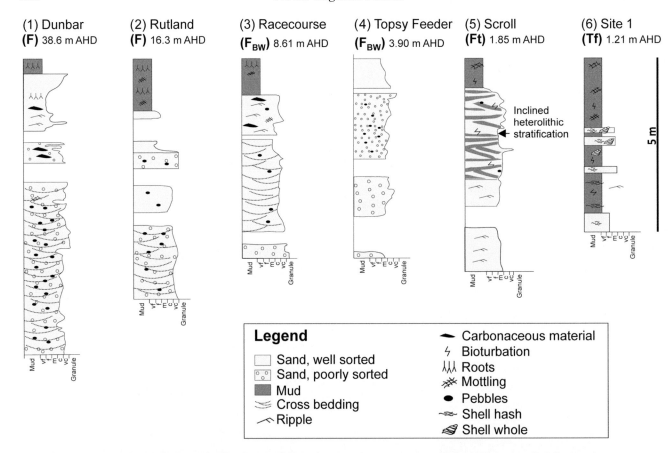

Figure 12.6 **Stratigraphic logs showing representative examples of depositional sequences**. These include fluvial-dominated (F) deposits on the megafan proper: (1) Dunbar and (2) Rutland; fluvial-dominated, backwater-influenced F_BW channel deposits on the upper delta plain: (3) Racecourse and (4) Topsy Feeder; fluvial-dominated, tide-influenced (Ft) channel deposits on the lower delta plain: (5) Scroll; and tide-dominated, fluvial-influenced (Tf) channel deposits on the lower delta plain (6): Site-1. See Figure 12.2 for location. A colour version of this figure is available in the SOD for Chapter 12.

lateral accretion deposits. Channel belts on the upper delta plain are typically 400–1,000 m wide (maximum: 1,600 m). Drilling indicates they are in the order of 7 m deep and are thus both shallower and narrower than those on the megafan (Fig. 12.5). Channel fill comprises fining-upwards medium to coarse sands, predominantly cross-bedded with some ripple cross-lamination observed in the top 3 m (Fig. 12.6). Rare pebbles were observed (up to $D_B = 30$ mm). F_{BW} channel fills are capped with mottled soil. Extensive augering across the Racecourse transect (Fig. 12.5) indicates that this cap varies from 0 to 2 m thick and is deeper in swales than on the ridges. Bioturbation is comparable to that observed on the open megafan surface and includes roots and a variety of burrows generated by reptiles, insects, and worms. The surrounding flood basin is composed primarily of mud, with some interbeds of fine to very fine sand (splay deposits).

12.6.1.3 Lower Delta Plain Channel Belts (Ft, Tf)

Channel belts on the lower delta plain exhibit morphologies influenced to various degrees by tides (Ft, Tf). They are characterised by inclined heterolithic stratification, and the sand to mud ratio is a function of their relative tide-to-fluvial influence, with tides introducing significant amounts of fine-grained material (Figs. 12.5–12.6). The reader is referred to Jones et al. (1993, 2003), Nanson et al. (2013a) and Lane et al. (2017) for detailed facies descriptions.

12.6.2 Channel Belt Characteristics

Mapping of palaeobelt planform characteristics has been extended from the work of Lane et al. (2017). The boundary between channel-belt deposits and low megafan interfluve areas is evidenced by abrupt changes in both vegetation and elevation (often as levees observed during field inspection). As fluvial sediment distribution on the Mitchell megafan is controlled primarily by avulsion, avulsion nodes were also recorded.

12.6.2.1 Avulsions

One hundred and nine major avulsion nodes were identified on the Mitchell megafan. They have been subdivided into two types (Fig.12.7a; Table 12.2; Lane et al. 2017): (i) avulsions on the megafan upstream of backwater influence (Fig. 12.7a, Table 12.2), and (ii) avulsions on the delta plain. Avulsions on the megafan plain are related to large-scale shifts in sediment deposition loci and define the delta depocentres: main Main Mitchell and Central Mitchell (Fig. 12.7a, Table 12.2).

The first type of avulsion occurred upstream of the backwater influence on the megafan at an average rate of 2.3/kyr. However, while models of fan-deltas predict nodal avulsions initiated from the topographically pinned apex of fan-deltas (e.g., Huanghe River megafan in China, Ganti et al. 2014), avulsions on the Mitchell megafan do not occur nodally on the megafan surface (*sensu* Slingerland and Smith 2004). Rather, the modern M_5 megafan is entrenched into the greater Mitchell Megafan, and avulsion nodes occur at irregularly spaced positions across the fan surface. These avulsions are likely to be an autogenic response to megafan aggradation (see Schumm et al. 1987), or to local breaks in slope related to shifting intersection points (i.e., position on the channel where the fan becomes unincised: Hooke 1967). The avulsion node locations can also be at least partially attributed to the indurated Quaternary alluvium that occurs on the megafan surface as it retards channel geometry adjustments, thereby promoting avulsion as an alternative adjustment mechanism (Nanson et al. 2005).

Avulsion nodes of the second type are concentrated in the backwater-influenced portion of the megafan, landward of the effective tidal limit. Avulsions mapped in the backwater zone are relatively frequent and based on the total number of mapped avulsion nodes in this region (Fig. 12.7a, Table 12.2); they have been

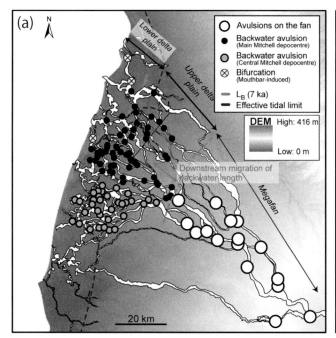

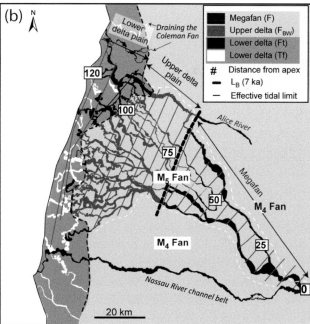

Figure 12.7 **Avulsion-related features on the Mitchell megafan**. (a) Mitchell M_5 megafan components and avulsion nodes. Avulsion is the primary control on sediment dispersal. Two distinct types are recognised: (1) avulsions on the megafan (2.3/kyr); (2) backwater-mediated avulsions (13.3/kyr). (b) Mitchell M_5 megafan channel belts (displayed over 30 arcsec DEM). Black: channel belts on the megafan proper, upstream of the backwater influence. Grey: fluvial-dominated, backwater-affected (F_{BW}) channel belts on the upper delta plain. Lower delta plain channel belts – black: fluvial-dominated, tide-influenced (Ft); white: tide-dominated, fluvial-influenced (Tf). Transverse/strike-oriented transects are shown at 5 km intervals from the M_5 fan apex to the tidal limit. Statistics in Figure 12.8 (discussed in text) are derived from measurements along these transects. Note: Nassau River channel is excluded from the analysis because part of the M_4 fan separates it from the remainder of fan M_5. A colour version of this figure is available in the SOD for Chapter 12.

Table 12.2 *Mitchell M₅ megafan mapped avulsion nodes: frequency and influence*

Mapped Avulsion Nodes divided by Depositional Zone		Minimum frequency (/kyr)		Enhanced by
Megafan (F)		16	2.3	Monsoonal flow regime, indurated sediment
	Central Mitchell depocentre	33	11	
Delta (F_{BW}, Ft, Tf)			13.3	Indurated sediment, backwater (base level, tides, seasonal increase in sea level)
	Main Mitchell depocentre	60	15	
Total Mapped M_5 Avulsion Nodes = 09		Overall frequency = 15.5/kyr		

occurring at an average rate of 13.3/kyr (Lane et al. 2017). This must be considered a minimum value, however, as channel reoccupation is common (see also Nanson et al. 1991; Jones et al. 2003). Avulsions are concentrated in the backwater zone because the deceleration of river flow in this zone promotes aggradation and avulsion. Since ~7 ka, the Mitchell River delta has prograded 17 km at its maximum extent and caused a similar lengthening of the backwater zone over this period (Lane et al. 2017). The most landward avulsion nodes on the delta are probably the oldest, but the landward-seaward avulsion trend is complicated by channel reoccupation.

Avulsion nodes within the backwater zone (delta plain) can also be linked to their major depocentres, and hence to sea-level trajectory (Lane et al. 2017). Thirty-three backwater-mediated avulsions are related to the formation of the Central Mitchell depocentre, which was operating between ~7–4 ka (a rate of 11/kyr; Fig. 12.7a; Table 12.2) under relatively high EP and probable highstand conditions (Fig. 12.3 Phase II; Lane et al. 2017). Sixty avulsion nodes are associated with the Main Mitchell depocenter, indicating an accelerated avulsion rate of 15/1000 years (Fig. 12.7a, Table 12.2). This acceleration is likely a function of rates of sea-level fall and the continuation of gradients offshore into the Gulf of Carpentaria (Nanson et al. 2013b), as follows. Fluvial-deltaic models link rates and trajectories of relative sea-level change impacts to delta channel avulsion frequency (e.g., Stouthamer and Berendsen 2007; Jerolmack 2009). Rapid transgression typically leads to accelerated aggradation and increased avulsion frequency compared with slower transgression

(Jerolmack 2009). In the case of a relative sea-level fall, fluvial incision and reduced rates of channel avulsion are generally expected (e.g., Slingerland and Smith 2004). However, models also indicate that avulsions will occur as a result of aggradation that leads to the superelevation of channels driven by sea-level fall (in those cases in which the fall in relative sea-level does not lead to significant channel incision, e.g., the Goose River delta – Nijhuis et al. 2015). Jones et al. (2003) have shown that the relative sea-level fall in the Gulf of Carpentaria did not cause significant channel incision. The apparent increased avulsion frequency observed in the Mitchell drainage, in spite of the decreased EP and lowered relative sea-level, is potentially related to the sea-level fall that caused the superelevation of channels.

12.6.2.2 Channel Belt Density, Width and Spacing

Channel-belt spacing and geometries were analysed along transverse transects set at 5 km intervals for the 115 km distance between the Mitchell megafan apex (M_5 Holocene apex) and the shoreline (Fig. 12.7b). There is a predictable increase in intersected channel belts in a downstream direction on the megafan (an average increase of ~1/10 km downstream) and upper delta plain (an average increase of ~1–2 km) as a result of the increased number of avulsions in this direction (Fig. 12.8a, Table 12.3). The sharp increase in the number of channel belts downstream of the backwater limit is attributed to increased avulsions caused by backwater processes; but the number of channel belts intersected decreases at the boundary between the upper and lower delta plains because

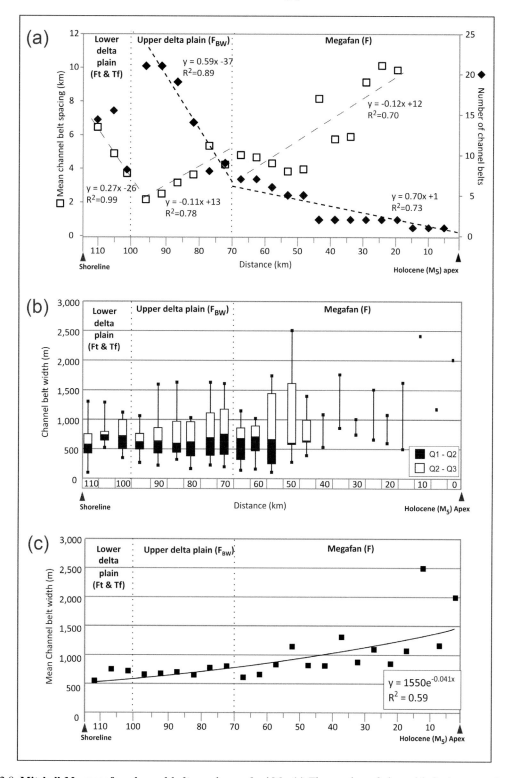

Figure 12.8 **Mitchell M$_5$ megafan channel-belt spacing and width**. (a) The number of channel belts increases downstream from the megafan apex as the system becomes more radial, with a rapid increase in channel-belt number evident downstream of the backwater limit. Conversely, channel-belt spacing reduces downstream on the megafan and upper delta plain as the system becomes more radial. Channel-belt spacing increases again on the lower delta plain as mouthbar-induced distributary channel formation is inhibited by waves. (b) Maximum, minimum, and interquartile ranges of channel belts downstream show significant spread in data. However, the widest belts (up to ~2,500 m) are located on the megafan. (c) The mean individual channel-belt width *vs.* distance from the Holocene (M$_5$) apex decreases downstream. Mean channel-belt widths for F channel belts on the megafan proper, for FBW channel belts on the upper delta plain, and for Ft and Tf channel belts on the lower delta plain are 931 m, 691 m, and 689 m, respectively.

Table 12.3 *Mitchell M5 megafan: channel belt statistics*

Distance from apex (km)	No. of channel belts	Mean average channel belt spacing (km)	Channel belt width (m)					
			Min	Q1	Q2	Mean	Q3	Max
5	1	NA	2004	2004	2004	2004	2004	2004
10	1	NA	1162	1162	1162	1162	1162	1162
15	1	NA	2500	2500	2500	2500	2500	2500
20	2	10	517	793	1069	1069	1345	1621
25	2	10	611	729	847	847	964	1082
30	2	9	662	878	1093	1093	1309	1524
35	2	6	750	813	875	875	938	1000
40	2	6	856	1081	1307	1307	1532	1757
45	2	8	532	672	811	811	951	1090
50	5	4	403	637	642	817	1003	1400
55	5	4	280	587	606	1117	1612	2630
60	6	4	109	254	664	832	1440	1740
65	7	5	162	481	707	659	895	991
70	7	5	141	325	679	619	859	1144
75	9	4	194	415	749	805	1173	1608
80	8	5	229	398	685	780	1111	1618
85	14	4	195	385	624	650	957	1026
90	19	3	334	443	597	700	971	1629
95	21	2	252	422	637	673	861	1600
100	21	2	278	499	617	643	760	1059
105	8	4	350	528	723	736	995	1136
110	15	5	553	642	734	769	794	1300
115	14	6	98	432	586	576	759	1300

Values combined by land unit (Megafan, Upper delta plain, Lower delta plain)								
	Total	Mean	Min	Q1	Q2	Mean	Q3	Max
Megafan proper (F)	45	6	109	521	800	931	1158	2500
Upper delta plain (F$_{BW}$)	92	3	194	414	623	691	951	1629
Lower delta plain (Ft, Tf)	37	5	98	572	680	689	788	1300
All belts (total)	174	5	98	451	655	753	979	2500

channels are funnelled into openings in the Phase I beach-ridge deposits (Fig. 12.3), which act as barriers.

Figure 12.8a and Table 12.3 show the channel-belt spacing (centreline to centreline) in three zones, the megafan proper, the upper delta, and the lower delta plains. Spacing on the megafan proper generally decreases downstream as the system rapidly avulses relative to radial spreading (average decrease of 100 m/km downstream). Mean channel-belt spacing on the fan is 6 km. On the upper delta plain, there is a more rapid decrease in spacing (average decrease of 80 m/km downstream) related to multiple backwater-mediated avulsion nodes. Mean channel-belt spacing on the upper

delta plain is approximately 3 km. However, on the lower delta plain mean average spacing increases to 5 km (Fig. 12.8a) because mouth-bar-induced bifurcation (and hence distributary channel splitting) is inhibited as waves rework mouth-bar sediment into laterally extended shoreline deposits (Lane et al. 2017).

Figure 12.8b and Table 12.3 also illustrate the relationship between channel-belt widths and distance from the megafan apex. Although there is a significant overlap in the interquartile ranges of channel belt widths, the widest channel belts occur on the megafan proper (up to 2,500 m wide). Mean channel-belt width decreases downstream (Fig. 12.5) from 931 m on the megafan to significantly smaller dimensions on the upper and lower delta plains (691 m and 689 m, respectively; Table 12.3).

Fluvial sediment distribution on the upper and lower delta occurs via the complicated network of modern and palaeochannel belts, and is controlled primarily by avulsion: OSL dating demonstrates significant lateral channel migration through time (Lane et al. 2017). However, ancient channel belts are seldom completely abandoned. During floods, additional discharge is not accommodated by an increase in downstream channel capacity of the Mitchell River, but is instead partitioned into distributary channels across the fan, particularly for floods with recurrence intervals greater than five years (Rustomji 2010).

Mean channel-belt width decreases downstream (Fig. 12.8b–c), which, combined with stratigraphic drilling data showing a decrease in depth, indicates a reduction in cross-sectional area (Fig. 12.5) of the most significant channel bodies. This trend accords with the DFS models of Weissmann et al. (2010, 2011, 2013). Although a proportion of flood flows is presumably lost to infiltration, as proposed in the Weissmann et al. model, in the Mitchell case this change is predominantly the result of flow partitioning through distributary channels. Aerial imagery of flood events and gauging station data provide evidence that during flood events with recurrence intervals or five years or more, flood losses to megafan distributary channels and floodplain is significant (Rustomji 2010).

The reduction in scale of the Mitchell channel belts is also strongly influenced by base level. Major palaeochannel belts landward of the backwater limit are up to 2.5 km wide, 9 to > 11 m deep, with width-to-thickness ratios typically between 47:1 and 128:1 (based on interquartile ranges of widths and stratigraphic coring). Fluvial channel belts in the backwater zone are up to 1.6 km wide and 5–7 m deep, with width-to-thickness ratios typically between 59:1 and 136:1. Backwater zone dynamics exert a strong control on avulsion, and the reduction in channel body size is also likely the result of partitioning of flow through distributary channels situated at backwater-mediated avulsion nodes. The vibracored channel-belt locations on the megafan proper (two F and two F_{BW} locations, Fig. 12.5) fall outside Gibling's (2006) predicted width-to-thickness envelope for channel bodies on

megafans, but fall within the envelope for meandering rivers or braided and low sinuosity rivers. This indicates that the Mitchell may be a useful analogue for low-gradient megafan-deltas.

Channel belts Ft and Tf on the lower delta plain are up to 1,300 m wide, and cored samples show depths of 5–6 m. However, scour depths of 13 m have been recorded at the channel mouth. Channel body width-to-thickness ratios thus fall between 95:1 and 158:1 (based on cored depths), though more work is needed to investigate the influence of tides on the lower delta plain, the tidal–fluvial transition, and their interactions with channel morphology and fill.

The downstream changes in channel-belt geometry, distribution, and gross sedimentary fill are shown in transverse transects (I–VI, Fig. 12.9a) and in longitudinal transects (VII–IX, Fig. 12.9b). Longitudinal transects show a break in slope along the fan, which appears to be related to the M_4 (Pleistocene) fan (transect IX, Fig. 12.9b). Transects VII and IX show a steepening of the fan surface at the backwater zone. This may be an autogenic response of the alluvial system, whereby it has aggraded to increase depositional slope. Indurated sediment is also known to have a localised effect on breaks in channel slope (Nanson et al. 2005).

Gross lithology logs were derived using vibracore (Lane et al. 2017), Queensland Geological Survey stratigraphic holes (Smart and Grimes 1971), water bore logs (The State of Queensland Department of Natural Resources and Mines 2014), and these data sources are located on the cross-sections in Figure 12.9. The errors inherent in these data sources are significantly less than the resolution of the stratigraphy that they have been used to inform. The depth of the underlying Wyaaba beds is derived from GQS Rutland Plains-1. The Wyaaba beds are probably Miocene to recent in age (Smart and Grimes 1971). The sand-to-mud ratio is provided for transects I–IV (Fig. 12.9a) and VII–IX (Fig. 12.9b) for the megafan down to the coastal plain. There is no notable change in sand-to-mud ratio down dip, but individual sand bodies are more compartmentalised (Fig. 12.9a). There is also a notable difference in sand-to-mud ratio from north to south (Fig. 12.9b). This may be attributed to the observed increase in sediment induration towards the north which has caused river channels to

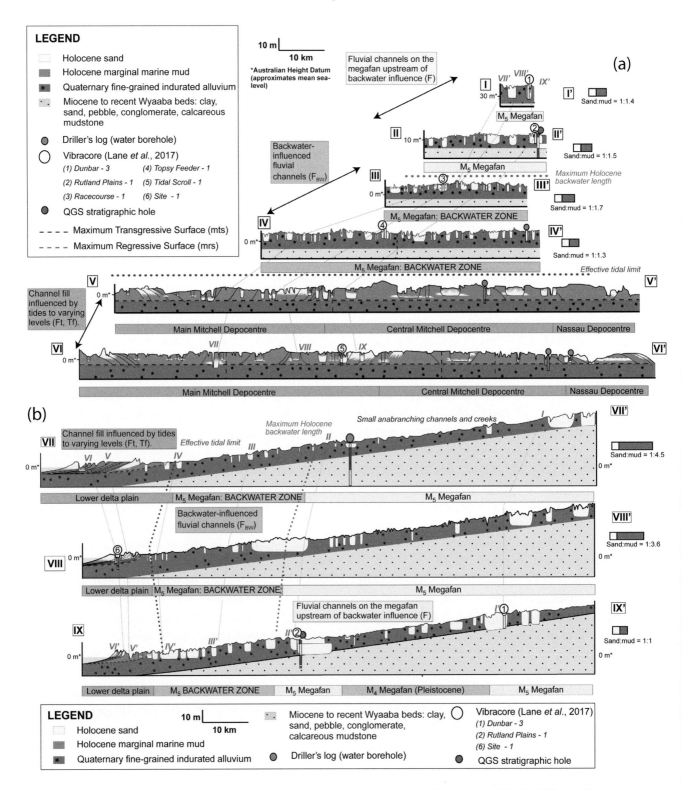

Figure 12.9 **The complex network of modern and ancient channels and channel belts on the Mitchell M$_5$ megafan, active subsequent to the Holocene post-glacial maximum marine transgression (~6 ka).** (a) Strike-oriented transects (I–IV) from near the apex to the effective tidal limit show the distribution of fluvial channel belts upstream of the backwater influence and those affected by backwater dynamics. (b) Dip-oriented transects (V–VI) from near the apex of the M$_5$ fan to the effective tidal limit; distribution and interpretation of fluvial channel belts upstream of the backwater-affected zones (see Figure 12.2 for transect locations). Modified from Lane et al. (2017). Asterisks (*): Australian Height Datum (AHD) approximates mean sea-level. A colour version of this figure is available in the SOD for Chapter 12.

more readily incise into the southern part of the fan and fill channel belts (subsequently abandoned) with sand and gravel.

12.7 Conclusions

The Mitchell River megafan is the largest megafan-delta system in Australia. It is characterised by a very low regional gradient (Holocene surface slope: ~0.0004) and has formed under a highly ephemeral flow regime. The megafan-delta has been subjected to repeated fluctuations in sea level and climate since its initiation in the Pliocene. Four major categories of palaeochannel belt are described on the megafan.

(i) Fluvial (F) channel belts upstream of the backwater influence are typically single-thread, have relatively large cross-sectional areas (up to 2.5 km wide, 9–11 m deep, width-to-depth Q1–Q3 interquartile range ratios 47:1–128:1) and coarse sand and gravel fills.
(ii) Backwater-influenced fluvial channel belts (F_{BW}) on the upper delta plain, which typically display scrolled lateral accretion morphology. Relative to F channels, F_{BW} occurrences have reduced cross-sectional dimensions (up to 1.6 km wide, 7 m deep and width-to-depth ratio Q1–Q3 interquartile range ratios 59:1–136:1) and reduced sediment calibre (medium and coarse sand).
(iii) Fluvial-dominated, tide-influenced (Ft) channels on the lower delta plain also typically display a scrolled lateral accretion morphology. They are up to 1.3 km wide and 6 m deep. Ft channels are characterised by inclined heterolithic strata, are medium-textured, and characteristically contain shell material.
(iv) Tide-dominated, fluvial-influenced (Tf) channels on the lower delta plain may have scrolled morphologies, inherited from the Ft channels from which they typically evolve (following upstream abandonment). Tf channels are filled with marine sediment (tidally derived muds and shell material), and reach 1.3 km wide and 5 m deep. Ft and Tf channels have width-to-depth ratios of 95:1–158:1 (Q1–Q3).

Avulsion is the primary mechanism of sediment dispersal. Two types have operated on the Mitchell River megafan: (i) frequent backwater-mediated avulsions cluster at the backwater limit, and occurred more than 13.3 times every 1000 years (once every 75 years). These more frequent avulsions usually result in local shifts in system depocentre. (ii) Less frequent

avulsions on the megafan occurred approximately 2.3 times every 1000 years (once every 434 years), and resulted in much more dramatic shifts in depositional loci. These avulsions do not cluster at the fan apex as they do in most other systems, but rather are distributed less systematically across and down the fan in response to the distribution of indurated substrates. Indurated sediment promotes avulsion by severely restricting channel migration whereby channels cannot expend their excess energy by lengthening their courses, and therefore avulse instead. This process is typical of fans across the Carpentaria Plains and, given the impact on the morphology of the fan and its channels, it should be considered when investigating similar ancient systems.

Avulsion rate on the upper delta plain (within the backwater limit) also appears to be influenced by sea-level trajectory. The estimated avulsion rate increased from 11/kyr in the Central Mitchell depocentre (under highstand conditions) to 15/kyr in the Main Mitchell depocentre (during a sea-level fall), in spite of decreased effective precipitation (EP) after ~3.7 ka. This may be attributed to the sea-level fall causing the superelevation of the channels and thereby promoting avulsive conditions.

In the complex modern and palaeochannel network that characterises the Mitchell megafan, sediment is partitioned radially, both in time through successive avulsions, and coevally through simultaneously active distributary channels. As a result, channel-belt numbers increase downstream from the apex. Correspondingly, channel-belt width, depth, and cross-sectional area also decrease downstream from the apex. These geometric attributes should provide useful analogue metrics for low-gradient ancient megafan systems that have been underrepresented in previous syntheses of channel-belt geometries.

Acknowledgements

We would like to thank the sponsors of Phases I and II of the WAVE Consortium (Apache, BAPETCO, BHPBP, BG, BP, Chevron, ConocoPhillips, Nexen, OMV, Shell, Statoil, Todd Energy, and Woodside Energy) for partial funding of this research. The Kowanyama Land Management Office, traditional owners of the Kowanyama region, and Viv Sinnamon

238   *Part II Regional Studies*

are thanked for their permission to undertake this research and for providing logistical support in this remote region. We thank Denise Smith, Joanne Lane, Kathryn Amos, Jessica Trainor, Rodney Whitfield, Dugald Smith, Gerald Nanson, Bartosz Cichonski, John Lane, and Andrew Fernie for assistance in the field, with special thanks to Brent Peterson from the University of Wollongong for drill rig operations. We also thank Frances Williams and Lee Arnold from the University of Adelaide for OSL contributions and Jorge Garcia Garcia, John Counts and Daniel Lane for assistance in core processing and OSL collection. Thanks are also due to Justin Wilkinson, Gerald Nanson and two anonymous reviewers for their comments that improved the quality of this work.

References

Ainsworth, R. B., Vakarelov, B. K., and Nanson, R. A. (2011). Dynamic spatial and temporal prediction of changes in depositional processes on clastic shorelines: Toward improved subsurface uncertainty reduction and management. *American Association of Petroleum Geologists Bulletin*, 95, 267–297.

BOM. (2018). Australian Bureau of Meteorology. *Tropical cyclones in the Northern Territory*. [online] Australian Government. Available at: www.bom.gov.au/cyclone/about/northern.shtml. [Accessed 05 February 2019]

Brooks, A. P., Spencer, J., Shellberg, J. G., and Knight, J. (2008). Using remote sensing to quantify sediment budget components in a large tropical river – Mitchell River, Gulf of Carpentaria. In J. Schmidt, T. Cochrane, C. Phillips, et al., eds., *Sediment Dynamics in Changing Environments*. International Association of Hydrological Sciences Press, Wallingford, 225–236.

Brooks, A. P., Shellberg, J. G., Knight, J., and Spencer, J. (2009). Alluvial gully erosion: an example from the Mitchell fluvial megafan, Queensland, Australia. *Earth Surface Processes and Landforms*, 34, 1951–1969.

Callen, R. A. and Bradford, J. (1992). Cooper Creek fan and Strzelecki Creek – hypsometric data, Holocene sedimentation, and implications for human activity. *Mines and Energy Reviews*, 158, 52–57.

Chappell, J. (1983). Evidence for smoothly falling sea level relative to north Queensland, Australia, during the past 6,000 yr. *Nature*, 302, 406–408.

Chatanantavet, P., Lamb, M. P., and Nittrouer, J. A. (2012). Backwater controls of avulsion location on deltas. *Geophysical Research Letters*, 39, DOI: 10.1029/2011GL050197.

Cohen, T. J., Nanson, G. C., Larsen, J. R., et al. (2010). Late Quaternary aeolian and fluvial interactions on the Cooper Creek Fan and the association between linear and source-bordering dunes, *Strzelecki Desert,*

Australia. Quaternary Science Reviews, 29, 455–471.

CSIRO, Commonwealth Scientific and Industrial Research Organisation, Australia. (2019). Sea-level Rise – Improving projections for the future. [online] CSIRO Available at: https://www.cmar.csiro.au/sealevel/sl_hist_last_decades.html. [Accessed 02 December 2019]

Croke, J. C., Magee, J. M., and Price, D. M. (1998). Stratigraphy and sedimentology of the lower Neales River, West Lake Eyre, Central Australia: from Palaeocene to Holocene. *Palaeogeography, Palaeoclimatology, Palaeoecology*, 144, 331–350.

Doutch, H. F. (1976). The Karamba Basin, northeastern Australia and southern New Guinea. Bureau of Mineral Resources (BMR) *Journal of Geology & Geophysics*, 1, 131–140.

Doutch, H. F., Ingram, J. A. Smart, J., and Grimes, K. G. (1970). *Progress Report on the Geology of the Central Carpentaria Basin*, Bureau of Mineral Resources (BMR) Australia Record 1972/64.

Doutch, H. F. and Nicholas, E. (1978). The Phanerozoic sedimentary basins of Australia and their tectonic implications. *Tectonophysics*, 48, 365–388.

Forbes, M., Jankowski, N., Cohen, T., et al. (2020). Palaeochannels of Australia's Riverine Plain – Reconstructing past vegetation environments across the Late Pleistocene and Holocene. *Palaeogeography, Palaeoclimatology, Palaeoecology*, 545, 109533.

Galloway, R. W., Gunn, R. H., and Story, R. (1970). The lands of the Mitchell–Normanby area, Queensland. *CSIRO Land Research Series*, 26.

Ganti, V., Chu, Z., Lamb, M. P., Nittrouer, J. A., and Parker, G. (2014). Testing morphodynamic controls on the location and frequency of river avulsions on fans versus deltas: Huanghe (Yellow River), China. *Geophysical Research Letters*, 41, 7882–7890.

Geoscience Australia and CSIRO Land & Water (2011). *1 Second SRTM derived Digital Elevation Models User Guide. Version 1.0.4*, Canberra, Geoscience Australia.

Gibling, M. R. (2006). Width and thickness of fluvial channel bodies and valley fills in the geological record: a literature compilation and classification. *Journal of Sedimentary Research*, 76, 731–770.

Grant, K. (1968). *Terrain classification for engineering purposes of the Rolling Downs Province, Queensland*. CSIRO Division of Soil Mechanics Technical Paper 3.

Grimes, K. G. (1974). *Mesozoic and Cainozoic geology of the Lawn Hill, Westmoreland, Mornington and Cape Van Diemen: Queensland*, Bureau of Mineral Resources (BMR) Australia, 1:250,000 sheet areas, Record 1974/106.

Grimes, K. G. and Doutch, H. F. (1978). The late Cainozoic evolution of the Carpentaria Plains, North Queensland. Bureau of Mineral Resources (BMR). *Journal of Geology & Geophysics*, 3, 101–112.

Hesse, P. P., Williams, R., Ralph, T. J., et al. (2018) Palaeohydrology of lowland rivers in the Murray-Darling Basin, Australia. *Quaternary Science Reviews*, 200, 85–105.

Hooke, R. (1967). Processes on arid-region alluvial fans. *The Journal of Geology*, 75, 438–460.

Isbell, R. F., Webb, A. A., and Murtha, G. G. (1968). *Atlas of Australian Soils, Sheet 7, North Queensland, with Explanatory Data*, Melbourne, CSIRO and Melbourne University Press.

Jerolmack, D. J. (2009). Conceptual framework for assessing the response of delta channel networks to Holocene sea-level rise. *Quaternary Science Reviews*, 28, 1786–1800.

Johnson, P. J. (1982). Sedimentary facies of an arid zone delta: Gascoyne Delta, Western Australia. *Journal of Sedimentary Research*, 52, 547–563.

Jones, B. G., Martin, G. R., and Senapati, N. (1993). Riverine–tidal interactions in the monsoonal Gilbert River fandelta, northern Australia. *Sedimentary Geology*, 83, 319–337.

Jones, B. G., Woodroffe, C. D., and Martin, G. R. (2003). Deltas in the Gulf of Carpentaria, Australia: forms, processes and products. In F. H. Sidi, D. Nummedal, P. Imbert, H. Darman, and H. W. Posamentier, eds., *Tropical Deltas of Southeast Asia: Sedimentology, Stratigraphy, and Petroleum Geology*, SEPM Special Publication, 76, 21–43.

Kemp, J. and Rhodes, E. J. (2010). Episodic fluvial activity of inland rivers in southeastern Australia: palaeochannel systems and terraces of the Lachlan River. *Quaternary Science Reviews*, 29, 732–752.

Kemp, J., Pietsch, T., Gontz, A., and Olley, J. (2017). Lacustrine-fluvial interaction in Australia's Riverine Plains. *Quaternary Science Reviews*, 166, 352–362.

Lamb, M. P., Nittrouer, J. A., Mohrig, D., and Shaw, J. (2012). Backwater and river plume controls on scour upstream of river mouths: implications for fluvio-deltaic morphodynamics. *Journal of Geophysical Research: Earth Surface*, 117, DOI:10.1029/2011JF002079.

Lane, T. I. (2016). *Evolution and architecture of the Holocene Mitchell River Megafan and Delta, Gulf of Carpentaria, Australia*. PhD thesis, University of Adelaide, Australia.

Lane, T. I., Nanson, R. A. Vakarelov, B. K. Ainsworth, R. B., and Dashtgard, S. E. (2017). Evolution and architectural styles of a forced-regressive Holocene Delta and Megafan, Mitchell River, Gulf of Carpentaria, Australia. In G. J. Hampson, A. D. Reynolds, B. Kostic, and M. R. Wells, eds., *Sedimentology of Paralic Reservoirs: Recent Advances*. Geological Society of London, Special Publication, 444, 305–334.

Lang, S. C., Payenberg, T. H. D., Reilly, M. R. W., et al. (2004). Modern and analogues for dryland sandy fluvial-lacustrine deltas and terminal splay reservoirs. *APPEA Journal*, 44, 329–356.

Leonhard, L., Burton, K., and Milligan, N. (2013). Gascoyne River, Western Australia; alluvial aquifer, groundwater management and tools. In C. Wetzelhuetter, ed., *Groundwater in the Coastal Zones of the Asia-Pacific*. Springer, Berlin, 359–378.

Massey, T. A., Fernie, A. J., Ainsworth, R. B. Nanson, R. A., and Vakarelov, B. K. (2014). Detailed mapping, three-dimensional modelling and upscaling of a mixed-influence delta system, Mitchell River delta, Gulf of Carpentaria, Australia. In A. W. Martinius, J. A. Howell, and T. Good, eds., *Sediment Body Geometry and Heterogeneity: Analogue Studies for Modelling the Subsurface*. Geological Society of London, Special Publication, 387, 131–151.

Murray-Wallace, C. V. and Woodroffe, C. D. (2014). *Quaternary Sea-level Changes: A Global Perspective*. Cambridge, Cambridge University Press.

Nanson, G. C. and Huang, H. Q. (2008). Least action principle, equilibrium states, iterative adjustment and the stability of alluvial channels. *Earth Surface Processes and Landforms*, 33, 923–942.

Nanson, G. C., Price, D. M., Short, S. A., and Young, R. W. (1991). Comparative uranium–thorium and thermo-luminescence dating of weathered Quaternary alluvium in the tropics of northern Australia. *Quaternary Research*, 35, 347–366.

Nanson, G. C., Jones, B. G., Price, D. M., and Pietsch, T. J. (2005). Rivers turned to rock: Late Quaternary alluvial induration influencing the behaviour and morphology of an anabranching river in the Australian monsoon tropics. *Geomorphology*, 70, 398–420.

Nanson, R., Ainsworth, B., Vakarelov, B., Fernie, A., and Massey, T. (2012). Geometric attributes of reservoir elements in a modern, low accommodation, tide-dominated delta. *APPEA Journal*, 52, 483–492.

Nanson, R. A., Lane, T. L, Ainsworth, R. B., and Vakarelov, B. K. (2013a). *Evolution and Architecture of a Low Accommodation, Mixed Influence Marginal Marine System: Holocene Mitchell River Delta, Queensland, Australia*. A Field Workshop, 1–6 September 2013: WAVE consortium.

Nanson, R. A., Vakarelov, B. K., Ainsworth, R. B., Williams, F. M., and Price, D. M. (2013b). Evolution of a Holocene, mixed-process, forced regressive shoreline: The Mitchell river delta, Queensland, Australia. *Marine Geology*, 339, 22–43.

Nijhuis, A. G., Edmonds, D. A., Caldwell, R. R., et al. (2015). Fluvio-deltaic avulsions during relative sea-level rise. *Geology*, 43, 719–722.

Page, K. J. and Nanson, G. C. (1996). Stratigraphic architecture resulting from Late Quaternary evolution of the Riverine Plain, south-eastern Australia. *Sedimentology*, 43, 927–945.

Page, K., Nanson, G., and Price, D. (1996). Chronology of Murrumbidgee river palaeochannels on the Riverine Plain, southeastern Australia. *Journal of Quaternary Science*, 11, 311–326.

Page, K. J., Dare-Edwards, A. J., Owens, J. W., et al. (2001). TL and stratigraphy of riverine source bordering sand dunes near Wagga Wagga, New South Wales, Australia. *Quaternary International*, 83–85, 187–193.

Page, K. J, Kemp, J., and Nanson, G. C. (2009). 1002 Late Quaternary evolution of Riverine Plain palaeochannels, southeastern Australia. *Australian Journal of Earth Sciences*, 56, 19–33.

Perry, R. A., Sleeman, J. R., Twidale, C. R., et al. (1964). *General report on the lands of the Leichhardt-Gilbert area, Queensland, 1953–54*. CSIRO, Land Research Series 11.

Pietsch, T. J., Nanson, G. C., and Olley, J. M. (2013). Late Quaternary changes in flow-regime on the Gwydir distributive fluvial system, southeastern Australia. *Quaternary Science Reviews*, 69, 168–180.

Porritt, E. L., Jones, B. G., Price, D. M., and Carvalho, R. C. (2020). Holocene delta progradation into an epeiric sea in northeastern Australia. *Marine Geology*, 422, 106114.

Prosser, I. P., Rustomji, P., Young, W. J., Moran, C. J., and Hughes, A. O. (2001). Constructing river basin sediment budgets for the National Land and Water Resources Audit, CSIRO, *Land and Water, Technical Report 15/01*.

Reeves, J. M., Chivas, A. R., García, A., and De Deckker, P. (2007). Palaeoenvironmental change in the Gulf of Carpentaria (Australia) since the last interglacial based on Ostracoda. *Palaeogeography, Palaeoclimatology, Palaeoecology*, 246, 163–187.

Reeves, J. M., Chivas, A. R., García, A., et al. (2008). The sedimentary record of palaeoenvironments and sea-level change in the Gulf of Carpentaria, Australia, through the last glacial cycle. *Quaternary International*, 183, 3–22.

Rhodes, E. G. (1980). *Modes of Holocene Coastal Progradation, Gulf of Carpentaria*, PhD thesis, Australian National University, Canberra.

Rhodes, E. G. (1982). Depositional model for a chenier plain, Gulf of Carpentaria, Australia. *Sedimentology*, 29, 201–221.

Rhodes, E. G., Polach, H. A, Thom, B. G., and Wilson, S. R. (1980). Age structure of Holocene coastal sediments: Gulf of Carpentaria, Australia. *Radiocarbon*, 22, 718–727.

Rustomji, P. (2010). *A statistical analysis of flood hydrology and bankfull discharge for the Mitchell River catchment, Queensland, Australia*, CSIRO: Water for a Healthy Country National Research Flagship [01/2010].

Rustomji, P., Shellberg, J. G., Brooks, A. P., Spencer, J., and Caitcheon, G. (2010). *A catchment sediment and nutrient budget for the Mitchell River, Queensland. A report to the Tropical Rivers and Coastal Knowledge (TRaCK) Research Program*. In Department of Sustainability, Water, Population and Communities, ed., CSIRO Water for a Healthy Country National Research Flagship, 119 p.

Schumm, S. A., Mosley, M. P., and Weaver, W. E. (1987). *Experimental Fluvial Geomorphology*. Wiley, New York, 413 p.

Shellberg, J. G. (2011). *Alluvial Gully Erosion Rates and Processes Across the Mitchell River Fluvial Megafan in Northern Queensland, Australia*. PhD thesis, Griffith University, Australian Rivers Institute.

Shellberg, J. G., Brooks, A. P. Spencer, J., and Ward, D. (2012). The hydrogeomorphic influences on alluvial gully erosion along the Mitchell River fluvial megafan, northern Australia. *Hydrological Processes*, 27, 1086–1104.

Shellberg, J. G., Brooks, A. P., and Rose, C. W. (2013). Sediment production and yield from an alluvial gully in northern Queensland, Australia. *Earth Surface Processes and Landforms*, 38, 1765–1778.

Shellberg, J. G., Spencer, J. Brooks, A. P., and Pietsch, T. J. (2016). Degradation of the Mitchell River fluvial megafan by alluvial gully erosion increased by post-European land use change, Queensland, Australia. *Geomorphology*, 266, 105–120.

Shulmeister, J. (1999). Australasian evidence for mid-Holocene climate change implies precessional control of Walker Circulation in the Pacific. *Quaternary International*, 57/58, 81–91.

Simpson, C. J. and Doutch, H. F. (1977). The 1974 wet-season flooding of the southern Carpentaria Plains, northwest Queensland. *BMR Journal of Australian Geology and Geophysics*, 2, 43–51.

Slingerland, R. L. and Smith, N. D. (2004). River avulsions and their deposits. *Annual Review of Earth and Planetary Sciences*, 32, 257–285.

Sloss, C. R., Nothdurft, L., Hua, Q., et al. (2018). Holocene sea-level change and coastal landscape evolution in the southern Gulf of Carpentaria, Australia. *The Holocene*, 28, 1411–1430.

Smart, J. and Grimes, K. G. (1971). Shallow stratigraphic drilling, eastern Carpentaria Basin. *Bureau of Mineral Resources, Geology and Geophysics*, BMR Record 1971/143.

Smart, J., Morrissey, J. A., and Hassan, S. E. (1975). Index to drill-hole data – Carpentaria, Laura, and Karumba Basins, *Bureau of Mineral Resources, Geology and Geophysics*, BMR Record 1975/167.

Smart, J., Grimes, K. G., Doutch, H. F., and Pinchin, J. (1980). The Carpentaria and Karumba Basins, North Queensland, Australia. *Bureau of Mineral Resources (BMR), Geology and Geophysics Bulletin*, 202, 73 pp.

Stouthamer, E. and Berendsen, H. J. A. (2007). Avulsion: the relative roles of autogenic and allogenic processes. *Sedimentary Geology*, 198, 309–325.

Suppiah, R. (1992). The Australian summer monsoon: A review. *Progress in Physical Geography*, 16, 283–218.

The State of Queensland Department of Natural Resources and Mines (2014). *Groundwater database bore report: Mitchell River area*, 141 pp.

Twidale, C. R. (1966). Chronology of denudation in northwest Queensland. *Geological Society of America Bulletin*, 67, 3–23.

Vakarelov, B. K. and Ainsworth, R. B. (2013). A hierarchical approach to architectural classification in marginal marine systems: bridging the gap between sedimentology and sequence stratigraphy. *American Association of Petroleum Geologists Bulletin*, 97, 1121–1161.

Wakelin-King, G. and Amos, K. (2016). A time-slice of the Lake Eyre Basin: sand/mud depositional geometries in a diverse lowstand endorheic drylands setting. *Eastern*

Australian Basins Symposium: Publication of Proceedings, 97–133.

Weissmann, G. S., Hartley, A. J. Nichols, G. J., et al. (2010). Fluvial form in modern continental sedimentary basins: Distributive fluvial systems. *Geology*, 38, 39–42.

Weissmann, G. S., Hartley, A. J. Nichols, G. J., et al. (2011). Alluvial facies distributions in continental sedimentary basins: distributive fluvial systems. In S. K. Davidson, S. Leleu, and C. North., eds., *From River to Rock Record: The Preservation of Fluvial Sediments and their Subsequent Interpretation*. SEPM Special Publication, 97, 327–355.

Weissmann, G. S., Hartley, A. J., Scuderi, L. A., et al. (2013). Prograding distributive fluvial systems: geomorphic models and ancient examples. In S. G. Driese and L. C. Nordt, eds., *New Frontiers in Paleopedology and Terrestrial Paleoclimatology, Paleosols and Soil Surface Analog Systems*. SEPM Special Publication, 104, 131–147.

Whitehouse, F. W. (1941). The surface of western Queensland. *Proceedings of the Royal Society of Queensland*, 53, 1–22.

Whitney, B. B. and Hengesh, J. V. (2015). Geomorphological evidence of neotectonics deformation in the Carnarvon Basin, Western Australia. *Geomorphology*, 228, 579–596.

Woodroffe, S. A. (2009). Testing models of mid to late Holocene sea-level change, North Queensland, Australia. *Quaternary Science Reviews*, 28, 2474–2488.

13

Megafans of the Northern Victorian Riverine Plains, SE Australia

KAREN A. KAPTEINIS, JOHN A. WEBB, and SUSAN Q. WHITE

Abstract

Low-angle megafans occur along the northern boundaries of the Victorian Uplands and extend into the Murray Basin. These include the Loddon River, Campaspe River, and Bullock Creek, which range in length from 90 km to 120 km from apex to toe. The Loddon and Bullock fans overlap significantly in their middle and northern extents. Because of their very low topographic gradients ($< 0.001°$), these depositional features had previously been classified as general channel and flood sediments of the Shepparton Formation, a Pliocene- to Holocene-aged floodplain formation. Airborne radiometric imaging has nonetheless allowed identification of distinct, fan-like features extending north into the Murray Basin. Radiocarbon dating of the Bullock Creek and Loddon River surface sediments has provided ages of 7,270 yr BP and 140 yr BP, respectively. Sediment textures progress down-fan from coarser to finer material, with individual sites dominated by silt or additions of sand-sized aggregates of clay and silt particles. The fans were formed largely by high-discharge, intermittent floods within a complex, interconnected distributive channel system, with smaller inputs from day-to-day channel deposition. Sediment sources include a combination of redeposited windblown silt and weathered material from basalt flows and Paleozoic metasediments in the upland catchments.

13.1 Introduction

Although several megafans have been identified in eastern Australia along the semi-arid inland side of the Great Dividing Range, they have not received as much attention as their global counterparts because their extremely low slope makes them difficult to identify, and they have often been classified as floodplains with a complex system of channels. It has required a combination of satellite imaging, airborne radiometric data, aerial photography and ground truthing of distributary channels for them to be recognised (Riley and Taylor 1978; Wray 2009; Pietsch and Nanson 2011; Pietsch et al. 2013). Only a few have been studied in detail, all in the central and northern parts of eastern Australia: the Gwydir fan (Riley and Taylor 1978; Pietsch and Nanson 2011; Pietsch et al. 2013), the Mitchell fan (Brooks et al. 2009; Shellberg et al. 2012; Nanson et al. 2013; Lane et al., Ch. 12), the Murrumbidgee fan (Schumm 1968; Page et al. 2009) and the Namoi fan (Riley and Taylor 1978; Wray 2009).

Megafans are also present to the south on the Victorian Riverine Plains, but previous studies in this area did not recognise their true extent due to their very low slope, and they were described as only ~30 km long (Macumber 1969, 1991; Macumber and Macumber 2010). Recently available satellite and airborne radiometric images, combined with radar and LiDAR topography, have shown that in fact three megafans >90 km long cover a large part of the

242

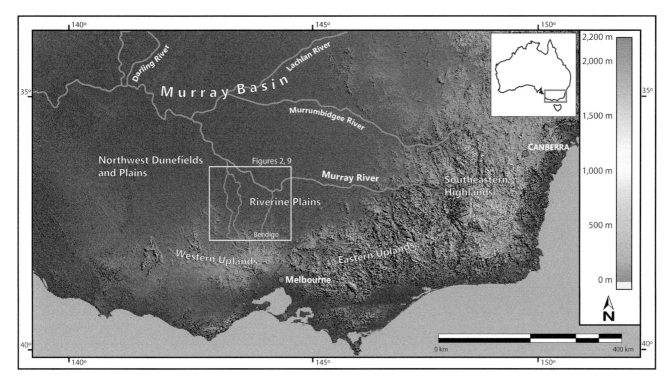

Figure 13.1 **Regional setting of the Victorian megafans, SE Australia**. Geomorphic regions of Victoria after Joyce et al. (2003). A colour version of this figure is available in the SOD for Chapter 13.

Victorian Riverine Plains (Figs. 13.1–13.3); these are described here, and the sedimentary environments associated with them are used to investigate the processes that formed the fans and their depositional history.

13.2 Regional Geological and Geomorphological Setting

The Victorian Riverine Plains comprise a large depositional plain with a very low northwards gradient (0.02–0.05°), crossed by rivers flowing towards the Murray River (Lawrence et al. 1976; Edwards et al. 2001). To the south are the Eastern and Western Victorian Uplands, that together form an east–west range increasing in elevation eastwards from an average of ~300 m in central Victoria to over 1,200 m in eastern Victoria (Fig. 13.1). The Eastern Uplands continue eastwards and increase in elevation to become the Southeastern Highlands, that extend northwards along the Australian coast as the Great Dividing Range.

The basement outcrops in the uplands, which also underlie the Riverine Plains, are largely composed of a thick sequence of Ordovician turbidites (Castlemaine

Group). They consist of medium- to fine-grained quartz-rich sandstones with interbedded siltstones, black shale, and chert. This sedimentary series has been folded, faulted, and subjected to low grade regional metamorphism (King 1986; VandenBerg et al. 2000; Edwards et al. 2001; Willman et al. 2002; Cayley et al. 2008). A series of granite bodies were intruded in Devonian time (King 1986; Edwards et al. 2001; Willman et al. 2002).

In the mid-Cretaceous (90 ± 5 Ma; Joyce et al. 2003) a major period of tectonism initiated uplift in the Southeastern Highlands of Australia, including the Victorian Uplands. This was followed by broad intra-cratonic downwarping in the Cenozoic, forming the inland Murray Basin, which accumulated up to 600 m of Paleocene to Holocene, mostly siliciclastic and calcareous sediments. Basal fluvio-lacustrine sands and clays are succeeded by shallow marine sediments in the western part of the basin, the latter inter-fingering with fluvial valley-fill sediments to the east (Brown and Stevenson 1991; Macumber 1991). The youngest unit is the Pliocene–Holocene Shepparton Formation (Lawrence 1966, 1975; Macumber and Macumber 2010), which is dominated by floodplain

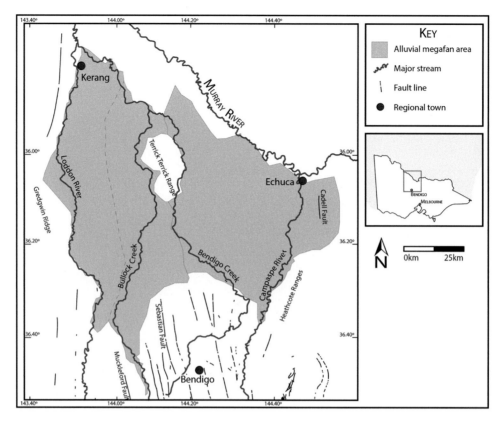

Figure 13.2 **Location map of megafans of the northern Victorian plains**.

and lacustrine clays and silty clays, interbedded with river channel sands and gravels (Calf et al. 1986; Ollier 1995; Cupper et al. 2003; Macumber and Macumber 2010). The average grain size of the Shepparton Formation sediments decreases with increasing distance from the upland source areas to the south (Calf et al. 1986).

In the Neogene, eruptions of the Newer Volcanics in the Victorian Uplands, emplaced basaltic lava flows with scoria cones (Price et al. 2003), partly infilling the valleys of the Loddon and Campaspe Rivers. A second period of active tectonism in the late Neogene further uplifted the Victorian Uplands (Hill 1999; Braun et al. 2009; Clark et al. 2011; Webb et al. 2011), causing incision of relatively narrow valleys. The Palaeozoic bedrock exposed on the slopes is deeply weathered. The late Neogene tectonism had a minor impact on the Victorian Riverine Plains by re-activating a series of north–south-aligned subvertical dip-slip faults, including the Muckleford and Sebastian Faults (Hills 1961; Clark et al. 2011), causing relatively small displacements (tens of metres; Robson and

Webb 2011). Tectonism has continued throughout the Quaternary, but with lower intensity (Clark et al. 2007).

The Victorian Riverine Plains occupy the southeastern part of the Murray Basin, and their fluvial landscape contains three components: river terraces and stream channels of the Shepparton Formation, recent alluvium deposited by perennial channels incised 2–5 m into the Shepparton Formation, and ephemeral lakes with lunette dunes on their eastern leeward sides (Macumber 1969; Edwards et al. 2001; Cupper et al. 2003). A distinction among the elevated channels of the Shepparton Formation has been made between 'prior streams' (with levee banks along their margins) and 'ancestral streams' (with wide meanders and upward-fining point bar deposits) (Butler 1950; Macumber 1969; Pels 1971; Butler et al. 1973; Edwards et al. 2001; Page et al. 2009; Macumber and Macumber 2010). The prior streams were active around 100–40 ka and the ancestral channels from 40 ka to 15–10 ka (Fried 1993). These channels carry flow only ephemerally under the current climatic

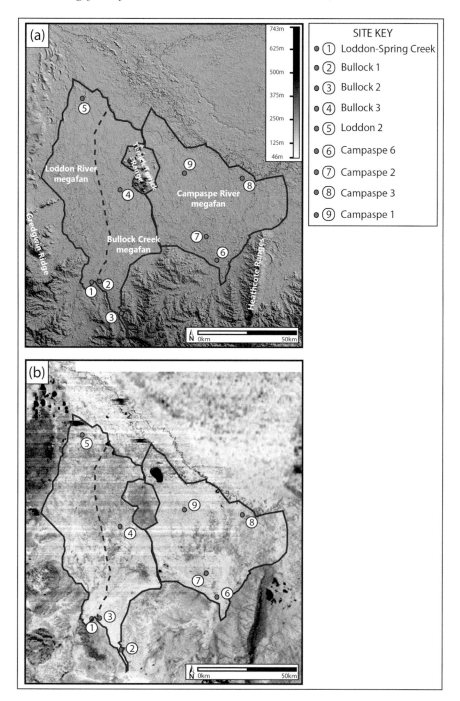

Figure 13.3 **Megafans of the northern Victorian plains**. (a) Shaded relief of the study area. (b) Airborne radiometric image showing fan-shaped features. Colour codes explained in text. A colour version of this figure is available in the SOD for Chapter 13.

regime (Bowler and Harford 1966; Edwards et al. 2001), and formed under a climate wetter than present (Page and Nanson 1996; Page et al. 2009; Kemp and Rhodes 2010; Pietsch et al. 2013).

A mantle of clayey Red-Brown Earth soils 80–120 cm thick, the Widgelli Pedoderm, covers the Victorian Riverine Plains (Butler and Hutton 1956; Sleeman 1975; Chen 2001; Chen et al. 2002; Bowler et al. 2006). This soil contains a component of aeolian clay, silt and silt-sized clay pellets, originally called 'parna' (Butler 1950; Lawrence et al. 1976; Butler and Hubble 1978) but now termed loess (following the broader definition of loess; Chen et al. 2002; Hesse and McTainsh 2003; Haberlah 2007; Cattle et al. 2009).

During periods of aridity in the late Quaternary, particularly the Last Glacial Maximum (LGM) at 22–18 ka (Kershaw and Nanson 1993; Bowler et al. 2006; Petherick et al. 2013), large amounts of aeolian dust were blown by prevailing westerly winds from dry lunette lakes, playas (i.e., salinas), and dry stream channels in inland Australia, onto the Riverine Plains and the slopes of the uplands to the south, contributing to the soils and creating mantles of quartz dust up to 20 cm thick on the uplands (Ash and Wasson 1983; Tickell and Humphrys 1987; Hesse and McTainsh 2003; Bowler et al. 2006; Cattle et al. 2009; Williams et al. 2009). The particle size of the aeolian material typically ranges from 20 to 50 μm, consistent with windblown loess (Cattle et al. 2002, 2009).

13.3 Climate

The present climate of the northern Victorian plains lies within two Köppen classes as revised by Peel et al. (2007): cold semi-arid (BSk) and warm oceanic/humid subtropical climates (Cfa). Average annual precipitation ranges from 1,000–600 mm in the upland catchments to 600–300 mm on the megafan plains (Bureau of Meteorology 2011). Most rainfall occurs during the cooler months (May–October) but shows high intra- and inter-annual variability. The evaporation rate is generally higher than precipitation, particularly in the warmer months. The high variability of precipitation results in equally variable stream flow, with low base flow interspersed with a few large discharge events (McMahon et al. 1987; Dettinger and Diaz 2000; McMahon and Finlayson 2003; Poff et al. 2006).

13.4 Methods

Due to the extremely low gradient of the fans, the location and extent of the alluvial megafans were initially delineated using the radiometric image of the area (Department of Primary Industries, DEPI 2003), rather than topography. This imagery uses gamma-ray spectrometry to record levels of the three common radiometric elements (potassium, thorium, and uranium) in the upper 1 metre of surface sediment (coloured red, green, and blue respectively), allowing soils and surface sediments of different sediment composition to be distinguished. The airborne radiometric

data was obtained at a resolution of 40–70 m and then gridded at 50 m mesh size, using a minimum curvature gridding method (DEPI 2003).

The megafans were identified by their light green-yellow colour, contrasting with the different colours evident in the uplands to the south: the pale pink-purple of weathered Ordovician sediments, the red or white of Devonian granites, the bright green of strongly ferruginised Ordovician sediments, and the blue-green of soils on the valley-filling basalts in the uplands (Fig. 13.3). The Campaspe megafan has some darker blue-green areas along the central and eastern parts (Fig. 13.3), probably reflecting the higher moisture content of soils where crop irrigation is most intense (water blocks the transmission of gamma radiation).

The fan outlines were overlain on the smoothed version of the Shuttle Radar Topography Mission DEM (digital elevation model) for northern Victoria (Geoscience Australia 2011), allowing the fan area, gradient, and catchment area to be measured and cross-sections to be constructed. Sediment volumes were calculated using the area (km^2) of the megafans and the average thickness of the Shepparton Formation. True colour satellite images from Google Earth and LiDAR data (DEPI 2009–2010) were used to map the distributary channels on the surface of the megafans. However, the distributary channels have been modified and obscured by laser grading for the installation of irrigation channels across much of the northern toes of the megafans, particularly around the Waranga irrigation channel; these channels carry water from the Murray River and regional reservoirs (Butler et al. 1973; Abernethy et al. 2004).

To determine the sediment characteristics of the three megafans studied, nine sites were sampled, one each in the proximal, middle, and distal parts of each fan. The sites were either undisturbed roadside verges or channel banks where streams had incised the fan. At the roadside sites, 50 × 50-cm test pits were dug by hand until the consistency of the sediment made it necessary to continue using a hand auger (diameter: 8 cm) to the maximum depth of the auger extension (1.2 m); sediment was removed from the auger holes in 10 cm lengths. At the stream sites, the channel walls were excavated to expose the undisturbed stratigraphy. All sites were sampled for sediment wherever changes in the stratigraphy were observed.

Sediment samples were analysed for particle size after using a 2 mm sieve to remove large pieces of organic material. The sediment was mixed into slurry using distilled water in 100 mL beakers and placed into the 1 L well of a Malvern Mastersizer 2000 particle-size instrument with 10 mL of 10% NaHMP until the laser obscuration reached 10–12%; the pump was set to 3,000 rpm. Most results showed that this method underestimated the amount of clay and silt, as these grain sizes were largely present as sand-sized pellets. Most of the samples were therefore reanalysed after dispersing the clay material by placing 50 g of sediment in 100 mL beakers with 80 mL of distilled water for four days, stirring vigorously once per day. The samples were then rerun on the Mastersizer 2000 using the same procedure, but with 20 mL of 10% NaHMP added as a dispersant and the ultrasonic displacement set to 20% to further disaggregate the pellets.

Charcoal was also collected from two samples manually under a Zeiss dissecting microscope. Given the small sample sizes, AMS radiocarbon dating was used.

13.5 Results: Megafan Morphology and Sedimentology

Using the radiometric image, the DEM, and satellite and LiDAR mapping of distributary systems, three megafans were identified in the central part of the Victorian Riverine Plains, stretching northwards from the Western Victorian Uplands to the Murray River, and built by the Loddon River, Bullock Creek, and Campaspe River (Table 13.1; Figs. 13.1 and 13.2). The three megafans are constrained to the east by the Heathcote ridge and to the west by the Gredgwin Ridge, which marks the eastern limit of the Pliocene marine transgression that inundated a large part of the Murray Basin (Figs. 13.1 and 13.2). The Bullock and Campaspe megafans are separated by the Terrick Terrick Range of Devonian granite to the north, and by a ridge of Ordovician metasediments in the south. The Loddon and Bullock megafans overlap (Figs. 13.2 and 13.3), so the boundary between the two is difficult to delineate in detail.

The fans vary in area (2,500–1,418 km^2), length (90–130 km from apex to toe), width (21–60 km), volume (84–165 km^3; Table 13.1) and gradient ($< 0.001°$; Table 13.1; Figs. 13.4 and 13.5), with their apices and toes at elevations of 160–220 m and 80–90 m, respectively (Figs. 13.4 and 13.5). Cross-profiles are gently convex, with the degree of transverse curvature decreasing down-fan, and the longitudinal profiles are very slightly concave (Fig. 13.5), typical of megafans (Hartley et al. 2010). The catchments in the uplands upstream of the apices have a maximum elevation of 746 m asl and vary in area and length from 4,732 km^2 and 310 km (Loddon River megafan) to 1,334 km^2 and 173 km (Bullock Creek megafan), respectively.

The trunk streams on the megafans are well-defined perennial streams that generally flow along the fan edges, apart from Bullock Creek, which flows through the centre of its megafan before it joins with the Loddon River megafan, whereupon it continues along the eastern margin of a possibly combined Loddon–Bullock megafan (as suggested by the dashed line dividing these two very similar surfaces; Figs. 13.2–13.5). A dense, interconnected network of shallow, sinuous channels is present along the Loddon and Campaspe trunk streams (Fig. 13.6). The trunk streams have incised 1–5 m, with depth of incision decreasing from apex to toe. Branching off the main streams is a network of ephemeral distributary channels that are generally single-thread and sinuous. These are commonly poorly defined or unconfined and consist of shallow depressions ranging in depth from several

Table 13.1 *Megafan and feeder-basin dimensions*

Megafan name	Length (km)	Area (km^2)	Sediment volume (km^3)	Gradient	Maximum width (km)	Upland catchment area (km^2)	Upland stream length (km)
Loddon	104	1,418	84	0.0006	22	4,732	310
Bullock	128	1,696	100	0.0009	21	1,334	173
Campaspe	91	2,506	165	0.0006	58	4,191	220

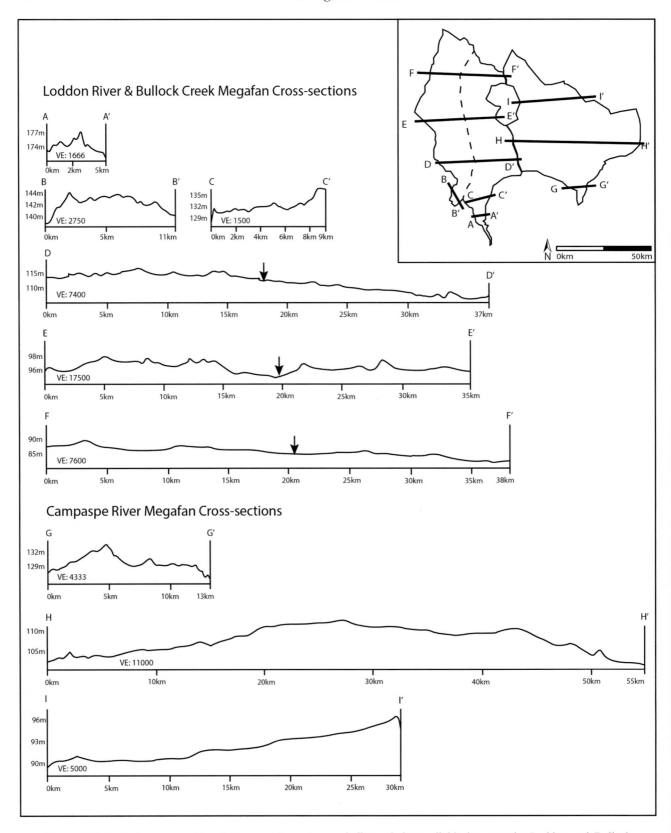

Figure 13.4 **Typical cross-profiles of the megafans.** Arrows indicate drainage divide between the Loddon and Bullock megafans.

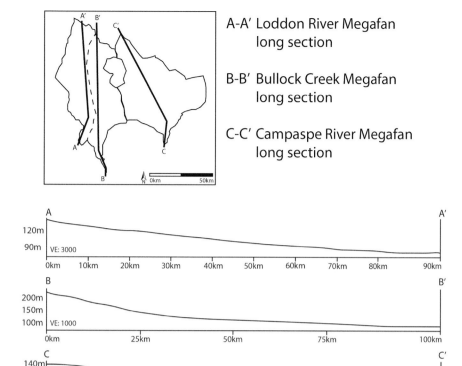

A-A' Loddon River Megafan
long section

B-B' Bullock Creek Megafan
long section

C-C' Campaspe River Megafan
long section

Figure 13.5 **Longitudinal profiles of the Loddon, Bullock and Campaspe megafans**.

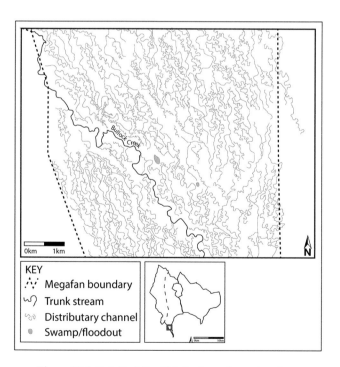

Figure 13.6 **Detail of the distributary channel system on Bullock Creek megafan**. Note complex system of interconnected sinuous channels.

centimetres to no more than 1 metre, and decrease in depth from south to north along the fans. They may not be visible in the field but can be identified using LiDAR imagery. On true-colour satellite images they are evident as discolourations on the surface of the fans or as a meandering line of trees. Shallow, connected channels sometimes contain gilgai along their courses. Distributary channels may be discontinuous; channel sections are separated by, or terminate in, floodouts and/or small swamps with characteristic vegetation such as rushes, sedges and swamp shrubs. Floodouts form where channels lose their ability to maintain a channelised course and distribute water and sediment across an expanded area; they are common in losing streams in arid and semi-arid environments (Tooth 1999).

The location of the Loddon River along the western margin of the Loddon megafan may be a consequence of westward tilting along the Muckleford Fault (Fig. 13.2), which lies between the apices of the Loddon and Bullock megafans (Fig. 13.2) and has recorded at least 50 m of post-Neogene vertical movement (Bowler and Harford 1966; Cherry and Wilkinson 1994). The Cadell Fault, which is located under the toe of the Campaspe megafan (Fig. 13.2), underwent up to 20 m of vertical movement at 80–60 ka (Clark et al. 2007), and this has caused a local offset of the megafan surface (Fig. 13.4, cross-profile I–I').

The megafan sediments, as exposed in the stream cuttings and in the auger holes, are dominated by thin, flat-lying beds of fine-grained material, mostly silt in the middle and toe areas of the fans; clay is a minor component, with a slight increase from apex to toe (Figs. 13.7 and 13.8; Table 13.2). Following disaggregation of samples, in many cases the amount of silt and clay increased, and the percentage of sand decreased, due to the presence of sand-sized pellets of clay and silt.

Sand is most abundant at the fan apices (Figs. 13.7 and 13.8; Table 13.2), where sand beds may show cross-bedding (Spring Creek stream site). Gravel is lacking. Similar sand-dominated sediments are common within the Shepparton Formation across the Riverine Plains, and represent channel deposits (Lawrence 1966; Macumber 1969).

Charcoal from near the apex of the Loddon and Bullock megafans (Fig. 13.7) gave AMS ages of 150 cal yr BP ± 30 from a depth of 1.26 m, and 7270 ± 40 yr BP from a depth of 90 cm.

13.6 Discussion

13.6.1 Sediment Source

The grain size and radiometric signature of the sediments of the three megafans are determined by the

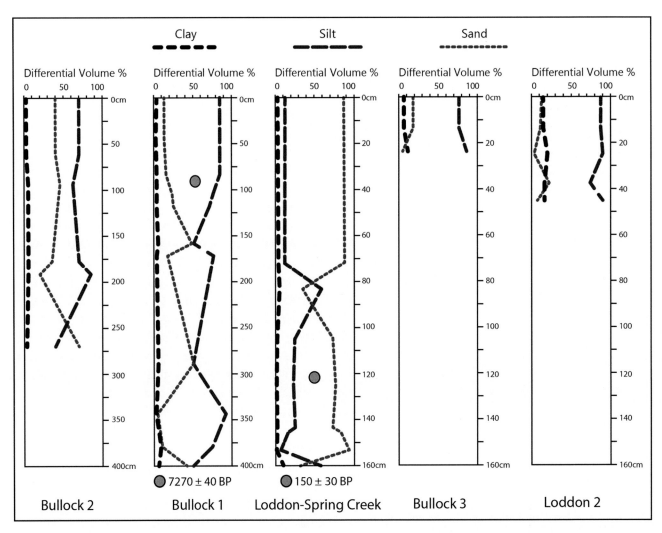

Figure 13.7 **Grain-size profiles on the Loddon and Bullock fans.** Test pit locations in Figure 13.3.

Table 13.2 *Sediment grain-size data*

Megafan name	Site location (see Fig. 13.3)	Distance from apex (km)	Disaggregated			Undisaggregated		
			Sand %	Silt %	Clay %	Sand %	Silt %	Clay %
Bullock	Bullock 1	25	23	73	4	30	64	7
	Bullock 2	12	36	62	2	39	59	1
	Bullock 3	49	10	80	10	n.a.	n.a.	n.a.
Loddon	Loddon 2	92	11	74	15	52	45	3
	Loddon-Spring Creek	29	69	27.9	3	73	24	3
Campaspe	Campaspe 6	12	20	75	4	18	78	5
	Campaspe 4	32	42	54	4	54	55	4
	Campaspe 3	53	6	78	16	17	78	4
	Campaspe 1	56	14	72	13	n.a.	n.a.	n.a.

Note. n.a.: not applicable (samples were not processed for aggregates)

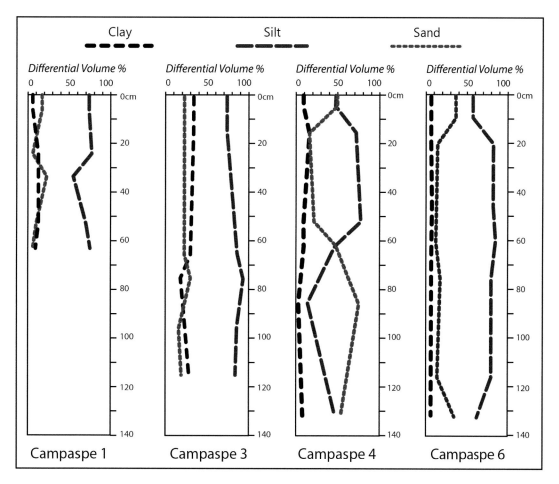

Figure 13.8 **Grain-size profiles on the Campaspe fan**. Test pit locations in Figure 13.3.

characteristics of the outcrops in the uplands source areas. The Ordovician sandstones and siltstones (Castlemaine Group) are deeply weathered, and because the siltstones are more easily eroded, the regolith is clay- and silt-dominated (Edwards et al. 2001), and therefore supplies mostly fine-grained sediment to the megafans. In addition, a component of aeolian quartz silt contains a substantial proportion of grains in the 20–50 μm size range. This is typical of the aeolian mantle of silt deposited across the uplands catchments during the dry, windy conditions of the cooler periods of the Quaternary, particularly the LGM (Cattle et al. 2002, 2009). The material was then washed onto the megafans by floods and mixed into the soil by illuviation (Tickell and Humphrys 1987; Chen et al. 2002; Harvey 2011).

The radiometric signature of the megafan sediments reflects that of the source materials. The dominant clay in the Ordovician sediments, and therefore the regolith developed on them, is illite (VandenBerg et al. 2000), which contains high levels of potassium (K) as well as moderate amounts of thorium (Th) and uranium (U). As a result, this material appears pale pink-purple in radiometric images (Fig. 13.3) (Edwards et al. 2001; Cayley et al. 2008). The Devonian granites have a red or pale-coloured radiometric signature due to, respectively, high levels of K or all three radiometric elements; the kaolinised sandy regolith on the granites retains this radiometric signature (Hill 1996; Edwards et al. 2001). The outcrops of Ordovician sediments on the northern flanks of the uplands are often strongly ferruginised and have a high Th signature with low K and U radiometric responses, i.e., green on the radiometric image (Fig. 13.3) (Kotsonis and Joyce 2003). The Newer Volcanics basalt flows present in the valleys of the Loddon and Campaspe rivers have weathered to clay-rich soils, which are blue-green on the radiometric images because they lack K (Fig. 13.3) (Kotsonis and Joyce 2003).

Overall, the radiometric colour of the megafans is a pale-yellow green (Fig. 13.3), reflecting the dominance of weathered (pale pink-purple) and ferruginised (green) Ordovician sediments in the catchments. The component of aeolian quartz silt in the megafan sediments does not affect their radiometric signature, as quartz lacks significant levels of any radiometric elements. The apex of the Bullock Creek megafan has a lighter radiometric colour than the other megafans due to the presence of Devonian granites in the catchment with a very pale-coloured radiometric signature.

13.6.2 Geomorphic and Sedimentary Processes

The Riverine Plains of northern Victoria contain three large, smooth, fan-like planforms that are 90–120 km long and have gradients of only 0.0006–0.0009 (Table 13.1); they comprise thin-bedded, fine-grained fluvial sediments, and can be classed therefore as megafans. The size of the Victorian megafans is largely a result of the virtually unconfined plain on which they have developed. Megafans elsewhere in the world typically lie adjacent to high relief mountain ranges, have steeper gradients and are dominated by sands or sands and gravels, with minor clays and silts (Ellery et al. 1993; Singh et al. 1993; Kar et al. 2014). The finer-grained sediments of the Victorian megafans reflect the dominance of weatherable siltstones and aeolian silt in the source areas (as discussed above). The very shallow gradients of the Victorian megafans are probably due at least partly to the relatively low elevation of the upland catchments, resulting from limited uplift during the intermittent Cenozoic tectonic history of Australia's Southeastern Highlands (Jansson 1988; Olive et al. 1994; Prosser et al. 2001; Braun et al. 2009; Webb et al. 2011). In addition, the megafans receive a comparatively small sediment supply because of low erosion rates in the uplands, which have limited relief.

The absence of gravel in the megafan sediments (Figs. 13.7 and 13.8) reflects both the low velocity of flood flow on the fans and the relatively small amount of gravel generated in the upland catchments, as a result of their low elevation and the absence of Quaternary glaciation. This contrasts with the large amounts of gravel in New Zealand and northern hemisphere fans, where glacial conditions have typified the Quaternary cool periods there (Hesse et al. 2004; Williams et al. 2009; Petherick et al. 2013).

Sediment transport on the megafans is intermittent, reflecting the flow regime of the feeder rivers, which is characterised by low base flow interspersed with a few large discharge events, typical of many Australian rivers (McMahon et al. 1987; Dettinger and Diaz 2000; Prosser et al. 2001; McMahon and Finlayson 2003). Higher-flow events (flood peaks) are represented in the stratigraphic profiles by individual peaks of fine and medium grained sand (Figs. 13.7 and 13.8). Sites dominated by sand (Loddon–Spring Creek; Fig. 13.7) are probably close to a perennial channel and received sediment during periods of lower stream discharge and more confined flow.

The general decrease in particle size from apex to toe of the megafans (Figs. 13.7 and 13.8) is probably ascribable to progressive depletion of the coarser silt fraction of the suspended load as it is deposited across the fan surface. This is related to the down-fan reduction in gradient and decrease in velocity of the water flow (Allen 1965) as flood pulses spread across the surfaces of the megafans, which lack the confinement of deeper channels at a local scale, and at a regional scale in mountain ranges.

The unconfined flood flow on these fans has also contributed to the merging of the Loddon River and Bullock Creek megafans, which are not separated by bedrock hills and ridges (Fig. 13.3). The boundary between these fans is unlikely to be stable temporally or geographically; coalescence of the two megafans at their midpoints has created a complex depositional feature that could be regarded as a type of bajada (a series of coalescing alluvial fans along a mountain front).

All the megafans of eastern and south-eastern Australia have formed where streams debouch from low elevation, low relief uplands onto broad, flat, poorly-confined plains (Bull 1977), including the Murrumbidgee, Lachlan, and Gwydir rivers (Langford-Smith 1960; Schumm 1968; Gingele and de Deckker 2005; Pietsch 2006; Page et al. 2009). Surface flow from the uplands decreases in velocity as it spreads across the extensive, very low gradient plains as shallow, virtually unconfined lateral flow, depositing sediment to form the megafans. This lateral movement of water is facilitated by the presence of shallow distributary channels, which, on the Victorian megafans, decrease in depth and width down-fan.

Floodouts are present along the ephemeral distributary channels as areas of lower gradient than the feeder stream, where water infiltration during floods creates a relatively stable wetland with swamp vegetation, including sedges and rushes (Wakelin-King and Webb 2007). Recurrent floods are required to maintain these features in the landscape, because the lack of water leads to dieback of the stabilising vegetation, thereby changing the floodout from a depositional feature to an erosional one as the flows are no longer slowed by vegetation (Wakelin-King and Webb 2007).

The sand-sized pellets of clay and silt present in the megafan sediments are common within Riverine Plain fluvial sediments and have also been recorded from aeolian regolith in Australia (Beattie 1970; Bowler 1973; Blackburn 1981). These aggregates form under seasonally hot, dry climates from high smectite clay soils such as vertisols, which are characterised by their capacity to swell and shrink upon wetting and drying, respectively (Rust and Nanson 1989; Maroulis and Nanson 1996; Wakelin-King and Webb 2007). Such soils occur commonly on the Victorian megafans. The pellets are generally robust and survive transport in the water column for several days; upon deposition and subsequent drying, they can merge to produce a massive mud deposit (Nanson et al. 1986; Maroulis and Nanson 1996; Wakelin-King and Webb 2007).

The depositional history of the Riverine Plains of northern Victoria has been strongly influenced by climate (De Caritat et al. 2007), and this is particularly true of the fans, where deposition is controlled in large part by processes occurring in the upland catchments where high rainfall events produce floods largely responsible for formation of the megafans. The intensity and frequency of these events determines the rate of sedimentation on the fans. Furthermore, the climate has also influenced sediment supply through control of vegetation density. Sparser vegetation cover in the upland catchments during Quaternary arid periods probably promoted seasonally larger stream discharges and therefore more sediment deposition (Page and Nanson 1996; Kemp and Rhodes 2010; Petherick et al. 2013; Pietsch et al. 2013). In addition, wildfire can dramatically reduce the amount of vegetation in the short term, increasing mass-wasting events on the catchment slopes and greatly increasing the transport of sediment onto the fans (Tomkins et al. 2007).

The influence of climate is also evident in the morphology of the elevated distributary channels on the megafan surfaces. These channels can be divided into prior streams (active around 100–40 ka) and ancestral streams (active from 40 to a 15–10 ka; Fried 1993). The change from prior to ancestral channels occurred when large mixed-bedload streams shifted to smaller suspended-load streams across much of the Riverine Plain in response to increasing aridity (Butler 1950; Schumm 1968; Fried 1993; Page et al. 2009). During this period of change, the paths of the sinuous distributary channels would have been in constant flux, with avulsion events changing the courses of streams until a complex network of channels evolved. Over time, as the streams responded to climatic forcing,

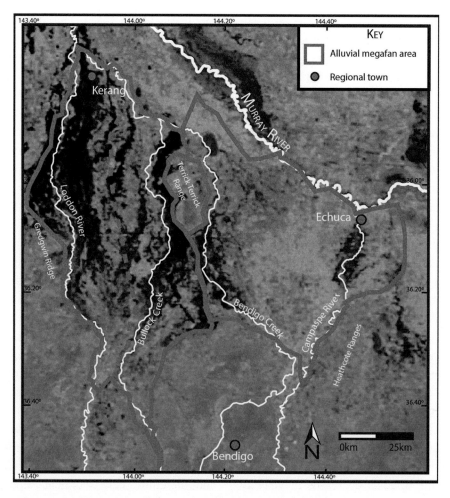

Figure 13.9 **MODIS/Terra satellite image of the January 2011 Loddon River floods**. Note the active distributary channels on the megafan (black) that are gathered into tributary channels towards the fan toe (NASA 2011). For location, see Figure 13.1. A colour version of this figure is available in the SOD for Chapter 13.

the distributary channels became ephemeral and the main rivers became entrenched (Macumber 1969; Edwards et al. 2001). The ephemeral streams now only carry water during periods of high discharge, as observed during the 2011 floods on the Loddon River, when the distributary channels transported large volumes of water through the fan system and into the Murray River (Fig. 13.9).

13.6.3 Depositional History

The predominantly fine-grained Shepparton Formation sediments of the megafans directly overlie, and in places interfinger with, the coarser-grained late Neogene Calivil Formation, a predominantly sand and gravel unit formed as river valley deposits (Brown and Stephenson 1991; VandenBerg 2009).

This change in sediment character records the transition from braided channels to more widespread megafan deposition in the late Pliocene, when the Shepparton Formation began to accumulate across the Victorian Riverine Plain (Lawrence 1966; Lawrence et al. 1976; Edwards et al. 2001).

Deposition of the Shepparton Formation has continued to the present, as shown by two charcoal ^{14}C dates obtained from the upper 2 m of sediments near the apices of the Loddon and Bullock megafans (Fig. 13.7). The older date (7270 ± 40 yr BP from Bullock Creek 1) represents sediment deposition during the Middle Holocene Climatic Optimum, a period of increased precipitation and stream discharge from 9 to 6 ka (Cohen and Nanson 2007). These conditions ended when aridity increased gradually from 4.5 ka to the present (Bowler et al. 1976; Calvo et al. 2007; Cohen and Nanson 2007; Nanson et al. 2003; Page et al. 2009; Petherick et al. 2013; Pietsch et al. 2013).

The much younger carbon date from Loddon-Spring Creek (150 ± 30 cal yr BP at 1.2 m depth; Fig. 13.7), at the apex of the Loddon megafan, indicates deposition after European settlement in Victoria. From 1830 onwards, much of the Victorian landscape, including the Riverine Plains, was cleared for grazing sheep and cattle (Butler et al. 1973). The shift from largely soft-footed marsupial herbivores to abundant hard-hooved mammal herbivores forced a significant modification of the vegetation, resulting in increased soil erosion, including along the main streams, despite no significant shift in climatic regime over the same time period (Butler et al. 1973; Dodson and Mooney 2002). This landscape instability was exacerbated by the advent of the Victorian Gold Rush during the 1850s and 1860s in the upper catchments of the Loddon and Campaspe Rivers, and to a lesser extent Bullock Creek (Dodson and Mooney 2002). The high discharge variability of rivers in central Victoria meant that water management in the form of water races, dams, and sluices was required for gold mining (Lawrence and Davies 2012). These large-scale modifications of the natural hydrological system, in combination with gold mining itself, resulted in the widespread stripping of vegetation from the landscape (Garden 2001; Abernethy et al. 2004; Lawrence and Davies 2012). Within a short period of time, runoff increased, transporting large amounts of soil to the streams (Garden 2001; Abernethy et al. 2004). Increased flood frequency was recorded in the Loddon River and Bullock Creek catchments between 1840 and 1860, with a large flood occurring in 1870 in the Loddon Catchment (Abernethy et al. 2004). During these years of landscape instability and higher-than-average precipitation, extensive siltation was recorded across the Loddon catchment (Abernethy et al. 2004). These processes resulted in >1 m of sediment deposition at the Loddon–Spring Creek site, and probably across large areas of the Loddon and Campaspe megafans.

13.7 Conclusions

Airborne radiometric data, distributary channel mapping, sediment grain size analysis, digital elevation model manipulation and AMS dating were used to determine the geometry and depositional processes of three large (90–120 km long), very low gradient ($<0.001°$) fluvial megafans in the northern Victorian Riverine Plains. Radiometric imaging of the relative proportions of K, Th and U in the upper parts of the sediment profile of the three megafans indicates that a mix of sediment inputs from exposed bedrock in the catchments of the megafans were deposited during episodic flooding. Weathering of the upland geology produced predominantly fine-grained material, which was further supplemented by aeolian inputs deposited on the slopes of the catchments in periods of aeolian activity (i.e., the Last Glacial Maximum). The transported sediments decrease in particle size down-fan as the coarser sand and silt particles are depleted, and are dominantly silt and clay; this differs from wedge-top megafans in foreland basins which are predominantly sand- and gravel-dominated facies. As with many eastern Australian examples, the megafans are not associated with high-relief mountain ranges or past glacial activity; instead, they occur on broad, flat plains adjacent to low elevation and low relief uplands which were not glaciated in the Cenozoic. Climatic controls have played a significant role in the formation of the megafans, in part due to vegetation dynamics across the catchments as a whole, and also because the intensity and frequency of high rainfall events determines the rate of sedimentation on the fans. Climate-controlled changes in discharge are also responsible for the change in distributary channel form, from large mixed bedload streams to smaller suspended-load streams, which occurred ca. 40 ka. The Shepparton Formation sediments of which the megafans are composed were deposited from the late Pliocene onward following a transition from braided channel deposition of the underlying Calivil Formation. Charcoal from the upper 2 m of sediment near the apices of the Loddon and Bullock megafans gave radiocarbon ages of 150 cal yr BP and 7270 yr BP. The younger Loddon megafan age is attributed to landscape modification following European settlement (including gold mining) in the fan catchment, while the older date from the Bullock Creek megafan resulted from higher discharge during the Holocene Climatic Optimum.

Acknowledgements

We wish to acknowledge all who offered their advice and time to making this research possible. Our thanks go to Andrew Cossey for his generosity in allowing us to compare his alluvial fan data to ours, and for showing us his study sites. Also, many thanks to Gresley Wakelin-King for her invaluable feedback on the draft version. We also wish to acknowledge the valuable effort of our field assistants, Margaret Reith, Fiona Glover, Susan Kapteinis, for giving up their weekends all the while wondering where the text-book examples of cone-shaped alluvial fans were in a very flat landscape. Our thanks also go to the La Trobe University Environmental Geoscience group for providing access to laboratory facilities.

References

Abernethy, B., Markham, A. J., Prosser, I. P., and Wansbrough, T. M. (2004). A sluggish recovery: the indelible marks of landuse change in the Loddon River catchment. In: *Fourth Australian Stream Management Conference: Linking Rivers to Landscapes*. Launceston, Tasmania, 19–22.

Allen, J. R. (1965). A review of the origin and characteristics of recent alluvial sediments. *Sedimentology*, 5, 89–11.

Ash, J. E. and Wasson, R.J. (1983). Vegetation and sand mobility in the Australian desert dunefield. *Zeitschrift für Geomorphologie*. 45, 7–25.

Beattie, J. A. (1970). Peculiar features of soil development in parna deposits in the Eastern Riverina, NSW. *Soil Research*, 8, 145–156.

Blackburn, G. (1981). Particle-size analyses of Widgelli parna in south-east Australia. *Australian Journal of Soil Research*, 19, 355–360.

Bowler, J. M. (1973). Clay dunes: their occurrence, formation and environmental significance. *Earth-Science Reviews*, 9, 315– 338.

Bowler, J. M. and Harford, L. B. (1966). Quaternary tectonics and the evolution of the riverine plain near Echuca, Victoria. *Journal of the Geological Society of Australia*, 13, 339–354.

Bowler, J. M., Hope, G. S., Jennings, J. N. Singh, G., and Walker, D. (1976). Late Quaternary climates of Australia and New Guinea. *Quaternary Research*, 6, 359–394.

Bowler, J. M, Kotsonis, A., and Lawrence, C. R. (2006). Environmental evolution of the Mallee region, Western Murray Basin. *Proceedings of the Royal Society of Victoria*, 118, 161–210.

Braun, J., Burbidge, D. R., Gesto, F. N., et al. (2009). Constraints on the current rate of deformation and surface uplift of the Australian continent from a new seismic database and low-T thermochronological data. *Australian Journal of Earth Sciences*, 56, 99–110.

Brooks, A. P., Shellberg, J. G., Knight, J., and Spencer, J. (2009). Alluvial gully erosion: an example from the Mitchell River fluvial megafan, Queensland, Australia. *Earth Surface Processes and Landforms*, 34, 1951–1969.

Brown, C. M. and Stevenson, A. E. (1991). *Geology of the Murray Basin, southeastern Australia*. Australian Government Publishing Service, Canberra.

Bull, W. B. (1977). The alluvial-fan environment. *Progress in Physical Geography*, 1, 222–270.

Bureau of Meteorology (2011). Average Annual Rainfall Map. www.bom.gov.au/jsp/ncc/climate_averages/rainfall/index.jsp?period=anandarea=vc#maps

Butler, B. E. (1950). Theory of prior streams as a causal factor of soil occurrence in the Riverine Plain of south-eastern Australia. *Australian Journal of Agricultural Research*, 1, 231–252.

Butler, B. E., Blackburn, G., Bowler, J. M., et al. (1973). *A Geomorphic Map of the Riverine Plain of South-Eastern Australia*. Australian National University Press, Canberra.

Butler, B. E. and Hubble, G. D. (1978). The general distribution and character of soils in the Murray-Darling River system. *Proceedings of the Royal Society of Victoria*, 90, 149–156.

Butler, B. E. and Hutton, J. T. (1956). Parna in the Riverine Plain of south-eastern Australia and the soils thereon. *Crop and Pasture Science*, 7, 536–553.

Calf, G. E, Ife, D., Tickell, S., and Smith, L. W. (1986). Hydrogeology and isotope hydrology of Upper Tertiary and Quaternary aquifers in Northern Victoria. *Australian Journal of Earth Sciences*, 33, 19–26.

Calvo, E., Pelejero, C., De Deckker, P., and Logan, G. (2007). Antarctic deglacial patterns in a 30 kyr record of sea surface temperature offshore South Australia. *Geophysical Research Letters*, 34, L130707.

Cattle, S. R., McTainsh, G. H., and Wagner, S. (2002). Aeolian dust contributions to soil of the Namoi Valley, northern NSW, Australia. *Catena*, 47, 245–264.

Cattle, S. R., Greene, R. S. B., and McPherson, A. A. (2009). The role of climate and local regolith-landscape processes in determining the pedological characteristics of aeolian dust deposits across south-eastern Australia. *Quaternary International*, 209, 95–106.

Cayley, R. A., Skladzien, P. B., Williams, B., and Willman, C. E. (2008). *Redesdale and part of Pyalong, 1:50,000 geological map report 128*. Geological Survey of Victoria, Melbourne, Australia.

Chen, X. Y. (2001). The red clay mantle in the Wagga Wagga region, New South Wales: evaluation of an aeolian dust deposit (Yarabee Parna) using methods of soil landscape mapping. *Australian Journal of Soil Research*, 39, 61–80.

Chen, X. Y., Spooner, N. A., Olley, J. M., and Questiaux, D. G. (2002). Addition of aeolian dusts to soils in southeastern Australia: red silty clay trapped in dunes bordering Murrumbidgee River in the Wagga Wagga region. *Catena*, 47, 1–27.

Cherry, D. P. and Wilkinson, H. E. (1994). *Bendigo, and part of Mitiamo, 1:100,000 geological map report 99.* Geological Survey of Victoria.

Clark, D., McPherson, A., and Collins, C. D. N. (2011). *Australia's Seismogenic Neotectonic Record: A Case for Heterogeneous Intraplate Deformation.* Geoscience Australia Record 2011/11. Geoscience Australia, Canberra.

Clark, D., Van Dissen, R., Cupper, M., Collins, C., and Prendergast, A. (2007). Temporal clustering of surface ruptures on stable continental region faults: a case study from the Cadell Fault scarp, southeastern Australia. In: *Proceedings of the Australian Earthquake Engineering Society Conference*, 23–25 November 2007, Wollongong, Paper 17.

Cohen, T. J. and Nanson, G. C. (2007). Mind the gap: an absence of valley-fill deposits identifying the Holocene hypsithermal period of enhanced flow regime in southeastern Australia. *The Holocene*, 17, 411–418.

Cupper, M. L., White, S., and Neilson, J. L. (2003). Quaternary: ice ages – environments of change. In W. D. Birch, ed., *Geology of Victoria*. Geological Society of Australia, Special Publication, 23, 337–360.

De Caritat, P. Lech, M. E., Jaireth, S., Pyke, J., and Fisher, A. (2007). *Riverina Region Geochemical Survey, Southern New South Wales and Northern Victoria.* CRC LEME Open File Report 234.

Department of Environment and Primary Industries (2009–2010). *2009–2010 Victorian State Wide Rivers Lidar Project.* Department of Environment and Primary Industries, Melbourne, Victoria, Australia.

Department of Primary Industries (2003). *Radiometric Ternary (K, Th, U) Image (1:1,000,000).* Department of Primary Industries, Melbourne, Victoria, Australia.

Dettinger, M. D. and Diaz, H. F. (2000). Global characteristics of stream flow seasonality and variability. *Journal of Hydrometeorology*, 1, 289–310.

Dodson, J. R. and Mooney, S. D. (2002). An assessment of historic human impact on south-eastern Australian environmental systems, using late Holocene rates of environmental change. *Australian Journal of Botany*, 50, 455–464.

Edwards, J., Slater, K. R., and McHaffie, I. W. (2001). *Bendigo 1:250 000 map area geological report.* Victorian Initiative for Minerals and Petroleum Report 72. Department of Natural Resources and Environment, Melbourne, Victoria, Australia.

Ellery, W. N., Ellery, K., Rogers, K. H., McCarthy, T. S., and Walker, B. H. (1993). Vegetation, hydrology and sedimentation processes as determinants of channel form and dynamics in the northeastern Okavango Delta, Botswana. *African Journal of Ecology*, 31, 10–25.

Fried, A. W. (1993). Late Pleistocene river morphological change, southeastern Australia: the conundrum of sinuous channels during the Last Glacial Maximum. *Palaeogeography, Palaeoclimatology, Palaeoecology*, 101, 305–316.

Garden, D., (2001). Catalyst or cataclysm? Gold mining and the environment. *Victorian Historical Journal*, 72, 28–44.

Geoscience Australia (2011). SRTM-derived 1 Second Digital Elevation Models Version 1.0. *Geoscience Australia*, Commonwealth of Australia.

Gingele, F. X. and De Deckker, P. (2005). Clay mineral, geochemical and Sr–Nd isotopic fingerprinting of sediments in the Murray–Darling fluvial system, southeast Australia. *Australian Journal of Earth Sciences*, 52, 965–974.

Haberlah, D. (2007). A call for Australian loess. *Area*, 39, 224–229.

Hartley, A. J., Weissmann, G. S., Nichols, G. J., and Warwick, G. L. (2010). Large distributive fluvial systems: characteristics, distribution, and controls on development. *Journal of Sedimentary Research*, 80, 167–183.

Harvey, A. (2011). Dryland alluvial fans. In D. S. G. Thomas, ed., *Arid Zone Geomorphology: Process, Form and Change in Drylands*. Wiley, Chichester, 333–371.

Hesse, P. P. and McTainsh G. H. (2003). Australian dust deposits: modern processes and the Quaternary record. *Quaternary Science Reviews*, 22, 2007–2035.

Hesse, P. P., Magee, J. W., and Van Der Kaars, S. (2004). Late Quaternary climates of the Australian arid zone: a review. *Quaternary International*, 118, 87–102.

Hill, S. M. (1996). The differential weathering of granitic rocks in Victoria, Australia. *AGSO Journal of Australian Geology and Geophysics*, 16, 271–276.

Hill, S. M. (1999). Mesozoic regolith and palaeolandscape features in southeastern Australia: significance for interpretations of denudation and highland evolution. *Australian Journal of Earth Sciences*, 46, 217–232.

Hills, E. S. (1961). Morphotectonics and the geomorphological sciences with special reference to Australia. *Quarterly Journal of the Geological Society*, 117, 77–90.

Jansson, M. B. (1988). A global survey of sediment yield. Geografiska Annaler. *Series A, Physical Geography*, 70, 81–98.

Joyce, E. B., Webb, J. A., Dahlhaus, P. G., et al. (with material by the late Jenkin, J. J.) (2003). Geomorphology: the evolution of Victorian landscapes. In W. D. Birch, ed., *Geology of Victoria*. Geological Society of Australia, Special Publication, 23, 533–561.

Kar, R., Chakraborty, T., Chakraborty, C., et al. (2014). Morpho-sedimentary characteristics of the Quaternary Matiali fan and associated river terraces, Jalpaiguri, India: Implications for climatic controls. *Geomorphology*, 227, 137–152.

Kemp, J. and Rhodes, E. J. (2010). Episodic fluvial activity of inland rivers in southeastern Australia: Palaeochannel systems and terraces of the Lachlan River. *Quaternary Science Reviews*, 29, 732–752.

Kershaw, A. P. and Nanson, G. C. (1993). The last full glacial cycle in the Australian region. *Global and Planetary Change*, 7, 1–9.

King, R. L. (1986). *Explanatory notes on the Ballarat 1:250,000 geological map*. Geological Survey of Victoria Report 75. Department of Industry and Resources, Melbourne, Victoria, Australia.

Kotsonis, A. and Joyce, E. B. (2003). *The regolith of the Bendigo 1:100 000 map area*. Victorian Initiative for Minerals and Petroleum Report 77. Department of Primary Industries, Melbourne, Victoria, Australia.

Langford-Smith, T. (1960). The dead river systems of the Murrumbidgee. *Geographical Review*, 50, 368–389.

Lawrence, C. R. (1966). Cainozoic stratigraphy and structure of the Mallee Region, Victoria. *Proceedings of the Royal Society of Victoria*, 79, 517–554.

Lawrence, C. R. (1975). *Geology, Hydrodynamics and Hydrochemistry of the Southern Murray Basin*. Geological Survey of Victoria Memoir, 30.

Lawrence, C. R., Macumber, P. G., Kenley, P. R., et al. (1976). Quaternary. In J. G. Douglas and J. A. Ferguson, eds., *Geology of Victoria*. Geological Society of Australia (Victorian Division), Special Publication, 5, 275–325.

Lawrence, S. and Davies, P. (2012). Learning about landscape: Archaeology of water management in colonial Victoria. *Australian Archaeology*, 74, 47–54.

Macumber, P. G. (1969). Interrelationship between physiography, hydrology, sedimentation, and salinization of the Loddon River Plains, Australia. *Journal of Hydrology*, 7, 39–57.

Macumber, P. G. (1991). *Interaction between Groundwater and Surface Systems in Northern Victoria*. Department of Conservation and Environment, Melbourne, Victoria, Australia.

Macumber, P. G. and Macumber, J. J. (2010). Groundwater flow in the Campaspe and Loddon Valleys of Northern Victoria: an enhanced role for the Shepparton Formation. *Proceedings of the Royal Society of Victoria*, 122, 43–69.

Maroulis, J. C. and Nanson, G. C. (1996). Bedload transport of aggregated muddy alluvium from Cooper Creek, central Australia: a flume study. *Sedimentology*, 43, 771–790.

McMahon, T. A. and Finlayson, B. L. (2003). Droughts and anti-droughts: the low flow hydrology of Australian rivers. *Freshwater Biology*, 48, 1147–1160.

McMahon, T. A., Finlayson, B. L., Haines, A., and Srikanthan, R. (1987). Runoff variability: a global perspective. In S. I. Solomon, M. Beran, and W. Hogg, eds., *The Influence of Climate Change and Climatic Variability on the Hydrologic Regime and Water Resources*. IAHS Publication, 168, 3–11.

Nanson, G. C., Cohen, T. J., Doyle, C. J., and Price, D. M. (2003). Alluvial evidence of major late-Quaternary climate and flow-regime changes on the coastal rivers of New South Wales, Australia. In K. Gregory and G. Benito, eds., *Palaeohydrology: Understanding Global Change*. Wiley, Chichester, 233–258.

Nanson, G. C., Rust, B. R., and Taylor, G. (1986). Coexistent mud braids and anastomosing channels in an arid zone river: Cooper Creek, Central Australia. *Geology*, 14, 175–178.

Nanson, R. A., Vakarelov, B. K., Ainsworth, R. B., Williams, F. M., and Price, D. M. (2013). Evolution of a Holocene, mixed-process, forced regressive shoreline: The Mitchell River delta, Queensland, Australia. *Marine Geology*, 339, 22–43.

Olive, L. J., Olley, J. M., Murray, A. S., and Wallbrink, P. J. (1994). Spatial variation in suspended sediment transport in the Murrumbidgee River, New South Wales, Australia. In L. J. Olive, R. J. Loughran, and J. A. Kesby, eds., *Variability in Stream Erosion and Sediment Transport*. IAHS Publication, 224, 241–250.

Ollier, C. D. (1995). Tectonics and landscape evolution in southeast Australia. *Geomorphology*, 12, 37–44.

Page, K. J., Kemp, J., and Nanson, G. C. (2009). Late Quaternary evolution of Riverine Plain paleochannels, southeastern Australia. *Australian Journal of Earth Sciences*, 56, S19–S33.

Page, K. J. and Nanson, G. C. (1996). Stratigraphic architecture resulting from Late Quaternary evolution of the Riverine Plain, south-eastern Australia. *Sedimentology*, 43, 927–945.

Peel, M. C., Finlayson, B. L., and McMahon, T. A. (2007). Updated world map of the Köppen–Geiger climate classification. *Hydrology and Earth System Sciences*, 4, 439–473.

Pels, S. (1971). River systems and climatic changes in southeastern Australia. In D. J. Mulvaney and J. Golson, eds., *Aboriginal Man and Environment in Australia*. Australian National University Press, Canberra, 38–46.

Petherick, L., Bostock, H., Cohen, T. J., et al. (2013). Climatic records over the past 30 ka from temperate Australia–a synthesis from the Oz-INTIMATE workgroup. *Quaternary Science Reviews*, 74, 58–77.

Pietsch, T. J. (2006). *Fluvial geomorphology and Late Quaternary geochronology of the Gwydir fan-plain*. PhD thesis, School of Earth and Environmental Sciences, University of Wollongong, Wollongong, New South Wales, Australia.

Pietsch, T. J. and Nanson, G. C. (2011). Bankfull hydraulic geometry; the role of in-channel vegetation and downstream declining discharges in the anabranching and distributary channels of the Gwydir distributive fluvial system, southeastern Australia. *Geomorphology*, 129, 152–165.

Pietsch, T. J., Nanson, G. C., and Olley, J. M. (2013). Late Quaternary changes in flow-regime on the Gwydir distributive fluvial system, southeastern Australia. *Quaternary Science Reviews*, 69, 168–180.

Poff, N. L., Olden, J. D., Pepin, D. M., and Bledsoe, B. P. (2006). Placing global stream flow variability in geographic and geomorphic contexts. *River Research and Applications*, 22, 149–166.

Price, R. C., Nicholls, I. A., and Gray, C. M. (2003). Cainozoic igneous activity. In W. D. Birch, ed., *Geology of Victoria*. Geological Society of Australia, Special Publication, 23, 361–375.

Prosser, I. P., Rutherfurd, I. D., Olley, J. M., et al. (2001). Large-scale patterns of erosion and sediment transport in river networks, with examples from Australia. *Marine and Freshwater Research*, 52, 81–99.

Riley, S. J. and Taylor, G. (1978). The geomorphology of the Upper Darling River System with special reference to the present fluvial system. *Proceedings of the Royal Society of Victoria*, 90, 89–102.

Robson, T. C. and Webb, J. A. (2011). Late Neogene tectonics in northwestern Victoria: evidence from the Late Miocene-Pliocene Loxton Sand. *Australian Journal of Earth Sciences*, 58, 579–586.

Rust, B. R. and Nanson, G. C. (1989). Bedload transport of mud as pedogenic aggregates in modern and ancient rivers. *Sedimentology*, 36, 291–306.

Schumm, S. A. (1968). River adjustment to altered hydrologic regimen - Murrumbidgee River and paleochannels, Australia. *Geological Survey Professional Paper*, 598, 65 pp.

Shellberg, J. G., Brooks, A. P., Spencer, J., and Ward, D. (2012). The hydrogeomorphic influences on alluvial gully erosion along the Mitchell River fluvial megafan. *Hydrological Processes*, DOI: 10.1002/hyp.9240.

Singh, H., Parkash, B., and Gohain, K. (1993). Facies analysis of the Kosi megafan deposits. *Sedimentary Geology*, 85, 87–113.

Sleeman, J. R. (1975). Micromorphology and mineralogy of a layered red-brown earth profile. *Australian Journal of Soil Research*, 13, 101–117.

Tickell, S. J. and Humphrys, W. G. (1987). *Groundwater resources and associated salinity problems of the Victorian part of the Riverine Plain*. *Geological Survey of Victoria*. Report 84, Department of Industry Technology and Resources, Victoria, Melbourne, Australia.

Tomkins, K. M., Humphreys, G. S., Wilkinson, M. T., et al. (2007). Contemporary versus long-term denudation along a passive plate margin: the role of extreme events. *Earth Surface Processes and Landforms*, 32, 1013–1031.

Tooth, S. (1999). Downstream changes in floodplain character on the Northern Plains of arid central Australia. In N. D. Smith and J. Rogers, eds., *Fluvial Sedimentology VI*. International Association of Sedimentologists, Special Publication, 28, 93–112.

VandenBerg, A. H. M. (2009). Rock unit names in western Victoria. Seamless Geology Project Report 130, Geological Survey of Victoria, Melbourne, Australia.

VandenBerg, A. H. M., Willman, C. E., Maher, S., et al. (2000). *The Tasman Fold Belt System in Victoria*. Geological Survey of Victoria, Special Publication, 134–154.

Wakelin-King, G. A. and Webb, J. A. (2007). Upper-flow-regime mud floodplains, lower-flow-regime sand channels: sediment transport and deposition in a drylands mud-aggregate river. *Journal of Sedimentary Research*, 77, 702–712.

Webb, J. A., Gardner, T. W., Kapostasy, D., Bremar, K. A., and Fabel, D. (2011). Mountain building along a passive margin: late Neogene tectonism in southeastern Victoria, Australia. *Geomorphology*, 125, 253–262.

Williams, M., Cook, E., van der Kaars, S., et al. (2009). Glacial and deglacial climatic patterns in Australia and surrounding regions from 35 000 to 10 000 years ago reconstructed from terrestrial and near-shore proxy data. *Quaternary Science Reviews*, 28, 2398–2419.

Willman, C. E., Bibby, L. M., Radojkovic, A. M., et al. (2002). *Castlemaine 1:100 000 map area, Geological report 121*. Geological Survey of Victoria, Melbourne, Australia.

Wray, W. A. L. (2009). Palaeochannels of the Namoi River Floodplain, New South Wales, Australia: the use of multispectral Landsat imagery to highlight a Late Quaternary change in fluvial regime. *Australian Geographer*, 40, 29–49.

Part III
Applications in Other Sciences

14

Geology of Fluvial-Fan Deposits

Facies Patterns, Architectural Organisation, and Implications for Economic Geology

DARIO VENTRA and ANDREA MOSCARIELLO

Abstract

Fluvial fans and megafans are being re-evaluated in terms of their importance for sediment distribution in present-day continental basins, with evidence that their thick alluvial successions have contributed to the aggradation of much larger fractions of the stratigraphic record than previously recognised. Research on active fans worldwide is illustrating processes and dynamics at system scale while also highlighting differences linked to climatic or basin-specific settings. Long-term progradation of these systems produces consistent architectural signatures and trends, such as vertical stacking of laterally extensive units of amalgamated channel fills and overbank fines with minimal erosional discontinuities; poorly developed pedogenic horizons; proximal-to-distal fining of channel fills, and reduction in their volume and connectivity. The predictable spatial organisation of facies associations in fluvial-fan successions, documented from numerous examples, suggests that depositional models for thick alluvial units may relate to their accumulation in fan settings, informing preliminary basin exploration and advanced phases of reservoir characterisation and enhanced recovery. Consideration of fluvial-fan stratigraphy may also improve development of coal, uranium and placer resources, as well as aquifer characterisation. More quantitative data on fluvial-fan sedimentology and architecture are needed to better apply concepts derived from the resurgence of interest in these systems.

14.1 Introduction

Recent developments in fluvial geomorphology and sedimentology have pushed the concept of fluvial (mega)fans to the forefront of debate, causing a resurgence of interest in these depositional systems not only from the perspective of continental basin analysis, but also regarding their application to economic geology. Fluvial fans are defined here as depositional landforms of great surface extent and subdued gradient, commonly originating from an apical zone at the topographic transition between a highland source area and an adjacent, low-relief sedimentary basin (Fig. 14.1), aggraded by rivers with a strong tendency to spread drainage and associated sediment load areally by means of frequent avulsion events and/or channel bifurcations. The term *megafan* is most often used for fluvial fans which attain radii in excess of 30–50 km or areas in the order of 10^3 km^2 or larger (Hartley et al. 2010; Fontana et al. 2014; Ventra and Clarke 2018), although some authors raise the lower boundary for radial extent up to ~80–100 km (e.g., Wilkinson et al. 2002, 2006). However, workers in the past used the term more liberally as a general synonym for fluvial fans, regardless of size.

Basic ideas on the distinctiveness of large, variably fan-shaped alluvial depositional landforms, on their construction by fluvial radial networks, and on their relevance for continental basin fills were originally proposed approximately forty years ago (Friend 1978) and elaborated in numerous works ranging in emphasis from alluvial geomorphology and sedimentology to continental stratigraphy and economic geology (Olsen 1987; Kelly and Olsen 1993; Nichols and

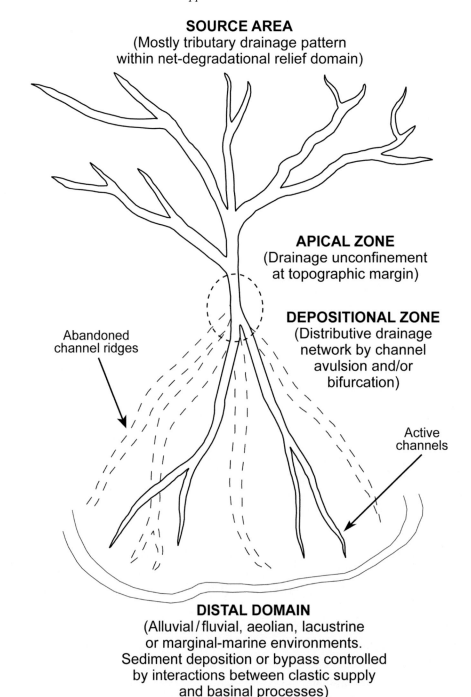

SOURCE AREA
(Mostly tributary drainage pattern
within net-degradational relief domain)

APICAL ZONE
(Drainage unconfinement
at topographic margin)

DEPOSITIONAL ZONE
(Distributive drainage
network by channel
avulsion and/or
bifurcation)

Abandoned
channel ridges

Active
channels

DISTAL DOMAIN
(Alluvial / fluvial, aeolian, lacustrine
or marginal-marine environments.
Sediment deposition or bypass controlled
by interactions between clastic supply
and basinal processes)

Figure 14.1 **Simplified, generalised model for drainage-pattern transition from tributive in a highland source area, to distributive in a lower-lying basinal area where sediment and runoff are dispersed in time by channel avulsions and/or bifurcations, ultimately forming a fan-shaped depositional landform**. Modified from Friend (1978).

Hirst 1998; DeCelles and Cavazza 1999; Horton and DeCelles 2001; Shukla et al. 2001; Moscariello 2005; Wilkinson et al. 2006; Nichols and Fisher 2007). Concepts were commonly applied to analyses of large, aggradational alluvial systems in modern sedimentary basins (e.g., Wells and Dorr 1987a,b; McCarthy et al. 1991, 1992; Iriondo 1993; Stanistreet and McCarthy 1993; Singh et al. 1993; Sinha and Friend 1994) and to equivalent systems more or less explicitly identified as such from thick stratigraphic successions of mostly fluvial origin (e.g., Graham 1983; Nichols 1987; Räsanen et al. 1992; Kumar 1993; Zaleha 1997a;

Schlunegger et al. 1997; Legarreta and Uliana 1998; Christophoul et al. 2002; Uba et al. 2005). However, the concept has gained major prominence in clastic sedimentology and basin analysis only in the last few years, reproposed as a systematic research programme by the British–American group advancing the concept of 'distributive fluvial systems' (hereafter DFS; Weissmann et al. 2010, 2011, 2015; Hartley et al. 2010; Davidson et al. 2013). Although considered partly contentious (Fielding et al. 2012; Latrubesse 2015), work by the 'DFS-group' has received a positive response and inspired new interpretations of outcrop-based and subsurface fluvial stratigraphy (e.g., Buehler et al. 2011; Rossetti et al. 2012; Gulliford et al. 2014; Rittersbacher et al. 2014; Quartero et al. 2015; Owen et al. 2015; Ribes et al. 2017; Moscariello 2018; Primm et al. 2018; Batezelli et al. 2019).

Fluvial-fan studies are opening a renewed approach to river geomorphology and sedimentology at the scale of entire drainage networks rather than of single channel patterns, and have increasing relevance for analyses of continental basins, where clastic fills might be represented by fan successions in much larger proportion than previously thought. This latter point provides a fresh perspective for fluvial sequence stratigraphy and economic geology, given that updated concepts on the development and architecture of thick fluvial-fan successions consent to set continental clastic deposits in a predictive framework applicable to numerous tectonic and (palaeo)climatic settings. In addition, fluvial (mega)fans develop under relatively high aggradation rates over geologically protracted timespans; this translates into relatively continuous sediment and proxy archives of environmental change in terrestrial settings, where erosion and non-deposition would otherwise prevail. Confident identification of clastic successions as products of fluvial-fan aggradation thus may entail retrieval of high-quality palaeoenvironmental records. For similar reasons, and given their relatively predictable stratigraphic organisation, fluvial-fan deposits are also interesting prospects for exploration and production of oil and gas resources (Moscariello 2018), and bear great potential also for coal deposits.

The latter aspects are the focus of this article, aiming to illustrate the general applicability of fluvial-fan

sedimentology to hydrocarbon geology and to explore potential for economic coal resources. From a terminological standpoint, we note that in spite of criticism (Fielding et al. 2012; Latrubesse 2015), the term 'distributive fluvial system' (Weissmann et al. 2010; Hartley et al. 2010), frequently adopted in recent literature, is here considered valid and unambiguous for a definition not only of the general planform but especially of the long-term dynamics of these landforms. However, the synonym 'fluvial (mega)fan' is favoured for reasons of historical precedence, and in order to stress the distinction with alluvial fans, which Hartley et al. (2010) subsume under 'distributive fluvial system' (Hartley et al. 2010), a conceptual and semantic position to which we do not subscribe.

14.2 General Morphology and Dynamics of Modern Fluvial Fans

Presently active fluvial fans and megafans feature worldwide distribution and a variety of surface processes and morphologies (Fig. 14.2) reflecting the broad range of climatic and geographic contexts in which they develop. Although research on modern fluvial fans has focused on a rather narrow subset in humid and arid tropical settings, useful generalisations can be drawn from large data compilations (Wilkinson et al. 2002; Hartley et al. 2010; Weissmann et al. 2011) which demonstrate that fans at present occur on all landmasses (except for those pervasively glaciated), in all tectonic settings, and under climatic conditions from subarctic to tropical hyper arid. The distinctiveness of these systems from a geomorphic perspective is given primarily by: (i) a drainage network that maintains an areally distributive configuration of shifting channel belts and/or bifurcating channels; (ii) a tendency to aggrade large fractions of sediment load locally, building up elongate depositional landforms that radiate away from an apical, elevated zone and merge topographically with distally contiguous, lower-relief landscapes (from alluvial mudflats to aeolian dunefields and sandsheets, from saline playas to lacustrine or marine coastlines; Fig. 14.2).

Such systems at present mostly originate from the establishment and progressive enlargement of wide, well-integrated catchments situated in active orogenic belts and highlands in a net-degradational geomorphic

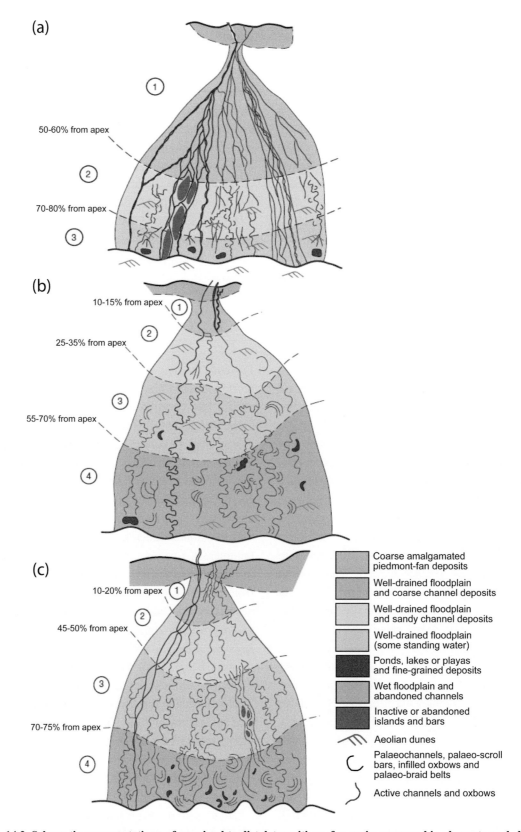

Figure 14.2 **Schematic representations of proximal-to-distal transitions for main geomorphic elements and drainage patterns in different kinds of fluvial fans**. (a) Braided bifurcating system. (b) Single-thread meandering and anabranching system. (c) Multi-thread (braided) anabranching system. Modified from Davidson et al. (2013). A colour version of this figure is available in the SOD for Chapter 14.

realm, acting as sources of terrigenous debris (Gupta 1997; DeCelles and Cavazza 1999; Friend et al. 1999). High or sustained rates of erosion often result in antecedent fluvial networks capable of maintaining their position by continuous incision into landscapes that are undergoing uplift (Gohain and Parkash 1990; Damanti 1993; Horton and DeCelles 2001; May 2011), conveying large volumes of sediment to areas of lower relief, possibly actively subsiding as sedimentary basins, where the balance between sediment supply and sediment-transport potential may switch to net aggradation. An excess of available sediment with respect to the potential to transfer it basinward is possibly the fundamental factor at the origin of river systems characterised by the key combination of long-term aggradation and frequent channel-belt avulsion, a process that ultimately results in the accretion of topographically convex fluvial fans with distributive drainage networks, rather than single channel belts extending downslope. This is confirmed by the direct relationship between high sediment loads and the avulsive tendency of fluvial channels (and of entire channel belts, over larger spatial and temporal scales), verified both experimentally and in field studies (Richards et al. 1993; Bryant et al. 1995; Slingerland and Smith 2004; Jerolmack and Mohrig 2007; Phillips 2011; Reits and Jerolmack 2012; Sinha et al. 2014; Galve et al. 2016). Avulsions in fluvial channels are favoured by high sediment loads and aggradation rates that rapidly build steeper local gradients between the channel (belt) and the surrounding floodplain, bringing the channel close to the threshold for redirection at a triggering event, such as a major flood, bank failure, obstruction by vegetation or by an overload of debris, etc. (Jones and Schumm 1999; Slingerland and Smith 2004). The probability and frequency of channel lateral migration and major relocation depends also on the frequency of exceptional hydrologic events capable of destabilising channels with excess discharge and/or sudden mobilisation of great sediment volumes. In fact, avulsion-prone, fan-shaped systems are commonly described from regions subject to seasonal or dramatic stochastic changes in discharge regime and sediment supply, such as monsoonal tropics, drylands, and subarctic latitudes (e.g., Boothroyd and Nummedal 1978; Wells and Dorr 1987b; McCarthy 1993; Jones et al. 1993; Leier et al. 2005; Pietsch and

Nanson 2011; Donselaar et al. 2013). Studies of Quaternary systems worldwide, for example, repeatedly demonstrate the genetic connection between phases of enhanced sediment supply from glaciated highlands and fluvial-fan aggradation (Leckie 1994; Weissmann et al. 2005; Fontana et al. 2014). Nodal avulsions from the proximal domain are commonly the basic mechanism by which a fan-like network is organised through time (Davidson et al. 2013; Weissmann et al. 2015; Ventra and Clarke 2018; Fig. 14.2). Paradigmatic in the literature is the example of the Kosi River megafan in northern India, on which the main channel belt has repositioned itself over a distance of > 100 km during the last two centuries (Wells and Dorr 1987b; Chakraborty et al. 2010).

Avulsion and bifurcation, however, also occur in medial and distal portions of distributive systems (Maizels 1990; McCarthy et al. 1992; Yonge and Hesse 2009; Donselaar et al. 2013; Assine et al. 2014), although true channel bifurcation is inherently unstable (Slingerland and Smith 2004; Kleinhans et al. 2013), and as such is probably only a transient configuration. Aerial and satellite images of fluvial-fan surfaces often show variable evidence for relict channels and channel clusters, preserved as moderately elevated, linear to sinuous ridges of coarse sediment delimiting overbank areas of finer-grained deposition and lower relief (Maizels 1990; Rannie 1990; Assine 2005; Blechshmidt et al. 2009; Assine et al. 2014; Van Dijk et al. 2016). Cross-fan topographic profiles highlight radially oriented, secondary aggradational lobes, several kilometres in lateral extent (Assine and Silva 2009; Chakraborty et al. 2010; Zani et al. 2012; Assine et al. 2014), representing discrete zones of aggradation corresponding to the activity of persistent channel clusters, subject in turn to repositioning in the context of system-scale avulsive events.

This avulsion-prone, dominantly aggradational regime generates basic morphological traits and drainage trends (Figs. 14.2 and 14.3a) that are common to fluvial fans in all settings, irrespective of secondary processes controlled by local climate or regional physiography. Independently of radius and surface area, almost all fluvial fans possess markedly *convex, lobate transverse profiles* clearly related to their long-term net-depositional regime, but *concave longitudinal concave profiles*, with a gradual reduction in gradient from proximal to distal sectors (Wilkinson et al. 2010; Hartley et al. 2010), likely related to the reduction in runoff competence downfan, particularly in systems subject to transmission losses in (semi)arid climates. A radially oriented drainage and morphosedimentary zonation is thus frequently recognised, whereby larger, deeper, bedload-dominated proximal channels with low-sinuosity braided patterns pass downfan into shallower, narrower, higher-sinuosity, locally anastomosed channels conveying finer bedload and a greater fraction of suspended load (Parkash et al. 1983; McCarthy et al. 1991; Räsanen et al. 1992; Singh et al. 1993; Blechschmidt et al. 2009; Yonge and Hesse 2009; Davidson et al. 2013; Fontana et al. 2014). Within this generalisation, however, the nature and spatial distribution of channel patterns are controlled by hydrology, sediment grade and types of soils and vegetation, and thus ultimately by local climate and catchment geology/morphometry. For example, fans emanating from tectonically active, high-gradient mountain ranges and/or set in periglacial or dryland climates tend to have greater proximal gradients aggraded by significant volumes of coarse bedload, transferred by braided channel systems (Hartley et al. 2010; Latrubesse 2015), whereas particularly high sediment supply with an irregular discharge regime can force braided channel patterns downfan to distal sectors. Dryland systems may further attain a limiting condition whereby runoff is reduced or becomes unconfined downfan to the point where practically neither water nor sediment leave the system distally ('terminal fans'; Parkash et al. 1983; Kelly and Olsen 1993; Donselaar et al. 2013).

From this proximal-to-distal zonation follow two fundamental trends (Fig. 14.3a). The typical tapering of fan planform toward the apical sector implies that shifting, wider, more competent channels on this relatively narrow zone effectively rework overbank fines because the zone is relatively narrow, thereby reducing the area occupied by stable floodplain domains. Conversely, the broader area of distal fan sectors allows for more stable, extensive overbank domains between narrower, less energetic channels. Depositional sub-environments superimposed on distal sectors depend also on climate and basinal geography (Wilkinson et al. 2010; Weissmann et al. 2015), ranging from distal mudflats and playas (Fisher et al.

2008) to dunefields (Cohen et al. 2010; Rossetti et al. 2012; Van Dijk et al. 2016), to semi-permanent wetlands (Tooth and McCarthy 2007; Latrubesse et al. 2007; Assine and Silva 2009; Yonge and Hesse 2009). In addition, it has been frequently documented that the grain size of channel-fill deposits on active fluvial fans tends to diminish downstream (Fig. 14.3a; Wells and Dorr 1987a; Singh et al. 1993; Horton and DeCelles 2001; Shukla et al. 2001; Browne 2004; Chakraborty et al. 2010; Ralph and Hesse 2010), although exceptions are known (Latrubesse 2015). The same trend, however, does not apply readily to floodplain fines (Davidson et al. 2013), although it might be reasonably predicted for deposits of more energetic overbank processes, such as crevasse-splay sands.

14.3 Geology of Fluvial-Fan Deposits

14.3.1 Facies Associations and General Architectural Trends

Facies associations in fluvial-fan successions are not different from those of most alluvial river systems, because processes at the scale of channel reaches and floodplain tracts are governed by the same interactions between fluid flow and mobile sediment grains. Discharge regime, gradients, sediment calibre and composition, and vegetative cover govern the planform patterns of channels and the organisation of their infills, as for all rivers. More characteristic architectural patterns emerge at larger scales, conferring on fluvial fan deposits a distinctive stratigraphic signature relevant both for identification of such depositional systems in the rock record and for economic geology.

Channel fills in most fluvial-fan deposits consist of vertically stacked, commonly amalgamated packages of dominantly coarse-grained facies associations featuring a recurrent vertical structure. They are commonly characterised by a basal erosional surface of generally low relief, overlain by variably thick intraformational (mud clasts, pedogenic nodules, plant debris, peat rafts) or clastic conglomerates, transitional upward to relatively coarse, massive to cross-stratified sandstones or gravelly sandstones. These in turn may be progressively overlain by finer, thinner-bedded sandstones and mudstones showing mostly planar

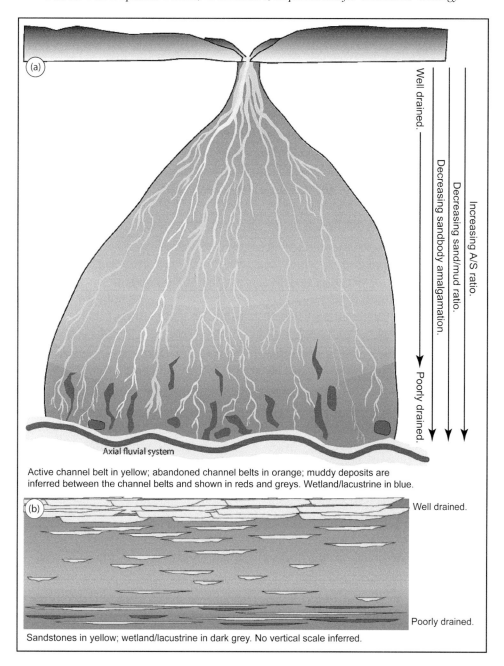

Active channel belt in yellow; abandoned channel belts in orange; muddy deposits are
inferred between the channel belts and shown in reds and greys. Wetland/lacustrine in blue.

Sandstones in yellow; wetland/lacustrine in dark grey. No vertical scale inferred.

Figure 14.3 **Models of fluvial-fan morphology and architecture**. (a) Generalised model (modified from Weissmann et al.
2013, with permission from the Society for Sedimentary Geology) of proximal-to-distal trends in drainage organisation and
sedimentology on a fluvial-fan surface. (A/S = ratio of accommodation to sediment supply). (b) Hypothetical cross-section
through a progradational fan succession, illustrating architectural transitions. No scale assumed for the fan radius or for the
thickness of the schematic stratigraphic interval. A colour version of this figure is available in the SOD for Chapter 14.

lamination, low-angle cross-bedding, ripple cross-
lamination or an absence of tractive structures, occasionally capped by weakly to moderately pedogenised
intervals (e.g., Campbell 1976; Willis and Bridge
1988; Evans 1991; Willis 1993; Zaleha 1997b; Ray
and Chakraborty 2002; Atchley et al. 2004; Kukulski
et al. 2013; Lawton et al. 2014; Owen et al. 2015).

These facies successions represent the gradual
decrease in energy and flow depth within a single
channel (Miall 1996; Atchley et al. 2004) subject to
progressive infill, and either channel abandonment or
repositioning on the alluvial surface. Specific facies
successions naturally vary depending on regional climate and dominant discharge trends (Fielding et al.

2018), with dryland rivers for example accumulating less well-organised deposits due to irregular, flashy discharge regimes (e.g., tractive sedimentary structures, downstream or laterally accreting macroforms, etc.; Olsen 1987; Maizels 1990; Hampton and Horton 2007; Cain and Mountney 2009). The tendency for fan channels to shift with high frequency over the aggrading surface induces their fills to amalgamate laterally as coarse-clastic sheets with characteristically low aspect ratios and great lateral continuity, up to kilometres at outcrop (Fig. 14.4a). These alternate vertically with mudstone-dominated packages that represent finer floodplain deposits accumulated over vast alluvial tracts distant from the active channel belt (Fig. 14.4b; Campbell 1976; Willis and Bridge 1988; Beer and Jordan 1989; Sadler and Kelly 1993; Willis 1993; Christophoul et al. 2002; Hampton and Horton 2007). This 'layer-cake' architecture of stacked coarse and fine clastic sheets with overall tabular geometry and minimal, local incision (Fig. 14.4a) is a hallmark of many outcrops of fluvial-fan successions, evolved in a dominant aggradational regime which allowed preservation of great volumes of overbank fines while main channels were repositioned continuously over the alluvial surface (Friend 1978; Bridge and Leeder 1979). Another frequent trait is the lack of thick, mature paleosols, although stratigraphic surfaces with some degree of pedogenic modification are common within overbank intervals or topping coarse-clastic sheets. This relates to the strong aggradational efficiency of fluvial fans, where shifting channel belts and sustained sedimentation rates imply that few areas are long unaffected by deposition (Friend 1996; Uba et al. 2005; Hampton and Horton 2007).

Considerable volumes of floodplain deposits are usually preserved in fluvial-fan successions, forming continuous, thick packages (Fig. 14.4b) of thinly bedded claystones and siltstones laid out during the waning and final ponding of unconfined flows in major floods. Depending on palaeoclimate, minor volumes of freshwater carbonates, evaporites or aeolian sandstones are associated (Allen et al. 1983; Martinius 2000; Hornung and Aigner 2002; Gibling et al. 2005, Cain and Mountney 2009; Owen et al. 2015). Variable volumes of very fine to fine sandstones and muddy sandstones are preserved within mudstone units, mostly as small, ribbon-shaped channel fills or

extensive sheets, up to hundreds of metres in length and from a few decimetres to a couple of metres thick, varying in structure from massive or pervasively pedoturbated to cross-laminated or (rarely) cross-stratified, with discontinuous basal intraclastic conglomerates. These represent emplacement of mixed bedload and suspended load dispersions by energetic unconfined floodwaters spreading onto the proximal floodplain or by shallow-channel networks (Allen et al. 1983; Willis and Bridge 1988; Bentham et al. 1993; Kumar 1993; Khan et al. 1997; Martinius 2000; Ray and Chakraborty 2002). The relative volume, bed thickness, and grain size of such sandstones decrease over several hundred metres away from the nearest channel fills or amalgamated channel-belt deposits, reflecting competence loss laterally for shallow, sediment-laden overbank flows (Bridge 1984).

The geometry and architecture of facies transitions between amalgamated channel fills and overbank deposits depend on the avulsion mechanism dominant either at local or system scale (Mohrig et al. 2000; Jones and Hajek 2007; Flood and Hampson 2014). Gradual transitions between floodplain mudstones and overlying coarse-clastic sheets occur through heterolithic intervals of sandstones and mudstones (Fig. 14.4c). The progressive thickening and coarsening upward of sandstone beds testifies to the progradational build-up of overbank splay complexes corresponding to permanent breaching zones from which parent-channel drainage was redirected (avulsion by progradation: Slingerland and Smith 2004). The process has been observed on medial and distal sectors of presently active fluvial fans (Buehler et al. 2011; Makaske et al. 2012; Assine et al. 2014). Conversely, avulsion by incision (Slingerland and Smith 2004; Jones and Hajek 2007) is triggered by abrupt erosion of a new drainage pathway from the parent channel, as observed in high-gradient, high-discharge proximal zones of the Kosi River fan (Sinha et al. 2014). The facies signature consists in coarse-grained channel fills erosionally superimposed on mudstones (Mohrig et al. 2000; Jones and Hajek 2007; Fig. 14.4d). However, studies on the architecture of overbank deposits in fluvial-fan successions are very rare (e.g., Gulliford et al. 2017), and it is not yet possible to make generalisations on architectural trends relating to the dominance of different avulsion

Figure 14.4 **Outcrop photographs of fluvial-fan deposits, Miocene Mariño Formation (central Andean foreland, ~20 km southwest of Luján de Cuyo, Mendoza Region, Argentina)**. (a) Typically aggradational architecture of alternating, tabular, coarse-grained, amalgamated channel-fills and mudstone-dominated overbank deposits (outcrop extent to farthest edge ~400 m). (b) Gradual transition between thin-bedded, heterolithic, proximal overbank strata up into a coarse-grained channel fill. (c), (d) Abrupt stratigraphic transitions between overbank mudstones and overlying, erosively based conglomeratic channel fill.

mechanisms, either spatially, from proximal to distal sectors, or through time during fan evolution.

Architectural trends at system scale are more consistently verified in the general volumetric and spatial relationships of overbank and channel strata, and the morphometry of coarse-grained channel fills throughout the stratigraphic column (e.g., Kelly and Olsen 1993; Horton and DeCelles 2001; Nichols and Fisher 2007; Weissmann et al. 2013). These are readily interpreted on the basis of patterns of geomorphic process on active fans (Fig. 14.3). The noted decrease in channel depth, width, and flow competence downfan, paralleled by an increase in relative floodplain area (and thus in the preservation potential of fine-clastic sediments), has prompted a comparison with vertical transitions in alluvial stratigraphic architectures ranging in thickness from a few hundred to a few thousand metres. Weissmann et al. (2013) described how long-term progradation of fluvial fans would lead to accumulation of alluvial successions in which deposits bearing a distal sedimentological signature are gradually overlain by stratal intervals of medial, then proximal fan domains (Fig. 14.3b). Such a genetic continuum of laterally migrating and aggrading sedimentary environments matches Walther's concept of latero-vertical facies superposition (Middleton 1973).

In distal sectors, the presence of rapidly migrating, distal distributary channels on a floodplain-dominated, low-gradient alluvial setting will be signalled by relatively small channel fills dispersed within an interval comprising mostly overbank mudstones and thin lenses and sheets of fine sandstones. Higher up through stratigraphy, alluvial architectures are expected to show a gradual increase in the size, texture, and relative volume of channel fills, so that the topmost intervals are characterised by a dominant volume of coarse-clastic units compared to overbank fines, with deeper and better amalgamated channel fills comprising the coarsest sediments. These reflect the activity of competent bedload-dominated runoff over the steepest, proximal part of the fan system, where limited 'lateral accommodation' would permit only minor preservation of overbank deposits.

A corollary to this predictive framework concerns the degree of pedogenic alteration at the fan surface and the expected position of related pedofacies through the sedimentary column (Atchley et al. 2013; Weissmann et al. 2013; Fig. 14.3). The sloping profiles of fluvial fans imply that distal surfaces are the lowest topographically, more likely affected by depth changes of the groundwater table in regions with a permanently humid or highly seasonal climate. Because the distal surface of a fan is dominated by low-energy floodplain environments with the least interference by channel processes, maximum pedogenic modification occurs in these distal sectors (e.g., Hampton and Horton 2007; Davidson et al. 2013; Wilson et al. 2014), where seasonal or (semi)perennial flooding, or groundwater saturation, are more likely. In humid climates, distal floodplains may also accumulate organic matter in the form of peat, whereas in seasonal per-humid to semi-arid climates they will be subject to periodic shifts in pedogenic processes governed by short- and long-term climate changes. This is reflected in the occurrence of relatively complex pedogenic signatures in the lowermost stratigraphic positions of progradational fan successions. In contrast, medial and proximal fan sectors, which lie topographically well above regional water tables and are affected by higher aggradation rates and more frequent channel reworking, will feature minor pedogenic modification and less hydromorphic pedofacies, in most cases represented by 'redbed' mudstones aggraded in well-drained conditions (Bentham et al. 1993; Hampton and Horton 2007; Gulliford et al. 2014; Marenssi et al. 2015; Hunger et al. 2018; Batezelli et al. 2019).

These general sedimentological and architectural trends, based on comparisons with spatial process distributions on active fans (Horton and DeCelles 2001; Weissmann et al. 2013), have been substantiated in numerous publications (Graham 1983; Willis and Bridge 1988; Beer and Jordan 1989; Evans 1991; Cuevas Gozalo and Martinius 1993; Kelly and Olsen 1993; Zaleha 1997a; DeCelles and Cavazza 1999; Martinius 2000; Shukla et al. 2001; Uba et al. 2005; Nichols and Fisher 2007; Sáez et al. 2007; Cain and Mountney 2009; Klausen et al. 2014; Rittersbacher et al. 2014; Owen et al. 2015, 2017; Hunger et al. 2018). It has also been demonstrated that such trends are substantially scale-invariant, as they apply to fans with radii ranging from a few tens to several hundreds of kilometres, and provide a robust conceptual base for evaluating the relevance of fluvial-fan successions to analyses of continental basin fills.

14.3.2 Significance of Fluvial-Fan Successions in the Context of Continental Basin Fills

Numerous fluvial successions described in the literature possess a vertical stratigraphic pattern matching the one described above (Weissmann et al. 2013), developed over thicknesses from a few hundred metres up to several thousands of metres. While the thickest examples recur in continental foreland settings (e.g., Beer and Jordan 1989; Zaleha 1997b; Horton and DeCelles 2001; Kumar et al. 2003; Uba et al. 2005; Zheng et al. 2006; Hampton and Horton 2007; Schlunegger and Norton 2015), thinner equivalents are recognised in rift-basin records (Hill 1989; Marquillas et al. 2005; McKie and Williams 2009). Explicit evidence from pull-apart basins is not yet reported, being minimal also in modern transtensional basins (Weissmann et al. 2015). Over thicknesses exceeding several hundreds of metres, it is commonplace for such fluvial successions to be assigned to distinct lithostratigraphic units, based on differences in dominant facies and architectures between specific intervals (e.g., Willis 1993; Kelly and Olsen 1993; Zaleha 1997b; Friend et al.

2001; Kukulski et al. 2013), whereas the conceptual model reported above convincingly makes the case for such thick sections to originate from long-term development of a single depositional system. Recognition of successions of potential fan origin however should not automatically lead to the inference of tectonically active margins, since modern fluvial fans prograde also through intracratonic depressions far from deformation belts (e.g., McCarthy 1993; Page and Nanson 1996; Lang et al. 2004; Wilkinson et al. 2010; Nageswara Rao et al. 2015).

This aspect of fluvial-fan geology recalls the basic geomorphic observations advanced by the DFS group, converging toward the conclusion that within modern continental basins most aggrading surfaces belong to fluvial fans/megafans or are affected by their activity (Hartley et al. 2010; Weissmann et al. 2010, 2015). This leads to the current hypothesis that the most efficient accumulation of alluvial sediments over geological timescales will be operated by these systems. Although supported by a substantial dataset, the hypothesis has generated a fair amount of controversy, with several authors (Fielding et al. 2012; Ashworth and Lewin 2012; Latrubesse 2015) noting that various modern large fluvial systems with tributary rather than distributary drainage patterns are also underlain by thick volumes of alluvial deposits. This objection does not consider the fact that most such rivers (Amazon, Paraná, Congo, Mississippi, Nile, among others) traverse regions lacking long-term subsidence, and thus have little potential for stratigraphic preservation over geological time. Such rivers are mostly corridors for sediment bypass basinward, where long-term accumulation occurs on deltas and turbidite-fed submarine fans (Figuereido et al. 2009; Bentley et al. 2016). Late Quaternary, glacio-eustatically paced base-level fluctuations favoured the creation and later infill of incised valleys, which have long been the most discussed (palaeo)geomorphic settings for aggradation of thick alluvial successions of all ages (e.g., Aitken and Flint 1995; Blum et al. 2013). However, in continental interiors remote from sea-level influence, sediment supply from well-integrated, net-erosional catchments expanding on highland areas can overwhelm the ability of local rivers to transfer terrigenous material basinward, triggering aggradation and consequently the onset of avulsive, distributive drainage systems. This

tendency is enhanced in (semi)arid climates, where transmission losses and high evapotranspiration rates further curtail effective sediment transport ('terminal fans'). The extent of coarse-clastic deposition on fluvial fans usually reaches tens of kilometres, greater than that of smaller alluvial fans dominated by mass-transport processes and unconfined runoff (Horton and DeCelles 2001; Moscariello 2005). In numerous continental basins, therefore, isolated or coalescent fluvial fans form extensive marginal belts of coarse-clastic storage. In contractional settings, however, the migration and uplift of thrust-belts usually impose a limit to persistent fluvial-fan aggradation up to a few tens of kilometres from the basin margin, and may lead to deformation and erosional recycling of proximal deposits of large megafans, and probably of entire fans with shorter radii (Latrubesse 2015).

Classical sequence-stratigraphic approaches to alluvial strata investigate relationships between base-level (i.e., sea-level) change and architectural patterns of fluvial and associated sediments (e.g., Shanley and McCabe 1991; Bohacs and Suter 1997). However, this perspective has long proven fruitless for thick fluvial sections in continental interiors, where eustatic forcing is suppressed either by distance from the palaeo-coastline or by an isolated, endorheic setting. The importance of other, less readily quantifiable controls, such as climate and tectonics, grows for terrestrial successions (Blum and Törnqvist 2000). An alternative approach to subdividing hinterland fluvial successions into genetically related units has developed, with a more objective, descriptive focus on alluvial architecture and less attention to potentially elusive, dominant controls. Depending on relative volumes of channel-fill *vs.* overbank sediments and on the morphometry of sandstone bodies, variable architectures are attributed to relatively high or low rates of accommodation creation *vs.* rates of sediment supply. The underlying principle has been applied by various workers, although presented with an inconsistent terminology (aggradational, transitional and degradational systems tracts of Currie 1997; forestepping, transitional and backstepping systems tracts of Legarreta and Uliana 1998; high-accommodation and low-accommodation systems tracts of Huerta et al. 2011).

Although applicable to a variety of basinal settings, the conceptual foundation to this genetic-stratigraphic approach has recently been questioned by a literature meta-analysis showing that channel-fill density, geometry and stacking patterns may not relate linearly to variations in accommodation (Colombera et al. 2015). This is because stratigraphic variations in channel-fill density and geometry and in the volume of overbank deposits can be controlled by other concomitant factors, such as climate-controlled discharge variations and extreme flood events (Schmitz and Pujalte 2007; Esposito et al. 2018), or changes in vegetation and soil cover (Smith 1995; Arche and López-Gómez 2005); therefore, it is conceptually simplistic (although not necessarily incorrect, in some instances) to attribute patterns of alluvial stratigraphy solely to tectonics/ accommodation variations even in landbound continental settings (Gibling et al. 2011). A possible solution to this uncertainty for some continental sequences is the above-mentioned consideration that fluvial-fan progradation inherently results in the stacking of alluvial successions with predictable architectural heterogeneity (Fig. 14.3b; Weissmann et al. 2013). In some instances, vertical variations through fluvial stratigraphy could relate simply to the autogenic stacking of proximal strata over progressively more distal alluvial strata (and facies associations from depositional systems adjacent to, or interfingering with, fan margins). The overall volume of deposits affected by such large-scale patterns would depend on regional subsidence rates and on fan radius, which ranges from a few tens up to several hundreds of kilometres, and therefore could amount to a considerable fraction of the basin fill.

Furthermore, it is possible that some thick fan successions were accumulated in proximity to marine coastlines. Most work so far has been carried out on inland basins originally adjacent to uplifting mountain chains, widely recognised by their analogy with examples of modern fluvial megafans in the Himalayan and Andean forelands. Numerous formations described in the literature however consist of thick, laterally extensive belts of fluvial and mixed fluvial and marginal-marine facies associations accumulated over broad, low-gradient coastal plains receiving abundant sediment from the hinterland and subject to complex interactions with shallow-marine environments controlled by eustatic base-level changes. This is a context for fluvial-fan development that finds no direct analogue

at present, and may have gone unnoticed in the geological record. Examples are the elongate epicontinental seaways that extended along Carboniferous and Late-Cretaceous foreland basins associated respectively with the Alleghenian and Sevier orogenic belts of eastern and mid-western North America. Both settings were occupied at different stages by thick alluvial wedges generated by highland catchments that prograde toward marine-flooded axial depocentres. Well-known stratigraphic units such as the middle Pennsylvanian upper Breathitt Group of southeast Kentucky and the Late Cretaceous Masuk, Blackhawk and Castlegate Sandstone formations of Utah feature distinct architectural trends such as proximal-to-distal (and downsection) fining, decrease in size and density of channel-fills, and an increase in the relative volume of associated floodplain strata, and especially of coal beds, the latter representative of extensive coastal peat mires (Aitken and Flint 1995; Corbett et al. 2011; Hampson et al. 2012; Rittersbacher et al. 2014).

These characteristics evidently match those recognised in large fans that dominate modern foreland basins. Eustatic fluctuations (frequent and pronounced for the Late Carboniferous Appalachian Basin, forced by glacio-eustatic cycles during the Gondwanan glaciation) paced the advance and retreat of coastal belts over low-gradient, distal alluvial plains traversed by numerous channel belts. Such fluctuations allowed for steadily rising groundwater tables that favoured peat accumulation during transgression, for marine flooding and establishment of marginal- to shallow-marine environments over areas previously occupied by distal fans, and for local valley incision during falling-stage (Aitken and Flint 1995; Olsen et al. 1995). While proximal sectors of large fans would have been less affected by base-level fluctuations, distal domains would have been subject to continuous eustatic forcing, aggrading highly heterogeneous facies belts from fluvial to transitional and shallow-marine, characterised by complex three-dimensional architectures and locally by the occurrence of incised valleys with fluvio-estuarine infills. A similar setting is envisaged here also for extensive Devonian fluvio-deltaic deposits known historically as 'Catskill Delta', associated with the Acadian Orogeny in the northeastern United States, throughout which stratal trends typical of avulsion-dominated fluvial fans have been recognised (Gordon and Bridge 1987; Willis and Bridge 1988); and for the areally widespread Triassic fluvial system that fed a shelf-to-slope system (Snadd Formation) along the rifted Boreal margin of the Panthalassan Ocean (Norwegian sector of the Barents Sea; Klausen et al. 2014). The latter formation accumulated in a greenhouse climate context that prevented high-frequency/high-magnitude eustatic fluctuations, reflected in greater lateral continuity of alluvial facies belts in the coastal-plain domain and absence of valley fills related to deep fluvial incision.

These perspectives on the origin of some distal-fluvial to marginal-marine clastic wedges suggest alternative hypotheses that may help to refine models for complex stratigraphies resulting from fluvial-fan progradation at the margins of marine basins. Further research is necessary at this stage to fine-tune our ability to identify stratigraphic successions of fluvial-fan origin (see Davidson et al. 2013, and Weissmann et al. 2015, for discussions), to understand how these systems evolved over long time scales under variable allogenic forcing, and to quantify their stratigraphic heterogeneity at levels comparable to those attained for other terrestrial depositional systems.

14.4 Implications for Economic Geology

The large volume and subcrop extent of fluvial-fan deposits along basin margins, both modern and ancient, establish these depositional systems as a fundamental, if overlooked until recently, target for exploration in economic geology (Moscariello 2018). Fluvial successions in general are characterised by complex distribution and rapid lateral variability of facies associations, and well data in conventional operations are generally too sparse and expensive to fully constrain sedimentological heterogeneity. However, most typical aspects of thick fluvial-fan successions are a recurrent stratal architecture and its consistent, often gradual downdip variation, as documented in numerous publications. These traits set fluvial fans apart as possessing a relatively predictable architecture, offering a high potential for successful characterisation at system scale. Counting recent evidence that such systems may be responsible for aggrading the largest volumes of fluvial deposits in stratigraphy, these considerations suggest a potential paradigm shift

in the analysis of fluvial strata for water resources and for geo-energy development.

The most common style of aggradation is represented by alternating, tabular, coarse-clastic- and mudstone-dominated sheets (Fig. 14.4a, b; Willis 1993; Sadler and Kelly 1993; Zaleha 1997a; Hampton and Horton 2007; Sinha et al. 2014; Marenssi et al. 2015) extending over areas determined by the original surface of the formative fan, thus potentially on the order of 10^3 to 10^5 km^2, i.e., equal to (and frequently exceeding) that of most oil and gas fields. The lack of deep incision in these stratal packages, which is related to the dominant aggradational regime typical of fluvial fans, implies that facies associations are vertically partitioned by relatively predictable surfaces over large scales, and their transitions follow a relatively regular geometry, affected mostly only by minor decimetric relief at the base of amalgamated channel-fills. Superposed on this simple, repetitive architecture are the larger-scale, proximal-to-distal trends discussed above (Figs. 14.3 and 14.5), reasonably predictable if a basin's palaeogeography is known to some degree.

The proximal facies belt of fluvial-fan successions is expected to comprise the coarsest, most laterally and vertically amalgamated channel fills, consequently presenting optimal reservoir potential in terms of primary porosity and permeability of constituent facies, as well as in terms of connectivity. Mudstone-dominated overbank units tend to represent small volumes of proximal-fan deposits, potentially minimising complications to reservoir performance due to stratigraphic buffers or barriers to fluid flow. Substantial volumes of relatively coarse sandstone bodies and fine conglomeratic channel fills will be preserved in medial fan sectors, but accompanied by increasing volumes of finer overbank deposits. Nonetheless, reservoir potential may still be significant (Campbell 1976) due to the great, often kilometric, lateral continuity of thick (up to ~10–15 m) amalgamated channel fills produced by channel migration and avulsion, coupled to a common lack of highly pedogenised horizons in floodplain strata. The latter implies relatively reduced early (bio) chemical cementation which commonly enhances impermeability. Possible high net-to-gross fairways with radial elongation and high continuity downfan (from tens to even hundreds of kilometres) may occur as clusters of channel fills (Hofmann et al. 2011; Hajek

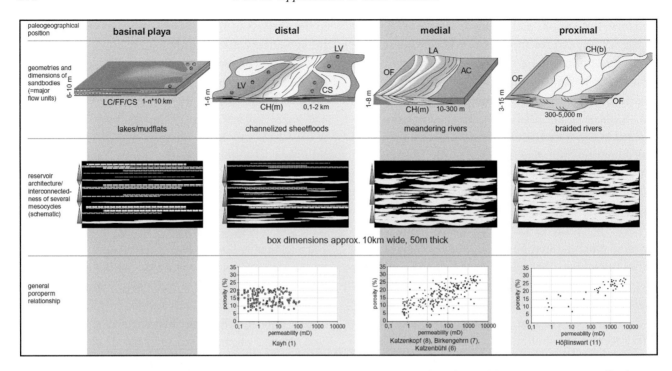

Figure 14.5 **Proximal-to-distal trends in dominant architectural elements, stratigraphic architectures, net-to-gross distributions and poro-permeability for representative deposits of the Triassic Stubensandstein Formation (southern Germany), an analogue for Permo-Triassic reservoirs of the North Sea** (modified from Hornung and Aigner 2002). The term 'net-to-gross' indicates the relative volumetric proportion of coarse-grained, potentially permeable facies associations to the total volume of a rock unit. Diagrams in the lowest row show the gradual increase in porosity and exponential increase in permeability values from relatively small, isolated channel fills and associated overbank sandstones of the distal sector to larger, increasingly amalgamated, sand-prone channel fills of proximal fan sectors (names underlying each diagram indicate the outcrop locality and number of sampled sections at each locality). Key to abbreviations: mD– millidarcy (permeability unit); LC– lacustrine carbonate deposits; FF– floodplain fines; CS– crevasse splay deposits; LV– levee deposits; CH– channel fill; LA– lateral accretion deposits; OF– overbank fines; AC– aggradational channel plug. A colour version of this figure is available in the SOD for Chapter 14.

et al. 2012; Van Dijk et al. 2016), many metres thick, hundreds of metres to a few kilometres wide, and up to many tens of kilometres long, being the expression of mesoscale aggradational lobes fed by major zonal avulsions (Chakraborty et al. 2010; Zani et al. 2012). However, the lateral, cross-fan connectivity of these coarse facies associations should be lower than the connectivity in a downfan direction, generating potential anisotropy in reservoir potential especially in systems where major avulsion timescales were long compared to system-wide aggradation rates. Sandstone volumes associated with overbank units in medial fan deposits may also be significant, although their distribution and facies properties have not yet been characterised sufficiently, let alone quantified.

The most distal sectors in fan successions are mudstone-prone (Fig. 14.5), expected to comprise significant volumes of poorly permeable facies associations deposited by biochemical sedimentation in ephemeral to permanent floodplain ponds and peat swamps. More mature pedogenic horizons in these sectors may enhance compartmentalisation. The low primary connectivity of ribbon-shaped, small channel fills dispersed in a mudstone-dominated stratigraphy translates into poorly predictable, 'labyrinthine' reservoir architectures in the distal sector (Weber and Van Geuns 1990; Silva et al. 1996). Here, sinuous channels with high suspended load imply also that corresponding channel fills will feature finer textures and heterolithic facies associations (i.e., lateral accretion strata, mud-dominated channel plugs, etc.) that may negatively affect reservoir quality. In contrast, typically bedload-dominated braided streams of proximal-fan domains generate less compartmentalised infills, with greater potential for flow transmission. However, in arid environments, medial to distal sectors of large fans subject to deflation generate large volumes of sand for accumulation of sand seas (e.g., McKie 2011;

Fryberger et al. 2011), which if preserved may represent interesting plays. These general trends have been recorded in stratigraphic and petrophysical analyses of reservoir potential in hydrocarbon provinces and regional aquifer systems comprising successions of fluvial-fan origin (Cuevas Gozalo and Martinius 1993; Silva et al. 1996; Hornung and Aigner 2002;

Hinds et al. 2004; Moscariello 2005; Hofmann et al. 2011; McKie 2011; Kukulski et al. 2013; Klausen et al. 2014; Keeton et al. 2015) and demonstrate a predictability at system scale that is uncommon for fluvial deposits (Fig. 14.6). Stratigraphic characterisation of fluvial-fan deposits is relevant also for prediction of aquifer units in dominantly alluvial successions and for

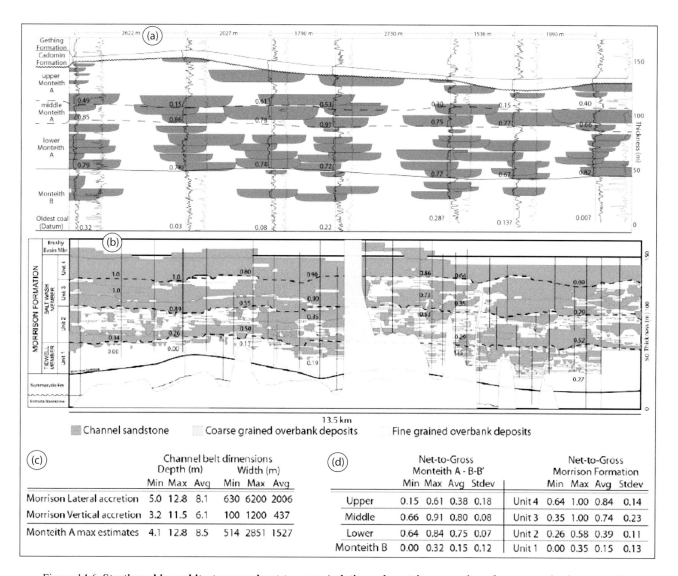

Figure 14.6 **Stratigraphic architectures and net-to-gross (relative volumetric proportion of coarse-grained, permeable facies associations over the total volume of a rock unit) potential through Mesozoic alluvial strata of the Cordilleran foreland belt, North America** (modified from Kukulski et al. 2013, with permission from the Canadian Society of Petroleum Geologists). (a) Cross-section perpendicular to regional palaeotransport direction (i.e., across-fan) through the 'B' unit of the Jurassic–Cretaceous Monteith Formation (Western Canadian Deep Basin, Alberta). Channel fills/belts interpreted from log data are highlighted as dark shapes. In panels a and b, decimal numbers positioned along the stratigraphy of individual well columns indicate estimated net-to-gross values for each stratigraphic interval. (b) Similar dataset and cross-fan section for the Tidwell and Salt Wash members of the Jurassic Morrison Formation (Utah). Note increasing volume and connectivity for channel-fills and channel belts upward through stratigraphy. (c) Channel-belt dimensions (depth and width) for the Morrison formation and the 'Monteith A' unit. (d) Stratigraphic net-to-gross trends through the Morrison Formation and 'Monteith B' unit.

management of groundwater resources (e.g., Hornung and Aigner 1999; Van Dijk et al. 2016). A relevant example is that of the High Plains Aquifer of the west-central United States, a complex of Cenozoic terrestrial sediments accounting for nearly one third of the groundwater resources extracted in the country (Dennehy 2000), subject to continuous depletion especially due to intensive agricultural exploitation (Rodell and Famiglietti 2002). Analyses of sparse outcrops and increasing availability of subsurface data suggest that a significant volume of the aquifer system, mostly represented by the Miocene to Pliocene Ogallala Group, may have been deposited by fluvial fans draining eastward from the Rocky Mountains, with stratal intervals representing also distinct phases of aeolian and volcaniclastic aggradation (Gustavson and Winkler 1988; Gustavson 1996; Harlow 2014).

More detailed work for reservoir characterisation should also consider possible deviations in stratigraphic architecture related to allogenically forced events of fan progradation and backstepping (with repetitive stacking of different facies associations caused by horizontal displacement of depositional belts over the aggrading surface) or autogenic switching of channel belts and major fan lobes (e.g., Martinius 2000; Kong et al. 2002; Trendell et al. 2012; Quartero et al. 2015). For example, exceptions to classical facies trends result from major interference of changing tectonic and environmental conditions or base-level changes during fan progradation. Climatic or hydro-geomorphic changes affecting the dynamics of highland catchments can induce disequilibrium between sediment supply and runoff transport capacity. In several active systems, deep radial incision has led to the establishment of longitudinal valleys (from several metres to a few tens of metres deep, up to many kilometres wide; Leckie 1994; Gibling et al. 2005; Weissmann et al. 2005; Assine et al. 2014; Van Dijk et al. 2016), moving the locus of sediment dispersal downfan, and separating perched, inactive proximal surfaces undergoing pedogenic modification from entrenched, topographically confined channel belts with coarse bedload deposition. Terraced, inset fan architectures also result from alternation of aggradation and degradation controlled by tectonic tilting and/or climate change (Maizels 1990; Gibling et al. 2005; Fontana et al. 2014). Valleys incised longitudinally

through fluvial fans alter the continuity of deposits, and add heterogeneity on a scale larger than facies transitions related to gradual progradation. Their dominant infill by highly amalgamated channel deposits renders them interesting exploration targets. At a larger scale, along basin margins, the occurrence of potentially large (palaeo)interfan areas hosting lesser, finer sandstone bodies and affected by greater pedogenic modification may also alter the lateral continuity and connectivity of proximal coarse-clastic facies belts. Other complications for system-scale architecture and compartmentalisation may derive from syndepositional or early post-depositional deformation, especially from salt tectonics and halokinetically driven deformation concomitant with fan development (e.g., McKie et al. 2010; Ribes et al. 2017), and must be assessed depending on the local context.

Of particular interest for geo-energy resources is also the repeated occurrence of regionally extensive, stratigraphically clustered coal seams in thick coastal-plain successions that are coeval with stratal intervals possibly representing more proximal domains of large fan systems. Important Palaeozoic and Mesozoic coal provinces are associated with heterolithic alluvial successions featuring tabular, aggradational architecture at the regional scale, and comprising numerous dispersed or amalgamated, mixed bedload and suspended-load channel fills with lateral accretion. Also displayed are local evidence for tidal influence, abundant floodplain fines, interbedded marginal-marine facies associations, and trends of textural coarsening, increased channel-fill amalgamation and 'terrestrialisation' up-section through stratigraphy (Gibling and Bird 1994; Aitken and Flint 1995; Allen and Fielding 2007; Fanti and Catuneanu 2009; Hampson et al. 2012; Rittersbacher et al. 2014). Whereas typical deltaic geometries and facies distributions are absent at system scale, evidence points to peat accumulation on broad, waterlogged, low-gradient alluvial plains fed by fluvial fans reaching the coastline and interacting with the shallow-marine realm. Distal-fan sectors would have been ideal for peat accumulation over vast rheotrophic mires, thereby resulting in extensive sheet-like coals given the low topography (which maintained the aggrading surface in contact with high groundwater tables), and given also the relatively large areas unaffected by active channel belts. Humid/perhumid palaeoclimates would have favoured peat formation (Bohacs and Suter 1997; Cecil 2013). Recurring base-level changes ranging from transgressive to early highstand might have been even more important (e.g., Gibling and Bird 1994; Aitken and Flint 1995), responsible for repeated episodes of rising water tables through the uppermost sedimentary column, which would have been otherwise subject to long-term aggradation and raised topographically out of shallow-groundwater range. A depositional model for large volumes of peat on distal fans has not yet been explored (but see Cecil 2013 for a possible example), but it would benefit the appraisal of coal quality and distribution at basin scale.

Similarly, investigations have been carried out and models developed for mineral resources such as placer ore deposits (including gold) and uranium precipitates in fan successions (e.g., McGowen and Groat 1971; Galloway 1977; Sanford 1982; Turner-Peterson and Fishman 1986; Owen et al. 2016). Again, the palaeo-hydraulics and connectivity of channel fills across fan sectors are relevant for evaluating potential mineral resources since they control accumulation of heavy-mineral grains (hydraulically equivalent to very coarse sand and fine gravel; Smith and Minter 1980) within proximal facies associations or specific architectural elements, and affect the chemistry and migration of subsurface fluids responsible for mobilisation and reprecipitation of uranium-bearing compounds along medial to distal permeability-transition zones and associated redox fronts. Economically interesting secondary accumulations of uranium and molybdenum minerals have been recognised in channel fills of the Permian Elliot Formation (Karoo Basin of South Africa; Le Roux 1993), in Jurassic channel strata of the Morrison Formation (Unites States Midwest; Turner-Peterson and Fishman 1986; Owen et al. 2016), and in amalgamated channel-belt deposits of fluvial-fan systems which aggraded the northern sectors of the Texas Gulf Coast (Galloway 1977, 1981). More prospectivity analyses for these resources are required to develop applicable models for primary and secondary ore emplacement in fan strata.

14.5 Conclusions

The current reprise of interest in fluvial (mega)fan geomorphology and sedimentology is increasing our understanding of processes and deposits of these systems at an unprecedented rate, as well as improving our ability to identify them from thick stratigraphic successions in the rock record. Numerous recent publications support the notion that large fan systems may be fundamental in controlling sediment distribution and aggradation within continental basins, and in dictating the organisation of alluvial facies belts and architectural patterns at length scales from tens to hundreds of kilometres. Advances in fluvial-fan stratigraphy are still lacking a more systematic, quantitative approach, but are opening new perspectives for prediction and characterisation of large volumes of subsurface deposits, especially along the margins of continental basin fills. Recurrent stratigraphic patterns within fluvial-fan successions offer good opportunities for subsurface exploration and reservoir development to be framed in a more predictable stratigraphic context than is commonly the case for thick fluvial successions. New depositional models for vast volumes of peat over distal, topographically low sectors of large fan systems may aid in the identification of major coal resources and in steering further development in producing coal provinces. The potential economic importance of fluvial-fan deposits, in turn, should open new opportunities for research on these as yet poorly explored depositional systems.

References

Aitken, J. F. and Flint, S. S. (1995). The application of high-resolution sequence stratigraphy to fluvial systems: a case study from the Upper Carboniferous Breathitt Group, eastern Kentucky, USA. *Sedimentology*, 42, 3–30.

Allen, J. P. and Fielding, C. R. (2007). Sedimentology and stratigraphic architecture of the Late Permian Betts Creek Beds, Queensland, Australia. *Sedimentary Geology*, 202, 5–34.

Allen, P. A., Cabrera, L., Colombo, F., and Matter, A. (1983). Variations in fluvial style on the Eocene-Oligocene alluvial fan of the Scala Dei Group, SE Ebro Basin, Spain. *Journal of the Geological Society, London*, 140, 133–146.

Arche, A. and López-Gómez, J. (2005). Sudden changes in fluvial style across the Permian-Triassic boundary in the eastern Iberian Ranges, Spain: analysis of possible causes. *Palaeogeography, Palaeoclimatology, Palaeoecology*, 229, 104–126.

Ashworth, P. J. and Lewin, J. (2012). How do big rivers come to be different? *Earth-Science Reviews*, 114, 84–107.

Assine, M. L. (2005). River avulsions on the Taquari mega-fan, Pantanal wetland, Brazil. *Geomorphology*, 70, 357–371.

Assine, M. L. and Silva, A. (2009). Contrasting fluvial styles of the Paraguay River in the northwestern border of the Pantanal wetland, Brazil. *Geomorphology*, 113, 189–199.

Assine, M. L., Corradini, F. A., Pupim, F. N., and McGlue, M. M. (2014). Channel arrangements and depositional styles in the São Lourenço fluvial megafan, Brazilian Pantanal wetland. *Sedimentary Geology*, 301, 172–184.

Atchley, S. C., Nordt, L. C., and Dworkin, S. I. (2004). Eustatic control on alluvial sequence stratigraphy: a possible example from the Cretaceous-Tertiary transition of the Tornillo Basin, Big Bend National Park, west Texas, U.S.A. *Journal of Sedimentary Research*, 74, 391–404.

Atchley, S. C., Nordt, L. C., Dworkin, S. I., et al. (2013). Alluvial stacking pattern analysis and sequence stratigraphy: concepts and case studies. In S. Driese, ed., *New Frontiers in Paleopedology and Terrestrial Paleoclimatology*, SEPM Special Publication, 104, 109–129.

Batezelli, A., Ladeira, F. S. B., Nascimento, D. L., and Silva, M. L. (2019). Facies and palaeosol analysis in a progradational distributive fluvial system from the Campanian-Maastrichtian Bauru Group, Brazil. *Sedimentology*, 66, 699–735.

Beer, J. A. and Jordan, T. E. (1989). The effects of Neogene thrusting on deposition in the Bermejo Basin, Argentina. *Journal of Sedimentary Petrology*, 59, 330–345.

Bentham, P. A., Talling, P. J., and Burbank, D. W. (1993). Braided stream and flood-plain deposition in a rapidly aggrading basin: the Escanilla Formation, Spanish Pyrenees. In J. L. Best and C. S. Bristow, eds., *Braided Rivers*. Geological Society of London, Special Publication, 75, 177–194.

Bentley, S. J., Blum, M. D., Maloney, J., Pond, L., and Paulsell, R. (2016). The Mississippi River source-to-sink system: perspectives on tectonic, climatic, and anthropogenic influences, Miocene to Anthropocene. *Earth-Science Reviews*, 153, 139–174.

Blechschmidt, I., Matter, A., Preusser, F., and Rieke-Zapp, D. (2009). Monsoon triggered formation of Quaternary alluvial megafans in the interior of Oman. *Geomorphology*, 110, 128–139.

Blum, M. D. and Törnqvist, T. E. (2000). Fluvial responses to climate and sea-level change: a review and look forward. *Sedimentology*, 47, 2–48.

Blum, M., Martin, J., Milliken, K., and Garvin, M. (2013). Paleovalley systems: insights from Quaternary analogs and experiments. *Earth-Science Reviews*, 116, 128–169.

Bohacs, K. and Suter, J. (1997). Sequence stratigraphic distribution of coaly rocks: fundamental controls and paralic examples. *Annals of the Association of Petroleum Geologists Bulletin*, 81, 1612–1639.

Boothroyd, J. C. and Nummedal, D. (1978). Proglacial braided outwash: a model for humid alluvial fan deposits. In A. D. Miall, ed., *Fluvial Sedimentology*. Canadian Society of Petroleum Geologists Memoir, 5, 641–668.

Bridge, J. S. (1984). Large-scale facies sequences in alluvial overbank environments. *Journal of Sedimentary Petrology*, 54, 583–588.

Bridge, J. S. and Leeder, M. R. (1979). A simulation model of alluvial stratigraphy. *Sedimentology*, 26, 617–644.

Browne, G. (2004). Downstream fining and sorting of gravel clasts in the braided rivers of mid-Canterbury, New Zealand. *New Zealand Geographer*, 60, 2–14.

Bryant, M., Falk, P., and Paola, C. (1995). Experimental study of avulsion frequency and rate of deposition. *Geology*, 23, 365–368.

Buehler, H. A., Weissmann, G. S., Scuderi, L. A., and Hartley, A. J. (2011). Spatial and temporal evolution of an avulsion of the Taquari River distributive fluvial system from satellite image analysis. *Journal of Sedimentary Research*, 81, 630–640.

Cain, S. A. and Mountney, N. P. (2009). Spatial and temporal evolution of a terminal fluvial fan system: the Permian Organ Rock Formation, south-east Utah, USA. *Sedimentology*, 56, 1774–1800.

Campbell, C. V. (1976). Reservoir geometry of a fluvial sheet sandstone. *American Association of Petroleum Geologists Bulletin*, 60, 1009–1020.

Cecil, C. B. (2013). An overview and interpretation of autocyclic and allocyclic processes and the accumulation of strata during the Pennsylvanian-Permian transition in the central Appalachian Basin, USA. *International Journal of Coal Geology*, 119, 21–31.

Chakraborty, T., Kar, R., Ghosh, P., and Basu, S. (2010). Kosi megafan: historical records, geomorphology and the recent avulsion of the Kosi River. *Quaternary International*, 227, 143–160.

Christophoul, F., Baby, P., and Dávila, C. (2002). Stratigraphic responses to a major tectonic event in a foreland basin: the Ecuadorian Oriente Basin from Eocene to Oligocene times. *Tectonophysics*, 345, 281–298.

Cohen, T. J., Nanson, G. C., Larsen, J. R., et al. (2010). Late Quaternary aeolian and fluvial interactions on the Cooper Creek Fan and the association between linear and source-bordering dunes, Strzelecki Desert, Australia. *Quaternary Science Reviews*, 29, 455–471.

Colombera, L., Mountney, N. P., and McCaffrey, W. D. (2015). A meta-study of relationships between fluvial channel-body stacking pattern and aggradation rate: implications for sequence stratigraphy. *Geology*, 43, 283–286.

Corbett, M. J., Fielding, C. R., and Birgenheier, L. P. (2011). Stratigraphy of a Cretaceous coastal-plain succession: Campanian Masuk Formation, Henry Mountains syncline, Utah, U.S.A. *Journal of Sedimentary Research*, 81, 80–96.

Cuevas Gozalo, M. C. and Martinius, A. W. (1993). Outcrop data-base for the geological characterization of fluvial reservoirs: an example from distal fluvial fan deposits in the Loranza Basin, Spain. In C. P. North and D. J. Prosser, eds., *Characterization of Fluvial and Aeolian Reservoirs*. Geological Society of London, Special Publication, 73, 79–94.

Currie, B. S. (1997). Sequence stratigraphy of nonmarine Jurassic-Cretaceous rocks, central Cordilleran foreland-basin system. *Geological Society of America Bulletin*, 109, 1206–1222.

Damanti, J. F. (1993). Geomorphic and structural controls on facies patterns and sediment composition in a modern foreland basin. In M. Marzo and C. Puigdefábregas, eds., *Alluvial Sedimentation*. International Association of Sedimentologists, Special Publication, 17, 221–233.

Davidson, S. K., Hartley, A. J., Weissmann, G. S., Nichols, G. J., and Scuderi, L. A. (2013). Geomorphic elements on modern distributive fluvial systems. *Geomorphology*, 180–181, 82–95.

DeCelles, P. G. and Cavazza, W. (1999). A comparison of fluvial megafans in the Cordilleran (Upper Cretaceous) and modern Himalayan foreland basin systems. *Geological Society of America Bulletin*, 111, 1315–1334.

Dennehy, K. F. (2000). High plains regional ground-water study. U.S.G.S. Fact Sheet, FS-091-00, 1-6.

Donselaar, M. E., Cuevas Gozalo, M. C., and Moyano, S. (2013). Avulsion processes at the terminus of low-gradient semi-arid fluvial systems: lessons from the Río Colorado, Altiplano endorheic basin, Bolivia. *Sedimentary Geology*, 283, 1–14.

Esposito, C. R., Di Leonardo, D., Harlan, M., and Straub, K. M. (2018). Sediment storage partitioning in alluvial stratigraphy: the influence of discharge variability. *Journal of Sedimentary Research*, 88, 717–726.

Evans, J. E. (1991). Facies relationships, alluvial architecture, and paleohydrology of a Paleogene, humid-tropical alluvial-fan system: Chumstick Formation, Washington state, U.S.A. *Journal of Sedimentary Petrology*, 61, 732–755.

Fanti, F. and Catuneanu, O. (2009). Stratigraphy of the Upper Cretaceous Wapiti Formation, west-central Alberta, Canada. *Canadian Journal of Earth Sciences*, 46, 263–286.

Fielding, C. R., Ashworth, P. J., Best, J. L., Prokocki, E. W., and Sambrook Smith, G. H. (2012). Tributary, distributary and other fluvial patterns: What *really* represents the norm in the continental rock record? *Sedimentary Geology*, 261–262, 15–32.

Fielding, C. R., Alexander, J., and Allen, J. P. (2018). The role of discharge variability in the formation and preservation of alluvial sediment bodies. *Sedimentary Geology*, 365, 1–20.

Figuereido, J., Hoorn, C., Van der Ven, P., and Soares, E. (2009). Late Miocene onset of the Amazon River and the Amazon deep-sea fan: evidence from the Foz do Amazonas Basin. *Geology*, 37, 619–622.

Fisher, J. A., Krapf, C. B. E., Lang, S. C., Nichols, G. J., and Payenberg, T. H. D. (2008). Sedimentology and architecture of the Douglas Creek terminal splay, Lake Eyre, central Australia. *Sedimentology*, 55, 1915–1930.

Flood, Y. S. and Hampson, G. J. (2014). Facies and architectural analysis to interpret avulsion style and variability: Upper Cretaceous Blackhawk Formation, Wasatch Plateau, central Utah, U.S.A. *Journal of Sedimentary Research*, 84, 743–762.

Fontana, A., Mozzi, P., and Marchetti, M. (2014). Alluvial fans and megafans along the southern side of the Alps. *Sedimentary Geology*, 301, 150–171.

Friend, P. F. (1978). Distinctive features of some ancient river systems. In A. D. Miall, ed., *Fluvial Sedimentology*. Canadian Society of Petroleum Geologists Memoir, 5, 531–542.

Friend, P. F. (1996). The development of fluvial sedimentology in some Devonian and Tertiary basins. *Cuadernos de Geologia Ibérica*, 21, 55–69.

Friend, P. F., Jones, N. E., and Vincent, S. J. (1999). Drainage evolution in active mountain belts: extrapolation backwards from present-day Himalayan river patterns. In N. D. Smith and J. Rogers, eds., *Fluvial Sedimentology VI*. International Association of Sedimentologists, Special Publication, 28, 305–313.

Friend, P. F., Raza, S. M., Geehan, G., and Sheikh, K. A. (2001). Intermediate-scale architectural features of the fluvial Chinji Formation (Miocene), Siwalik Group, northern Pakistan. *Journal of the Geological Society, London*, 158, 163–177.

Fryberger, S. G., Knight, R., Hern, C., Moscariello, A., and Kabel, S. (2011). Rotliegend facies, sedimentary provinces, and stratigraphy, Southern Permian Basin UK and Netherlands: a review with new observations. In J. Grötsch and R. Gaupp, eds., *The Permian Rotliegend of the Netherlands*. SEPM Special Publication, 69, 51–88.

Galloway, W. E. (1977). *Catahoula Formation of the Texas Coastal Plain: depositional systems, composition, structural development, ground-water flow, history, and uranium distribution.* Texas Bureau of Economic Geology, Investigation Report 87.

Galloway, W. E. (1981). Depositional architecture of Cenozoic Gulf coastal plain fluvial systems. In F. G. Ethridge and R. M. Flores, eds., *Recent and Ancient Nonmarine Depositional Environments: Models for Exploration*. SEPM Special Publication, 31, 127–155.

Galve, J. P., Alvarado, G. E., Pérez-Peña, J. V., et al. (2016). Megafan formation driven by explosive volcanism and active tectonic processes in a humid tropical environment. *Terra Nova*, 28, 427–433.

Gibling, M. R. and Bird, D. J. (1994). Late Carboniferous cyclothems and alluvial paleovalleys in the Sydney Basin, Nova Scotia. *Geological Society of America Bulletin*, 106, 105–117.

Gibling, M. R., Tandon, S. K., Sinha, R., and Jain, M. (2005). Discontinuity-bounded alluvial sequences of the southern Gangetic Plains, India: aggradation and degradation in response to monsoonal strength. *Journal of Sedimentary Research*, 75, 369–385.

Gibling, M. R., Fielding, C. R., and Sinha, R. (2011). Alluvial valleys and alluvial sequences: towards a geomorphic assessment. In S. K. Davidson, S. Leleu, and C. P. North, eds., *From River to Rock Record: The Preservation of Fluvial Sediments and their Subsequent Interpretation*. SEPM Special Publication, 97, 423–447.

Gohain, K. and Parkash, B. (1990). Morphology of the Kosi Megafan. In A. H. Rachocki and M. Church, eds., *Alluvial Fans: A Field Approach*. Wiley, Chichester, 151–178.

Gordon, E. A. and Bridge, J. S. (1987). Evolution of Catskill (Upper Devonian) river systems: intra- and extrabasinal controls. *Journal of Sedimentary Petrology*, 57, 234–249.

Graham, J. R. (1983). Analysis of the Upper Devonian Munster Basin, an example ofa fluvial distributary system. In J. D. Collinson and J. Lewin, eds., *Modern and Ancient Fluvial Systems*. International Association of Sedimentologists, Special Publication, 6, 473–483.

Gulliford, A. R., Flint S. S., and Hodgson, D. M. (2014). Testing applicability of models of distributive fluvial systems or trunk rivers in ephemeral systems: reconstructing 3-D fluvial architcture in the Beaufort Group, South Africa. *Journal of Sedimentary Research*, 84, 1147–1169.

Gulliford, A. R., Flint, S. S., and Hodgson, D. M. (2017). Crevasse splay processes and deposits in an ancient distributive fluvial system: the lower Beaufort Group, South Africa. *Sedimentary Geology*, 358, 1–18.

Gupta, S. (1997). Himalayan drainage patterns and the origin of fluvial megafans in the Ganges foreland basin. *Geology*, 25, 11–14.

Gustavson, T. C. (1996). *Fluvial and eolian depositional systems, paleosols, and paleoclimate of the upper Cenozoic Ogallala and Blackwater Draw formations, southern High Plains, Texas and New Mexico*. Texas Bureau of Economic Geology, Report of Investigations, 239.

Gustavson, T. C. and Winkler, D. A. (1988). Depositional facies of the Miocene-Pliocene Ogallala Formation, northwestern Texas and eastern New Mexico. *Geology*, 16, 203–206.

Hajek, E. A., Heller, P. L., and Schur, E. L. (2012). Field test of autogenic control on alluvial stratigraphy (Ferris Formation, Upper Cretaceous–Paleogene, Wyoming). *Geological Society of America Bulletin*, 124, 1898–1912.

Hampson, G. J., Gani, M. R., Sahoo, H., et al. (2012). Controls on large-scale patterns of fluvial sandbody distribution in alluvial to coastal plain strata: Upper Cretaceous Blackhawk Formation, Wasatch Plateau, central Utah, USA. *Sedimentology*, 59, 2226–2258.

Hampton, B. A. and Horton, B. K. (2007). Sheetflow fluvial processes in a rapidly subsiding basin, Altiplano Plateau, Bolivia. *Sedimentology*, 54, 1121–1147.

Harlow, R. H. (2014). *Depositional and paleoclimatic evolution of the Cenozoic High Plains succession from core: Haskell County, Kansas*. MSc thesis, University of Kansas.

Hartley, A. J., Weissmann, G. S., Nichols, G. J., and Warwick, G. L. (2010). Large distributive fluvial systems: characteristics, distribution, and controls on development. *Journal of Sedimentary Research*, 80, 167–183.

Hill, G. (1989). Distal alluvial fan sediments from the Upper Jurassic of Portugal: controls on their cyclicity and channel formation. *Journal of the Geological Society, London*, 146, 539–555.

Hinds, D. J., Aliyeva, E., Allen, M. B., et al. (2004). Sedimentation in a discharge dominated fluvial-lacustrine system: the Neogene Productive Series of the South Caspian Basin, Azerbaijan. *Marine and Petroleum Geology*, 21, 613–638.

Hofmann, M. H., Wroblewski, A., and Boyd, R. (2011). Mechanisms controlling the clustering of fluvial channels and the compensational stacking of cluster belts. *Journal of Sedimentary Research*, 81, 670–685.

Hornung, J and Aigner, T. (1999). Reservoir and aquifer characterization of fluvial architectural elements: Stubesandstein, Upper Triassic, southwest Germany. *Sedimentary Geology*, 129, 215–280.

Hornung, J. and Aigner, T. (2002). Reservoir architecture in a terminal alluvial plain: an outcrop analogue study (Upper Triassic, southern Germany). Part 1: sedimentology and petrophysics. *Journal of Petroleum Geology*, 25, 3–30.

Horton, B. K. and DeCelles, P. G. (2001). Modern and ancient fluvial megafans in the foreland basin system of the central Andes, southern Bolivia: implications for drainage network evolution in fold-thrust belts. *Basin Research*, 13, 43–63.

Hunger, G., Ventra, D., Moscariello, A. Veiga, G., and Chiaradia, M. (2018). High-resolution compositional analysis of a fluvial-fan succession: the Miocene infill of the Cacheuta Basin (central Argentinian foreland). *Sedimentary Geology*, 375, 268–288.

Iriondo, M. (1993). Geomorphology and late Quaternary of the Chaco (South America). *Geomorphology*, 7, 289–303.

Jerolmack, D. J. and Mohrig, D. (2007). Conditions for branching in depositional rivers. *Geology*, 25, 463–466.

Jones, B. G., Martin, G. R., and Senapati, N. (1993). Riverine-tidal interactions in the monsoonal Gilbert River fandelta, northern Australia. *Sedimentary Geology*, 83, 319–337.

Jones, H. L. and Hajek, E. A. (2007). Characterizing avulsion stratigraphy in ancient alluvial deposits. *Sedimentary Geology*, 202, 124–137.

Jones, L. S. and Schumm, S. A. (1999). Causes of avulsion: an overview. In N. D. Smith and J. Rogers, eds., *Fluvial Sedimentology VI*. International Association of Sedimentologists, Special Publication, 28, 171–178.

Keeton, G. I., Pranter, M. J., Cole, R. D., and Gustason, E. R. (2015). Stratigraphic architecture of fluvial deposits from borehole images, spectral-gamma-ray response and outcrop analogs, Piceance Basin, Colorado.

American Association of Petroleum Geologists Bulletin, 99, 1929–1956.

Kelly, S. B. and Olsen, H. (1993). Terminal fans – A review with reference to Devonian examples. *Sedimentary Geology*, 85, 339–374.

Khan, I. A., Bridge, J. S. Kappelman, J., and Wilson, R. (1997). Evolution of Miocene fluvial environments, eastern Potwar Plateau, northern Pakistan. *Sedimentology*, 44, 221–251.

Klausen, T. G., Ryseth, A. E., Helland-Hansen, W., Gawthorpe, R., and Laursen, I. (2014). Spatial and temporal changes in geometries of fluvial channel bodies from the Triassic Snadd Formation of offshore Norway. *Journal of Sedimentary Research*, 84, 567–585.

Kleinhans, M. G., Ferguson, R. I., Lane, S. N., and Hardy, R. J. (2013). Splitting rivers at their seams: bifurcations and avulsions. *Earth Surface Processes and Landforms*, 38, 47–61.

Kong, F., Jalfin, G., Lukito, P., and Sarkawi, I. (2002). Sequence stratigraphic framework in a humid alluvial fan complex, Quiriquire Oil Field, Venezuela. In J. M. Armentrout and N. C. Rosen, eds., *Sequence Stratigraphic Models for Exploration and Production*. GCS, SEPM Special Publication, 601–631.

Kukulski, R. B., Moslow, T. F., and Hubbard, S. M. (2013). Tight gas sandstone reservoir delineation through channel-belt analysis, Late Jurassic Monteith Formation, Alberta Deep Basin. *Bulletin of Canadian Petroleum Geology*, 61, 133–156.

Kumar, R. (1993). Coalescence megafan: multistorey sandstone complex of the late-orogenic (Mio-Pliocene) sub-Himalayan belt, Dehra Dun, India. *Sedimentary Geology*, 85, 327–337.

Kumar, R., Ghosh, S. K., Mazari, R. K., and Sangode, S. J. (2003). Tectonic impact on fluvial deposits of Plio-Pleistocene Himalayan foreland basin, India. *Sedimentary Geology*, 158, 209–234.

Lang, S. C., Payenberg, T. H. D., Reilly, M. R. W., et al. (2004). Modern analogues for dryland sandy fluvial-lacustrine deltas and terminal splay reservoirs. *Journal of the Australian Association of Petroleum Production and Exploration*, 44, 329–356.

Latrubesse, E. M. (2015). Large rivers, megafans and other Quaternary avulsive fluvial systems: a potential "who's who" in the geological record. *Earth-Science Reviews*, 146, 1–30.

Latrubesse, E. M., Da Silva, S. A. F., Cozzuol, M., and Absy, M. L. (2007). Late Miocene continental sedimentation in southwestern Amazonia and its regional significance: biotic and geological evidence. *Journal of South American Earth Sciences*, 23, 61–80.

Lawton, T. F., Schellenbach, W. L., and Nugent, A. E. (2014). Late-Cretaceous fluvial-megafan and axial-river systems in the southern Cordilleran foreland basin: Drip Tank Member of Straight Cliffs Formation and adjacent strata, southern Utah, U.S.A. *Journal of Sedimentary Research*, 84, 407–434.

Leckie, D. A. (1994). Canterbury Plains, New Zealand – Implications for sequence stratigraphic models.

American Association of Petroleum Geologists Bulletin, 78, 1240–1256.

Legarreta, L. and Uliana, M. A. (1998). Anatomy of hinterland depositional sequences: Upper Cretaceous fluvial strata, Neuquen Basin, west-central Argentina. In K. W. Shanley and P. J. McCabe, eds., *Relative Role of Eustasy, Climate and Tectonism in Continental Rocks*. SEPM Special Publication, 59, 83–92.

Leier, A. L., DeCelles, P. G., and Pelletier, J. D. (2005). Mountains, monsoons, and megafans. *Geology*, 33, 289–292.

Le Roux, J. P. (1993). Genesis of stratiform U–Mo deposits in the Karoo Basin of South Africa. *Ore Geology Reviews*, 7, 485–509.

Maizels, J. (1990). Long-term palaeochannel evolution during episodic growth of an exhumed Plio-Pleistocene alluvial fan, Oman. In A. H. Rachocki and M. Church, eds., *Alluvial Fans: A Field Approach*. Wiley, Chichester, 271–304.

Makaske, B., Maathuis, B. H. P., Padovani, C.R., Stolker, C., Mosselman, E. et al., (2012). Upstream and downstream controls of recent avulsions on the Taquari megafan, Pantanal, south-western Brazil. *Earth Surface Processes and Landforms*, 37, 1313–1326.

Marenssi, S. A., Ciccioli, P. L., Limarino, C. O., Schencman, L. J., and Diaz, M. Y. (2015). Using fluvial cyclicity to decipher the interaction of basement- fold-thrust-belt tectonics in a broken foreland basin: Vinchina Formation (Miocene), northwestern Argentina. *Journal of Sedimentary Research*, 85, 361–380.

Marquillas, R. A., Del Papa, C., and Sabino, I. F. (2005). Sedimentary aspects and paleoenvironmental evolution of a rift basin: Salta Group (Cretaceous-Paleogene), northwestern Argentina. *International Journal of Earth Sciences*, 94, 94–113.

Martinius, A. W. (2000). Labyrinthine facies architecture of the Tórtola fluvial system and controls on deposition (Late Oligocene – early Miocene, Loranca Basin, Spain). *Journal of Sedimentary Research*, 70, 850–867.

May, J.-H. (2011). The Rio Parapetí – Holocene megafan formation in the southernmost Amazon Basin. *Geographica Helvetica*, 66, 193–201.

McCarthy, T. S. (1993). The great inland deltas of Africa. *Journal of African Earth Sciences*, 17, 275–291.

McCarthy, T. S., Stanistreet, I. G. and Cairncross, B. (1991). The sedimentary dynamics of active fluvial channels on the Okavango fan, Botswana. *Sedimentology*, 38, 471–487.

McCarthy, T. S., Ellery, W. N., and Stanistreet, I. G. (1992). Avulsion mechanisms on the Okavango fan, Botswana: the control of a fluvial system by vegetation. *Sedimentology*, 39, 779–795.

McGowen, J. H. and Groat, C. G. (1971). *Van Horn Sandstone, West Texas: an alluvial fan model for mineral exploration*. Texas Bureau of Economic Geology, Investigation Report 72.

McKie, T. (2011). A comparison of modern dryland depositional systems with the Rotliegend Group in the

Netherlands. In J. Grötsch and R. Gaupp, eds., *The Permian Rotliegend of the Netherlands*. SEPM Special Publication, 69, 89–103.

McKie, T. and Williams, B. (2009). Triassic palaeogeography and fluvial dispersal across the northwest European basins. *Geological Journal*, 44, 711–741.

McKie, T., Jolley, S. J., and Kristensen, M. B. (2010). Stratigraphic and structural compartmentalization of dryland fluvial reservoirs: Triassic Heron Cluster, central North Sea. In S. J. Jolley, Q. J. Fisher, R. B. Ainsworth, P. Vrolijk, and S. Delisle, eds., *Reservoir Compartmentalization*. Geological Society of London, Special Publication, 347, 165–198.

Miall, A. D. (1996). *The Geology of Fluvial Deposits*. Springer, Berlin, 582 pp.

Middleton, G. V. (1973). Johannes Walther's law of the correlation of facies. *Geological Society of America Bulletin*, 84, 979–988.

Mohrig, D., Heller, P. L., Paola, C., and Lyons, W. J. (2000). Interpreting avulsion process from ancient alluvial sequences: Guadalope-Matarranya system (northern Spain) and Wasatch Formation (western Colorado). *Geological Society of America Bulletin*, 112, 1787–1803.

Moscariello, A. (2005). Exploration potential of the mature Southern North Sea basin margins: some unconventional plays based on alluvial and fluvial fan sedimentation models. In A. G. Doré and B. A. Vining, eds., *Petroleum Geology: North-West Europe and Global Perspectives*. Geological Society of London, Petroleum Geology Conference Series, 6, 595–605.

Moscariello, A. (2018). Alluvial fans and fluvial fans at the margins of continental sedimentary basins: geomorphic and sedimentological distinction for geo-energy exploration and development. In D. Ventra and L. E. Clarke, eds., *Geology and Geomorphology of Alluvial and Fluvial Fans: Terrestrial and Planetary Perspectives*. Geological Society of London, Special Publication, 440, 215–243.

Nageswara Rao, K., Saito, Y., Nagakumar, K .C. V., et al. (2015). Palaeogeography and evolution of the Godavari delta, east coast of India during the Holocene: an example of wave-dominated and fan-delta settings. *Palaeogeography, Palaeoecology, Palaeoclimatology*, 440, 213–233.

Nichols, G. J. and Hirst, J. P. (1998). Alluvial fans and fluvial distributary systems, Oligo-Miocene, northern Spain: contrasting processes and products. *Journal of Sedimentary Research*, 68, 879–889.

Nichols, G. J. and Fisher, J. A. (2007). Processes facies and architecture of fluvial distributary system deposits. *Sedimentary Geology*, 195, 75–90.

Olsen, H. (1987). Ancient ephemeral stream deposits: a local terminal fan model from the Bunter Sandstone Formation (L. Triassic) in the Tønder-3, -4 and -5 wells, Denmark. In L. Frostick and I. Reid, eds., *Desert Sediments: Ancient and Modern*. Geological Society of London, Special Publication, 35, 69–86.

Olsen, T., Steel, R., Høgseth, K., Skar, T., and Røe, S.-L. (1995). Sequential architecture in a fluvial succession: sequence stratigraphy in the Upper Cretaceous Mesaverde Group, Price Canyon, Utah. *Journal of Sedimentary Research*, B65, 265–280.

Owen, A., Nichols, G. J., Hartley, A. J., Weissmann, G. S. and Scuderi, L. A. (2015). Quantification of a distributive fluvial system: the Salt Wash DFS of the Morrison Formation, SW U.S.A. *Journal of Sedimentary Research*, 85, 544–561.

Owen, A., Nichols, G. J., Hartley, A. J., and Weissmann, G. S. (2017). Vertical trends within the prograding Salt Wash distributive fluvial system, SW United States. *Basin Research*, 29, 64–80.

Owen, A., Hartley, A. J., Weissmann, G. S., and Nichols, G. J. (2016). Uranium distribution as a proxy for basin-scale fluid flow in distributive fluvial systems. *Journal of the Geological Society, London*, 173, 569–572.

Page, K. J. and Nanson, G. C. (1996). Stratigraphic architecture resulting from Late Quaternary evolution of the Riverine Plain, south-eastern Australia. *Sedimentology*, 43, 927–945.

Parkash, B., Awasthi, A.K., and Gohain, K. (1983). Lithofacies of the Markanda terminal fan, Kurukshetra District, Haryana, India. In J. D. Collinson and J. Lewin, eds., *Modern and Ancient Fluvial Systems*. International Association of Sedimentologists, Special Publication, 6, 337–344.

Phillips, J. D. (2011). Universal and local controls of avulsions in southeast Texas rivers. *Geomorphology*, 130, 17–28.

Pietsch, T. J. and Nanson, G. C. (2011). Bankfull hydraulic geometry; the role of in-channel vegetation and downstream declining discharges in the anabranching and distributary channels of the Gwydir distributive fluvial system, southeastern Australia. *Geomorphology*, 129, 152–165.

Primm, J. W., Johnson, C. L. and Stearns, M. (2018). Basin-axial progradation of a sediment supply driven distributive fluvial system in the Late Cretaceous southern Utah foreland. *Basin Research*, 30, 249–278.

Quartero, E. M., Leier, A. L., Bentley, L. R., and Glombick, P. (2015). Basin-scale stratigraphic architecture and potential Paleocene distributive fluvial systems of the Cordilleran Foreland Basin, Alberta, Canada. *Sedimentary Geology*, 316, 26–38.

Ralph, T. J. and Hesse, P. P. (2010). Downstream hydro-geomorphic changes along the Macquarie River, southeastern Australia, leading to channel breakdown and floodplain wetlands. *Geomorphology*, 118, 48–64.

Rannie, W. F. (1990). The Portage La Prairie 'floodplain fan'. In A. H. Rachocki and M. Church, eds., *Alluvial Fans: A Field Approach*. Wiley, Chichester, 179–193.

Räsänen, M., Neller, R., Salo, J., and Jungner, H. (1992). Recent and ancient fluvial deposition systems in the Amazonian foreland basin, Peru. *Geological Magazine*, 129, 293–306.

Ray, S. and Chakaborty, T. (2002). Lower Gondwana fluvial succession of the Pench-Kanhan valley, India: stratigraphic architecture and depositional controls. *Sedimentary Geology*, 151, 243–271.

Reits, M. D. and Jerolmack, D. J. (2012). Experimental alluvial fan evolution: channel dynamics, slope controls, and shoreline growth. *Journal of Geophysical Research*, 117, F02021, http://dx.doi.org/10.1029/2011JF002261.

Ribes, C., Kergaravat, C., Crumeyrolle, P., et al., (2017). Factors controlling stratal pattern and facies distribution of fluvio-lacustrine sedimentation in the Sivas mini-basins, Oligocene (Turkey). *Basin Research*, 29, 596–621.

Richards, K., Chandra, S., and Friend, P. (1993). Avulsive channel systems: characteristics and examples. In J. L Best and C. S. Bristow, eds., *Braided Rivers*. Geological Society of London, Special Publication, 75, 195–203.

Rittersbacher, A., Howell, J. A., and Buckley, S. J. (2014). Analysis of fluvial architecture in the Blackhawk Formation, Wasatch Plateau, Utah, U.S.A., using large 3D photorealistic models. *Journal of Sedimentary Research*, 84, 72–87.

Rodell, M. and Famiglietti, J. S. (2002). The potential for satellite-based monitoring of groundwater storage changes using GRACE: the High Plains aquifer, central US. *Journal of Hydrology*, 263, 245–256.

Rossetti, D. F., Zani, H., Cohen, M. C. L., and Cremon, É. H. (2012). A Late Pleistocene-Holocene wetland megafan in the Brazilian Amazonia. *Sedimentary Geology*, 282, 276–293.

Sáez, A., Anadón, P., Herrero, M. J., and Moscariello, A. (2007). Variable style of transition between Palaeogene fluvial fan and lacustrine systems, southern Pyrenean foreland, NE Spain. *Sedimentology*, 54, 367–390.

Sadler, S. P. and Kelly, S. B. (1993). Fluvial processes and cyclicity in terminal fan deposits: an example from the Late Devonian of southwest Ireland. *Sedimentary Geology*, 85, 375–386.

Sanford, R. F. (1982). Preliminary model of regional Mesozoic groundwater flow and uranium deposition in the Colorado Plateau. *Geology*, 10, 348–352.

Schlunegger, F. and Norton, K. P. (2015). Climate *vs.* tectonics: the competing roles of Late Oligocene warming and Alpine orogenesis in constructing alluvial megafan sequences in the North Alpine foreland basin. *Basin Research*, 27, 230–245.

Schlunegger, F., Matter, A., Burbank, D. et al. (1997). Sedimentary sequences, seismofacies and evolution of depositional systems of the Oligo/Miocene Lower Freshwater Molasse Group, Switzerland. *Basin Research*, 9, 1–26.

Schmitz, B. and Pujalte, B. (2007). Abrupt increase in seasonal extreme precipitation at the Paleocene–Eocene boundary. *Geology*, 35, 215–218.

Shanley, K. W. and McCabe, P. J. (1991). Predicting facies architecture through sequence stratigraphy – An example from the Kaiparowits Plateau, Utah. *Geology*, 19, 742–745.

Shukla, U. K., Singh, I. B., Sharma, M., and Sharma, S. (2001). A model of alluvial megafan sedimentation: Ganga Megafan. *Sedimentary Geology*, 144, 243–262.

Silva, F. P., Silva, A., Martinius, A. W., and Weber, K. J. (1996). The Tórtola fluvial system: an analogue for the Upper Gharif of the Sultanate of Oman. *GeoArabia*, 1, 325–342.

Singh, H., Parkash, B. and Gohain, K. (1993). Facies analysis of the Kosi megafan deposits. *Sedimentary Geology*, 85, 87–113.

Sinha, R. and Friend, P. F. (1994). River systems and their sediment flux, Indo-Gangetic plains, Northern Bihar, India. *Sedimentology*, 41, 825–845.

Sinha, R., Sripriyanka, K., Jain, V., and Mukul, M. (2014). Avulsion threshold and planform dynamics of the Kosi River in north Bihar (India) and Nepal: a GIS framework. *Geomorphology*, 216, 157–170.

Slingerland, R. and Smith, N. D. (2004). River avulsions and their deposits. *Annual Review of Earth and Planetary Sciences*, 32, 257–285.

Smith, N. D. and Minter, W. E. (1980). Sedimentological controls of gold and uranium in two Witwatersrand paleoplacers. *Economic Geology*, 75, 1–14.

Smith, R. M. H. (1995). Changing fluvial environments across the Permian-Triassic boundary in the Karoo Basin, South Africa, and possible causes of tetrapod extinctions. *Palaeogeography, Palaeoclimatology, Palaeoecology*, 117, 81–104.

Stanistreet, I. G. and McCarthy, T. S. (1993). The Okavango Fan and the classification of subaerial fan systems. *Sedimentary Geology*, 85, 115–133.

Tooth, S. and McCarthy, T. S. (2007). Wetlands in drylands: geomorphological and sedimentological characteristics, with emphasis on examples from southern Africa. *Progress in Physical Geography*, 31, 3–41.

Trendell, A. M., Atchley S. C., and Nordt, L. C. (2012). Depositional and diagenetic controls on reservoir attributes within a fluvial outcrop analog: Upper Triassic Sonsela Member of the Chinle Formation, Petrified Forest National Park, Arizona. *American Association of Petroleum Geologists Bulletin*, 96, 679–707.

Turner-Peterson, C. E. and Fishman, N. S. (1986). Geologic synthesis and genetic models for uranium mineralization in the Morrison Formation, Grants uranium region, New Mexico. In C. E. Turner-Peterson, E. S. Santos, and N. S. Fishman, eds., *A Basin Analysis Case Study: The Morrison Formation, Grants Uranium Region*. AAPG Studies in Geology, 22, 357–388.

Uba, C. E., Heubeck, C., and Hulka, C. (2005). Facies analysis and basin architecture of the Neogene Subandean synorogenic wedge, southern Bolivia. *Sedimentary Geology*, 180, 91–123.

Van Dijk, W. M., Densmore, A. L., Singh, A., et al. (2016). Linking the morphology of fluvial fan systems to aquifer stratigraphy in the Sutlej-Yamuna plain of northwest India. *Journal of Geophysical Research: Earth Surface*, 121, 201–222.

Ventra, D. and Clarke, L. E. (2018). Geology and geomorphology of alluvial and fluvial fans: current progress and research perspectives. In D. Ventra and L. E. Clarke, eds., *Geology and Geomorphology of Alluvial and Fluvial Fans: Terrestrial and Planetary*

Perspectives. Geological Society of London, Special Publication, 440, 1–21.

Weber, K. J. and Van Geuns, L. (1990). Framework for constructing clastic reservoir simulation models. *Journal of Petroleum Technology*, 42, 1248–1297.

Weissmann, G. S., Bennett, G. L., and Lansdale, A. L. (2005). Factors controlling sequence development on Quaternary fluvial fans, San Joaquin Basin, California, USA. In A. M. Harvey, A. E. Mather, and M. Stokes *Alluvial Fans: Geomorphology, Sedimentology, Dynamics.* Geological Society of London, Special Publication, 251, 169–186.

Weissmann, G. S., Hartley, A. J., Nichols, G. J., et al. (2010). Fluvial form in modern continental sedimentary basins: Distributive fluvial systems. *Geology*, 38, 39–42.

Weissmann, G. S., Hartley, A. J., Nichols, G. J., et al. (2011). Alluvial facies distributions in continental sedimentary basins – Distributive fluvial systems. In S. K. Davidson, S. Leleu, and C. P. North, eds., *From River to Rock Record: The Preservation of Fluvial Sediments and their Subsequent Interpretation.* SEPM Special Publication, 97, 325–356.

Weissmann, G. S., Hartley, A. J., Scuderi, L. A., et al. (2013). Prograding distributive fluvial systems – Geomorphic models and ancient examples. In S. Driese, ed., *New Frontiers in Paleopedology and Terrestrial Paleoclimatology.* SEPM Special Publication, 104, 131–147.

Weissmann, G. S., Hartley, A. J., Scuderi, L. A., et al. (2015). Fluvial geomorphic elements in modern sedimentary basins and their potential preservation in the rock record: a review. *Geomorphology*, 250, 187–219.

Wells, N. A. and Dorr, J. A. (1987a). A reconnaissance of sedimentation on the Kosi alluvial fan of India. In R. M. Flores and M. D. Harvey, eds., *Recent Developments in Fluvial Sedimentology.* SEPM Special Publication, 39, 51–61.

Wells, N. A. and Dorr, J. A. (1987b). Shifting of the Kosi River, northern India. *Geology*, 15, 204–207.

Wilkinson, M. J., Cameron, N. R., and Burke, K. (2002). Global geomorphic surveys of large modern subaerial fans: distribution and exploration implications. *Houston Geological Society Bulletin*, 44, 11–13.

Wilkinson, M. J., Marshall, L. G., and Lundberg, J. G. (2006). River behavior on megafans and potential influences on diversification and distribution of aquatic organisms. *Journal of South American Earth Sciences*, 21, 151–172.

Wilkinson, M. J., Marshall, L. G., Lundberg, J. G., and Kreslavsky, M. H. (2010). Megafan environments in northern South America and their impact on Amazon Neogene aquatic ecosystems. In C. Hoorn and F. P. Wesselingh, eds., *Amazonia, Landscape and Species Evolution: A Look into the Past.* Blackwell, London, 162–184.

Willis, B. (1993). Ancient river systems in the Himalayan foredeep, Chinji Village area, northern Pakistan. *Sedimentary Geology*, 88, 1–76.

Willis, B. J. and Bridge, J. S. (1988). Evolution of Catskill River systems, New York state. In N. J. McMillan, A. F. Embry, and D. J. Glass, eds., *Devonian of the World.* Canadian Society of Petroleum Geologists Memoir, 14, 85–106.

Wilson, A., Flint, S., Payenberg, T., Tohver, E., and Lanci, L. (2014). Architectural styles and sedimentology of the fluvial lower Beaufort Group, Karoo Basin, South Africa. *Journal of Sedimentary Research*, 84, 326–348.

Yonge, D. and Hesse, P. P. (2009). Geomorphic environments, drainage breakdown and channel and floodplain evolution on the lower Macquarie River, central-western New South Wales. *Australian Journal of Earth Science*, 56, S35–S53.

Zaleha, M. J. (1997a). Intra- and extrabasinal controls on fluvial deposition in the Miocene Indo-Gangetic foreland basin, northern Pakistan. *Sedimentology*, 44, 369–390.

Zaleha, M. J. (1997b). Fluvial and lacustrine palaeoenvironments of the Miocene Siwalik Group, Khaur area, northern Pakistan. *Sedimentology*, 44, 349–368.

Zani, H., Assine, M. L., and McGlue, M. M. (2012). Remote sensing analysis of depositional landforms in alluvial settings: method development and application to the Taquari megafan, Pantanal (Brazil). *Geomorphology*, 161–162, 82–92.

Zheng, H., Huang, X., and Butcher, K. (2006). Lithostratigraphy, petrography and facies analysis of the Late Cenozoic sediments in the foreland basin of the West Kunlun. *Palaeogeography, Palaeoclimatology, Palaeoecology*, 241, 61–78.

15

An Investigation of Lithology, Hydrogeology, Bioturbation, and Pedogenesis in a Borehole through the Cubango Megafan, Northern Namibia

ROY MCG. MILLER, STEPHEN T. HASIOTIS, CHRISTOPH LOHE,
FALK LINDENMAIER, and MARTIN QUINGER

Abstract

The 260 m thick Andoni Formation, deposited by an ancestor to the Cubango/Okavango River, forms the symmetrical, ultra-low gradient (~0.017°) Cubango Megafan, 350 km long and 300 km wide. As part of the thick, Cenozoic 'Kalahari succession' of northern Namibia, the formation consists of fine-grained, unsorted, silt- to clay-rich sands; fine-grained, well-sorted and unconsolidated aquifer sands; and minor aquitard clays. Avulsive deposition under seasonal, semi-arid conditions in a grassland savanna environment was highly intermittent. Periodic flood events were separated by long intervals of non-deposition. Extensive post-depositional bioturbation and pedogenesis of each layer to depths of >1 m took place during subaerial exposure, destroying bedding and sediment sorting. Bioturbation consists of meniscate backfilled burrows likely produced by beetle larvae, soil bug nymphs, and other invertebrates in the A, B, and C horizons. As climate became drier, these features were overprinted by pedogenic calcrete and phreatic carbonate nodules. Unsorted clay- and silt-rich sands suggest weak depositional currents, but aquifer sands point to wetter conditions with greater runoff. The pedogenic calcretes are indicative of limited annual rainfall (< 550 mm) and extremely stable land surfaces upon which no new sediment was deposited for centuries or even millennia.

15.1 Introduction

Namibia is a semi-arid country that relies largely on artificial reservoirs and groundwater resources for its water supply. In the western part of the Owambo Basin (Fig. 15.1), local summer rains and flooding of the Cuvelai Drainage System replenish shallow, hand-dug wells, generally not much deeper than 5–10 m, which have supported the local population for centuries (Marsh and Seely 1992; Mendelsohn et al. 2000, 2002). This area is the most populous of Namibia. Waters in the Cenozoic Kalahari Group deposits in the Cuvelai drainage system are highly saline, however, and the wells become brackish towards the end of the dry season, which lasts seven months or more. Similar hand-dug wells further east tap a perched aquifer (KOH 0) in thin aeolian-sheet sands covering and supporting an open, broad-leaf forest on the Cubango megafan, the uppermost unit of the Kalahari succession in the Owambo Basin. These wells also start to give out before the summer rains begin.

To ensure a better supply of potable water for the populous north, storage dams, south-easterly flowing canals, a purification plant, and pipelines were constructed in the 1970s for water pumped from the Kunene River, near Ruacana (Fig. 15.1). One of numerous relatively shallow boreholes drilled, borehole WW 37070 in the saline western part of the Cubango megafan, intersected a freshwater aquifer at a depth of 190 m b.g.l. (below ground level) (Boussard and Bittner 2008). This was the spark that set in motion a systematic deep groundwater exploration

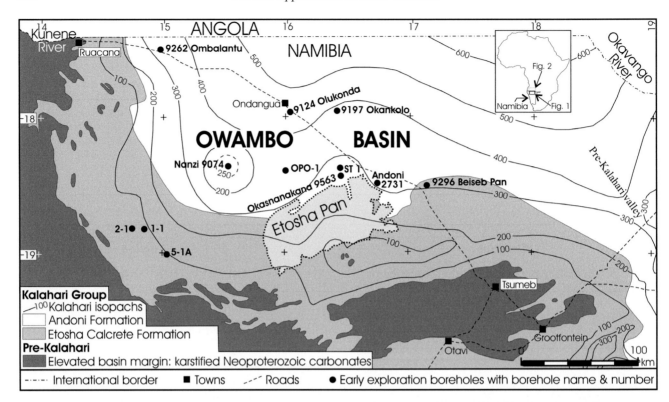

Figure 15.1 **The Owambo Basin of northern Namibia with its elevated margins of karstified Neoproterozoic carbonates and fill of Kalahari sediments**. Kalahari Group isopachs from Hedberg (1979) and Miller (1997) based on borehole data and seismic interpretation. The locations of early exploration boreholes are shown: Andoni 2731 (Coal Commission 1961); 1-1, 2-1, 5-1A, OPO-1, ST 1 (Etosha Petroleum: Fulkerson 1964; Hedberg 1979; Miller 1997, 2008; 9074 Nanzi, 9124 Olukonda, 9197 Okankolo, 9262 Ombalantu, 9296 Beiseb Pan, 9563 Okasnanakana: Hugo 1969). Modified after Miller (2008) and Miller et al. (2010) and reproduced with permission from the Director of the Geological Survey of Namibia and the Geological Society of South Africa.

programme in the western part of the Cubango mega-fan. That programme is still ongoing and has been largely sponsored and supervised by the German Federal Institute for Geosciences and Natural Resources (BGR) in coordination with the Department of Water Affairs of the Namibian Ministry of Agriculture, Water, and Forestry (in official records, all boreholes have the prefix WW; this prefix is not rendered in the figures to reduce cluttering). Three cored boreholes drilled by the BGR, especially the 400-m core of WW 203302 (Miller et al. 2016), has given us a much better understanding of the lithostratigraphy of the Andoni Formation and Olukonda Formation beneath it.

This study presents an overview of the geological and evolutionary record of the Andoni Formation, the formation that forms the Cubango megafan, as documented by these and other deep groundwater exploration boreholes.

15.2 Geological Setting

15.2.1 Pre-Cenozoic Geology of the Owambo Basin

A Palaeoproterozoic granitic basement and overlying Neoproterozoic successions form the floor of the Owambo Basin and are exposed along its elevated margins. Mineral exploration boreholes established the younger basin stratigraphy consisting of the Paleozoic–Mesozoic Karoo Supergroup and the Cenozoic Kalahari Group (SACS 1980; Miller 1997, 2008). The thickest Kalahari Group interval totalled 475 m in borehole 9124 but it may be up to 600 m thick (Fig. 15.1; Hedberg 1979; Miller 1997).

The humid climate of the late Mesozoic produced the African Erosion Surface (King 1963; Partridge 1985, 1998; Partridge and Maud 1989), an extensive, erosional land surface over much of the African continent (Burke and Gunnell 2008) and a deeply weathered regolith up to 50 m thick. High rainfall produced deeply incised river networks in the interior of southern Africa (McCarthy 1983; Partridge 1998; de Wit 1999; Moore 1999; Moore and Moore 2004; Miller 2008), now buried by the Kalahari succession. Only remnants of that regolith remain where they have been protected from Cenozoic erosion by weathering-resistant caps of silcrete, calcrete-cemented Kalahari gravels, sheets of groundwater calcrete, and early Kalahari gravel deposits (Partridge and Maud 1989; Ward and Corbett 1990; Partridge 1998; Miller 2008). In the Owambo Basin, remnants of the regolith are still preserved below the Kalahari sediments in the weathered Karoo and Late Neoproterozoic shales (Miller 2008).

A deeply incised, pre-Kalahari river system under 300 m of Kalahari deposits southeast of Grootfontein extends in a northeasterly direction, then bends to the northwest (Miller and Schalk 1980; Miller 2008) and enters the Owambo Basin (Fig. 15.1) where it cannot be followed far due to inadequate deep drilling. It may have drained originally into a palaeo-Zambezi River system. End-Cretaceous and episodic Cenozoic epeirogenic axial uplift of the southern and northern margins of the basin (Du Toit 1933; Partridge 1998; Moore 1999) resulted in a relative deepening of the Owambo Basin (Figs. 15.1 and 15.2). The Owambo Basin is a marginal subbasin of the greater Kalahari Basin of central and southern Africa.

The time-equivalent offshore succession provides evidence of the intense onshore Cretaceous fluvial erosion. The 7-km-thick, Cretaceous, delta-like succession off the Orange River mouth (McLachlan and McMillan 1979) was fed by the palaeo-Orange River and its tributaries from the interior of southern Africa, up until 67 Ma. At that point the interior of southern Africa became semi-arid (Partridge 1993), and the palaeo-Orange River and its tributaries supplied far less sediment to the offshore (Dingle et al. 1983; Muntingh and Brown 1993; Brown et al. 1995). The limited rainfall left sediments stranded in the basin interiors of southern and central Africa. This marked the start of deposition of the Kalahari succession (Miller 2008).

15.2.2 The Kalahari Group in the Owambo Basin

15.2.2.1 Kalahari Group Stratigraphy

Within the Owambo Basin, the Kalahari Group consists of four superposed formations, namely, from the base upwards: the Ombalantu, Beiseb, Olukonda and Andoni formations (Miller 1997, 2008). It is only in the deep coal boreholes drilled by the Namibian Geological Survey (Hugo 1969; Miller 1997) that all four units have been intersected and recognised. The two upper units have also been intersected in several shallower boreholes. A fifth unit, the Etosha Calcrete Formation (Miller 2008), or Etosha Limestone (Buch 1996; Buch and Rose 1996), forms an inner, ~80-km-wide apron to the elevated southern basin margins and is a lateral equivalent of the above four siliciclastic formations. Deposited as a groundwater (phreatic) calcrete from innumerable springs sourced in the karstified Neoproterozoic platform carbonates of the elevated basin margins, its deposition began at the same time as the earliest Kalahari siliciclastic sediments began, continued throughout the Cenozoic, and is still being deposited at the water table (Miller 2008). The extensive subsurface karst features of these platform carbonates provide the towns, villages, and farms along the basin margin with their water as well as the capital city of Windhoek 400 km to the south via an open canal and pipelines.

At the start of Kalahari deposition, the African Surface regolith must have been extensive and the source of much of the early Kalahari sediment. Consequently, the Ombalantu Formation is red and clay rich, but locally has a conglomeratic base. It has a maximum thickness of 80 m. A similar clay-rich base to the Kalahari is present in Botswana (Boocock and van Straten 1962; Haughton 1969). The Beiseb Formation is a thin (max. thickness 60 m), almost basin-wide, matrix-supported conglomerate that is both red and white in colour. It contains extra-basinal clasts and intra-basinal Karoo clasts. The sources for these two formations are assumed to have been both internal and external to the basin. The Olukonda Formation is red throughout and also clay rich. In contrast, the Andoni Formation is largely pale-yellow, with colour variations resulting from extensive bioturbation. The few deep exploration boreholes scattered across the basin show the Andoni Formation to be areally more extensive than the underlying Olukonda Formation. It also forms the uppermost unit in the Kunene megafan further west (Miller 1997).

15.2.2.2 The Cubango Megafan

The symmetrical, north–south-trending Cubango megafan, first recognised by Wilkinson et al. (2008), straddles the Namibia-Angola border and forms the upper part of the Cenozoic Kalahari Group fill in the eastern part of the Owambo Basin (Figs. 15.1 and 15.2). This megafan, 350 km long and 300 km wide, has a convex cross section and, below the apex, a broadly convex form down its axis, with a gradient of

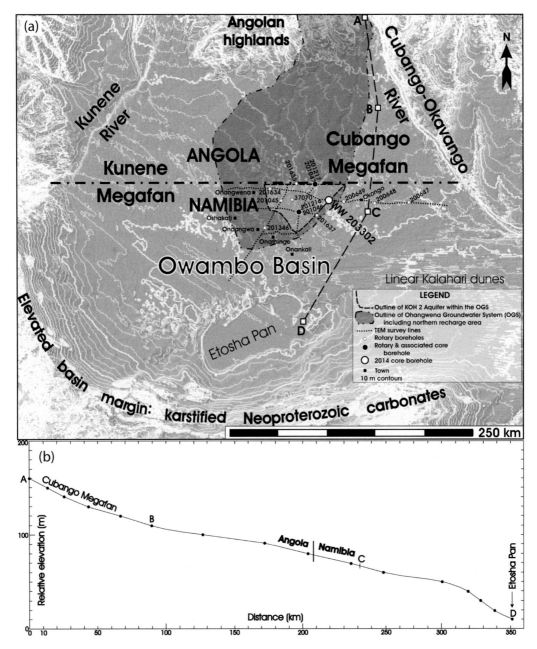

Figure 15.2 **Aerial extent and gradient of the Cubango megafan**. (a) Sun-shaded digital terrain model (DTM) of the Owambo Basin showing the well-defined Cubango megafan and the less obvious Kunene megafan. The Cubango River in Angola becomes the Okavango River in Namibia and Botswana. Palaeolake Etosha, now termed the Etosha Pan, was the Cenozoic terminal collector for both the Cubango and Kunene megafans (Miller 2008; Miller et al. 2010). The northwest- and northeast-trending lineaments on the Kunene megafan are artifacts from image processing. DTM image provided by Corner Geophysics, Swakopmund, Namibia. (b) Axial gradient of the Cubango megafan along line ABCD constructed from 10 m contours on a 3-arcsec Shuttle Radar Topography Mission image (a and b modified after Miller et al. 2010, and reproduced with permission of the Geological Society of South Africa).

0.017° over much of its length (Miller et al. 2010). It was formed by the palaeo-Cubango–Okavango River. The fan is bounded to the west by the smaller, less well-defined and almost imperceptibly convex surface of the Kunene megafan (Miller 2008; Miller et al. 2010). The Kunene megafan is incised by a wide network of broad, shallow channels, the Cuvelai Drainage System, fed by seasonal floods that rise in the western highlands of central Angola and flow southwards (Miller 2008). Both megafans terminated in, and fed, Palaeolake Etosha (Miller 2008, Miller et al. 2010). The 50 m thick, pale-green, saline, smectite- and analcime-bearing clays of the palaeolake bed now form the floor of the present-day Etosha Pan. Fossil evidence suggests that Palaeolake Etosha began to dry up some 6 Myr ago and that fluvial accumulation of both megafans and drainage therefrom into the palaeolake ceased at about 4 Ma as southern and central Africa became ever more arid (DeMenocal 1995, 2004; Miller et al. 2010; Miller 2014).

15.3 Methodology

A comprehensive account of investigations and results up to 2012 has been presented by Lindenmaier et al. (2014). Investigations in the field began in 2008 by conducting a total of 440 transient electromagnetic soundings (TEM) along several traverses (Schildknecht 2012) (Fig. 15.2). These revealed a deep-seated, freshwater aquifer, the Ohangwena 2 Aquifer (KOH 2), sandwiched between saline sediments (Fig. 15.3). Between 2009 and 2011 several rotary boreholes (to 400 m depth) and two shallower core boreholes were drilled to better constrain the extent and nature of the aquifer. Additional monitoring boreholes were drilled in 2011, revealing a shallower, fresh to brackish – but laterally variable aquifer – the Ohangwena 1 Aquifer (KOH 1). This was investigated by the Department of Water Affairs by means of many boreholes to depths of 150 m (van Wyk 2009). East of the KOH 2 Aquifer outlined in Figure 15.2, the water in the aquitard between the two aquifers becomes much fresher and the TEM survey simply revealed a

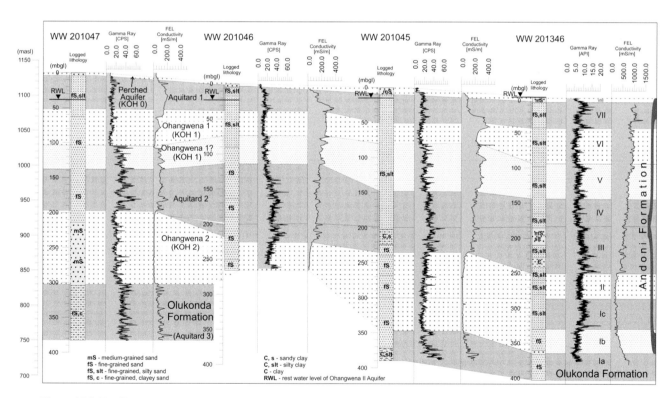

Figure 15.3 **Profile across the western part of the Cubango megafan highlighting the aquifer horizons.** Constructed from descriptions of rotary borehole samples collected every metre and down-hole gamma-ray and conductivity logs. Note that the geophysical logs define lithological changes and individual layers better than the logging of the rotary samples. Ia–VIII: lithological units of Lindenmaier et al. 2014 (modified after Lindenmaier et al. 2014 and reproduced with permission from Springer Nature).

thick section of fresh water down to a depth of 200 m. Water samples were taken for analysis and pump tests were carried out on selected boreholes.

All boreholes were logged for sedimentological characteristics, with particular attention to particle sorting. Selected cores were analysed for mineralogy, micropaleontology, geochemistry, and vertical hydraulic conductivity (Fenner 2010; Walzer 2010; Lindenmaier et al. 2014). Downhole geophysical logging was carried out on 20 old and recent boreholes, obtaining gamma-ray, apparent resistivity, magnetic susceptibility and focused electrical logs. Numerous logs from pre-2008 government drilling campaigns (Bittner and Kleczar 2006; van Wyk 2009) were digitised, reinterpreted and incorporated into a sedimentological database. From this data, a 3-D image of the layered aquifer–aquitard system was constructed.

In 2014 a core borehole (WW 203302, Fig. 15.2) was drilled to 400 m b.g.l. with only 8% core loss, resulting mainly from the totally unconsolidated, water-saturated aquifers. The 10-cm-diameter core was retrieved in 3-m plastic sleeves which were then cut into 1-m lengths. Each metre length was logged visually (Miller et al. 2016) and photographed with a core scanner. As far as the authors are aware, this valuable core is the first almost continuous core from the unconsolidated Kalahari succession in the greater Kalahari Basin. The core reveals sedimentological details that permit interpretations of depositional processes, environmental conditions, climate, and modification of the sediments by pedogenesis, bioturbation, and phreatic carbonate cementation and nodule growth. This contribution provides a preliminary description of that core, an interpretation of the depositional conditions, and hydrological conductivity of intersected aquifers. Reported compositions of carbonate cements and nodules are based on their responses to cold 10% HCl. Detailed sedimentological studies are still in progress. Determination of aquifer parameters and calculations of safe sustainable yields are nearing completion. These will guide future drilling to tap the aquifers.

15.4 Results: Lithostratigraphy of the Cubango Megafan

15.4.1 Andoni Formation (275–0 m)

The contact between the Olukonda and Andoni formations is taken 15 m below the base of the KOH 2

Aquifer (Fig. 15.4) (Houben et al. 2020). Although the sediments on either side of the contact are red, the continuous aquifer sands represent a major change in depositional conditions from those that prevailed during deposition of the red, clay-rich, generally well-bedded Olukonda Formation, which lacks aquifers. Three aquifers in the Andoni Formation, the surficial KOH 0 aquifer (Lindenmaier et al. 2014), the fresh to brackish KOH 1 Aquifer, and the deep, freshwater KOH 2 Aquifer, are separated by aquitards. The formation, which reaches 275 m in thickness elsewhere in the basin, is described below from bottom to top.

15.4.1.1 KOH 2 Aquifer (259.75–159.8 m)

The KOH 2 Aquifer pinches out westwards, southwards, and apparently eastwards (Fig. 15.3). The aquifer is clearly defined in the gamma-ray and conductivity logs. It is unconsolidated, massive, unbedded, and consists of well sorted, medium- to fine-grained quartz sand with well-rounded sand grains and only minor amounts of clay. It is red in the lower 50 m, varicoloured in the middle, and pale yellow to pale grey in the upper part. It reaches a maximum thickness of 120 m but is less well defined towards the southwest where the contained water becomes slightly brackish (Fig. 15.3). The upper surface falls from 200 m to 250 m b.g.l. in a southwesterly direction (Fig. 15.3). Apart from a cluster of diagenetic carbonate nodules at 190 m (Fig. 15.4), very few occur in the aquifer. There is patchy bioturbation (Fig. 15.4) in which individual layers with a pale-yellow base contain a few distinct grey burrows. The upper parts of such layers are either a uniform grey, or mottled in shades of grey due to intense bioturbation (Fig. 15.6b). There are no sharp stratal boundaries in the upper part of the aquifer due to intense bioturbation (grey colour).

15.4.1.2 Aquitard between the KOH 2 and 1 Aquifers (159.8–99.35 m)

This aquitard consists mainly of pale-yellow to grey, poorly sorted, fine-grained, highly silty to clayey sands with several thin, interbedded lenses of variously silty to sandy clay. Clay minerals are smectite and authigenic clinoptilolite (Dill et al. 2012; Lindenmaier et al. 2014). There is an almost total lack of bedding up to a depth of 128 m b.g.l. Randomly scattered coarse-grained sand

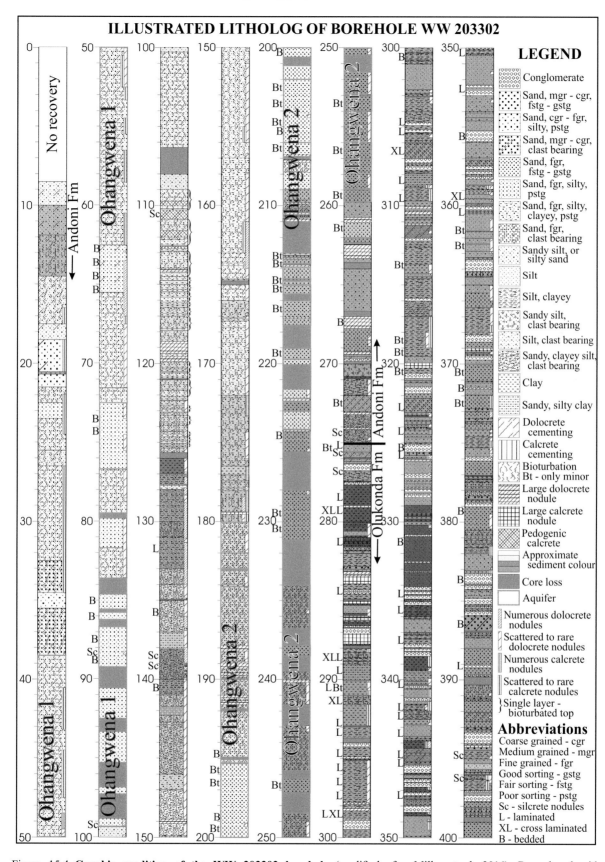

Figure 15.4 **Graphic rendition of the WW 203302 borehole** (modified after Miller et al. 2016). Reproduced with permission from the Director of the Geological Survey of Namibia. A colour version of this figure is available in the SOD for Chapter 15.

Figure 15.5 **Randomly scattered coarse sand grains (up to 2 mm in size) in core sections of pale-yellow, poorly sorted, carbonate-cemented, clayey, fine-grained sand from the aquitard between the KOH 1 and KOH 2 aquifers**. Core from borehole WW 201217, which was drilled 100 m away from WW 201047 close to the Angolan border (Fig. 15.2); depths – (a) 128.9 m; (b) 139.1 m. A colour version of this figure is available in the SOD for Chapter 15.

grains, some the size of small pebbles, occur in all cores throughout this aquitard (Fig. 15.5). None occur concentrated in a distinct bed. Individual layers above 128 m have been distinguished, particularly those with pale-brown bases and green, bioturbated tops (125–109 m: Fig. 15.4), but no sharp bedding planes have been identified. The only preserved bedding planes are those bounding clay layers and even these appear gradational over 1 cm or more, or consist of patches of clay and sand that clearly have been mixed by bioturbation over a thickness of 1–5 cm. Bioturbation is evident throughout almost all the fine-grained, silty, clayey sands of this aquitard, the most obvious being where the original pale yellow or pale brown colour of the sediment has been changed to pale grey in the lower part of the aquitard, and to shades of green higher up.

Intergranular cementation by calcrete or dolocrete is locally evident. Dolocrete nodules occur almost throughout the core between 126 m and the base of the aquitard. In places these are highly concentrated (Fig. 15.4). Locally, only calcrete nodules are present, or both calcrete and dolocrete nodules occur together. The nodules have also disrupted many stratal contacts. Above 126 m, nodules are rare or absent. Four closely spaced, nodular, pedogenic calcrete layers that lack hardpan tops occur at the top of the green silts (Fig. 15.4).

15.4.1.3 KOH 1 Aquifer (99.35–38.5 m)

The KOH 1 Aquifer is lithologically heterogeneous. The whole section consists almost entirely of

moderately to well sorted, fine-grained, slightly silty or clayey sands or almost only of poorly sorted, fine-grained, silty to clayey sands, or by an irregular interbedding of these two end members (van Wyk 2009; Lindenmaier et al. 2014). As in the underlying aquitard, a few randomly scattered coarse grains occur through the unit. The correlation between boreholes is usually poor (Fig. 15.3). The effective aquifer ranges from being only a few metres to almost 60 m thick. Borehole geophysical logs reflect this variability (Fig. 15.3). Intergranular calcrete cementation is also present. Dolocrete nodules are rare and largely confined to the upper part of the aquifer (Fig. 15.4).

15.4.1.4 Aquitard between the KOH 1 and the KOH 0 Aquifers (38.5–8.5 m)

This aquitard consists of fine-grained silty to clayey sands with a gradual upward increase in the proportion of well rounded, medium-grained sand. As below, a few randomly scattered and well-rounded grains of coarse sand and granules are present even in the most silty zones. Pale yellow is the dominant colour, but a 4.5 m-thick, red paleosol horizon occurs towards the top. Much of the upper 25 m of the core is cemented by dolocrete or calcrete. Most of the scattered nodules are calcrete, even in the dolocrete-cemented zones.

15.4.1.5 KOH 0 Aquifer (8.5–0 m)

The KOH 0 Aquifer is 1- to 40-m thick and consists of white, well-rounded, well-sorted, unconsolidated, fine- to medium-grained aeolian sheet sand that forms the present surface of the Cubango megafan and the uppermost unit of the Andoni Formation. Its average thickness is 11 m. Open, broad-leaf forest covers much of the Cubango megafan and is rooted in this aquifer.

15.4.1.6 Distal Facies of the Andoni Formation

Numerous layers of green clay and sandy clay occur through the Andoni Formation in the southern and western parts of the Owambo Basin from depths of 200 m almost up to the surface (Miller 1997, 2008). Evaporitic gypsum occurs in association with these clays in the west (Miller et al. 2010). In addition, the green, 50 m thick, saline clays with up to 13% halite that accumulated in Palaeolake Etosha, the end point of both the Cubango and Kunene megafan drainage systems, also form part of this distal facies. The clays contain smectite and authigenic analcime (Miller et al. 2010).

15.4.2 Sedimentary Petrology: Some Remarkable Features

15.4.2.1 Micropalaeontology

Samples for micropalaeontological studies were collected from all types of sediments within the KOH 2 to the KOH 1 Aquifers. Siliceous specimens were preserved in sandy sediments but not in clay, due to greater diagenetic mobilisation of silica in clays which are invariably saline and generally contain authigenic clinoptilolite and analcime (Dill et al. 2012). Most abundant in all samples were phytoliths of C4 grasses, typical of savanna environments with low to moderate soil moisture during the growing season. Scattered algal diatoms and sponge spicules indicate periods of open standing water, often somewhat saline (Fenner 2010; Lindenmaier et al. 2014).

15.4.2.2 Coarse Sand Grains and Granules

Enigmatically, a few very coarse-grained sand grains, granules, and even rare small pebbles (Wentworth scale; Pettijohn 1957) occur completely randomly scattered in all cores in the mainly fine-grained deposits of the Andoni Formation (Fig. 15.5), from the surface down to a depth of up to 175 m (borehole WW

203302) – almost the base of the aquitard above the KOH 2 Aquifer. There is not a single layer in which these coarse-grained grains are concentrated. This random scattering is ascribed to extensive postdepositional bioturbation which is described below and considered further in the discussion.

15.4.2.3 Evidence for Soil Formation and Bioturbation

In the Andoni Formation extensive bioturbation begins to become apparent in the KOH 2 Aquifer. It remains minor, subtle, and faint in the lower red aquifer sands between the depths of 260 to 250 m b.g.l., and consists of a few scattered, slightly darker red, centimetre to decimetre spots or burrows, or as paler red to whitish spots or burrows. This unconsolidated sand is so soft that it clearly has been disturbed by the drilling, obscuring some of the evidence for bioturbation. From the depths of 250 to 234 m the colour contrasts become slightly more obvious, but bioturbation is still minor. Nevertheless, faint, irregular, and patchy colour mottling in the core suggest that bioturbation may have been more extensive (e.g., from 234 to 235 m). For most of the interval from 231 to 195.3 m there is limited evidence of bioturbation. The deepest, intensely bioturbated layer occurs between 207.25 and 207.76 m (Fig. 15.6a). More abundant evidence for burrows appears in the pale-yellow sands above 195.3 m. The earliest forms are pale purplish grey burrows and patches followed by similar, but darker, purplish grey and then deep rust-red forms, the latter with the least abundant burrows (Fig. 15.6b,c). Upwards, the yellow sands become patchy and pale grey becomes the background colour to the burrows. Individual beds, where recognisable, have a pale-yellow background colour at the base becoming pale-grey at the top (Fig. 15.6b).

In the aquitard between the KOH 1 and 2 Aquifers, bioturbation is extensive almost throughout. The abundance of burrows is variable but, in general, increases upwards. The same three generations of burrows are fairly evenly distributed through the yellow and grey sections. Many burrows in the grey sands between depths of 168 and 176.5 m have half-centimetre-thick yellow halos around them. In the pale-yellow sands higher up, the greatest concentration of burrows occurs

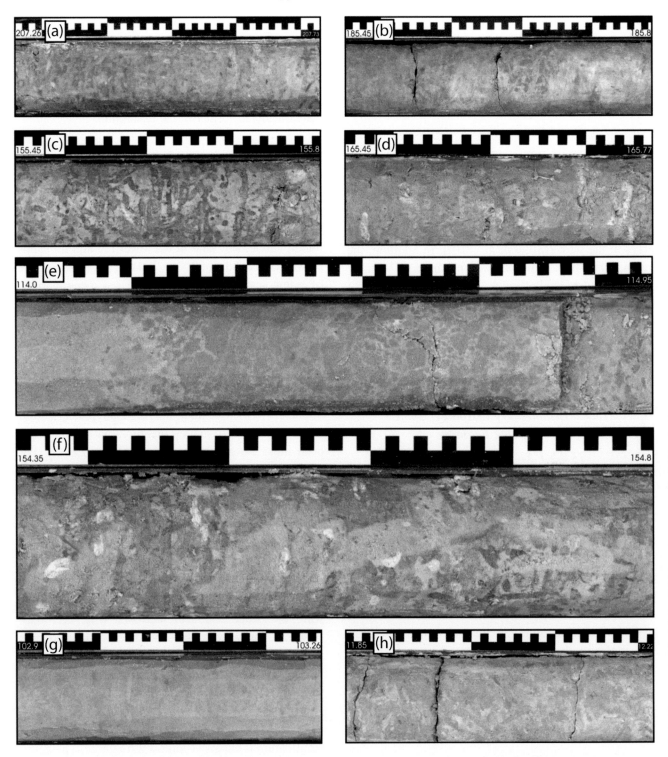

Figure 15.6 **Bioturbation in the Andoni Formation**. Depths shown on each image; top of core to the left of each image. (a) The deepest intensely bioturbated layer; pale yellow aquifer sand, red bioturbation; bioturbation is minor immediately above and below this layer. (b) Pale-yellow, well-sorted, fine-grained sand with two generations of purplish grey bioturbation features which become more abundant towards the top of the layer. The younger generation is darker than the older. (c) Pale-yellow, fine-grained, silty, clayey sand with two generations of pale-purple bioturbation features (older paler) and a third generation of rust-red bioturbation. Note the internal lamination in some of the burrows. (d) Intense, pale green bioturbation of a pale-yellow, silty, fine-grained sand; the latter visible tiny, irregularly shaped remnants. (e) Pale-brown, unsorted, fine-grained, clayey sand with pale-green bioturbation features which increase in concentration towards the top of the layer and then are smothered by deposition of the next pale brown layer. (f) Large-scale blocky disturbance of pale-yellow, poorly sorted, fine-grained, silty

between 158 and 155 m (Fig. 15.6c). Between 146 and 134 m depth, the greenish silts contain a few small, irregularly shaped remnants of pale-brown or yellow silt (Fig. 15.6d), suggesting that the green colour, itself somewhat patchy, is a later feature possibly related to extensive bioturbation with some circular, burrow-like features. From 134 to 110.78 m depth, there are only two generations of burrows, the earlier being a pale green in colour, the second darker green. In this section of the core are several pale-brown layers with a green bioturbated overprint that increases in intensity upwards (Fig. 15.6e). With the deposition of each new pale-brown layer, the underlying biota that caused the bioturbation appears to have been smothered. With time, the new bed became bioturbated in its turn. Larger-scale disturbance of the sand (possibly by mammals) is rare (Fig. 15.6f). Above 110.78 m depth, bioturbation is subtle and recognised as a faint, irregular colour mottling of the silts (Fig. 15.6g). The bioturbation manifests itself as a pervasive, early-phase, faint, uneven, patchy speckling of the core. Later burrows are not common but are sharply outlined and grey in colour. Some grey burrows have a broad, whitish outer rim. Calcretised rhizoliths, some with a sandy core, are present but rare. Calcrete and dolocrete nodules are scattered through this unit (Fig. 15.4).

Above the pedogenic calcretes at a depth of 109.8 m, bioturbation is mostly subtle and indistinct, represented by faint colour variations in patches approximately the size of single burrows. Nevertheless, individual burrows or rhizoliths are occasionally distinct. Burrows are more distinct and numerous only between the depths of 66–71 m and in the carbonate-cemented section from 11 to 17 m depths (Fig. 15.6g).

15.4.2.4 Carbonate Cement and Nodules

Phreatic intergranular carbonate cementation and nodule growth postdate bioturbation and probably occurred after considerable burial. Dolocrete and less common calcrete cements occur but are absent from the well-sorted, water-saturated aquifer sands.

Carbonate nodules from 0.1 to 80 cm in diameter are far more widespread than the intergranular cements in the Kalahari succession (Fig. 15.4), occurring in cemented and uncemented beds as well as in impermeable clays and within totally unconsolidated, water-saturated aquifer sands. In places, the nodules are located at the contacts between layers. The majority of nodules are dolocrete or calcareous dolocrete. Some layers contain both dolocrete and calcrete nodules. In contrast to the carbonate-cemented layers, most nodules contain a much lower proportion of clastic grains.

15.4.3 Hydrogeology

Lindenmaier et al. (2014) described eight lithofacies units based on sorting and permeability of core samples, and on geochemistry and geophysical borehole logs (Fig. 15.3). The difficulty with older core boreholes (WW 201216, WW 201217; see Figure 15.2 for location) was core loss, especially from aquifer units. A total of 56 undisturbed core samples from WW203302 have enabled a much more detailed analysis of hydraulic properties of the lithofacies units (Fig. 15.4).

15.4.3.1 Hydraulic Conductivity Test on Core Samples

Hydraulic conductivity of the KOH 2 Aquifer ranges from 1.3×10^{-5} to 1.1×10^{-9} m/s. The aquifer is sub-artesian since the water rises 164 m from the top of the aquifer at a depth of 180 m to 16 m below ground level, which is the rest water level (Lindenmaier et al. 2014).

Hydraulic conductivity of the aquitard between the KOH 2 and 1 Aquifers ranges from 3.3×10^{-11} to 4.1×10^{-12} m/s (Lindenmaier et al. 2014). Pore water is saline.

The KOH 1 Aquifer is confined to semi-confined and contains both fresh and brackish water. Hydraulic conductivity ranges from 1.1×10^{-6} to 4.0×10^{-10} m/s. Rest water level, at 37 m b.g.l., is significantly deeper than that for the KOH 2 Aquifer. Eastwards

Caption for Figure 15.6 (*cont.*) sand that followed formation of pale purple and red bioturbation features. (g) Very pale-grey to yellow, fine-grained, slightly silty sand with fair sorting. Patchy coloration suggests bioturbation. (h) Pale-red, poorly sorted, fine-grained, silty, dolocrete-cemented sand with deeper-red bioturbation features. Sample locations: (a), (b) from the KOH 2 Aquifer; (c)–(g) from the aquitard between the KOH 1 and KOH 2 aquifers; (h) from just below the KOH 0 Aquifer.

(boreholes 200647, 200648 and 200649; Fig. 15.2), the sands of both aquifers shown in Fig. 15.3 become less clayey, salinity decreases and water quality improves (Lindenmaier et al. 2014, fig. 12 therein).

15.4.3.2 Single-Well Pump Tests

Results of the laboratory permeability tests after DIN 18130-1 for each lithofacies unit could be determined and compared to single-well pump tests conducted in 36 boreholes (12 in the KOH 2 Aquifer and 24 in the KOH 1 Aquifer):

- Lithofacies unit (LU) Ia and Ib (Fig. 15.3, Olukonda Formation) at depths of 400–280 m b.g.l. have an average hydraulic conductivity of 8.0×10^{-8} m/s;
- Results of the permeability tests for the KOH 2 Aquifer (LU II, Fig. 15.3) range between 5.40×10^{-8} and 7.60×10^{-6} m/s (vertical conductivity), with a median value 5.00×10^{-6} m/s. By comparison, single well tests delivered an average hydraulic conductivity of 4.00×10^{-5} m/s;
- Core samples from the KOH 1 Aquifer (LU V and VI, Fig. 15.3) yield permeabilities between 1.10×10^{-9} and 1.30×10^{-6} m/s and an average hydraulic conductivity of 3.15×10^{-5} m/s;
- The aquitards have low average hydraulic conductivities of 8.0×10^{-8} m/s (LU Ia, Aquitard 3, Fig. 15.3), 1.5×10^{-9} m/s (LU III and IV, Aquitard 2, Fig. 15.3) and 4.28×10^{-10} m/s (LU VII, Aquitard 1, Fig. 15.3). This suggests very limited connectivity between the aquifers on the Namibian side of the groundwater system.

15.5 Interpretation and Discussion

The tectonic setting of the Owambo Basin is that of a remnant late-Neoproterozoic foreland basin. Apart from faulting of probable Mesozoic age, associated mafic lava and dyke emplacement (Miller 1997, 2008) and minor end-Cretaceous and later epeirogenic uplift of the southern and northern margins (Du Toit 1933; Moore 1999), the basin has been tectonically inactive since the end of the Neoproterozoic. The sedimentary fill of the basin began during the late Cretaceous and may be as much a 600 m thick in the deepest parts (Hedberg 1979; Miller 1997, 2008; Miller et al. 2010).

The primary depositional pattern that formed and produced the fan shape of the Cubango megafan is that of a distributive fluvial system (DFS) as described by Weissmann et al. (2010). Although channel-fill deposits have not been identified within the formation either within cores or by interpretation of the regional aeromagnetic data, the fan must have had channel and floodplain deposits radiating outwards from an apex at the upper end of the low-gradient, unconfined sedimentary receiving basin. The apex is identified where the palaeo-Cubango/Okavango River exited its confined valley in the Angolan highlands to the north (Miller et al. 2010).

Bioturbation and pedogenesis can destroy bedding and sedimentary structures, and thereby hamper interpretations of depositional conditions of sediments that accumulated under a subaerial environment. Varying intensities of bioturbation are evident in the Andoni Formation. Consequently, the effects of bioturbation and pedogenesis are important to consider and understand before proceeding to discuss the depositional environment of the sediments.

15.5.1 Soil Formation, Bioturbation, and Carbonate Deposition

The stratigraphic patterns of palaeosols and bioturbation preserved in the core of WW 203302 provide a rich record of (i) soil development and types, (ii) groundwater profile and hydrology, (iii) sediment accumulation rates through time, (iv) palaeoclimatic setting, including seasonality, and (v) broad palaeoenvironments. Such interpretations are possible because soil (paleosol) formation is the result of the interplay between climate, organisms, relief, parent material, and time (Jenny 1941; Thorp 1949; Hole 1981; Johnston et al. 1987; Birkeland 1999), all of which modify sediments deposited in subaerial conditions and those that become subaerially exposed after deposition in aquatic settings (Hasiotis et al. 2007; Hasiotis and Platt 2012). The architectural and surficial morphologies of modern plant and animal traces (Hasiotis and Mitchell 1993; Bromley 1996) record information about the trace maker, its behaviour, and the physicochemical conditions of its surroundings (Hasiotis 2002, 2003, 2007; Hasiotis and Platt 2012). Organisms in terrestrial and aquatic environments are distributed vertically in sediments and soils based on biological and physicochemical characteristics with respect to their affinity for water, i.e., soil moisture and water table levels (Cloudsley-Thompson 1962;

Wallwork 1970; Whittaker 1975; Glinski and Lipiec 1990; Hasiotis and Mitchell 1993; Hasiotis 2002, 2003, 2007; Counts and Hasiotis 2009; Hasiotis and Platt 2012).

The understanding that the soil moisture regime is the major control of the depth, diversity, and abundance of soil fauna and flora from the surface to the subsurface is particularly well established. This regime is controlled mostly by the groundwater profile and climate, which, in turn, are measured by temperature, precipitation, seasonality, and solar insolation. These are all an expression of continentality, latitude, wind patterns, and orographic effects (Jenny 1941; Thornthwaite and Mather 1955; Lydolph 1985; Hasiotis et al. 2007). In the rock record, root traces and animal burrows are preserved as trace fossils in outcrop and core (Hasiotis and Mitchell 1993; Bromley 1996; Hasiotis 2002, 2003, 2004, 2008; Smith et al. 2008a). Deciphering the occurrence, depth, diversity, and tiering of bioturbation in paleosols is paramount to interpreting the sedimentological, pedological, hydrological, and climatic conditions under which paleosols formed (Bown and Kraus 1983; Johnson et al. 1987; Hasiotis 2007; Hasiotis et al. 2007; Hembree and Hasiotis 2008; Smith et al. 2008a; Fischer and Hasiotis 2018). The vertical succession of facies associations and trace fossils (using Walther's law and sequence stratigraphic concepts) can be used to interpret the depositional and climatic history of continental deposits (Marriott and Wright 1993; Hasiotis 2002; Smith et al. 2008a; Fischer and Hasiotis 2018).

Intervals dominated by horizontally oriented burrows point to periods of higher soil moisture and higher groundwater tables and/or sedimentation rates. These features do not allow organisms to burrow deeply due to surface sediment accumulation associated with high water content. Intervals dominated by vertically oriented burrows indicate periods of better drainage and lower water table levels, and/or slower sedimentation rates (Hasiotis 2002, 2003, 2004, 2007, 2008; Hembree and Hasiotis 2008; Smith et al. 2008a; Hasiotis et al. 2012; Hasiotis and Platt 2012; Fischer and Hasiotis 2018). Rhizoliths – root traces preserved as casts, moulds, rhizohaloes, and rhizocretions – indicate the presence of surface plants and their association with the soil biota which fed on them (Klappa 1980; Hasiotis 2002, 2004,

2008; Kraus and Hasiotis 2006; Hembree and Hasiotis 2008; Hasiotis and Platt 2012) (see SOD 15.5.1).

Paleosols can occur singly in a sedimentary succession when they form during a period of landscape stability or after a period of landscape degradation (i.e., formation of an unconformity). The latter are commonly thick and exceptionally well developed (i.e., very mature). Paleosols occur most commonly as vertically stacked or multistoried intervals of pedogenic modification because they often formed in sedimentary systems undergoing net aggradation. On the basis of the rate of post-depositional pedogenesis resulting from the interplay of all the above, paleosols are classified as compound, composite, or cumulative (Jenny 1941; Birkeland 1999; Kraus 1999; Hasiotis 2007; Hasiotis et al. 2007). Compound paleosols generally form when sedimentation is rapid, erosion insignificant, and pedogenesis short-lived. If pedogenesis outcompetes deposition for extended periods of time, composite paleosols can develop where a profile is welded to or merged onto an underlying profile (when present). Thick, cumulative paleosols indicate that sedimentation was relatively steady and erosion was insignificant, with pedogenesis outcompeting sedimentation. The types of paleosols that formed in a sedimentary succession, therefore, can be used to interpret the autogenic and allogenic processes that influenced the depositional and erosional history of a sedimentary basin (e.g., Kraus 1999; Birkeland 1999; Hasiotis and Platt 2012; Layzell et al. 2015).

Pedogenic calcretes record climates with potential evapotranspiration greater than precipitation ($Ep > P$). The early stages of development thereof with limited carbonate precipitation take 10^1 to 10^3 years (Stages I–III of Gile et al. 1966; Machette 1985). The later stages, which culminate in intense carbonate cementation and include the typical laminated hardpan facies that develops at the top, form compound to cumulative sedimentation patterns and take 10^3–10^6 years to develop (Stages III–VI of Gile et al. 1966; Machette 1985).

15.5.2 The Andoni Formation

15.5.2.1 KOH 2 Aquifer

The fine-grained, yet well sorted sands of the KOH 2 Aquifer at the base of the Andoni Formation indicate a major change in depositional conditions that were possibly climate driven. The good rounding of sand grains, however, is more suggestive of aeolian dunes as a sediment source. The gradual upward change in colour of the aquifer sands (Fig. 15.4) could be ascribed either to changing colour of the source sands or to more reducing conditions after deposition, or both.

15.5.2.2 Aquitard between the KOH 2 and 1 Aquifers

The general lack bedding planes, the few diffuse bedding contacts that are present, and the lack of sedimentary structures are all ascribed to processes of pedogenesis by extensive bioturbation and high concentrations of calcrete and dolocrete nodules that grew in the vadose (unsaturated) zone and phreatic (saturated) zone. Consequently, a satisfactory analysis of depositional conditions of the aquitard is impossible but there are features that permit some conclusions to be made.

The upward change in colour from grey to pale yellow then green, and then back to pale yellow, is probably due to a combination of such factors as mineralogy, degree and nature of pedogenesis and bioturbation, redox conditions during bioturbation and after burial (note the grey, green, and red burrows) and the composition of the fluids from which the calcrete and dolocrete nodules grew.

The randomly scattered coarse-sand grains in the otherwise fine-grained sediments need explaining. The waning strength of depositional currents down the extremely low gradient of the megafan over 100 km from the apex (point B in Fig. 15.2b) to core locations (Fig. 15.2), whether channelised or as sheet floods, should have produced some degree of differential settling and downfan particle grading. Sedimentological evidence in all three cores does not support this, however. This lack of evidence, together with the extremely rare preservation of bedding planes and sedimentary structures in the Andoni Formation as a whole, is ascribed to extensive post-depositional bioturbation. This points to pedogenesis and bioturbation outpacing sedimentation for most of the Andoni succession above the KOH 2 Aquifer (see below).

The four pedogenic calcretes that occur over a thickness of 2.92 m near the top of the aquitard record a major climatic change coincident with a change in the initial colour of the sediments from pale brown to pale greenish yellow, and with the complete disappearance

of bioturbation for a brief, 1-m thick interval. Pedogenic calcretes require a few thousand to tens of thousands or even hundreds of thousands of years to form in arid environments and in the absence of carbonate-rich sediment sources (Gile et al. 1966; Machette 1985; Durand et al. 2016). They can form from a few centimetres to 10–20 cm below the soil surface on stable surfaces in environments with less than 550 mm of annual rainfall in equatorial regions (Goudie 1983; Wright and Tucker 1991). Each pedogenic calcrete horizon is hosted within a new layer of soil/sediment following surface aggradation. Thus, hundreds of thousands of years could separate the individual calcrete layers from each other. Consequently, this 2.92 m thick section could represent a prolonged period during which very little sediment was deposited.

In much of the aquitard, the smectite, even when making up only 5% of the fine-grained, silty and clayey sands, swells to such an extent when wet that porosity and permeability are significantly reduced. This factor and the generally poor sorting of sediments are responsible for the low hydraulic transmissivity (Lindenmaier et al. 2014) (Table 15.1). Furthermore, Figure 15.3 shows both aquitards to be saline. The presence of clinoptilolite in both of them (Dill et al. 2012; Lindenmeier et al. 2014) supports this as it requires an alkaline environment in which to grow (Hay 1966; Renault 1993; Mees et al. 2005).

15.5.2.3 The KOH 1 Aquifer

In this confined aquifer, the sorting of fine-grained, silty sands and the hydraulic conductivity vary significantly vertically and along strike towards the west, changing the aquifer quality through space (Fig. 15.3). Eastwards, however, (boreholes 200647 to 200649, Fig. 15.2), sand sorting and water quality improve significantly, together with the aquifer quality (Lindenmaier et al. 2014).

15.5.2.4 Pedogenesis and Bioturbation

The degree of bioturbation in the Andoni Formation can be attributed to deposition dominated by overland flow, sheet floods, and fluvial overbank processes. These alluvial plain environments supported flora, epifauna, and infauna, and are displayed in the core by weak to strong degrees of pedogenesis that resulted in compound, composite, and cumulative palaeosols

Table 15.1 *Links between lithofacies units, subsurface hydraulics and mean hydraulic conductivity in the Olukonda and Andoni formations.*

Formation	Lithofacies units[1]	Hydro-stratigraphy	Hydraulic conductivity (m/s)			
			Triaxial WW201216 DIN 18130	Triaxial WW201217 DIN 18130	Average (permeameter) WW203302 DIN 18130	Average[2]
Andoni Formation	VIII	KOH 0 Aquifer	–	–	–	–
	VII	Aquitard 1	–	–	$4.28 \cdot 10^{-10}$	
	VI	KOH 1 Aquifer	–	$2.2 \cdot 10^{-7}$ $6.3 \cdot 10^{-8}$	$5.3 \cdot 10^{-7}$	$3.15 \cdot 10^{-5}$
	V	KOH 1 Aquifer and secondary Aquitard	$4.0 \cdot 10^{-10}$	$1.1 \cdot 10^{-6}$ $9.1 \cdot 10^{-12}$ $2.9 \cdot 10^{-11}$		
	IV	Aquitard 2	$1.2 \cdot 10^{-11}$	$4.10 \cdot 10^{-12}$ $3.30 \cdot 10^{-11}$	$1.5 \cdot 10^{-9}$	
	III	Aquitard 2	$1.4 \cdot 10^{-12}$ $5.9 \cdot 10^{-12}$	–		
	II	KOH 2 Aquifer	$1.1 \cdot 10^{-9}$ $3.3 \cdot 10^{-9}$ $4.9 \cdot 10^{-8}$	$6.0 \cdot 10^{-6}$	$1.3 \cdot 10^{-5}$	$4.00 \cdot 10^{-5}$
	Ib	KOH 2 Aquifer	–	–		
Olukonda Formation	Ia	Aquitard 3	–	–	$8.0 \cdot 10^{-8}$	

[1] Definition of lithofacies units is based on the combination of the investigation of heavy, light, and clay minerals plus geochemical and geophysical logging (modified after Lindenmaier et al., 2014).
[2] Mean hydraulic conductivity values derived from triaxial flow and permeability tests on undisturbed core sections (one value for each tested section) and average values of single-well pump tests.

developed across the wide areas of an aggrading megafan landscape. Here the climatic interpretations show consistently continental climates ranging from semi-arid to humid. Such reconstructions of landscape history require multidisciplinary approaches that not only incorporate but integrate interpretations from geomorphology, sedimentology, petrology, ichnology, paleontology, paleobotany, and geochemistry.

Interpretations of the bioturbation based on modern experimental research suggest that backfilled meniscate burrows were most likely produced by beetle larvae (Coleoptera) and soil bug nymphs (Hemiptera) in the vadose zone (e.g., Smith and Hasiotis 2008; Counts and Hasiotis 2009). These trace fossils are assigned to the ichnotaxon *Naktodemasis* (e.g., Smith

et al. 2008b; Counts and Hasiotis 2009; Fischer and Hasiotis 2018). Bioturbation of this style suggests periods of better drainage conditions with lower to moderate soil moisture levels.

Based on the dominant colour of the paleosols (Vepraskas 2015), the depth of the groundwater table (i.e., the boundary between the vadose and phreatic zones) is interpreted to have varied widely throughout the time of deposition of the Andoni Formation sediments. Gray- and green-dominated paleosols reflect groundwater tables at or above the surface as well as subaqueous depositional environments. Blue-grey-green-dominated paleosols reflect poorly drained conditions with high soil moisture and high water-table conditions. Yellow, greyish purple, and red mottled paleosols reflect imperfectly drained conditions associated with seasonal precipitation. Red and orange-dominated paleosols reflect well-drained conditions, usually associated with seasonal precipitation.

The compound and cumulic, pedogenically modified sediment in the red section of core from depths of 234–275 m are each likely to have taken 10^0–10^2 years to form. Bioturbation features preserved as burrow mottles or ghost mottles were probably made by *Naktodemasis* and *Planolites*. The few identifiable root traces appear as rhizohaloes with associated *Naktodemasis* and *Planolites*.

The core section from ~140 to 234 m depths displays colour variations of purples, reds, greens, greys, pale yellow, and white (Ca); and compound, composite, and cumulic intervals – some with, others without, pedogenic and/or phreatic carbonate. Gleyed intervals attest to standing water above the soil or at the soil–air interface. Both horizontal (high water table) and vertical burrows (better drained vadose zone) are present. Individual paleosol horizons would likely have taken between 10^0 and 10^4 years to form, with greater development representing a greater number of years to form.

The upper 140 m of core contains compound, composite, and cumulic palaeosols formed within the same time range as above. Pedogenic calcrete layers that are nodular but lack laminated hardpan tops, however, represent 10^2–10^5 years of formation time (e.g., Gile et al. 1966; Machette 1985). Significant pedogenic features in this section include *Naktodemasis*, *Planolites*, rhizoliths, rhizocretions, nodules, autoclastic brecciation, plugged horizons, fractures, and laminae.

The patterns of bioturbation and soil formation observed in the Cubango megafan core are now known to be typical of well-studied continental deposits preserved in late Paleozoic, Mesozoic, and Cenozoic deposits. Examples are the Lower Permian Council Grove Group, Kansas (e.g., Counts and Hasiotis 2009), the Upper Triassic Chinle Formation, Utah, USA (e.g., Fischer and Hasiotis 2018), the Upper Jurassic Morrison Formation, Utah, USA (e.g., Hasiotis 2004, 2008), the Eocene Willwood Formation, Wyoming, USA (e.g., Bown and Kraus 1983; Smith et al. 2008a), the Eocene-Oligocene White River Formation, Colorado, USA (Hembree and Hasiotis 2007), and the Pawnee Creek Formation, Colorado, USA (Hembree and Hasiotis 2008).

15.5.3 Carbonate Cements and Nodules

In addition to bioturbation, the growth of non-pedogenic carbonate cements and nodules are sediment-modifying, post-depositional events. These processes produce major volume changes within the sediment at or below the water table and after significant burial (Wright and Tucker 1991). The volumetric expansion of cemented layers can erase primary bedding planes or render them very indistinct (Ward 1987; Wright and Tucker 1991; Goudie 1993; Miller 2008). Since these cements occur specifically in the saline aquitards, pore-water chemistry must also have played a role. The source of the cement in the fine-grained sediments is unknown at present. The origin of dolocretes is also as yet unresolved, whether primary (Watts 1980, 1991) or diagenetic (Watson and Nash 1997), distal (Wright and Tucker 1991) or proximal (Gevers 1930; Miller 2008). The section in WW 20330 between the depths of 8.5 and 22.5 m is pervasively cemented by massive dolocrete or calcrete horizons without any nodular texture (Fig. 15.4). This represents an impervious barrier to flow at the base of the KOH 0 Aquifer, and formed as a groundwater calcrete immediately below the aquifer over the last ~2 Ma and subsequent to the aeolian reworking of the top of the Kalahari succession between ~4 and 2 Ma (Miller et al. 2010). There is some evidence of karstification and recementation near the top of this barrier.

The high carbonate content of the nodules and their occurrence in both cemented and uncemented layers also raises questions as to the carbonate source and the reason for high carbonate availability. The isolated occurrence of many of the nodules throughout most of the succession, and the fact that impervious clay layers contain either dolocrete or calcrete nodules, or sometimes both, all suggest that carbonate sources for some of the nodules must be extremely local, i.e., internal to the sediment.

15.5.4 Groundwater Dynamics and Recharge

Monitoring of the rest water level over the last four years suggests that the Ohangwena Groundwater System (OGS) (Fig. 15.2) is recharged during the rainy season. There was good recharge of the KOH 2 Aquifer during the 2011 and 2012 rainy seasons. The

2014–2015 drought, in contrast, resulted in a slight drop in the water table. Because of the thick aquitards that separate the KOH 1 and KOH 2 from the Perched Aquifer (KOH 0) on the Namibian side of the OGS, local recharge is very limited. Groundwater flow patterns point to the northern part of the OGS as the main recharge area for the KOH 1 and KOH 2 aquifers. The recharge area is located not far south of the foot of the Angolan highlands in the catchments of the Tchimporo, Cuvate, and Cumbati/Chipolo rivers (IGA and GEOSA 1998; Mendelsohn and Weber 2011). KOH 1 and KOH 2 are confined aquifers, resulting in subartesian conditions in the southernmost part of the OGS and a potentiometric surface only several metres below ground level in the central part of the OGS. Groundwater ages (^{14}C) for KOH 1 and KOH 2 range from 16,000 to 25,000 years; Lohe 2019).

Isotope tracer tests indicated that local recharge into the KOH 0 aquifer on the Namibian side of the system was 45 mm per year during the 2013 and 2014 rainy seasons (Beyer et al. 2015).

There are no river gauges or weather stations on the Angolan side of the OGS. Consequently, an estimation of recharge to the groundwater system and determination of the sustainable water yield can only be approximated by referring to fluctuations in the rest water level. Based on data from the years 2012 and 2013, inflow to the Namibian part of the KOH 2 was estimated to be between 2.5 and $5 \times 10^6 \, \text{m}^3/\text{yr}$. The stored volume of water in the KOH 2 on the Namibian part of the OGS is approximately $5 \times 10^9 \, \text{m}^3$.

15.6 Conclusions

Palaeoenvironments recorded by the Andoni Formation include alluvial plains with fluvial and sheetflood deposition, with or without palustrine conditions with seasonal floodplain ponds (lacustrine). The majority of intervals with soil formation reflect unsteady sediment accumulation that are represented by compound paleosols developed between seasonally dominated depositional events. There were correspondingly fewer intervals with composite and cumulic paleosols, the latter produced by less frequent periods of sediment accumulation, which allowed for more protracted pedogenesis. More specifically, time periods over which the paleosols would have formed are:

- weak soil development in compound paleosols without pedogenic carbonate: $1–10^2$ years;
- cumulic paleosols with limited pedogenic carbonate deposition during stages I–II of Gile et al. (1966) and Machette (1985): $10–10^3$ years;
- compound paleosols with pedogenic carbonate stages I–III: $10–10^3$ years;
- composite paleosols with pedogenic carbonate stages III–V: $10^3–10^5$ years.

The groundwater profile and hydrology varied with local to regional water input and sediment permeability to form well-drained to poorly drained conditions, with seasonality recorded by colour mottling, and both biotic and abiotic pedogenic features. Palaeoclimate ranged from semiarid to humid, likely with strong seasonality, which, during more humid pluvials, supported abundant soil biota, plant life, and soil development but resulted in the deposition of pedogenic carbonate during interpluvials. Intense evapotranspiration over the long term will have gradually resulted in a build up of salinity in the aquitards (e.g., McCarthy et al. 1991).

Flooding and sediment supply to the system were strong enough and regular enough to ensure that avulsive flow processes aggraded a symmetrical, convex megafan cone. The sedimentological and diagenetic evidence highlight the contrast over the long term between higher rainfall in the source of the feeder river, the Angolan highlands, and the drier receiver basin, between pluvial and interpluvial periods and even annually between relatively short, wet summers feeding fresh water into the system and long dry winters during which surface water and the humid vadose zone became progressively more saline as evapotranspiration intensified. The phytoliths of C4 grasses (Fenner 2010; Lindenmaier 2014) and rhyzoliths attest to the latter. The 4- and 6-Ma fossils in the uppermost clays of Palaeolake Etosha, the end-point depression of the Cubango and Kunene megafans, reflect the gradual salinisation of freshwater influxes. These include hippopotami that lived in and got washed down the Ekuma River as it flowed into the lake, and flamingos that thrived in the lake as its waters became more saline (Miller et al. 2010). Given such extensive pedogenesis and bioturbation in the Andoni Formation the fluvial construction of the Cubango megafan will have taken many millions of years.

Acknowledgements

This work was funded by the Federal Ministry of Economic Cooperation and Development, Germany (BMZ), the Ministry of Agriculture, Water and Forestry, Namibia (MAWF) [Project: Groundwater Management in the north of Namibia; Project No. 2013.2472.2] and the European Union [Project: Integrated Water Resource Management (IWRM) in the Cuvelai–Etosha Basin; Project No. 9 ACP RPR 5044]. These projects were supported and executed by the Federal Institute for Geosciences and Natural Resources, Hannover, Germany (BGR) and the Landesamt für Bergbau, Energie und Geologie, Hannover, Germany (LBEG). Anna Nguno, Ralph Muyamba and Regina Joseph assisted with core handling and preparation.

References

Beyer, M., Gaj, M., Hamutoko, J. T., et al. (2015). Estimation of groundwater recharge via deuterium labelling in the semi-arid Cuvelai-Etosha Basin, Namibia. *Isotopes in Environmental and Health Studies*, 15, 533–552.

Birkeland, P. (1999). *Soils and Geomorphology*. Oxford University Press, New York, 448 pp.

Bromley, R. G. (1996). *Trace Fossils: Biology, Taphonomy and Applications*. Chapman and Hall, London, 361 pp.

Bittner, A. C. W. and Kleczar, M. L. (2006). *Desk study report: Cuvelai-Etosha groundwater investigation. Department of Water Affairs and Forestry and Bittner Water Consult*, Windhoek, Namibia, 48 pp.

Boocock, C. and van Straten, C. J. (1962). Notes on the geology and hydrogeology of the central Kalahari region, Bechuanaland Protectorate. *Transactions of the Geological Society of South Africa*, 65, 125–171.

Boussard, C. and Bittner, A. 2008). *Drilling of water boreholes for resettlement farms in the Ohangwena region. Unpublished Report for Ministry of Lands, Resettlement and Rehabilitation*, Windhoek, Namibia, 63 pp.

Bown, T. M. and Kraus, M. J. (1983). Ichnofossils of the alluvial Willwood Formation (Lower Eocene), Bighorn Basin, northwest Wyoming, U.S.A. *Palaeogeography, Palaeoclimatology, Palaeoecology*, 43, 95–128.

Brown, L. F.., Benson, J. M., Brink, G. J. van den, et al. (1995). Orange Basin. In J. M., Benson, G. J. van den Brink, and L F. Brown, eds., *Sequence Stratigraphy in Offshore South African Divergent Basins. AAPG Studies in Geology*, 41, 139–184.

Buch, M. W. (1996). Geochrono-Geomorphostratigraphie der Etoscha Region, Nord-Namibia. *Erde*, 127, 1–22.

Buch, M. W. and Rose, D. (1996). Mineralogy and geochemistry of the sediments of the Etosha Pan region in northern Namibia: a reconstruction of the depositional environment. *Journal of African Earth Science*, 22, 355–378.

Burke, K. and Gunnell, Y. (2008). *The African Erosion Surface: A Continental-scale Synthesis of Geomorphology, Tectonics, and Environmental Change over the Past 180 Million Years*. Geological Society of America Memoir, 201, 66 pp.

Cloudsley-Thompson, J. L. (1962). Microclimates and the distribution of terrestrial arthropods. *Annual Review of Entomology*, 7, 199–222.

Coal Commission (1960). Interim report of the Coal Commission. *Unpublished internal Report of the Geological Survey of Namibia*, 124 pp.

Counts, J. W. and Hasiotis, S. T. (2009). Neoichnological experiments documenting burrowing behaviors and traces of the masked chafer beetle (Coleoptera: Scarabaeidae: *Cyclocephala* sp.): Behavioral significance of extant soil-dwelling insects to understanding backfilled trace fossils in the continental realm. *Palaios*, 24, 75–92.

DeMenocal, P. B. (1995). Plio-Pleistocene African climate. *Science*, 270, 53–59.

DeMenocal, P. B. (2004). African climate change and faunal evolution during the Pliocene-Pleistocene. *Earth and Planetary Science Letters*, 220, 3–24.

de Wit, M. C. J. (1999). Post-Gondwana drainage and the development of diamond placers in western South Africa. *Economic Geology*, 94, 721–740.

Dill, H., Kaufhold, S., Lindenmaier, F., et al. (2012). Joint clay-heavy-light mineral analysis: a tool to investigate the hydrographic-hydraulic regime of Late Cenozoic deltaic inland fans under changing climatic conditions (Cuvelai-Etosha Basin, Namibia). *International Journal of Earth Science*, 102, 1–40.

Dingle, R. V., Siesser, W. G., and Newton, A. R. (1983). *Mesozoic and Tertiary Geology of Southern Africa*. A.A. Balkema, Rotterdam, 375 pp.

Durand, N., Hamelin, B., Deschamps, P., Gunnell, Y., and Curmi, P. (2016). Systematics of U–Th disequilibria in calcrete profiles: lessons from southwest India. *Chemical Geology*, 446, 54–69.

Du Toit, A. L. (1933). Crustal movement as a factor in the geographical evolution of South Africa. *South African Geographical Journal*, 16, 2–20.

Ewart, A., Marsh, J. S., Milner, S. C., et al. (2004b). Petrology and geochemistry of early Cretaceous bimodal continental flood volcanism of the NW Etendeka, Namibia. Part 2: characteristics and petrogenesis of the high-Ti latite and low-Ti voluminous quartz latite eruptives. *Journal of Petrology*, 45, 107–138.

Fenner, J. (2010). *Silt fraction analysis of the boreholes NAM 201216 and 201217, Namibia and its indications for palaeoenvironment and sediment age: a pilot study*. Bundesanstalt für Geowissenschaften und Rohstoffe, Hannover, Germany, 27 pp.

Fischer, S. A. and Hasiotis, S. T. (2018). Stratigraphic changes in ichnopedofacies of the Upper Triassic

Chinle Formation, Utah: implications to paleohydrology and paleoclimate. International Congress on Continental Ichnology, *Annales Societatis Geologorum Poloniae, Special Publication*, 88, 127–162.

Fulkerson, D. H. (1964). Log of Stratigraphic Test No. 1 well. *Unpublished Report of the Etosha Petroleum Company*, 139 pp.

Gevers, T. W. (1930). Terrestrer Dolomit in der Etoscha-Pfanne, Südwestafrika. *Zentralblad für Mineralogie, Geologie und Paläontologie, Section B*, 224–230.

Gile, L. H., Peterson, F. F., and Grossman, R. B. (1966). Morphological and genetic sequences of carbonate accumulation in desert soils. *Soil Science*, 101, 347–360.

Glinski, J. and Lipiec, J. (1990). *Soil Physical Conditions and Plant Roots*. CRC Press, Boca Raton, 250 pp.

Goudie, A. S. (1983). Calcrete. In A. S. Goudie and K. Pye, eds., *Chemical Sediments and Geomorphology*. Academic Press, London, 93–131.

Hasiotis, S. T. (2002). *Continental Trace Fossils*. SEPM, Short Course Notes Number 51, Tulsa, Oklahoma, 132 pp.

Hasiotis, S. T. (2003). Complex ichnofossils of solitary to social soil organisms: understanding their evolution and roles in terrestrial paleoecosystems. *Palaeogeography, Palaeoclimatology, Palaeoecology*, 192, 259–320.

Hasiotis, S. T. (2004). Reconnaissance of Upper Jurassic Morrison Formation ichnofossils, Rocky Mountain region, USA: environmental, stratigraphic, and climatic significance of terrestrial and freshwater ichnocoenoses. *Sedimentary Geology*, 167, 277–368.

Hasiotis, S. T. (2007). Continental ichnology: fundamental processes and controls on trace-fossil distribution. In W. Miller III, ed., *Trace Fossils – Concepts, Problems, Prospects*. Elsevier, Amsterdam, 268–284.

Hasiotis, S. T. (2008). Reply to the Comments by Bromley et al. of the paper "Reconnaissance of the Upper Jurassic Morrison Formation ichnofossils, Rocky Mountain Region, USA: paleoenvironmental, stratigraphic, and paleoclimatic significance of terrestrial and freshwater ichnocoenoses" by Stephen T. Hasiotis. *Sedimentary Geology*, 208, 61–68.

Hasiotis, S. T. and Mitchell, C. E., (1993). A comparison of crayfish burrow morphologies: Triassic and Holocene fossil, paleo- and neo-ichnological evidence, and the identification of their burrowing signatures. *Ichnos*, 2, 291–314.

Hasiotis, S. T. and Platt, B. F. (2012). Exploring the sedimentary, pedogenic, and hydrologic factors that control the occurrence and role of bioturbation in soil formation and horizonation in continental deposits: an integrative approach. *The Sedimentary Record*, 10, 4–9.

Hasiotis, S. T., Kraus, M. J., and Demko, T. M. (2007). Climate controls on continental trace fossils. In W. Miller, III, ed., *Trace Fossils – Concepts, Problems, Prospects*. Elsevier, Amsterdam, 172–195.

Hasiotis, S. T., Reilly, M., Amos, K., et al. (2012). Actualistic studies of the spatial and temporal distribution of terrestrial and aquatic traces in continental environments to differentiate lacustrine from fluvial, eolian, and marine environments. In O. W. Berganz, Y. Bartov, K. Bohacs, and D. Nummedal, eds., *Lacustrine Sandstone Reservoirs and Hydrocarbon Systems*. American Association of Petroleum Geologists Memoir, 95, 433–489.

Haughton, S. H. (1969). *Geological History of Southern Africa*. Geological Society of South Africa, 535 pp.

Hay, R. L. (1996). *Zeolites and zeoloitic reactions in sedimentary rocks. Geological Society of America Special Paper*, 85, 126 pp.

Hedberg, R. M. (1979). Stratigraphy of the Owamboland Basin, South West Africa. *Bulletin of the Precambrian Research Unit, University of Cape Town*, 24, 325 pp.

Hembree, D. I. and Hasiotis, S. T. (2007). Paleosols and ichnofossils of the White River Formation of Colorado: Insight into soil ecosystems of the North American midcontinent during the Eocene-Oligocene transition. *Palaios*, 22, 123–142.

Hembree, D. I., Hasiotis, S. T. (2008). Miocene vertebrate and invertebrate burrows defining compound paleosols in the Pawnee Creek Formation, Colorado, U.S.A. *Palaeogeography, Palaeoclimatology, Palaeoecology*, 270, 349–365.

Hole, F. D. (1981). Effects of animals on soil. *Geoderma*, 25, 75–112.

Houben, G. J., Kaufhold, S., Miller, R. McG., et al. (2020). Stacked megafans of the Kalahari Basin as archives of paleogeography, river capture, and Cenozoic paleoclimate of southwestern Africa. *Journal of Sedimentary Research*, 90, 980–1010.

Hugo, P. J. (1969). *Report on the core-drilling programme in Owamboland 1967–1968*. Unpublished Report of the Geological Survey of South West Africa, Windhoek, Namibia, 46 pp.

IGA & GEOSA (1988). Carta Geologica de Angola, 1:1 000 000, Instituto Nacional de Geologia de Angola, Luanda.

Jenny, H. (1941). *Factors of Soil Formation: A System of Quantitative Pedology*. McGraw-Hill, New York, 281 pp.

Johnson, D. L., Watson-Stegner, D., Johnson, D. N., and Schaetzl, R. J. (1987). Proisotropic and proanisotropic processes of pedoturbation. *Soil Science*, 143, 278–291.

King, L. C. (1963). *South African Scenery*. Oliver and Boyd, London, 308 pp.

Klappa, C. F. (1980). rhizoliths in terrestrial carbonates: classification, recognition, genesis and significance. *Sedimentology*, 27, 613–629.

Kraus, M. J. (1999). Paleosols in clastic sedimentary rocks: their geologic applications: *Earth-Science Reviews*, 47, 41–70.

Kraus, M. J. and Hasiotis, S. T. (2006). Significance of different modes of rhizolith preservation to interpreting paleoenvironmental and paleohydrologic settings: examples from Paleogene paleosols, Bighorn basin, Wyoming. *Journal of Sedimentary Research*, 76, 633–646.

Lindenmaier, F., Miller, R., Fenner, J., et al. (2014). Structure and genesis of the Cubango Megafan in northern Namibia: implications for its hydrogeology. *Hydrogeology Journal*, 22, 1307–1328.

Lohe, C. (2019). *The Ohangwena Groundwater System; Final Technical Report, Technical Cooperation Project "Groundwater Management in the North of Namibia"*. Department of Water Affairs and Forestry (DWAF), Namibia & Federal Institute for Geosciences and Natural Resources (BGR), Germany, BMZ-No.: 2009.2096.7, 101 pp.

Lydolph, P. E. (1985). *The Climate of Earth*. Rowman & Allanheld Publishers, Totowa (NJ), 386 pp.

Machette, M. N. (1985). Calcic soils of the southwestern United States. In D. L. Weide, ed., *Soils and Quaternary Geology of the Southwestern United States. Geological Society of America, Special Paper*, 203, 1–21.

Marriott, S. B. and Wright, V. P. (1993). Palaeosols as indicators of geomorphic stability in two Old Red Sandstone alluvial suites, South Wales. *Journal of the Geological Society, London*, 150, 1109–1120.

Marsh, A. and Seely, M. (1992). *Oshanas, Sustaining People, Environment and Development in central Owambo, Windhoek: Desert Research Foundation of Namibia*, 56 pp.

McCarthy, T. S. (1983). Evidence for the former existence of a major, southerly flowing river in Griqualand West. *Transactions of the Geological Society of South Africa*, 86, 37–49.

McCarthy, T. S, McIver, J. R., and Verhagen, B. T. (1991). Groundwater evolution, chemical sedimentation and carbonate brine formation on an island in the Okavango Delta swamps, Botswana. *Applied Geochemistry*, 6, 577–595.

McLachlan, I. R. and McMillan, I. K. (1979). Microfaunal biostratigraphy, chronostratigraphy and history of Mesozoic and Cenozoic deposits on the coastal margin of South Africa. *Geological Society of South Africa, Special Publication*, 6, 161–181.

Mees, F., Stoops, G., Van Ranst, E., Paepe, R., and Van Overloop, E. (2005). The nature of zeolite deposits of the Olduvai Basin, northern Tanzania. *Clays and Clay Mineralogy*, 53, 659–673.

Mendelsohn, J. and Webber, B. (2011). CUVELAI: The Cuvelai Basin, its water and people in Angola and Namibia. *RAISON*, Windhoek, Namibia, 64 pp.

Mendelsohn, J., El Obeid, S., and Roberts, C. (2000). *A Profile of North-Central Namibia*. Gamsberg Macmillan, Windhoek, 79 pp.

Miller, R. McG. (1997). The Owambo Basin of northern Namibia. In R. C. Selley, ed., *Sedimentary Basins of the World: African Basins*. Elsevier, Amsterdam, 237–317.

Miller, R. McG. (2008). *The Geology of Namibia (3 vols.). Geological Survey of Namibia*, Windhoek, 1564 pp.

Miller, R. McG. (2014). Evidence for the evolution of the Kalahari dunes from the Auob River, southeastern Namibia. *Transactions of the Royal Society of South Africa*, 69, 195–204.

Miller, R. McG. and Schalk, K. E. L. (1980, reprinted 1990). *Geological Map of Namibia*, 1:1 million scale. Windhoek: Geological Survey of Namibia.

Miller, R. McG., Frimmel, H. E., and Will, T. M. (2009b). Geodynamic synthesis of the Damara Orogen sensu lato. Neoproterozoic to Early Palaeozoic evolution of Southwestern Africa. In C. Gaucher, A. N. Sial, G. P. Halverson, and H. E. Frimmel, eds., *Neoproterozoic-Cambrian Tectonics, Global Change and Evolution: A Focus on Southwestern Gondwana*. Developments in Precambrian Geology, 16, Elsevier, Amsterdam, 231–235.

Miller, R. McG., Pickford, M., and Senut, B. (2010). The geology, palaeontology and evolution of the Etosha Pan, Namibia: implications for terminal Kalahari deposition. *South African Journal of Geology*, 113, 307–334.

Miller, R. McG., Lohe, C., Hasiotis, S. T., et al. (2016). The Kalahari Group in the 400-m deep core borehole WW 203302, northern Owambo Basin. *Communications of the Geological Survey of Namibia*, 17, 143–238.

Moore, A. E. (1999). A reappraisal of epeirogenic flexure axes in southern Africa. *South African Journal of Geology*, 102, 363–376.

Moore, J. M. and Moore, A. E. (2004). The roles of primary kimberlitic and secondary Dwyka glacial sources in the development of alluvial and marine diamond deposits in southern Africa. *Journal of African Earth Science*, 38, 115–134.

Muntingh, A. and Brown, L. F. (1993). Sequence stratigraphy of petroleum plays, post-rift Cretaceous rocks (Lower Aptian to Upper Maastrichtian), Orange Basin, western offshore, South Africa. In P. Weimer and H. Posamentier, eds., *Siliciclastic Sequence Stratigraphy: Recent Developments and Applications*. American Association of Petroleum Geologists Memoir, 58, 71–98.

Partridge, T. C. (1985). The palaeoclimatic significance of Cainozoic terrestrial, stratigraphic and tectonic evidence from Southern Africa: a review. *South African Journal of Science*, 81, 245–247.

Partridge, T. C. (1993). The evidence for Cenozoic aridification in Southern Africa. *Quaternary International*, 17, 105–110.

Partridge, T. C. (1998). Of diamonds, dinosaurs and diastrophism: 150 million years of landscape evolution in southern Africa. 25th Alex du Toit Memorial Lecture, *South African Journal of Geology*, 101, 167–184.

Partridge, T. C. and Maud, R. R. (1989). The end-Cretaceous event: new evidence from the southern hemisphere. *South African Journal of Science*, 85, 428–430.

Pettijohn, F. J. (1957). *Sedimentary Rocks*. Harper, New York, 718 pp.

SACS, South Africa Committee for Stratigraphy (1980). Stratigraphy of South Africa. Kent, L. E., (Comp.);

Part 1. Lithostratigraphy of the Republic of South Africa, South West Africa/Namibia, and the Republics of Bophuthatswana, Transkei and Venda. *Geological Survey of South Africa, Handbook*, 8, 690 pp.

Schildknecht, F. (2012). *Groundwater for the north of Namibia: groundwater exploration with TEM soundings in the Cuvelai-Etosha-Basin*, vol. Id. Bundesanstalt für Geowissenschaften und Rohstoffe, Hannover (Germany) 222 pp.

Smith, J. J. and Hasiotis, S. T. (2008). Traces and burrowing behaviors of the cicada nymph *Cicadetta calliope*: Neoichnology and paleoecological significance of extant soil-dwelling insects. *Palaios*, 23, 503–513.

Smith, J. J., Hasiotis, S. T., Kraus, M. J., and Woody, D. T. (2008a). Relationship of floodplain ichnocoenoses to paleopedology, paleohydrology, and paleoclimate in the Willwood Formation during the Paleocene–Eocene Thermal Maximum (PETM) at Polecat Bench, Bighorn Basin, Wyoming. *Palaios*, 23, 696–711.

Smith, J. J., Hasiotis, S. T., Kraus, M. J., and Woody, D. (2008b). Morphology and paleoenvironmental implications of adhesive meniscate burrows (AMB), Paleogene Willwood Formation, Bighorn Basin, Wyoming. *Journal of Paleontology*, 82, 267–278.

Thornthwaite, C. W. and Mather, J. R. (1955). *The Water Balance: Publications on Climatology*, v. VIII. Drexel Institute of Technology, Centerton, 104 pp.

Thorp, J. (1949). Effects of certain animals that live in soils. *Scientific Monthly*, 68, 180–191.

van Wyk, B. (2009). *Rural Water Supply Drilling Program 2009: Ohangwena Region*. Bittner Water Consult, Windhoek, Namibia, 18 pp.

Vepraskas, M. J. (2015). *Redoximorphic Features for Identifying Aquic Conditions*. North Carolina State University (Raleigh, N.C.), Tech. Bull., 301, 29 pp.

Wallwork, J. A. (1970). *Ecology of Soil Animals*. McGraw-Hill, New York, 283 pp.

Walzer, A. (2010). *Multilayered aquifers in the central-north of Namibia and their potential use for water supply*. Technische Universität Dresden, Dresden; Bundesanstalt für Geowissenschaften und Rohstoffe, Hannover, Germany, 114 pp.

Ward, J. D. (1987). The Cenozoic succession in the Kuiseb Valley, central Namib Desert. *Memoir of the Geological Survey of Namibia*, 9, 124 pp.

Ward, J. D. and Corbett, I. (1990). Towards and age for the Namib. In: M. K. Seely, ed., *Namib Ecology: 25 Years of Namib Research. Transvaal Museum, Monograph*, 7, 17–26.

Watson, A. and Nash, D. J. (1997). Desert crusts and varnishes. In D. S. G. Thomas, ed., *Arid Zone Geomorphology*. Wiley, New York, 69–107.

Watts, N. L. (1980). Quaternary pedogenic calcrete from the Kalahari (southern Africa): mineralogy, genesis and diagenesis. *Sedimentology*, 27, 661–686.

Watts, N. L. (1991). Quaternary pedogenic calcrete from the Kalahari (southern Africa): mineralogy, genesis and diagenesis. In V. P. Wright and M. E. Tucker, eds., *Calcretes*. Blackwell, London, 69–74.

Whittaker, R. W. (1975). *Communities and Ecosystems*. Macmillan, New York, 385 pp.

Wilkinson, M. J., Kreslavsky, M. H., and Miller, R. McG. (2008). Megafans of the Northern Kalahari Basin. *Third Southern Deserts Conference (Oxford University, School of Geography), Molopo Lodge, Northern Cape, South Africa, 16–19 September, 2008*.

Weissmann, G. S., Hartley, A. J., Nichols, G. J., et al. (2010). Fluvial form in modern continental sedimentary basins: distributive fluvial systems. *Geology*, 38, 39–42.

Wright, V. P. and Tucker, M. E., eds. (1991). *Calcretes*. Blackwell, London, 352 pp.

16

Megafans on Mars
A Fluvial Analogue for the Sinus Meridiani Layered Sediments

M. JUSTIN WILKINSON, MARK SALVATORE, and RICARDO VILALTA

Abstract

The layered sediments at Sinus Meridiani, Mars, ~1 km thick and covering 300,000 km², have been probed by the rover *Opportunity*. Numerous observations on these rocks are reevaluated through the poorly-known model of the fluvial megafan. We conclude that at least some sections of the Meridiani stack are vestiges of large fluvial megafans. Our reasons include the following: the southern uplands of Mars are a feasible sediment source; sediment was likely delivered via rivers that cut the extensive valley network that drain the upland toward Meridiani; the units cover large areas commensurate with terrestrial megafan landscapes, and display the same very low slopes; megafan landscapes lie directly adjacent to upland sediment sources, as seen at Meridiani; megafans require neither closed basins nor waterbodies for sedimentation to occur; numerous examples of fluvial channels appear in some units; and morphologies of the widespread raised ridges of the ridge-forming unit (RFU) are suggestive of indurated channel networks seen on megafans in Oman. Features of vast aggradational landscapes as encapsulated in the novel megafan analogue thus provide answers to several key observations, whereas existing fluvial analyses usually apply the classic attributes of erosional landscapes, leading to significant difficulties in interpretation of the Meridiani units.

16.1 Introduction

The sedimentary units in the Sinus Meridiani region of Mars are situated on a bench at an intermediate altitude between the southern highlands and northern lowlands (Fig. 16.1). The Meridiani units are 'the most areally extensive exposures of layered units (...) in the intercrater plains...' (Malin and Edgett 2000:1928). The units are ~1 km thick (Hynek and Di Achille 2017) and comprise four units, one of which is the *ridge-forming unit* (RFU)[1], which is the main focus of our study because it is the most widely exposed. Its surface is also described as etched because of the rough texture of the numerous ridges exposed on its surface. The RFU and two underlying units are friable, light-toned and sulphate-bearing, with high thermal inertias. The RFU is overlain by the hematite-bearing, plains-forming unit (e.g., Hynek 2002; Arvidson et al. 2003; Edgett 2005; Hynek and Phillips 2008; Hynek and Di Achille 2017) that was under direct scrutiny by the Mars Exploration Rover (MER) *Opportunity*.

This chapter is a re-evaluation of numerous strands of observation and interpretation, although some new observations are included. The re-evaluation is undertaken in light of a now better developed fluvial model for the formation of this unit, namely the megafan analogue. We examine the viability of the hypothesis that at least some parts of the Meridiani layered units are the depositional remnant of large fluvial megafans. At regional scales we focus on the setting of the Meridiani units, and at the more local scale on morphologies of the widespread raised ridges of the RFU.

The origin of the Meridiani layered units has been considered enigmatic (Edgett and Parker 1997; Edgett and Malin 2002; Edgett 2005; Hynek and Phillips 2008; Hynek and Di Achille 2017), having been ascribed to several layer-forming processes. Many

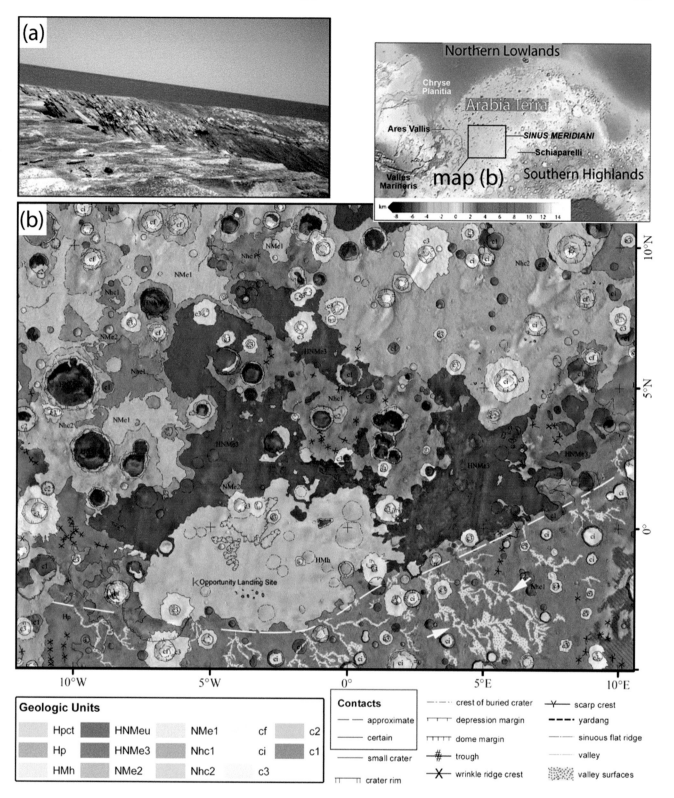

Figure 16.1 **Sinus Meridiani layered units, southwestern Arabia Terra** (inset).
(a) Layered units exposed on the southeast rim of Endurance Crater, western Sinus Meridiani, near the MER Opportunity landing site shown in panel b. MER Opportunity Left Navigation Camera image 1N138388315EFF2700P1994L0M1, taken Sol 115, 21 May 2004, looking southwest.
(b) Geological map of Sinus Meridiani layered units. Etched units (north of the dashed line) are the ridge-forming unit (RFU, purple, mapped as HNMe$_3$); they overlie units NMe$_1$, NMe$_2$. The smooth, hematite-rich unit (HMh) overlies the RFU. Immediately south of a linear margin (dashed line), ancient rocks of the southern highlands (Nhc$_1$) are incised by dendritic drainage patterns (valley surfaces: blue) that are oriented northwest toward the RFU and associated layered units. Large areas covered by valley surfaces are shown between arrows. Modified from Hynek and Di Achille (2017). Inset modified from Davis (2017).

existing lines of evidence point to a fluvial origin, and the fluvial megafan analogue meets not only the many requirements listed by Hynek and Phillips (2008) for acceptable hypotheses (see Section 16.5.2), but many other observations as well. It is beyond the scope of this chapter to evaluate other hypotheses that have been advanced, but we present the latest arguments in support of a fluvial megafan hypothesis for the formation of the RFU.

Megafans have been seen as a larger version of classic, small alluvial fans, although the dominant depositional processes are quite different (Wilkinson, Ch. 17). Megafan depositional processes are related to classic floodplain environments, but evolve into landscapes of significantly larger areas. Megafans are now known to be ubiquitous, with more than two hundred terrestrial examples identified (Fig. 16.2d) (Wilkinson and Currit, Ch. 2).

16.2 Layered Units of Meridiani Planum

16.2.1 Description and Emplacement

The Meridiani sedimentary units are plains-forming deposits that occupy parts of western, northern and eastern Sinus Meridiani, stretching 1,100 km E–W, and 700 km NNW–SSE with areal estimates of 300,000 km^2 or larger (Hynek and Phillips 2008; Grotzinger and Milliken 2012) and an estimated volume of $> 10^5$ km^3 (Hynek et al. 2003). The widely exposed RFU (HNMe$_3$; M–Meridiani Planum, e–etched) is underlain by etched units NMe$_1$ and NMe$_2$, both of which display smoother exposed surfaces (Edgett 2005). The smaller hematite-rich unit (HMh) (Fig. 16.1b) overlies part of the RFU (Hynek and Di Achille 2017) along the southern boundary of the layered units (boundary rendered as dashed lines, Figs. 16.1b and 16.2a–c), a boundary that divides the layered rocks from the more ancient Noachian cratered terrain (Nhc$_1$). The ancient terrain is incised by dendritic valley networks (VN, used here in the plural), with features indicating an undoubted fluvial origin (Hoke and Hynek 2008) and whose flow directions are broadly oriented northwest toward the Meridiani layered units (Hynek and Di Achille 2017) (lower right, Figs. 16.1b and 16.2a). Outliers of the layered units occupying some larger craters are mapped

separately (HNMe$_u$). Grotzinger and Milliken (2012) have classed the Meridiani units as Laterally Continuous Sulfate (LCS) orbital facies.

The late Noachian–Early Hesperian layered units at Meridiani entomb larger impact craters and exhibit unconformities, observations that suggest 'a long, dynamic history' (Edgett and Malin 2002:4; Edgett 2005). High-resolution topographic data of fourteen exposed horizons show a regional dip to the northwest at low angles, between 0.1° and 1° (Hynek and Phillips 2008). Numerous outliers of similar layered deposits occur over a wide area beyond the main Meridiani body (Hynek and Di Achille 2017), even including much of Arabia Terra (Zabrusky et al. 2012).

The entire Meridiani stack includes significant quantities of sulphates emplaced by regional, fluctuating groundwater (e.g., Squyres and Knoll 2005; Squyres et al. 2009; Andrews-Hanna et al. 2010). Subsequent extensive exhumation has exposed the rocks over thousands of km^2, with progressive removal of inter-ridge material in many localities (e.g., Edgett 2005; Hynek and Di Achille 2017) resulting in the etched topography of the RFU ridge networks (Figs. 16.3 and 16.4). The overlying hematite-rich unit covers $\sim 9 \times 10^4$ km^2 (Hynek and Di Achille 2017) as a lag concentration of hematite nodules, hypothesised to be derived from the underlying RFU (Hynek 2004; Squyres et al. 2005).

16.2.2 Theories of Origin

The many hypotheses for the origin of the etched units illustrate the enigmatic nature of the rocks. Hypotheses include volcanic ash flows and air fall materials (Arvidson et al. 2003; McCollom and Hynek 2005; Michalski and Bleacher 2013), cemented aeolian deposits (Squyres and Knoll 2005), impact-generated materials (Knauth et al. 2005), subaqueous deposits (Edgett and Parker 1997; Christensen and Ruff 2004), subaerial and/or subaqueous deposits (Malin and Edgett 2000; Edgett and Malin 2002; Edgett 2005), ice-related deposits (Niles and Michalski 2009), and 'dirty evaporites' emplaced by long-duration regional groundwater upwelling (Andrews-Hanna et al. 2007, 2010). Hynek and Phillips (2008) suggested hybrid processes, namely either aeolian

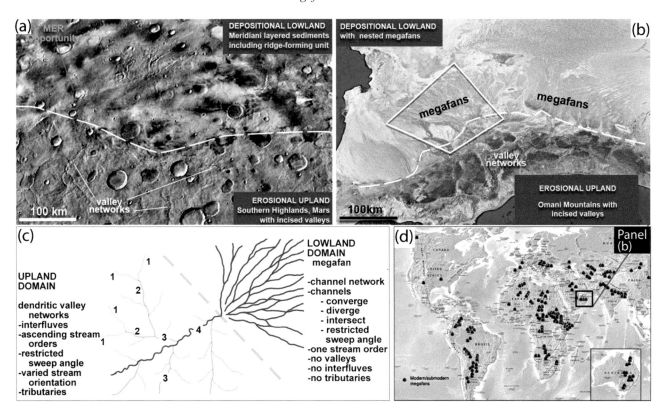

Figure 16.2 **Upland *vs.* lowland fluvial domains** (separated by dashed lines in panels (a), (b), (c), and by straight lines in panels (e) and (f)).
(a) Sinus Meridiani: upland valley networks (lower half) oriented toward smoother Meridiani Planum lowland (upper half) with layered sediments. North to top. (b) Oblique image of Oman at similar scale to panel (a): upland valleys in dark rocks of the coastal mountains (lower centre of image) are sediment source for adjacent megafans (upper centre). North to right. Box shows locations of Figs. 16.5 and 16.6. (c) Distinctions between adjacent upland and lowland domains – diagrammatic rendering. Upland drainage net: dendritic stream pattern; Lowland (megafan cone) drainage net: radial stream pattern (see text). (d) World megafan map (see Wilkinson and Currit, Ch. 2), indicating location of Omani megafans. Box indicates area of astronaut image in panel (b). (e) Diagrammatic section of megafan sedimentation, showing adjacent upland/lowland domain configuration and absence of waterbodies: upland river profiles (I, II, III) are incised valleys (right side); in the lowland these profiles develop extensive megafan deposits (shaded center and left). Upward migration of stream profiles (I to II to III) increases accommodation space as stratigraphic base level shifts distally (a) to (b) to (c); fan apex zone can shift horizontally (arrows) with small changes in profile altitude. Long profile can display convexity distally (dashed line) (see text). (f) Okavango megafan, Botswana (150 km radius), showing Okavango River (flow left to right) that supplies the megafan apex with permanent near-surface water (within semi-circle); near-surface water declines downfan; valley-confined Okavango River floodplain ('panhandle') projects 100 km into upland domain (flanked by linear palaeo-dunes). Inset: linear dune invasion of distal megafan. Dunes emplaced by easterly winds. A colour version of image (f) is available in the SOD for Chapter 16. Imagery sources: Google Mars CTX composite (a); astronaut images ISS006-E-25047 (c) and STS036-E-7726 (f), courtesy Earth Science and Remote Sensing Unit, NASA–Johnson Space Center.

sediments, or pyroclastic ash airfall, but both influenced by a fluctuating groundwater table.

Malin and Edgett (2000) outlined the constraints that must have operated in the past for the emplacement of the extensive Meridiani sedimentary units, involving 'enormous volumes of material' (p. 1930). Processes of emplacement

'had to create…layered units of similar thickness, physical properties, and great areal extent that are not confined within a specific crater or chasm. … [T]he scale and extent of outcrops…are reminiscent of…sedimentary rocks of the Colorado Plateau. … By analogy…the martian outcrop materials are most likely to have been deposited under subaerial and/or subaqueous conditions…' (Malin and Edgett 2000:1931).

Malin and Edgett (2000:1932) define subaqueous processes broadly, encompassing 'those associated with flowing water (alluvial), standing water (submarine and lacustrine), and the mixing of flowing water and standing water (deltaic)'.

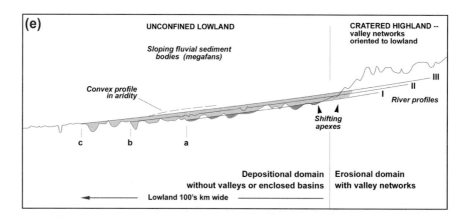

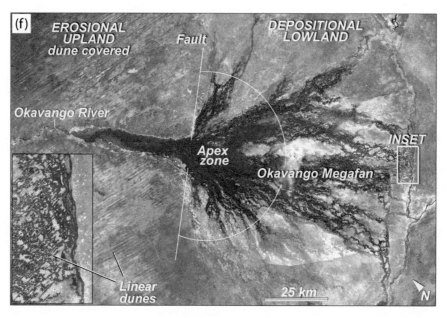

Figure 16.2 (*cont.*)

Mars has an extensive fluvial history, especially in the Meridiani region (e.g., Edgett and Malin 2002; Edgett 2005; Hynek and Phillips 2008; Hoke and Hynek 2009). Our examination pursues a major conclusion drawn by Malin and Edgett (2000:1932), that low-energy 'alluvial processes may have played a major role in the formation of the martian outcrop materials'. As evidence, they pointed to the valley networks immediately upslope in the cratered terrain. Malin and Edgett (2000:1932) identified a difficulty raised by this conclusion, namely the absence of the corresponding evidence of 'overland transport systems of gullies, channels, streams, and valleys'.

In weighing preferred models, Malin and Edgett (2000:1935) favoured an Earth-like model, invoking 'a planet and environment capable of sustaining liquid water on its surface, and the movement by this water of substantial amounts of eroded rock material'. Subsequent evaluation of the Meridiani sediments, partly from evidence for fluvial channels in the Meridiani units, led Edgett (2005) to refine the subaqueous interpretation by emphasising fluvial processes and describing the Meridiani units as a 'valley-ed volume' (Edgett 2005: 5, 14, 54). Edgett (2005) gave as support the existence of valley networks upstream in the cratered terrain, but noted as a difficulty the lack of obvious valley networks *within* the Meridiani

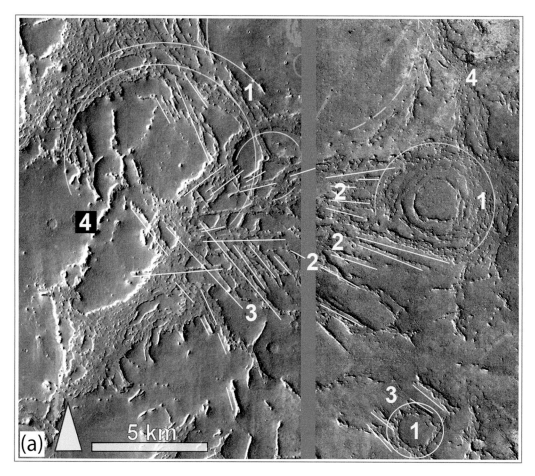

Figure 16.3 **Ridge types in the RFU (ridge-forming unit), Sinus Meridiani**.
(a) Multi-origin RFU ridge complexes, enhanced with white lines: crater-related features are circles, partial circles, concentric circles (rim-related – e.g., 1, 1, 1; linear radial fractures 2). Parallel linear fractures apparently not crater-related (3, 3). Sinuous ridges, single and multiple, are part of the complex pattern (4, 4). Image centre: 4.1 N, 355.7 E.
(b) to (d): RFU sinuous ridges. Most coherent ridges are outlined; several others are apparent. None of the sinuous ridges appear to be crater-related. (b) Simple sinuous ridge is ~30 km long (lower right to upper left). Compound sinuous ridge at top right. (c) Sinuous ridges appear to circumvent impact crater (top right). Linear ridges lower left appear to be fault-controlled. (d) Two compound sinuous ridges appear to intersect. This intersection is reminiscent of rivers on megafans; divergence of the lower ridge into narrower arms (arrowed) is reminiscent of scroll bars developed by lateral shift of a stream channel.
Image centres 4.1 N, 355.7 E (a); 4.05 N, 355.41 E (b); 3.88 N, 355.45 E (c); 6.97 N, 354.83 E (d). Imagery data in all panels derived from the CTX Global Mosaic (beta01) (Dickson et al. (2018)).

sediments. The raised ridges of relict fluvial channels in interior Oman (Figs. 16.2b and 16.5) were mentioned briefly but specifically by Edgett (2005) as possible analogues for the raised ridges of the RFU.

16.3 The Megafan Analogue and Its Implications

We take the fluvial reasoning of these pioneering interpretations a step further by means of the megafan analogue. We illustrate how the megafan analogue provides coherent explanations for several major constraints and problems with interpreting Meridiani's layered units.

16.3.1 The Megafan Analogue

Megafans have been defined as partial cones of river-laid sediment with radii > 80 km. Cones displaying very low slopes ($\ll 1°$) are built by the switching action (avulsion) of the fan-forming river (e.g., Wilkinson, Ch. 17). More than 270 terrestrial mega-fans have been documented in the most recent study (Fig. 16.2d, and Wilkinson and Currit, Ch. 2).

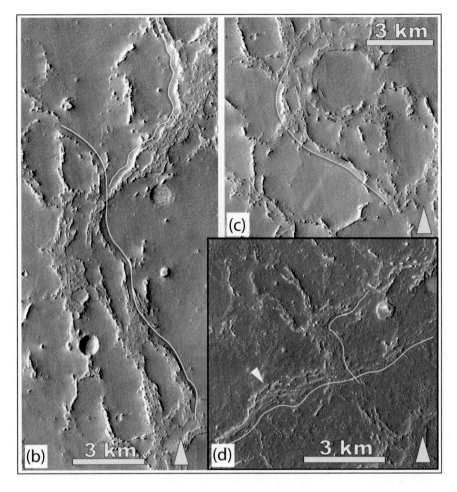

Figure 16.3 (*cont.*)

Indeed, in a new perspective, megafans have been referred to as a geomorphic norm in continental basins undergoing active fluvial deposition (Wilkinson et al. 2010; Weissmann et al. 2010). Many are mapped in the Sahara Desert and its fringes (Wilkinson et al., Ch. 3). Megafans underlie the Chaco Plains (Iriondo 1993) of central South America, where contiguous megafans extend along strike for more than 1,100 km (Wilkinson et al. 2006, 2010) and cover a vast area of 750,000 km^2 (Wilkinson et al. 2006). Many other parts of the world, such as the southeastern Arabian peninsula (Maizels 1990a, b) (Fig. 16.2b), also display landscapes dominated by nested megafans. South America and Africa boasts the largest number (almost 90 each) (Wilkinson and Currit, Ch. 2).

Few overt megafans are documented on Mars. This may reflect the difficulty of interpreting low, flat landscapes as megafans, as has been notably the case in Earth geomorphology until recently. This difficulty may be more pronounced in martian landscapes that are billions of years old and disrupted by numerous impact craters. However, possible examples are Amazonis Planitia, which displays divergent fluvial drainage lines and lies within the dimensional, slope and smoothness ranges of major megafans (> 700 km long, $< 0.5°$, high smoothness signatures: Fuller and Head 2002; Wilkinson and McGovern 2010). A coniform feature on the southern margin of Isidis is one of several coalesced alluvial fans that reaches our criterion for large fans (> 80 km long), as does the fanlike sediment body at the mouth of Hypanis Valles (Fawdon et al. 2018). In the hierarchy of landforms, megafan alluvial systems are larger than almost all coastal deltas and even very large floodplains – and notably, multimegafan landscapes reach the dimensions of the

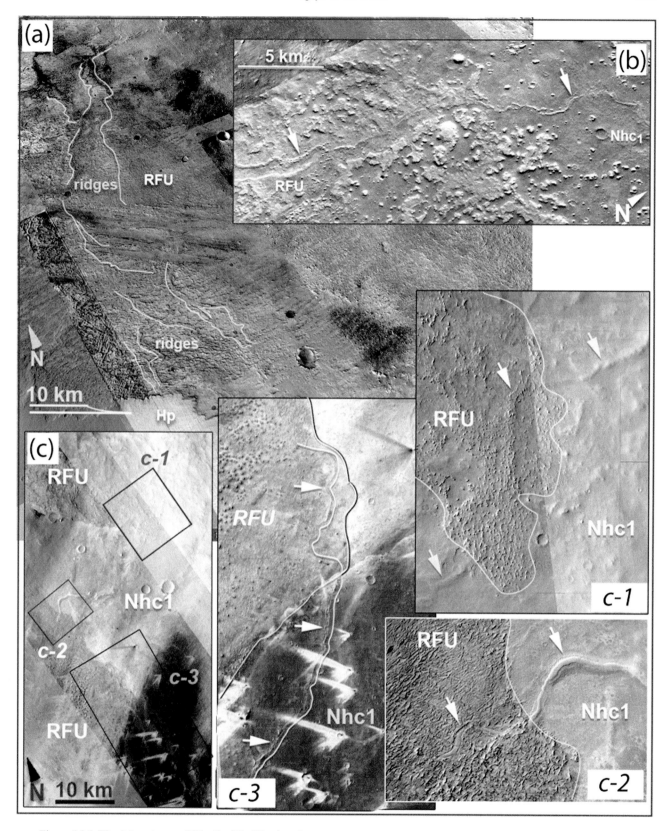

Figure 16.4 **Fluvial systems within the Meridiani rocks**.
(a) Western Sinus Meridiani. Oblique view of the largest accepted fluvial system in characteristically etched RFU rocks (from Edgett 2005); channel lengths (outlined) total > 100 km. Drainage net centred on 0.06 N, 351.61 E.
(b) Northern Sinus Meridiani. A ridge in RFU etched material (left arrow) joins a fluvial channel on a smoother surface cut in Nhc$_1$ rocks (right arrow). Image centre: 8.43 N, 357.01 E.
(c) Eastern Sinus Meridiani. One river channel crosses from Nhc$_1$ rock surfaces into RFU rocks with a visible channel form (channel mapped by Hynek and Di Achille, 2017, in the extreme north of the image). Image centre: 3.07 N, 9.26 E. Detail in panels c-1, c-2, c-3. Unit margins as mapped by Hynek and Di Achille (2017). Imagery data in all panels derived from the CTX Global Mosaic (beta01) (Dickson et al. (2018)).

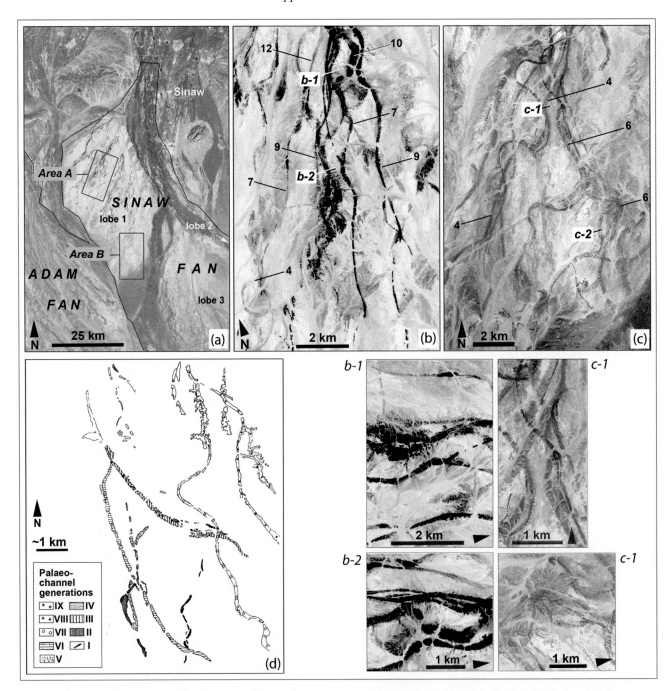

Figure 16.5 **Generations of sinuous channels (dark lines) on Omani megafans**. (a) Portions of Adam and Sinaw megafans, Oman, showing the Sinaw megafan apex and parts of two constituent lobes. Omani Mountains at top margin of image are sediment source. Flow direction to the south. Megafan surfaces show numerous channels of different generations (numbered after Maizels 1990a). Areas A and B are shown in panels (b) and (c). Image centre: 22.22 N, 57.95 E. (b), (c) Detail of numerous superimposed dark channels as narrow ridges of varying sinuosity. Locations of excerpt panels indicated as *b-1*, *b-2*, and *c-1*, *c-2*. Enlarged excerpts below show complexity of network detail. (d) Detail of nine generations of megafan channels on lobe 3 (from Maizels 1990a, (b)). Image centre 21.72 N, 58.1 5 E. A colour version of this figure is available in the SOD for Chapter 16.

Meridiani layered sediments (we exclude outflow deposits such as the fan-like feature at the mouth of Maja Valles, centered ~18.5 N, 307.2 E, because the dynamics of outflow events are different from those of common terrestrial rivers; the latter are

closer in size/discharge to those of the channels of VN – e.g., Irwin et al. 2005a).

In contrast, subenvironments within megafan landscapes also hold specific interest for interpretations of the detailed stratigraphy documented by MER

Table 16.1 *Attributes of channels and sinuous ridges on Earth and at Sinus Meridiani. Attributes of sinuous ridge types (shaded) track closely with megafan attributes*

Setting	Composite channel/ridge pattern	Slope variability[a]	Orientation variability[a]	Channel/ridge width variability	Junctions Convergent	Junctions Divergent	Intersections
EARTH – Upland valley channel network	*strictly dendritic*	*high*	*high*	*high*	*frequent*	*absent*	*absent*
Lowland channels: megafans	*'floodplain'*	*low*	*low*	*high*	*frequent*	*frequent*	*frequent*
MARS – Upland valley channel network (VN)	*strictly dendritic*	*high*	*medium to high*	*high*	*frequent*	*absent*	*rare*
Lowland RFU sinuous ridges of Meridiani[b]	*'floodplain'*	*low*	*low to medium*	*high*	*frequent*	*frequent*	*frequent*

Notes: [a] Relates to stream-order variety in the multi-order channel networks of dendritic uplands *vs.* single-order fan-forming river channels on megafans. [b] Excludes overt crater-related and fracture-related ridges.

Opportunity. Apart from overt fluvial depositional styles, these depositional subenvironments include not only dune fields, but also small lakes that can exist in their hundreds – within deflation depressions, in deflation hollows within dune sheets, and in depressions between alluvial ridges. These subenvironments are briefly examined in Section 16.5.4.

16.3.2 Geomorphic Domains: Regional Context of Megafans

The following three terrestrial perspectives have been central to the interpretation of the Meridiani sedimentary sequence developed in this chapter.

16.3.2.1 Erosional Upland vs. Depositional Lowland Domains

The classic regional/subregional-scale upland is an eroding landscape in which single river channels flow along valley axes that are arranged in strictly dendritic patterns. Interfluves are the topographic highs between valleys (Fig. 16.2c). In contrast, the depositional lowland domain displays dense networks of

channels without valleys; alluvial ridges are the highest parts of the megafan landscape[2] (Fig. 16.2c, Table 16.1).

16.3.2.2 Multi-Domain Geomorphic Models vs. Schumm's Three-Domain System

The familiar subcontinental-scale fluvial model of Schumm (1977) envisages a Mississippi-like system with an upland erosional domain, a (long) sediment transport sector, and a continent-margin depositional domain (deltaic zone). But six other multi-domain geomorphic models of similar scale are now known that include megafan zones (Wilkinson, Ch. 17, Fig. 17.9). In five of these, an upland erosional domain lies *immediately adjacent* to a major depositional domain of clustered megafans, usually located distant from shorelines (Fig. 16.2b, d) (Wilkinson, Ch. 17, Fig. 17.9). This kind of juxtaposition is unfamiliar compared with the Schumm configuration, but is nevertheless common in Africa, South America, Asia, and Australia. Parallels with alluvial fans are apparent, but the scale of the megafan landscapes is orders of magnitude larger.

Schumm's model not only lacks a major continental (non-coastal) fluvial depositional domain, but does not envision an aggrading lowland adjacent to an eroding upland. Juxtaposition of domains could well apply in the Meridiani region; and further, continental fluvial sedimentation in the form of multi-megafan landscapes is a simpler model for Meridiani because large waterbodies are not a necessary condition for sedimentation to occur (Fig. 16.2e).

16.3.2.3 Accommodation Space

A persistent problem for a Meridiani fluvial analogue has been visualising the way in which accommodation for the volume of almost horizontally-laid sediments could be achieved in the absence of a major water body such as an ocean. Sequence stratigraphic models posit accommodation based on control by base levels and the vertical fluctuation of base levels. There is a less well-known continental variant of sequence stratigraphy models. This posits accommodation *without* control of base levels by a distal waterbody – the *stratigraphic base level* (Shanley and McCabe 1994). In its simplest form, as sedimentation progresses by downlap onto a preexisting surface (*a → b → c*, Fig. 16.2e), the accommodation envelope expands upward as shown by the stream profile sequence *I → II → III* (see Wilkinson, Ch. 17, and SOD 16.1; also discussion of low-angle distal slopes forcing upward migration of equilibrium profiles in Miall 1996; and Kapteinis et al., Ch. 13). Such depozones display sloping surfaces but are nevertheless topographically and hydrologically open, and differ specifically from an 'ultimate topographic low' such as Mars's northern plains; they also differ from 'the 'open' basins [formed by] craters or topographic depressions with both an inlet and outlet channel' (Grotzinger and Milliken 2012:27, 29).

Another, more subtle control of accommodation is the downfan decline in discharge, one of the dominant trends on megafans (Weissmann et al. 2010, 2015). A major implication of this observation is that many megafans are demonstrably sediment sinks (discharge loss controlled by infiltration and evaporation), at times even becoming closed sedimentary systems – not only as individual megafans but as an entire multi-megafan landscape (see SOD 16.2).

These three perspectives from the global study of megafans (Wilkinson and Currit, Ch. 2) all point towards a direct connection between the VN in the Sinus Meridiani region and the layered units, a conclusion examined in detail in the Discussion.

Global data modelling of fluvial discharge that would be required to excavate the VN systems on Mars implies the possibility that an ocean may have occupied the northern plains of Mars (Luo et al. 2017). Even though megafan sedimentation specifically does not require the existence of waterbodies, such waterbodies only make fluvial hypotheses more viable in general. Deltas and smaller fan-like landforms such as terminal fans, fan-deltas and bajadas (and prograding sequences of these) may well have acted as components of fluvial sediment dispersal systems. We focus on megafans because they appear to be the simplest fluvial form (i) to apply to the accumulation of the layered deposits, (ii) of appropriate dimensional scale, and (iii) to examine sedimentation in terms of the non-necessity of waterbodies. We do not exclude such associated depositional environments in focusing on the megafan analogue.

16.4 Ridge Patterns of the RFU

16.4.1 Description and Hypotheses

The RFU varies from ~200 to 300 m thick (cross-section, Hynek and Di Achille 2017). Ridges of the etched terrain rise 10–40 m above the surrounding plains (Arvidson et al. 2003; Edgett 2005; Hynek and Phillips 2008) with 'rugged, criss-crossing' patterns of no preferred orientation (Edgett 2005:23) described earlier as polygonal (Arvidson et al. 2003). There is general agreement that the ridges are the indurated remnants after the selective erosion of less coherent inter-ridge material. Edgett (2005) suggested that the ridges may represent so-called *inverted landscapes* that are documented in terrestrial deserts – e.g., Miller (1937) in Australia, Rhodes (1987) in California, and Maizels (1990a, b) in the Arabian Peninsula. Mars displays numerous examples of undoubted inverted channels (e.g., Irwin et al. 2005b; Pain et al. 2007; Burr et al. 2010). Based on terrestrial examples, Hayden et al. (2019) emphasised the distinction between sinuous ridges that represent inverted channels

Table 16.2 *Type (i) sinuous RFU ridge characteristics suggestive of fluvial forms (as known from terrestrial megafan surfaces)*

Channel-like attributes	Description
Length	Numerous ridges up to tens of kms long (Figs. 16.3, 16.4, 16.6c)
Sinuosity	Numerous examples; cryptic influence of buried or unrecognisable impact craters remains a significant interpretational impediment; also appear in regions where overt cratering is low (Figs. 16.3, 16.4, 16.6c)
Single and braid-like habits	Very common, reflecting known variability in terrestrial rivers
Junctions	
• Downslope convergence	Numerous in some places (Figs. 16.3, 16.4, 16.6c)
• Downslope divergence & ridge crossing	Numerous in some places (Figs. 16.3, 16.4, 16.6c)
Obstacle interaction	Numerous ridges track around craters (suggesting a single system influenced by, but independent of, crater location) (Figs. 16.3a, c)
Meander migration	Curved parallel ridge sets suggestive of meander scroll bars; infrequent
Splay-like features	Oriented orthogonally to larger sinuous ridges; present but uncommon
Delta-like distributaries in craters	Present but uncommon

versus those that represent entire channel belts (channels, levees, and overbank floodplain components).

From their planet-wide review of polygonal ridge networks, Kerber et al. (2017:206–207) described *Meridiani type* morphologies as 'a patchy network with occasional circular shapes, …[that] tend to be wider than ridges elsewhere'; and as 'arcuate and sometimes near circular…rather than…strictly rectilinear' (Figs. 16.3 and 16.4). The ridges often appear layered. Kerber et al. (2017:206) summarised competing hypotheses that have been employed to account for the patterns of exposed RFU ridges:

(i) Polygonally-fractured lava flows, with fractures subsequently filled by later lava or volcaniclastic deposits; with final preferential erosion of the non-fracture material (Hynek et al. 2002; Arvidson et al. 2003);

(ii) irregular, poddy precipitation and leaching of iron-rich compounds endowing fracture zones with greater resistance to erosion (Ormö et al. 2004);

(iii) sediment filling of giant linear troughs, which become more resistant and hence inverted after erosion (Edgett 2005), leading to patterns reminiscent of 'giant polygons' seen in Acidalia Planitia and Elysium Planitia;

(iv) inverted channels from … megafans.

16.4.2 Ridge Patterns and the Fluvial Hypothesis

Existing hypotheses concerning the origin of the ridges broadly envisage a single origin. But more detailed examination reveals that several morphological types (Fig. 16.3a) can be distinguished.

Sinuous ridges: appear to be a coherent subtype, some of which have been interpreted as fluvial channels (Fig. 16.4); ubiquitous in the RFU, and of particular interest in this chapter, displaying bifurcating patterns and other attributes reminiscent of fluvial patterns (Table 16.2) such as single-thread and braid-like channels, crevasse splays and occasional small deltas entering craters.

(i) Probable crater-related ridges: the most prominent, ubiquitous and coherent group, namely circular, apparently rim-related outlines (complete and partial), central peaks, radial dikes, radial fluvial channels, radial ejecta ridges, and infrequent sinuous margins of ejecta blankets.

(ii) Parallel linear ridges: probable fractures, small in number.

(iii) Polygonal networks: (multi)ridge patterns with straight or curved planform, possibly related to one or more polygon-forming processes.

(iv) Other ridge patterns.

The origin of type (i), and its apparent interactions with type (ii), are discussed in Section 16.5.5.

Compared with those of the erosional domain, terrestrial channel patterns of the depositional geomorphic domain (Section 16.3.2) are entirely distinctive. Figures 16.2c, 16.6a and b, and Table 16.1 (upper rows) show the dendritic, convergent, multi-order stream pattern of the upland domain; these streams also display *higher* variability not only of slope, but also of sinuosity, radius of curvature, and orientation. Channels on megafans present the opposite attributes: *lower* variability of stream slopes (all extremely low), sinuosity, radii of curvature, and channel orientation, more diverse junction types (convergent, divergent, crossing), but

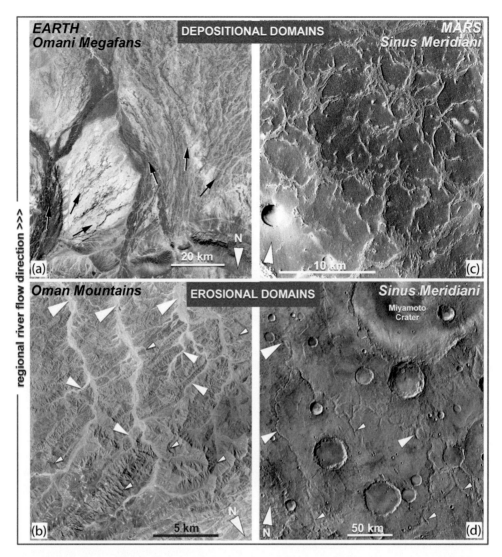

Figure 16.6 **Erosional and depositional domains (lower and upper panels respectively) on Earth and Mars (left and right panels respectively)**. (a) Omani megafans. Arrows show radial stream flow from apexes of the Sinaw (lower left) and Adam (lower right) megafans (centre: 22.22 N, 57.95 E). (b) Omani Mountains, source of megafan sediment. Arrows show stream order, from low to high order (smallest arrows, largest arrows respectively) (centre: 22.22 N, 57.95 E). (c) RFU ridge networks (centre: 22.22 N, 57.95 E). (d) Southern Highlands, Sinus Meridiani, sediment source. Arrows show stream order, from low to high (smallest arrows, largest arrows respectively) (centre: 22.22 N, 57.95 E). Imagery data in panels (c) and (d) derived from the CTX Global Mosaic (beta01) (Dickson et al. (2018)). A colour version of this figure is available in the SOD for Chapter 16.

single-order stream dimensions (Fig. 16.5). Indeed, statistical comparison between attributes of the two terrestrial domains (Table 16.1, upper rows) shows significant differences at high confidence levels (L. Yunxi, pers. comm. to MJW). Martian VN patterns mirror the dendritic patterns of terrestrial erosional uplands (Table 16.1, compare upland rows 1 and 3). More importantly, most attributes of the sinuous ridges mirror those of terrestrial megafan channels and their intertwined networks of ridge junctions and networks (Table 16.1, compare lowland rows 2 and 4).

We test the hypothesis that the close similarity between the sinuous ridge morphologies and terrestrial channel morphologies *specifically of the type found on megafan surfaces* represent the sinuous ridges that are the inverted remnants of single-thread and braided streams.

16.4.3 Known and Suspected Fluvial Channels

Demonstrable fluvial channels have been mapped by Hynek and Di Achille (2017) in units above (Hp) and

Table 16.3 ***Ridges in RFU recognised as fluvial or connected to fluvial channels in other units*** (geological units from Hynek and Di Achille 2017)

Location	Description	Comment
	Definite fluvial ridges within the RFU, with lengths	
0.14 N, 8.34 W (Fig. 16.4a)	Extended ridge complex in western Meridiani (Edgett 2005) > 85 km	Three or four parallel tributaries (converge northward); northern half mapped by Hynek and Di Achille (2017) (length of all tributaries > 160 km)
8.33 N, 357.01 E (Fig. 16.4b)	Ridge in valley ~25 km	Continues into subjacent unit Nhc1 as channel (channel widens into network with possible fan)
3.28 N, 9.41 E (Fig. 16.4c-1)	Ridge > 4.6 km	Connects to fluvial channel mapped by Hynek and Di Achille (2017)
3.11 N, 9.05 E (Fig. 16.4c-2)	Ridge > 3.6 km	Both ends connect to recognisable channels in subjacent unit Nhc1 (several other short fluvial dewatering channels cross this Nhc1 sector)
2.87 N, 9.11 E (Fig. 16.4c-3)	Ridge > 9.5 km	Continues south as a recognisable channel in unit subjacent Nhc1 for another 20 km (joining a wider network of channels)
	Possible fluvial ridges	
3.75 N, 8.53 E	Ridge > 12 km	Forms margin between RFU and unit Nhc1; possible northward continuation for 6.5 km
5.93 N, 358.21 E	Ridge 2.5 km	Continuous into unit Nhc1 rocks (Nhc1 here displays an entire channel network > 58 km long)
2.19 N, 5.89 E	Ridge > 8.3 km	Crosses narrow Nhc1 zone; hundreds of similar ridges within the RFU
1.32 N, 353.92 E	Ridge ~7 km	From Edgett (2005); resembles hundreds of sinuous ridges in the vicinity

below (NMe_{1-2}, Nhc_{1-2}) the RFU. We have also observed overt channels that are directly continuous with some sinuous ridges (Fig. 16.4, Table 16.3). Only five undoubted examples are accepted from the RFU itself, the most prominent of which is a network of channels (Edgett 2005) totalling more than 160 km in length in western Meridiani (Fig. 16.4a), and oriented with the slope of the Meridiani units which has been shown to be generally between north and northwest (Hynek and Phillips 2008). Another example shows an RFU ridge that is continuous with a channel cutting the underlying Nhc_1 unit (Fig. 16.4b). One other example from this unit shows distinct megafan characteristics of convergent and divergent fluvial channels (Edgett 2005, his Fig. 16a). Davis et al. (2016:847) mapped regionally extensive networks of sinuous ridges in a large swath of Arabia Terra, which they interpreted as 'inverted fluvial channels developed on extensive aggrading flood plains'.

16.5 Discussion

Two of the most persuasive lines of evidence accounted for by the megafan analogue are regional characteristics of the layered units and, at a smaller scale, the ridge patterns of the RFU. At the regional scale, the megafan analogue accounts for the following facts: (i) layered units lie adjacent to a highland, downstream of well-developed valley networks in the highland; (ii) layered unit dimensions are on a scale of hundreds of kilometres; (iii) the slope of these units is very low and oriented away from the highland; and (iv) these units display thicknesses of hundreds of metres. (v) At the local scale, the analogue suggests that the sinuous ridges of the RFU may represent at least a part of the sediment transport network.

These lines of evidence are discussed below in terms of what we consider a viable hybrid hypothesis for the delivery of large quantities of material to Meridiani Planum.

16.5.1 A Hybrid Hypothesis for Sediment Supply to Meridiani

A difficulty of the fluvial hypothesis lies in the fact that the volumes of rock eroded during incision of the VN *per se* were calculated to be insufficient to account for the volume of the Meridiani layered sediments (Hynek and Phillips 2008). However, we suggest that several other sources of supply probably operated in the long period of VN activity. Indeed, Hynek and Phillips (2008:220) had suggested that a 'hybrid' hypothesis might best account for the existence of the Meridiani sediments; and Malin and Edgett (2000:1931) suggested the same idea, that the layered units 'are most likely to have been deposited under subaerial and/or subaqueous conditions'. Such multi-faceted explanations are consistent with historical geomorphic approaches to landscape evolution.

We thus envisage areally extensive sedimentation in the style of fluvial megafans, with sediments likely derived from various sources that could render calculations of the (smaller) volume of the existing VN less accurate as an indication of material supplied to Meridiani Planum. In addition, the mapped zones of slope erosion (*vs.* linear channel/VN erosion) along the courses of some VN (arrows, Fig. 16b) widens the contributary zone significantly, and therefore the volume of transported materials that needs to be considered.

Likely sources are well known: sediment supplied by VN rivers over geologically lengthy periods from both the linear VN forms but also from wider erosional stripping of southern highland landscapes; aeolian and volcanic airfall materials from nearer and more distant sources; and significant volumes of evaporite supplied by groundwater pumping at the surface.

Rainfed stream processes, erosional and depositional, would then have operated in both domains. In the cratered terrain, where drainage nets are hundreds of kilometres long (Hoke and Hynek 2009), slopes would have shed these airborne materials to the streams, combined with finer fractions of materials such as impact ejecta and regolith. By contrast, in the downstream aggradational terrain of Meridiani, sedimentation would have comprised the abovementioned fluvially transported sediment from highland sources augmented by airborne materials, both as primary and

fluvially reworked deposits. Evaporative processes also appear to have operated. Accumulation of sulphate materials predicted by global hydrologic models ('playa hypothesis': Andrews-Hanna 2007, 2010, 2011) would have further augmented the clastic and evaporitic accumulations (the lag deposit of the Burns Formation in the overlying HMh unit contains up to ~40% sulphate precipitates – Glotch et al. 2006; Squyres et al. 2006). Local sub-megafan-scale evaporite accumulation processes known from megafans in arid lands likely also contributed (see SOD 16.3). Fig. 16.7 gives a simplified, diagrammatic rendering of layered sedimentation by hybrid processes – fluvial, aeolian and fluvially reworked materials and evaporites – in a megafan setting, without the influence of a topographic basin or waterbody.

Several considerations support the suggested hybrid hypothesis. Precipitation and surface runoff have been major players in VN development (Craddock and Howard 2002; Howard et al. 2005; Hynek and Phillips 2003; Irwin et al. 2005a; Zabrusky et al. 2012, among others). With higher-resolution imagery, greatly extended drainage densities in southern highland basins are now documented by Hynek et al. (2010:1), who concluded that their data 'reveal characteristics of sustained precipitation and surface runoff'. Added to this, entire swaths of the drainage nets may have been rendered unrecognisable by resurfacing processes (Irwin and Howard 2002; Craddock and Howard 2002; Hynek and Phillips 2003; Irwin et al. 2008), suggesting that drainage systems could have been significantly more extensive. It is possible that the hundreds of metres of stripping, as documented by Hynek and Phillips (2001), produced at least some material that was transported to the nearest sink, namely Meridiani.

Furthermore, newly mapped valley surfaces (Hynek and Di Achille 2017) greatly expand zones directly drained by the VN. Of several such zones mapped by them, we estimate the largest as $>10,000\,\mathrm{km}^2$ in southeastern Meridiani (arrows, Fig. 16.1b, centred 3 S, 5.5 E), and a similar area in eastern Meridiani (centred 3.5 N, 12.5 E). The quantities of material from these areas and transported downstream are unknown, but they may have been significant considering the time spans involved.

Erosion rates also may well have been as much as three orders of magnitude higher during the Noachian

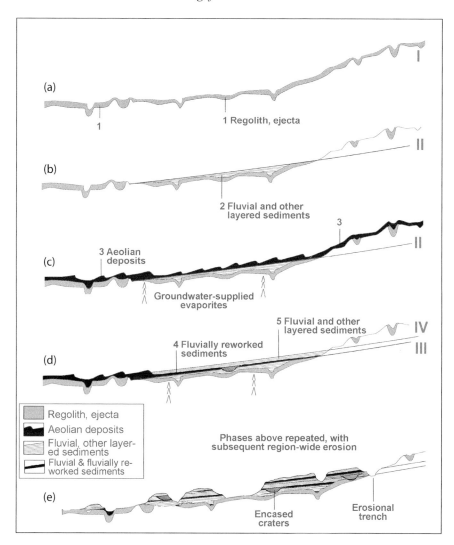

Figure 16.7 **Simplified diagrammatic rendering of the hybrid hypothesis of accumulation of the Meridiani layered units (vertically exaggerated).** Cratered uplands on right; Meridiani Planum zone centre and left. Phases (b) to (d) probably repeated in different combinations at different locations.
(a) Early cratered surface I. Surface layer (1) derived from aeolian, airfall, regolith, ejecta materials.
(b) Early river profile II. Fluvial sedimentation (2) derived especially from uplands, occurs as sloping units onlapping onto prior surfaces without the influence of a waterbody; other depositional processes may contribute.
(c) Airborne/aeolian materials (3) emplaced during phases of reduced fluvial activity. Arrows suggest impregnation and volume augmentation by evaporitic materials.
(d) Fluvial activity resumes along surface III. Fluvial reworking/partial reworking of airborne/aeolian materials (4); various constituents comprise subsequent deposition of layered units (5), surface IV; impregnation and volume augmentation by evaporitic materials continues; impact craters are engulfed by the layered units.
(e) Prior phases b through d repeated. Accumulation of ~1-km-thick layered units; layers enclose numerous craters; subsequent widespread erosion truncates beds; trench eroded along the southern margin of the layered units at the contact zone (Newsom et al. 2003).

(Craddock and Maxwell 1993; Hynek and Phillips 2001; Craddock and Howard 2002; Golombek et al. 2006), an increase interpreted from evidence of surface water flow, long continued, and specifically in the form of precipitation (Hynek et al. 2010). Indeed, low-latitude precipitation persisted actively as late as the early Hesperian (Andrews-Hanna et al. 2010; Hynek et al. 2010, 2015; Zabrusky et al. 2012).

Fluvial reworking of desert sediments is a well-established phenomenon, and is included as a realistic component of the hybrid processes. Aeolian sand sheets and dunes can be completely redeposited as fluvially laid units, as seen for example in the central Namib Desert (Wilkinson et al., Ch. 4; Wilkinson 1990). *Partial* fluvial reworking of aeolian units is an associated phenomenon, evident in the field as truncated dunal structures capped by fluvial units. Examples are the complex stratigraphy of the Tsondab Sandstone underlying the Namib Sand Sea (Hüser 1976; Ollier 1977; Besler 1984; Ward 1984).

Locally derived volcanic materials, mentioned as a source for the friable Meridiani deposits (Michalski and Bleacher 2013), would have been significant. Calculated from *minimum* volumes given by Michalski and Bleacher (2013), total erupted materials from seven paterae in the Meridiani region could have exceeded 18,400–28,800 km^3, making it possible that erupted tephra contributed a substantial fraction of the estimated 100,000 km^3 volume of the existing etched units (Hynek et al. 2003).

All these considerations suggest that locally derived and more distant materials could have provided significant, if as yet incalculable, sediment volumes to Meridiani, emplaced episodically over tens of millions of years, and could be sufficient to account for the proposed depositional origin of the sediments.

16.5.2 The Megafan Analogue and Hypothesis Requirements

Apart from the points raised above, a series of observations and interpretations of the Meridiani rocks (e.g., Malin and Edgett 2000; Edgett and Malin 2002; Edgett 2005; Hynek and Phillips 2008) have yielded eight requirements that any explanatory hypothesis ought to satisfy (requirement statements quoted from Hynek and Phillips 2008:219).

Requirements 1 and 2 – *'coherent deposition over at least the hundred kilometer scale; exposures of these units over hundreds of thousands of square kilometers'*. Individual units within megafans can extend as coherent units over large portions of a megafan (sub-fan-scale lobes evident on many modern megafans, e.g., Santos et al., Ch. 6). The extent of known terrestrial multi-megafan landscapes (up to 7×10^5 km^2; Hartley et al. 2010; Wilkinson et al. 2010; Wilkinson, Ch. 17) easily equates to the exposures of the areas covered by the Meridiani etched units. The megafan analogue thus obviates the necessity for regional-scale processes associated with larger water-bodies as an explanation for the dimensions of the deposits (recent calculations developed by Luo et al. 2017, of river runoff volumes needed to excavate the

VN cavities of Mars ultimately imply an open body of water on the northern plains. This conclusion does not preclude sedimentation by the agency of water in the specific fluvial style of extensive, low-energy lowland-domain (megafan) aggradation).

Requirement 3 – *'generally planar strata'*. Megafan strata are widely accepted as planar at the subregional scale. For example, extensive subsurface probes into modern megafans have demonstrated the stratiform architecture of buried units (Van Dijk et al. 2016; Miller et al., Ch. 15; Sinha et al., Ch. 10; Ventra and Moscariello, Ch. 14).

Requirement 4 – *'development conformally to the (likely) preexisting shallow regional slope'*. The present very low NW slope of the Meridiani sediments is close to that of the underlying regional surface (Hynek and Phillips 2008). The megafan analog, however, suggests that a preexisting slope may be an unnecessary requirement to explain the slope of the layered units, because megafans generate their own extensive, very low slopes that reach hundreds of kilometres in extent, independent of the underlying slope. For example, modern megafans of the Kalahari Group of layered rocks in southern Africa's Kalahari Desert overlie a deeply incised, pre-Kalahari river system (Miller et al., Ch. 15), burying valleys and interfluves of the prior landscape (Haddon and McCarthy 2005). The smooth, low slope aggradational surfaces of the modern megafans therefore do not reflect the erosional topographies buried beneath ~ 300 metres of sediment. The megafan analogue thus allows de-coupling of the layered units from the underlying surface, rather than implying a necessary connection. In addition, we note that existing terrestrial megafan slopes are universally $< 0.5°$ (Wilkinson, Ch. 17), matching the range of slopes calculated for the layered units (0.1–1.0°; Hynek and Phillips 2008).

Requirement 5 – *'geographical association with both older and younger valley networks'*. The possible association between the layered sediments and the VN has drawn attention (e.g., Edgett 2005). Hynek and Phillips (2008:219) reasoned that 'valleys are either buried by the layered deposits or cut down into them (indicating pre- or post-dating the materials but with no clear observation that they were coeval with deposition)'. They concluded that a fluvial hypothesis proposing a connection to the VN was therefore not feasible. The analogue, however, indicates that mega-fan apex locations shift vertically and horizontally as streams adjust to fluctuating conditions (arrows, Fig. 16.2e), per the *complex stream response* of Schumm (1977, 1981). Tongues of sediment can extend at least tens of kilometres upstream from mega-fan apexes into upland valleys. The 'panhandle' flood-plain upstream of the apex of Botswana's Okavango megafan, for example, projects more than 100 km into the erosional domain (Fig. 16.2f). Such geometries reflect shifting connections rather than a lack of connection between VN and the Meridiani units.

Requirement 6 – *'formation beginning toward the end of the Noachian epoch and continuing into the Hesperian'*. In their recent detailed evaluation of age estimates, Hynek and Di Achille (2017) gave a combined age for all the Meridiani units as 3.88 Ga (giga-years, or billion years). They estimated an age of 3.92 Ga for units underlying the Meridiani units, and two ages (3.86 and 3.66 Ga, computed on different bases) for the overlying Hp unit (Hynek and Di Achille 2017). They considered the 3.66 Ga age to be more consistent with the stratigraphic sequence. Deposition likely took a few tens of millions of years to build the > 800 m of the Meridiani rock mass (evidenced *inter alia* by numerous buried craters, some very large, and an area covered by the Meridiani units that was probably far larger than it is today; Hynek et al. 2002; Zabrusky et al. 2012). Thus, the RFU may well be younger than the 3.88 Ga age of the combined Meridiani units, but older than the 3.66 Ga age for the overlying unit Hp.

We note that the more robust dates bracketing the RFU (3.92 and 3.66 Ga) span the two dated periods of fluvial activity in the upland VN. The ages of valley formation *per se* (as opposed to the surfaces cut by the VN) centre at ~3.74 and ~3.70 Ga (Fassett and Head 2008; Hynek and Phillips 2008), or 'roughly 3.8 to 3.6 Ga' (Hynek et al. 2010:6). This bracket itself therefore suggests a *prima facie* connection between the RFU and VN activity. Indeed, the connection appears more likely if (i) fluvial activity acted periodically over long periods of time, as is now accepted, and (ii) if deposition of the layered sediments spanned long periods.

From examination of remnants of Meridiani-like layered rocks, widely distributed across ~3×10^6 km² of Arabia Terra, Zabrusky et al. (2012) calculated

that the units were laid down and then subsequently eroded (to the point when erosion ceased) between 3.83 + 0.02/–0.03 and 3.56 + 0.19/–0.40 Ga. Zabrusky et al. (2012) argued that the VN erosion phase occurred earlier than the evaporite-driven accumulation phase of the widely scattered layered units. However, we note that the given age range spans the period in which the VN were active, suggesting a direct connection between VN excavation and deposition of at least some of the layered units.

Requirement 7 – *'lack of evidence that deposition took place in a closed basin'*. Malin and Edgett (2000:1335) queried the phenomenon of 'regional layering without obvious basin control'. But this phenomenon is accounted for by *regional* fluvial sedimentation evidenced by many multi-megafan terrestrial landscapes (Section 16.3.2.2) that has taken place in open basins. This explanation is inherently more parsimonious than hypotheses that require waterbodies as the medium of deposition. The lack of delta deposits at the mouths of the VN led Hynek and Phillips (2008) to argue for a lack of connection between the VN and the Meridiani layers. Here too, the megafan analogue supplies an explanation: the layered units mapped at the mouths of two VN (1.7 S, 2.6 E and 0.77 S, 3.7 E) may be sediments laid down in a non-deltaic but nevertheless fluvial (megafan) setting.

Arabia Terra is one of the few places along the dichotomy that acts as a 'bench' between the higher cratered terrain and the northern lowlands (inset, Fig. 16.1b). This bench displays very low slopes (Andrews-Hanna et al. 2007). In such terrain, megafan dynamics would see the prograding fan deposits terminating on the bench (as much as hundreds of kilometres from the source upland, if what Zabrusky et al. 2012, consider to be the layered remnants of the Meridiani body is an indication – their Fig. 9a, c). There are various reasons to conclude that the proposed megafans terminated somewhere in Arabia Terra, the first being the very common observation that river discharge on megafans frequently declines down-fan, even in climates that are not arid, with the fan gradient intersecting the subjacent surface (Figs. 16.2e and 16.7). The second is the effect of increasing aridity: we note below (Section 16.5.3) that the Meridiani units were laid down close to the termination of the wetter early phase of Mars' history, and probably overlapped with the ensuing phase of aridification. Discharge decline on arid fans is known to translate into *convex* long profiles (dashed line, Fig. 16.2e), to produce a marked distal termination (Wilkinson, Ch. 17), quite independent of a waterbody. The implied existence of a waterbody, as seems to be indicated by recent calculations (Luo et al. 2017), is not inconsistent with megafan fluvial dynamics (Lane et al., Ch. 12, have shown that fluvial dynamics *without* coastal influence operate in cases where fan apexes lie hundreds of kilometres inland from the coastline).

Requirement 8 – *'significant physical compositional differences between layers'*. Fluvial dynamics are generally recognised as generating compositional differences in the course of megafan deposition (e.g., Weissmann et al. 2013). The advocated hybrid hypotheses (Section 16.5.1) provide for fluvial sediments derived from three distinct sources in the southern highlands (eroded bedrock, entrained aeolian and airfall materials, and yet other layers composed of mixtures of these), all of which are likely to have contributed to the observed compositional differences. In addition, aeolian and airfall materials deposited directly onto Meridiani Planum surfaces might have been buried with primary structures, intact or reworked fluvially. Post-depositional selective induration by sulphate precipitates would likely further enhance layer distinctions. All these processes would produce units with distinctive physical properties.

16.5.3 The Late Noachian–Early Hesperian Wet-to-Dry Climatic Shift

Quite apart from our argument for the *prima facie* connection between VN excavation and deposition of the layered deposits, is the possibility that fluvial processes continued into the more arid climatic phase of groundwater upwelling and evaporite accumulation at the surface of Meridian Planum. Far from ending fluvial activity, it seems likely that such upwelling may have *extended* fluvial activity into the period of the drying phase: 'The transition to more arid conditions would have led to the pre-eminence of the global-scale deep groundwater flow over the waning shallow hydrology…, the *groundwater interacting with the local surface environment* to form the deposits at Meridiani Planum' (emphasis added) (Andrews-Hanna et al.

(2007:165). Restated, the fluvial contribution to the layered units then appears as being feasibly prolonged into at least the early phases of groundwater-driven sediment accumulation – bridging the wet-to-dry shift when channels would transition from higher to lower discharge – in a period acknowledged by Zabrusky et al. (2012:312) as probably 'complicated'. The Meridiani units may thus represent one of the best preserved and most complete sedimentary records of this dynamic phase of Mars' history.

Aspects of the megafan analogue closely parallel the following stated strengths of the 'playa hypothesis', namely (i) the great extent of the areas apparently affected by this global hydrological phenomenon (epitomised in Arabia Terra); (ii) the implied long-continued supply of water (late Noachian to early Hesperian) 'with time scales on the order of millions of years'; and (iii) the 'steady flux of groundwater to the surface… *in the absence of a topographic basin*' (Andrews-Hanna et al. 2010:5, 16; emphasis added).

16.5.4 Megafan Analogue: Other Phenomena

Several other characteristics of the Meridiani rocks are accounted for by the megafan analogue. These concern the location of the layered unit mass, mineralogic distributions within it, the nature of sediment transport systems, and the origin of thick, layered units in large craters. Questions of unit thickness and grain-size trends have also been raised (and amplify considerations of Requirement 8 above).

Hynek and Phillips (2008) concluded that the location of the Meridiani units is an unsolved problem. The megafan analogue, however, suggests that the location is connected to both the north-facing scarp termed a 'break of slope' by Andrews-Hanna et al. (2010:16) and to the VN systems on the upland side of the scarp. This topographic discontinuity (dashed lines in Fig. 16.2a–c) matches exactly the abrupt margin between upland and lowland geomorphic domains common in megafan landscapes (Section 16.3.2.2; and Table 17.1, Wilkinson, Ch. 17) – and also matches the juxtaposition of ancient and younger rocks that differ strongly in appearance, attitude and geomorphic expression – but which can be genetically connected.

Hynek and Phillips (2008:218) noted that distal points of the VN now appear at altitudes lower than parts of the layered units, an observation reasonably interpreted by them to indicate the lack of connection between the VN and layered units. However, this observation does not disqualify the megafan analogue because a megafan apex necessarily rises in the course of fan building, either in relative terms (in many mountain-front locations), or in absolute terms as suggested by the profiles in models illustrated in Figure 16.2e (profiles I–III) and Figure 16.7 (profiles III and IV). Indeed, layered sediments are found extending into the lower courses of some VN. Hynek and Phillips (2008:218) noted that the VN 'are in contact with the layered sequence'). The cut-and-fill behaviour at megafan apexes is a geomorphic norm (relating either to autogenic system adjustments described by Schumm 1981, or to imposed allogenic factors) – in which cases raised terrace remnants lie above the modern feeder river and megafan surface (e.g., Iriondo 1993). Such megafan units can extend tens of kilometres into the upland domain along the valley floor (Fig. 16.2f). Adjustments of this kind are to be expected. They illustrate the complex character of source-river altitude with respect to juxtaposed sedimentary units, and do not necessarily call into question the connection of the river as a source of the sediments in the aggradational domain.

Overt connections between the highlands and Meridiani have been suggested: for example, similarities between derived surface spectra for the RFU and rocks of the cratered terrain indicate that these units 'are related mineralogically' at a relatively high level of confidence (Arvidson et al. 2003:14). This connection readily fits a scenario of sediment transport to Meridiani via the VN.

The fact that the layered units are thickest at locations farthest from the southern boundary of Meridiani led Hynek and Phillips (2008) to argue that a continental shelf setting was unlikely, and that the VN were an unlikely source for the layered sediments. By the megafan analogue, the slope of sedimentary units is mainly controlled by fluvial processes operating at the upland-lowland domain scale, so that trends in thickness of the Meridiani units may not depend on topography of the underlying surface, nor on distance from Meridiani's southern boundary. In the abovementioned example (Requirement 4) from the Kalahari Basin, the ~300 m-thick units that culminate with megafan facies

not only mask prior topography entirely, but also thicken with distance from the upland-lowland margin (Haddon and McCarthy 2005).

Mapped outliers of the layered units were probably continuous with the major outcrops at Meridiani (e.g., Malin and Edgett 2000; Edgett 2005; Hynek and Phillips 2008; Hynek and Di Achille 2017) and further northeast in the rest of Arabia Terra (Zabrusky et al. 2012). Such connections are accounted for by the regional extent of sedimentation typical of that in multi-megafan environments, and are consistent with the regional groundwater upwelling 'playa hypothesis' of Andrews-Hanna et al. (2007, 2010, 2011).

A likely regional northward decline in sediment grain-size in the Meridiani units was identified by Edgett and Parker (1997), a trend that conforms to the widely documented downstream fining of sediment textures in megafan depositional environments (e.g., Weissmann et al. 2010, 2011, 2013, 2015; Ventra and Moscariello, Ch. 14). Here, too, the megafan analogue obviates the existence of a water body.

At the scale of the MER *Opportunity's* observations, the hematite-rich HMh unit is interpreted as being at least partly associated with water (e.g., Grotzinger et al. 2005; Grotzinger and Milliken 2012; Squyres and Knoll 2005). At the wider regional scale, two perspectives from the megafan analogue have bearing. Circumstantial evidence suggests a connection between the Meridiani sediments and river systems: the highest channel frequencies occur in southern Sinus Meridiani, a region that also displays one of the highest drainage densities (Hynek et al. 2010). This conclusion relates to the distribution of hematite abundances mapped by Christensen and Ruff (2004) who see no systematic change in hematite abundances. However, it may be significant that the highest abundances, as mapped, are located along the southern margin of the hematite unit – that is, geographically adjacent to termini of the VN channels, in what we suggest is the proximal zone of a megafan landscape. Accepting a water-mediated mode of hematite accumulation, it is striking that this zone of apparent maximum concentration overlaps the zone where fluvial systems concentrate runoff from large upland drainages. Terrestrial examples of preferential water supply in proximal megafan zones are numerous (types (i) and (ii), SOD 16.4), and immediately apparent in the active

Okavango megafan of the Kalahari Desert (Fig. 16.2f) where surface moisture is restricted to the apex zone as a persistent annual pattern (e.g., McCarthy 2013).

Furthermore, the mapped hematite abundances also show a general decline northward (Christensen and Ruff (2004). This trend, if real, also mimics two arid-zone megafan patterns: (i) the decline in near-surface water occurrences with distance downstream from megafan apexes (Weissmann et al. 2015), one of the commonest patterns identified in the global megafan survey; and (ii) the trend from fluvial to aeolian environments downfan. The location of a wetter (likely brine-rich) proximal zone may supply the 'special circumstance' called for by Christensen and Ruff (2004) to explain not only the hematite concentrations in their general location (along the southern boundary of Meridiani Planum), but also the detail of higher concentrations nearer the southern boundary.

Small lakes are a common subenvironment within megafan landscapes. Terrestrial examples are known in which megafans host hundreds of small dry lakes – in depressions between alluvial ridges, as deflation hollows on megafan surfaces, and counterintuitively, even in deflation hollows *within* megafan dune fields (Wilkinson, Ch. 17). Such subenvironments provide at least one plausible analogue for the units examined by MER *Opportunity* in southwestern Meridiani that have been interpreted as a 'facies [that] marks where the groundwater table breached the surface and wind-driven subaqueous currents transported the sediment' (Grotzinger and Milliken 2012:21). An example, one of many that could be quoted, is known from the Cubango megafan, sourced from Angola's dune-covered upland (Wilkinson et al., Ch. 4; Miller et al., Ch. 15). The northern flank of this megafan near the apex hosts a deflational depression 50×20 km (centred $16.33°$ S, $17.01°$ E). Interfan streams from the upland flow into this depression, the floor of which is streaked with small linear dunes. Such depressions appear to be an analogue for the kind of complex aeolian–subaqueous depositional environment of the Burns formation (Grotzinger and Milliken 2012).

At the other extreme, some megafan surfaces hundreds of kilometres long are almost entirely covered with dune fields, such as those at the foot of the Altun Shan in China's Taklimakan desert. Here, the presently

weak fan-forming rivers debouch into the dune field, ceasing to flow in interdune hollows. Such rivers feed lakelets on the Yawatongguz He megafan, 40 km from the apex (centred at $37.9°$ N, $83.5°$ E); on the Andir He megafan apex (centred at $37.9°$ N, $83.7°$ E); and also on the Keriya He megafan, in a zone 70 km-long situated 140 km from the apex (centred at $38.5°$ N, $82°$ E). The common co-location of fluvial systems and the aeolian systems to which they give rise on arid megafans is more fully presented in SOD 16.5, with further examples from arid lands.

16.5.5 Ridge Patterns and the Fluvial Hypothesis

A few undoubted fluvial channels are documented in the Meridiani units, underlying, within, and overlying the RFU (Section 16.4.3). These give circumstantial evidence for the role of rivers in the emplacement of these units. Furthermore, several channel systems in units that underlie the RFU are continuous with the sinuous ridges of the RFU – demonstrating the connection of some ridges to fluvial channels (Fig. 16.4). This fact alone suggests that Meridiani was a region in which rivers were active during sedimentation, at least in some locations.

Other circumstantial evidence suggests a connection between the RFU and river systems: the distribution on Mars of sinuous ridges that have been interpreted as fluvial features shows high concentrations in the Meridiani region (Williams 2007), which coincides with the region that also displays the highest fluvial-channel drainage densities (Hynek et al. 2010). Zabrusky et al. (2012:312) specifically suggested that evaporitic cementation driven by groundwater upwelling is one of the processes that contributed to the evolution of inverted valley networks across Arabia Terra, although these researchers did not include the Meridiani raised ridges in their sample of inverted networks.

Megafans are a normal component of terrestrial fluvial systems, allowing one to argue further that megafans may have been a more dominant landform on an arid, episodically fluvial and fluvially etched planet. Indeed, one reason that megafans are not encountered more widely on Mars may relate to less effective induration and therefore poorer preservation of such deposits in the absence of Meridiani-like evaporitic cements.

A major problem of Meridiani geology has been the identification of a sediment transport system (e.g., Malin and Edgett 2000; Edgett and Malin 2002; Edgett 2005). Fluvial channel networks typical of the megafan surfaces, however, suggest that a feasible transport system lies precisely in the widely exposed, raised-ridge networks of the RFU, if the sinuous ridges, one of several types we identify (Section 16.4.2), are interpreted as the exhumed relief of indurated river channels.

Despite expressing doubts about fluvial hypotheses, Edgett (2005) specifically pointed to the sinuous raised ridges of relict river channels in interior Oman as possible analogues for the RFU ridges. The Omani channels appear today as erosionally resistant ridges up to 30 metres high (Maizels 1990a, b) that display networks of numerous intersecting ridges of variable sinuosity (Fig. 16.5). What was not realised, or not regarded as significant by Edgett (2008), was the wider setting of the ridges, namely a megafan depositional environment. Here eight contiguous large fans, including three megafans, are clustered along 200 km of the hyperarid western flanks of the Oman Mountains, extending >200 km into the Arabian Peninsula (Fig. 16.2b). Oman's raised ridges indeed appear to be a good analogue for the RFU ridge networks. But the analogue supplies the wider domain-level context: it accounts for fluvial deposition, in an aggrading lowland, juxtaposed to an upland sediment source, separated by an abrupt topographic margin.

Furthermore, a detailed study of one of the Omani megafans (Maizels 1990a, b) has revealed a long history of at least twelve generations of superimposed palaeochannel systems (Fig. 16.5 shows nine generations), estimated to extend back to humid phases in the Pliocene–early Pleistocene. The terrestrial megafan analogue indicates slow landscape evolution from sequential deposition to exhumation over at least a few million years. The implication of a holistic megafan analogue is for long-continued, if episodic, fluvial activity at Meridiani.

Edgett and Malin (2002:2) have noted the 'affinity' of ridges to be co-located with crater forms. A major difference between terrestrial and the proposed martian megafan landscapes is the large number of crater obstacles that a fluvial dispersal system would have had to negotiate. Sinuous ridges in Meridiani are frequently seen to circumvent crater forms, suggesting the kind of interactions that are common between megafan river courses and bedrock obstacles (Fig. 16.3a, c). Other ridge types such as the apparently crater-related type (i) (Section 16.4.2) are ubiquitous and do not conflict with interpretations based on the proposed megafan analogue.

The presence of polygonal ridge patterns (Edgett and Parker 1997; Edgett 2005; Arvidson et al. 2003) has been suggested as evidence for subaqueous emplacement of unit NMe_1, which underlies sectors of the RFU (Edgett 2005). In the RFU, multiple-ridged features in polygon-like conformation may well have originated by non-fluvial processes (e.g., Arvidson et al. 2003; Edgett 2005). But it is also true that other ridges, of sinuous conformation, track closely with channel attributes of the terrestrial lowland domain (Table 16.1, compare lowland rows), and that they appear quite distinct from attributes of the upland VN (Table 16.1, compare upland rows).

Diachronic 'before-or-after' scenarios are usually employed to account for the existence of the RFU ridges, as for example offered by Arvidson et al. (2003) and Edgett (2005). Scenarios considered by Edgett (2005:31) were 'whether the ridges formed and then the spaces between them were filled, or whether the ridges consist of material that filled cracks or voids within a host rock'. However, the megafan analogue suggests a synchronic scenario: individual layers may be simply non-homogeneous – i.e., containing coarser channel sediments more prone to induration, encased in low-permeability fines, at least as regards ridge type (i).

A circumstantial argument supporting a fluvial interpretation of the RFU origins relates to the distribution of raised curvilinear features (RCFs). Ridge type (i) is characterised primarily by a sinuous, bifurcating morphology (Section 16.4.2), and thus conforms to the criteria for RCFs (Burr et al. 2010). Williams (2007) studied RCFs that included long sinuous ridges (10–200 km long, 0.3–3 km wide) and shorter bifurcating ridges. Williams (2007) identified 175 separate locales on Mars that show these features, including some in Meridiani, and concluded that RCFs can be interpreted as remnants of fluvial channels.

16.6 Conclusions

16.6.1 Domain Perspectives: The Megafan Analogue and a Meridiani Sediment Dispersal System

Grotzinger and Milliken (2012:7) have observed that 'classification [of martian layered terrains] reaches its fullest potential if regional topographic/structural context is considered...'. The megafan analogue advocated here emphasises not only the existence of purely fluvial depositional domains but also an appropriate scale, that is, a terrestrial domain of subcontinental extent whose significance has been recognised only recently. Depositional or aggradational fluvial domains are contrasted with upland fluvial domains that are dominantly erosional, with characteristic dendritic drainage patterns. Fluvial interpretations of the RFU have been rejected for reasons usually related to the lack of recognition of geomorphic, hydrological and sedimentologic patterns that typify aggradational landscapes. Our interpretations have proposed a primarily fluvial origin for the Meridiani layered units, based on a variety of evidence assembled by prior researchers and here interpreted in terms of the megafan analogue. Our conclusion supports that of Malin and Edgett (2000:1932) in their rigorously argued but guarded opinion: 'Despite the absence of well-defined overland transport systems of gullies, channels, streams and valleys, *alluvial processes may have played a major role* in the formation of the martian outcrop materials' (emphasis added). The evidence seems strong that the layered units likely relate at least in part to an active fluvial phase as the *depositional correlative* of the incisional VN phases. Overlap of the dates for both further suggest this connection.

16.6.2 Hybrid Sediment Sources

Hybrid sediment sources may account for the volume of material contained in the extant Meridiani sediment mass. External sources include sediment eroded from rocks in the adjacent southern highlands, especially during episodes of resurfacing, and repeated influxes of fluvially-entrained airborne materials (aeolian and volcanic) that probably blanketed the region episodically. Airborne materials (aeolian and volcanic) would also have been deposited directly onto Meridiani

Planum surfaces. Playa-style evaporitic accumulation was also significant, promoting selective induration of impact features and ridge-forming materials. The length of time evidently spanned by the Meridiani layered units that entomb numerous impact craters adds feasibility to such 'hybrid' hypotheses.

16.6.3 Minimum Requirements

The megafan analogue meets several specific requirements identified by prior researchers for the acceptability of hypotheses that address origins of the Meridiani units. (i) Megafan dimensions, especially when clustered, easily account for the wider former extent of the Meridiani sediments. (ii) Widespread, relatively low-energy fluvial activity occurred in periods spanning deposition of the units (i.e., spanning the Noachian–Hesperian boundary). (iii) The geographic association of the layered units with the VN exactly parallels the many terrestrial 'domain' settings of an eroding upland immediately upstream of an aggrading lowland. (iv) Megafan fluvial systems account for the planar and coherent clastic character and very low slope of the Meridiani units. (v) Employing a 'hybrid' hypothesis, the megafan analogue also accounts for marked differences in composition of the Meridiani layers, which are seen as derived from fluvial emplacement of eroded sediment, and fluvial reworking of aeolian and airfall materials. Later episodic sulphate impregnation would have further served to vary compositional attributes of units. (vi) The megafan analogue accounts for the similarity in derived spectra for rocks of the cratered terrain and the Meridiani materials, but also (vii) accounts for details of mapped hematite abundances, and (viii) the observed downslope diminution in grain-size.

Two requirements fall away in light of the megafan analogue: the need for a topographic/closed basin to contain the large mass of Meridiani sediment; and the related argument for the possible existence of shallow seas, although the possible existence of waterbodies is consistent with the megafan analog. The requirement for the layered units to conform to the underlying regional surface also falls away because megafan systems impose their own gradients on sedimentary units. The conundrum of the geographic location of the layered units at Sinus Meridiani seems adequately explained by geomorphic patterns now well established in megafan landscapes, namely the juxtaposition of fluvial units immediately downstream of valleys eroded in older upland rocks, with the proximal zone of the fluvial units anchored specifically along a well-defined topographic scarp separating the two landscapes.

16.6.4 Prolonged and Persistent Fluvial Activity

Delivery of sediment to the layered units and enhancement of surface flow during active groundwater upwelling, both favour prolonged and persistent surface water activity, even after precipitation had begun to wane (but before the final phase of aridity set in).

The length of time required to lay down the Omani megafans and exhume 12 stacked generations of channels has been estimated at a few million years (Maizels 1990a, b) – implying that fluvial activity may also have operated on similar time scales, or longer, during deposition of the Meridiani units, if indeed some ridges of the RFU represent fluvial channels.

16.6.5 Burns Formation and Other Phenomena

The co-location of water-filled depressions and aeolian subenvironments is documented on terrestrial megafans in arid/semiarid climates. Such subenvironments encompass the depositional settings subsumed in the Burns Formation examined by MER *Opportunity* in southwestern Meridiani (e.g., Grotzinger et al. 2005; Squyres et al. 2006). Several examples are given of terrestrial arid and semi-arid megafans that describe the numerous dune-related depressions that hold ephemeral water and episodic aeolian deposits.

The megafan analogue provides explanations for other problems relating to the Meridiani sediments, especially the noted absence of gullies, channels, streams and valleys. All are attributes of erosional domains and should therefore not be expected to dominate in fluvial aggradational domains. The analogue supports Malin and Edgett's (2000:1935) conclusion that an environment existed which involved the 'movement by…water of substantial amounts of eroded rock material'.

VN have been mapped more widely in Arabia Terra to the north and east of Meridiani (Davis et al. 2016, 2019) on terrains dated as middle to late Noachian (Hynek and Di Achille 2017). We surmise that fluvial

activity continued in the cratered terrain south of Meridiani, delivering sediment not only to Meridiani, but probably to most of Arabia Terra.

16.6.6 Sub-Megafan-Scale Features: Ridge Morphologies

One subset of RFU ridge morphologies that are long, sinuous and widespread arguably may be exhumed fluvial channels. These represent a major component of the extant ridge networks on the exposed RFU surfaces. The sinuous ridges do not mimic dendritic valley (upland) patterns, which may explain why the fluvial hypothesis has failed to be persuasive (versus the dendritic VN patterns to the south which are indeed persuasive as fluvial features). But the lowland megafan analogue mirrors many attributes of the RFU sinuous ridges (Table 16.2). The undoubted fluvial character of some ridges demonstrates that fluvial processes have played some part in the evolution of the vast and complex ridge network at Meridiani.

The megafan analogue suggests a plausible transport and dispersal system for sediment supply to Meridiani, one that accommodates a sediment mass covering hundreds of thousands of km^2, in the same way as do palaeochannel networks in megafan landscapes. The upland VN and Meridiani ridges may be correlatives of an *integrated sediment transport and dispersal system*, operating in different but associated geomorphic domains.

We conclude that erosional-domain terms and patterns appear to have been insufficient or misleading for the task of analysing very large depositional landscapes, not only on Earth (Wilkinson et al., Ch. 4) but also, not unreasonably, on Mars. An example is the unsuccessful search for valleys within the patently aggradational Meridiani sediment mass so that the lack of valleys then appeared to argue against a fluvial interpretation. This, combined with possibly unfamiliar fluvial patterns in aggrading lowlands – in the form of channel ridge networks that are both widespread and specifically do not occupy valleys – has resulted in an incomplete toolkit of features and patterns for the analysis of the Meridiani rocks. Ideas concerning regional-scale sediment accommodation and subcontinental configurations of drainage and landscape domains are also generally applied without reference to the full array of depositional domains that is now known to include vast megafan landscapes – thereby missing analogues that might give explanations for sediments laid down in fluvial environments.

In several ways, the megafan analogue is more parsimonious than other hypotheses in explaining the lack of closed depressions, the slope of the Meridiani units, the apparent lack of waterbodies associated with the layered units, the prominent linear boundary where the layered units meet the cratered terrain, and the very low slope of the Meridiani units. The Omani raised sinuous fluvial ridges, proposed by Edgett (2005) as a potential analogue for the sub-population of sinuous raised ridges of the RFU, seem an entirely appropriate analogue – especially when viewed in their wider megafan context, as proposed in this chapter.

We suggest that the fluvial megafan analogue, especially in the form of hybrid hypotheses as proposed by Hynek and Phillips (2008), answers these and other problems of interpretation and geographical relationship that have dogged scientific examination of the Meridiani sediments – possibly solving what Hynek and Phillips (2008:220) termed the 'mystery' that has indeed attached to understanding these units.

Acknowledgements

MJW has benefitted from discussions on issues dealt with in this chapter over the years with Carl Allen, Dorothy Oehler, Devon Burr, Misha Kreslavsky, and Mike Carr. Jacobs Engineering provided logistical support for field research in Australia, South America, and southern Africa. L. Yunxi is thanked for providing expert statistical comparisons of terrestrial domain attributes. The authors thank Alan Howard for his valuable critique of this chapter and for helpful suggestions.

References

Andrews-Hanna, J. C., Phillips, R. J., and Zuber, M. T. (2007). Meridiani Planum and the global hydrology of Mars. *Nature*, 446, 3–6.

Andrews-Hanna, J. C., Zuber, M. T., Arvidson, R. E., and Wiseman, S. M. (2010). Early Mars hydrology: Meridiani playa deposits and the sedimentary record of Arabia Terra. *Journal of Geophysical Research*, 115, E06002.

Andrews-Hanna, J. C. and Lewis, K. W. (2011). Early Mars hydrology: 2 – Hydrological evolution in the Noachian and Hesperian epochs. *Journal of Geophysical Research*, 116, 02007. doi: 10.1029/2010JE003709.

Arvidson, R. E., Seelos, F. P., Deal, K. S., et al. (2003). Mantled and exhumed terrains in Terra Meridiani, Mars. *Journal of Geophysical Research*, 108, 8073, doi:10.1029/2002JE001982.

Besler, H. (1984). The development of the Namib dune field according to sedimentological and geomorphological evidence. In J. C. Vogel, ed., *Late Cainozoic Palaeoclimates of the Southern Hemisphere*. Balkema, Rotterdam, 445–454.

Burr, D. M., Williams, R. M. E., Wendell, K. D., Chojnacki, M., and Emery, J. P. (2010). Inverted fluvial features in the Aeolis/Zephyria Plana region, Mars: Formation mechanism and initial paleodischarge estimates. *Journal of Geophysical Research*, 115, E07011, doi:10.1029/2009JE003496, 2010

Christensen, P. R. and Ruff, S. W. (2004). Formation of the hematite-bearing unit in Meridiani Planum: Evidence for deposition in standing water. *Journal of Geophysical*, 109, E08003 doi:10.1029/2003JE002233.

Craddock, R. A. and Howard, A. D. (2002). The case for rainfall on a warm, wet early Mars. *Journal of Geophysical Research*, 107, 5111, doi:10.1029/2001JE001505.

Craddock, R. A. and Maxwell, T. A. (1993). Geomorphic evolution of the Martian highlands through ancient fluvial process. *Journal of Geophysical Research*, 98, 3453–3468.

Davis, J. (2017). Unraveling a Martian enigma: The hidden rivers of Arabia Terra. Planetary Society 21 March 2017 [https://www.planetary.org/blogs/guest-blogs/2017/the-river-plains-of-mars-arabia-terra.html, accessed 10 May 2020]

Davis, J. M., Balme, M., Grindrod, P. M., Williams, R. M. E., and Gupta, S. (2016). Extensive Noachian fluvial systems in Arabia Terra: Implications for early Martian climate. *Geology*, 44, 847–850.

Davis, J. M., Gupta, S., Balme, M., et al. (2019). A diverse array of fluvial depositional systems in Arabia Terra: Evidence for mid-Noachian to early Hesperian rivers on Mars. *Journal of Geophysical Research: Planets*, 124, 1913–1934. doi:10.1029/2019JE005976

Dickson, J. L., Kerber, L. A., Fassett, C. I., and Ehlmann, B. L. (2018). A global, blended CTX mosaic of Mars with vectorized seam mapping: A new mosaicking pipeline using principles of non-destructive image editing. *Lunar Planetary Science Conference*, 49, Abstract 2083.

Edgett, K. S. (2005). The sedimentary rocks of Sinus Meridiani: Five key observations from data acquired by the Mars Global Surveyor and Mars Odyssey orbiters. *Mars*, 1, 5–58.

Edgett, K. S. and Malin, M. C. (2002). Martian sedimentary rock stratigraphy: Outcrops and interbedded craters of northwest Sinus Meridiani and southwest Arabia Terra. *Geophysical Research Letters*, 29, 2179.

Edgett, K. S. and Parker, T. J. (1997). Water on early Mars: Possible subaqueous sedimentary deposits covering ancient cratered terrain in western Arabia and Sinus Meridiani. *Geophysical Research Letters*, 24, 2897–2900.

Fassett, C. I. and Head, J. W. (2008). The timing of martian valley network activity: Constraints from buffered crater counting. *Icarus*, 195, 61–89.

Fawdon, P., Gupta, S., Davis, J., et al. (2018). Hypanis Valles delta: The last high stand of a sea on early Mars. *Lunar Planetary Science Conference*, 49, Abstract 2839.

Fuller, E. R. and Head, J. W. (2002). Amazonis Planitia: The role of geologically recent volcanism and sedimentation in the formation of the smoothest plains on Mars. *Journal of Geophysical Research*, 107, 5081, doi:10.1029/2002JE001842.

Glotch, T. D., Bandfield, J. L., Christensen, P. R., et al. (2006). Mineralogy of the light-toned outcrop at Meridiani Planum as seen by the Miniature Thermal Emission Spectrometer and implications for its formation. *Journal of Geophysical Research*, 111, E12S03. doi:10.1029/2005JE002672.

Golombek, M. P., Grant, J. A., Crumpler, L. S., et al. (2006). Erosion rates at the Mars Exploration Rover landing sites and long-term climate change on Mars. *Journal of Geophysical Research*, 111, E12S10, doi:10.1029/2006JE002754.

Grotzinger, J. P. and Milliken, R. E. (2012). The sedimentary rock record of Mars: Distribution, origins, and global stratigraphy. In J. P. Grotzinger and R. E. Milliken, eds., *Sedimentary Geology of Mars*. SEPM Special Publication, 102, 1–48.

Grotzinger, J. P., Arvidson, R. E., Bell, J. F., et al. (2005). Stratigraphy and sedimentology of a dry to wet eolian depositional system, Burns formation, Meridiani Planum, Mars. *Earth and Planetary Science Letters*, 240, 11–72.

Haddon, I. G. and McCarthy, T. S. (2005). The Mesozoic–Cenozoic interior sag basins of Central Africa: The Late-Cretaceous–Cenozoic Kalahari and Okavango basins. *Journal of African Earth Sciences*, 43, 316–333.

Hartley, A. J., Weissmann, G. S., Nichols, G. J., and Warwick, G. L. (2010). Large distributive fluvial systems: Characteristics, distribution, and controls on development. *Journal of Sedimentary Research*, 80, 167–183.

Hayden, A. T., Lamba, M. P., Fischer, W. W., et al. (2019). Formation of sinuous ridges by inversion of river-channel belts in Utah, USA, with implications for Mars. *Icarus*, 332, 92–110.

Hoke, M. R. T. and Hynek, B. M. (2009). Roaming zones of precipitation on ancient Mars as recorded in valley networks. *Journal of Geophysical Research*, 114, E08002, doi:10.1029/2008JE003247.

Howard, A. D., Moore, J. M., and Irwin R. P. (2005). An intense terminal epoch of widespread fluvial activity on early Mars: 1. Valley network incision and associated deposits. *Journal of Geophysical Research*, 110, E12S15, doi:10.1029/2005JE002460.

Hüser, K. (1976). Kalkkrusten im Namib-Randbereich des mittleren Südwestafrika. *Basler Afrika Bibliographien*, 15, 51–77.

Hynek, B. M. and Di Achille, G. (2017). Geologic Map of Meridiani Planum, Mars. Pamphlet, Scientific Investigations, Map No. 3356, United States Geological Survey, Washington DC.

Hynek, B. M. and Phillips, R. J. (2001). Evidence for extensive denudation of the martian highlands. *Geology*, 29, 407–410.

Hynek, B. M. and Phillips, R. J. (2003). New data reveal mature, integrated drainage systems on Mars indicative of past precipitation. *Geology*, 31, 757–760.

Hynek, B. M. and Phillips, R. J. (2008). The stratigraphy of Meridiani Planum, Mars, and implications for the layered deposits' origin. *Earth and Planetary Science Letters*, 274, 214–220.

Hynek, B. M. Arvidson, R. E., and Phillips, R. J. (2002). Geologic setting and origin of Terra Meridiani hematite deposit on Mars. *Journal of Geophysical Research*, 107. doi:10.1029/2002JE001891.

Hynek, B. M., Phillips, R. J., and Arvidson, R. E. (2003). Explosive volcanism in the Tharsis region: Global evidence in the Martian geologic record. *Journal of Geophysical Research*, 108. doi:10.1029/2003JE002062.

Hynek, B. M., Beach, M., and Hoke, M. R. T. (2010). Updated global map of Martian valley networks and implications for climate and hydrologic processes. *Journal of Geophysical Research*, 115, E09008, doi:10.1029/2009JE003548.

Hynek, B. M., Osterloo, M. K., and Young, K. S. (2015). Late stage formation of martian chlorides. *Geology*, 43, 787–790.

Iriondo, M. H. (1993). Geomorphology and late Quaternary of the Chaco (South America). *Geomorphology*, 7, 289–303.

Irwin, R. P. and Howard, A. D. (2002). Drainage basin evolution in Noachian Terra Cimmeria, Mars. *Journal of Geophysical Research*, 107 (E7), 5056. doi: 10.1029/2001JE001818.

Irwin, R. P., Howard, A. D., Craddock, R. A., and Moore, J. M. (2005a). An intense terminal epoch of widespread fluvial activity on early Mars: 2. Increased runoff and paleolake development. *Journal of Geophysical Research*, 110, E12S15. doi: 10.1029/2005JE002460.

Irwin, R. P., Craddock, R. A., and Howard, A. D. (2005b). Interior channels in Martian valley networks: Discharge and runoff production. *Geology*, 33, 489–492.

Irwin, R. P., Howard, A. D., and Craddock, R. A. (2008). Fluvial valley networks on Mars. In S. Rice, A. Roy, and B. Rhoads, eds., *River Confluences, Tributaries, and the Fluvial Network*. Wiley, Chichester, 419–452.

Kerber, L., Dickson, J. L., Head, J. W., and Grosfils, E. B. (2017). Polygonal ridge networks on Mars: Diversity of morphologies and the special case of the Eastern Medusae Fossae Formation. *Icarus*, 281, 200–219.

Knauth, L. P., Burt, D. M., and Wohletz, K. H. (2005). Impact origin of sediments at the Opportunity landing site on Mars. *Nature*, 438, 1123–1128.

Luo, W., Cang, X., and Howard, A. D. (2017). New Martian valley network volume estimate consistent with ancient ocean and warm and wet climate. *Nature Communications*, 8:15766. doi: 10.1038/ncomms15766

Maizels, J. (1990a). Raised channel systems as indicators or palaeohydrologic change: a case study from Oman. *Palaeogeography, Palaeoclimatology, Palaeoecology*, 76, 241–277.

Maizels, J. (1990b). Long-term paleochannel evolution during episodic growth of an exhumed Plio-Pleistocene alluvial fan, Oman. In A. H. Rachocki and M. Church, eds., *Alluvial Fans: A Field Approach*. Wiley, Chichester, 271–304.

Malin, M. C. and Edgett, K. S. (2000). Sedimentary rocks of early Mars. *Science*, 290, 1927–1937.

McCarthy, T. S. (2013). The Okavango Delta and Its place in the geomorphological evolution of Southern Africa. *South African Journal of Geology*, 116, 1–54.

McCollom, T. M. and Hynek, B. H. (2005). A volcanic environment for bedrock diagenesis at Meridiani Planum on Mars. *Nature*, 438, 1129–1131.

Miall, A. D. (1996). *The Geology of Fluvial Deposits*. Springer, New York, 582 pp.

Michalski, J. R. and Bleacher, J. E. (2013). Supervolcanoes within an ancient volcanic province in Arabia Terra, Mars. *Nature*, 502, 47–52.

Miller, R. P. (1937). Drainage lines in bas-relief. *Journal of Geology*, 4, 432–438.

Newsom, H. E., Barber, C. A., Hare, T. M., et al. (2003). Paleolakes and impact basins in southern Arabia Terra, including Meridiani Planum: Implications for the formation of hematite deposits on Mars. *Journal of Geophysical Research*, 108(E12), 8075, doi:10.1029/2002JE001993, 2003.

Niles, P. B. and Michalski, J. (2009). Meridiani Planum sediments on Mars formed through weathering in massive ice deposits. *Nature Geoscience*, 2, 215–220.

Ollier, C. D. (1977). Outline geological and geomorphic history of the Central Namib Desert. *Madoqua*, 10, 207–212.

Ormö, J., Komatsu, G., Chan, M. A., Beitler, B., and Parry, W. T. (2004). Geological features indicative of processes related to the hematite formation in Meridiani Planum and Aram Chaos, Mars: A comparison with diagenetic hematite deposits in southern Utah, USA. *Icarus*, 171, 295–316.

Pain, C. F., Clarke, J. D. A., and Thomas, M. (2007). Inversion of relief on Mars. *Icarus*, 190, 478–491.

Rhodes, D. D. (1987). *Table Mountain of Calaveras and Tuolumne counties, California*. In *Centennial Field Guide Volume 1, Cordilleran Section*. Geological Society of America, 269–272.

Schumm, S. A. (1977). *The Fluvial System*. Wiley Interscience, New York, 338 pp.

Schumm, S. A. (1981). Evolution and response of the fluvial system: sedimentologic implications. In E. G. Ethridge and R. M. Flores, eds., *Recent and Ancient Nonmarine Depositional Environments*. SEPM Special Publication, 31, 19–29.

Shanley, K. W. and McCabe, P. J. (1994). Perspectives on the sequence stratigraphy of continental strata. *American Association of Petroleum Geologists Bulletin*, 78, 544–568.

Squyres, S. W. and Knoll, A. H. (2005). Sedimentary rocks at Meridiani Planum: origin, diagenesis, and implications for life on Mars. *Earth and Planetary Science Letters*, 240, 1–10.

Squyres, S. W., Arvidson, R. E., Bollen, D., et al. 2006). Overview of the Opportunity Mars Exploration Rover Mission to Meridiani (Planum: Eagle Crater to Purgatory Ripple. *Journal of Geophysical Research*, 111, E12S12. doi:10.1029/2006JE002771.

Squyres, S. W., Knoll, A. H., Arvidson, R. E., et al. (2009). Exploration of Victoria Crater by the Mars Rover Opportunity *Science*, 324, 1058–1061.

Van Dijk, W. M., Densmore, A. L., Singh, A., et al. (2016). Linking the morphology of fluvial fan systems to aquifer stratigraphy in the Sutlej-Yamuna plain of northwest India. *Journal of Geophysical Research: Earth Surface*, 121, 201–222.

Ward, J. D. (1984). A Reappraisal of the Cenozoic stratigraphy in the Kuiseb valley of the central Namib desert. In J. C. Vogel, ed., *Late Cainozoic Palaeoclimates of the Southern Hemisphere*. Balkema, Rotterdam, 455–463.

Weissmann, G. S., Hartley, A. J., Nichols, G. J., et al. (2010). Fluvial form in modern continental sedimentary basins: distributive fluvial systems. *Geology*, 38, 39–42.

Weissmann, G. S., Hartley, A. J., Nichols G. J., et al. (2011). Alluvial facies distributions in continental sedimentary basins – distributive fluvial systems. In S. K. Davidson, S. Leleu, and C. P. North, eds., *From River to Rock Record: The Preservation of Fluvial Sediments and their Subsequent Interpretation*. SEPM Special Publication, 97, 327–355.

Weissmann, G. S., Hartley, A. J., Scuderi, L. A., et al. (2013). Prograding distributive fluvial systems— Geomorphic models and ancient examples. In S. G. Driese and L. C. Nordt, eds., *New Frontiers in Paleopedology and Terrestrial Paleoclimatology*, SEPM Special Publication, 104, 131–147.

Weissmann, G. S. Hartley, A. J., Scuderi, L. A., et al. (2015). Fluvial geomorphic elements in modern sedimentary basins and their potential preservation in the rock record: A review. *Geomorphology*, 250, 187–219.

Wilkinson, M. J. (1990). *Palaeoenvironments in the Namib Desert: The Lower Tumas Basin in the Late Cenozoic*. The University of Chicago, Geography Research Paper, 231, 196 pp.

Wilkinson, M. J. and McGovern, P. J. (2010). Megafans and paleo-megafans in Amazonis and northwest Tharsis: implications for fluvial processes, surface geology, and spreading of the Olympus Mons volcano. *Lunar and Planetary Science Conference*, 41, Abstract 2253.

Wilkinson, M. J., Marshal, L. G., and Lundberg, J. G. (2006). River behavior on megafans and potential influences on diversification and distribution of aquatic organisms. *Journal of South American Earth Science*, 21, 151–172.

Wilkinson, M. J., Marshal, L. G., Lundberg, J. G., and Kreslavsky, M. H. (2010). Megafan environments in northern South America and their impact on Amazon Neogene aquatic ecosystems. In C. Hoorn and F. P. Wesselingh, eds., *Amazonia, Landscape and Species Evolution: A Look into the Past*. Blackwell, London, 162–184.

Williams, R. M. E. (2007). Global spatial distribution of raised curvilinear features on Mars. *Lunar Planetary Science Conference* 38, Abstract 1821.

Zabrusky, K., Andrews-Hanna, J. C., and Wiseman, S. M. (2012). Reconstructing the distribution and depositional history of the sedimentary deposits of Arabia Terra, Mars. *Icarus*, 220, 311–330.

Notes

1 We adopt Edgett's (2005) descriptive term 'ridge-forming unit' ('unit R'; also unit 'HNMe$_3$', Hynek and Di Achille 2017; 'Unit E', Arvidson et al. 2003)

2 To avoid rising flood waters in the vast megafan landscapes in central South America, local inhabitants flee toward the river, to gain the higher ground of the levees. As quoted from local inhabitants, 'when the [Pilcomayo] river floods we go to the river – we and the snakes' (Martín Iriondo, 1989, pers. comm. to Wilkinson).

Part IV

Megafans in World Landscapes

17

Megafans in World Landscapes
Results of a Global Survey

M. JUSTIN WILKINSON

Abstract

Major morphological characteristics of megafans and the associated drainage networks of multi-megafan landscapes are outlined. Such landscapes differ significantly from well-known erosional landscapes with dendritic drainage in eroded valleys: megafan landscapes display partial cone morphology; longitudinal profiles can be convex, concave or both; interfluves and tributaries are lacking; the fan-forming river behaviour is highly avulsive; fan-margin rivers display three discharge regimes; convergent drainage zones of various types are widespread despite the broad partial-cone morphology; avulsions often occur at subapexes distant from the prime apex; perirheic zones differ significantly from the model developed for valley-confined floodplains. These and other attributes bespeak the complexity that arises as fan-like features increase to megafan proportions. Controls of megafan formation are presented. Megafans are distinguished from small alluvial fans, midsize distributive fluvial systems (DFS) (30–80 km long), valley-confined floodplains, deltas, major avulsive fluvial systems (MAFS), and large accretionary fluvial systems (LAFS) – the latter two relatively new to the geomorphic lexicon. Megafan nesting patterns and several wider continental lowland drainage models that encompass megafans (e.g., central South America) are described. Megafans show striking similarities with attributes of large axial floodplains despite being formed by smaller rivers. Terms with confusing meanings are clarified.

17.1 Introduction

This chapter describes the major characteristics of megafans from the perspective of recent research. It is now recognised that megafans lie at the intersection of sets of auto- and allogenic controls (Ventra and Clarke 2018). The origin of megafan landscapes is frequently ascribed to allogenic factors, partly because few megafans are well known. However, autogenic factors are emphasised in this chapter. The chapter also attempts to dispel some of the confusion that has arisen around megafans as relatively recently recognised features that are now known to be widely distributed on Earth. Megafans have been viewed, for example, simply as a variant of the confined floodplain (Blair and McPherson 1994), or as a version of the small, well-known alluvial fan (e.g., Stanistreet and McCarthy 1993; Miall 1996), despite being orders of magnitude larger. Megafans are a subset of the fan-like landforms of all dimensions that have been termed 'distributive fluvial systems' (DFS: Hartley et al. 2010a, b; Weissmann et al. 2010, 2011). DFS are described in the Introductory chapter (Wilkinson and Gunnell, Ch. 1) and discussed briefly in the section on terminology below (Section 17.7.2).

The examples used here are drawn from the global map presented in Wilkinson and Currit (Ch. 2), developed as part of an ongoing project to map active and relict megafans that can be identified remotely in which our criteria for identifying megafans are discussed (Wilkinson and Currit, Ch. 2). Unless otherwise specified, the data are based on megafans that lie distant from modern shorelines in order to avoid any confusion with deltas. The data on which this chapter is based are partly drawn from presentations and internal reports (Wilkinson 1996, 1998, 1999, 2001, 2002, 2003, 2005, 2006, 2015; Wilkinson and Cameron 2002; Wilkinson et al. 2002, 2006, 2010).

17.2 Megafans in World Landscapes and the Hierarchy of Fluvial Landforms

17.2.1 Significance of Megafans in World Landscapes

The extent of fluvial fans and megafans now takes on regional significance, especially when two or more are nested together (Fig. 17.1a, b). Larger megafans measure up to $1–2 \times 10^5$ km^2: the area of the Pilcomayo megafan approaches a quarter of a million km^2 (Iriondo 1987; Latrubesse et al., Ch. 5), which is the largest known on Earth (Wilkinson 2001; Wilkinson et al. 2006). Contiguous fan surfaces reach a total of ~ 0.75 M km^2 in South America with a total for the entire continent of ~ 1.25 M km^2 (Wilkinson et al. 2006). Iriondo (1984, 1987) emphasised the flatness and great extent of *depositional landscapes* of modern nested megafans in the Chaco Plains of northern Argentina and western Paraguay. These extend for $> 1,200$ km along the Andean mountain front (Wilkinson et al. 2010, their fig. 10.3). Wilkinson et al. (2010) also emphasised the smooth, flat morphology of such landscapes which are now known to display almost unique roughness signatures (Fig. 17.1a) comparable at subregional scales with very extensive and better researched pedimented (erosional) surfaces with which they are known to be confused.

In the Chaco Basin, modern megafans occupy 85% of lowlands, with the axial, valley-confined Paraguay-Paraná floodplain occupying the remnant. Similar megafan-tributary *vs.* axial proportions are apparent in parts of the Gangetic foreland basin of northern India.

17.2.2 Megafans in the Hierarchy of Fluvial Deposits

The state of definitional confusion and lack of precision in applying appropriate processes at the different scales can be partly mitigated by using the hierarchical schema of sedimentary bodies proposed by Miall (1996, 2014, 2015). Twelve *Sedimentation Rate Scales* (*SRS* 1–12, Fig. 17.2) are based on normal time scales of sediment-body evolution, which translates into a geometrically-scaled hierarchy of fluvial architectural units. The processes that act at each level change as dimension changes. The schema is intended by Miall (1996, 2014, 2021) as a flexible arrangement of fluvial architectural elements, which are modified here to accommodate megafans and other major avulsive fluvial systems (Section 17.6.1.4).

Smaller bodies become components or building blocks of larger bodies. An example specific to megafans is the familiar channel-belt (*SRS* 8, Fig. 17.2), a critical building block of several larger landforms, namely valley-confined floodplains, megafans (Fig. 17.2), and deltas. The fact of the channel belt acting as a component of both valley-confined floodplains and megafans was specifically one of the reasons why floodplains and megafans were classified as the *same* feature by Blair and McPherson (1994:481): they described megafans as 'rivers with locally expanding reaches'. For reasons discussed in this chapter, megafans in fact appear to be quite distinct from valley-confined floodplains.

Because of varying definitions, Miall (1996, 2014) appeared ambivalent about the place of alluvial fans in his hierarchy, classifying them as features of *SRS* levels 8 and 9 (Fig. 17.2, and also in his tables 3.2, 4.2, and 9.1, Miall 1996). In this writer's view, and partly following the dimensional scale proposed by Weissmann et al. (2011), features discussed here (shaded, Fig. 17.2) fit best at the following *SRS* levels: classic alluvial fans (*sensu stricto*) with dimensions < 20 km at the *SRS* 7 level, fluvial fans (30–80 km long) at the *SRS* 8 level and megafans (> 80 km long) at the *SRS* 9 level. Sets of nested megafans then fit best as *SRS* 10 bodies (Wilkinson et al. 2006, 2010). Smaller architectural elements can be components of all the larger architectural elements, namely floodplains, deltas and megafans.

The sedimentological distinction between small alluvial fans (*sensu stricto*) and fans > 30 km long is now generally recognised. However, the well-described processes on small fans are often invoked uncritically as being responsible for the development of large fans. This applies especially with respect to the familiar 'expansion of flow' dynamic, documented as a key process for alluvial fan (*sensu stricto*) development:

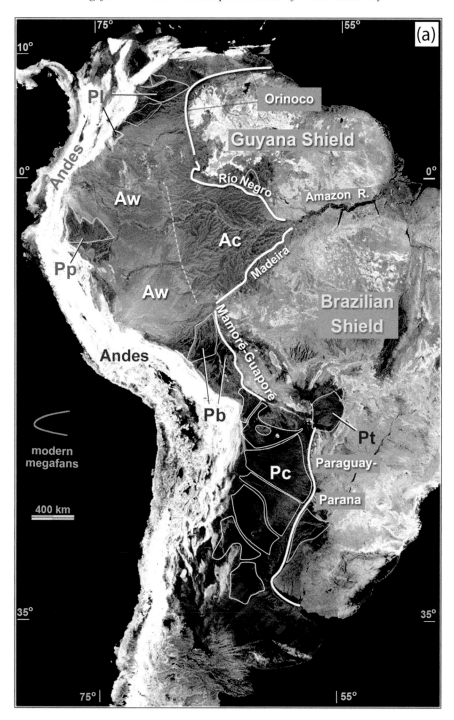

Figure 17.1 **Megafan landscapes**. (a) Roughness map of most of South America. Darkest tones: smoothest, lowest-sloped landscapes, dominated by megafan plains especially well developed in the Chaco Plains (Pc). Other major megafan plains are: Pastaza–Marañón (Pp), Llanos (Pl), Pantanal (Pt) and Beni (Pb). Larger megafans are outlined: Chaco (Pc) megafans from Iriondo (1984, 1987, 1993). Lightest tones: roughest, steepest landscapes of the Andean mountains. Intermediate tones: Brazilian and Guiana shields. From Wilkinson et al. (2010). (b) Gangetic Plain and Himalaya catchments. Himalaya-sourced Kosi and Tista rivers have generated the triangular Kosi megafan and diamond-shaped Mahananda-Tista megafan. The axial Ganga River is displaced southward by megafan bodies against the craton margin. Astronaut handheld image STS027–39-27, courtesy of the Earth Science and Remote Sensing Unit, NASA Johnson Space Center. From Wilkinson et al. (2010). A colour version of image (b) is available in the SOD for Chapter 17.

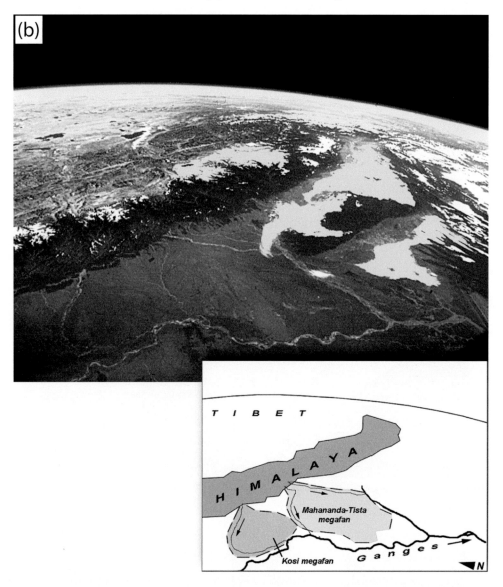

Figure 17.1 *(cont.)*

'The expansion of flows in response to their instantaneous unconfinement upon passage beyond either the apex or the intersection point initiates sedimentation typically over a pie-piece-shaped fan segment called the active depositional lobe... The optimal conditions that promote sedimentation through flow expansion on the semiconical form [of the alluvial fan] create radii that rarely exceed 10 km in length...' (Blair and McPherson 1994:454).

Although repeatedly invoked as the initiation process for fluvial and megafan formation, this process is specifically not responsible for forming cones of sediment of megafan dimensions. On larger fans, the lack of confinement operates at a different scale that involves avulsion, namely the repeated lateral shift of an *entire*

channel belt (individual channels confined *locally* by levees) to different orientations across vastly larger landscapes not confined by valley walls. 'Expansion of flow' thus means different things at different scales, and provides a prime example of a process erroneously applied to landforms at different levels of the fluvial hierarchy.

The examples above emphasise two provisos: first, that the similarity of system components such as channel belts should not lead to confusion between features such as megafans, valley-confined floodplains and deltas, of which they are constituents. Nor should features at different levels in the hierarchy, even if they show fan-like morphology such as alluvial fans *(sensu*

SRS[a] Group	Time scale (yr)	Architectural Units of Fluvial Deposits		
		Rivers and alluvial fans (Miall 1996, 2014, 2015; DeCelles et al. 1991)	Megafan system -- radii > 80 km	Major Avulsive Fluvial System (MAFS)[b], Large Accretionary Fluvial System (LAFS)[c]
1	10^{-6}	Lamina		
2	10^{-6}–10^{-4}	Ripple (microform)		
3	10^{-3}	Diurnal dune increment		
4	10^{-2}–10^{1}	Dune (mesoform)		
5	10^{0}–10^{1}	Macroform growth increment		
6	10^{2}–10^{3}	Shallow channel, large stream-bed macroform (point bar, levee, splay), alluvial fan channels		
7	10^{3}–10^{4}	Channel, delta lobe, alluvial fan trench backfill, avulsion events		
8	10^{4}–10^{5}	Channel belt, alluvial fan		
9	10^{5}–10^{6}	Floodplain, alluvial fan, alluvial fan tract (bajada), delta, major depositional system axis, e.g., G. Mexico coast depositional axes	Megafan – this volume	*Major Avulsive Fluvial Systems (MAFS)*, e.g., Bananal (Araguaia R., Brazil), Cumberland Marshes (Saskatchewan, Canada); *Large Accretionary Fluvial Systems (LAFS)*, e.g., Channel Country floodplains of Australia
10	10^{6}–10^{7}	Smaller basin-fill complexes, e.g., Tertiary fms., G. Mexico coast	Nested megafans, e.g., Chaco Plains, N. Argentina— W. Paraguay	
11	10^{6}–10^{7}	Basin-fill complex		
12	10^{6}–10^{7}	Basin-fill complex		

Figure 17.2 Megafans and grouped megafans within the hierarchy of fluvial deposits. Rivers, alluvial fans, and deltas– left column, from Miall (1996, 2014) and DeCelles et al. (1991). Midscale forms are shaded. Adapted from Wilkinson et al. 2006, 2010. [a]*SRS –Sedimentation Rate Scale*, per Miall 2014, 2015; [b,][c]see Section 17.5.1.

stricto) and megafans, automatically be attributed to the same processes (e.g., expansion of flow).

17.3 Megafan Morphologies and Controls: Area, Slope, Drainage Styles, Planform

17.3.1 Megafan Surface Morphology: Surface Character, Drainage Styles, Apex Types, and Planform

The extremely low gradients of individual megafans can be judged from the fact that despite widths up to hundreds of kilometres, the convexity in transverse profile is always apparent despite being usually less than 10 m in relief. In global terms – according to digital-elevation-model-based roughness measurements (3 arcsec SRTM) – megafan surfaces form one of the largest and smoothest surface types on Earth, comparable to lake floors (Fig. 17.1a, also Wilkinson et al., Ch. 3, Fig. 3.5).

Seen remotely, a dominant characteristic of megafan surfaces is the pattern of densely packed paleochannels and channel belts. This characteristic is so widespread that it is often possible to identify megafans by this means alone. Complex networks such as these are related to the various river patterns associated with avulsion phenomena enumerated by Slingerland and Smith (2004) (nodal, random, local, and regional, with variants termed partial and full avulsion), as well as other phenomena reviewed in sections below. Processes responsible (avulsion by annexation, incision, and progradation) add further complexity to the patterns. Numerous examples exist on any one megafan surface. A burgeoning literature on the types and mechanics of avulsion has arisen (e.g., see reviews in Miall 2014). Avulsions are placed in the broader megafan context below (Section 17.4.1.1).

Microtopographic features show extreme variety compared with that of smaller alluvial fans. Detailed descriptions of megafan surfaces (Cordini 1947 for the Pilcomayo surface; several for the Kosi megafan, e.g., Sinha et al., Ch. 11; and McCarthy 2013, who summarises a large literature on the Okavango megafan) include a wide range of features in addition to paleochannel belts, such as: underfit channels, depressions of various origins (open and closed), some occupied by lakes or wetland, aeolian forms (both depositional and erosional), and even positive, mound-like features a few km^2 in area produced partly by evaporite precipitation (McCarthy 2013).

Very common features found on megafans are zones of fluvial incision. The best known is a proximal incision cut by the megafan-forming river, analogous to that developed on smaller alluvial fans. An 'intersection point' marks the distal end of the incised zone, typified by a subapex from which an individual 'lobe' of sediment forms as a new, active surface with locally radial drainage; the abandoned, older surface flanking the incision is subject to pedogenesis and fluvial reworking. A well described example is the modern distal lobe of the Taquari megafan in SW Brazil (Section 17.4.3) (Assine 2005; Weissmann et al. 2015; Pupim et al. 2017; Santos et al., Ch. 6). Fan-forming rivers can incise the entire length of a megafan, rendering the fan surface relict as far as the fan-forming river is concerned – the Cubango relict megafan of northern Namibia being a prime example (Wilkinson et al., Ch. 4; Miller et al., Ch. 15).

17.3.1.1 Apexes and Avulsion, Sediment Lobes, Drainage Planform

The existence of an *apex*, as the topographically highest point of a cone, is regarded as a primary criterion for the recognition of a megafan (e.g., Weissmann et al. 2011) and is directly associated with channels of fan-forming rivers that radiate from the catchment outlet at the margin of an upland source (defined as regional, nodal avulsions *sensu* Slingerland and Smith 2004) (Figs. 17.3a, c, d). The resulting diverging channel pattern is readily observed remotely. Radiating channels do not imply simultaneously active channels, as suggested by the term *distributary* (see *distributive/distributary*, Section 17.7.2) that is frequently applied to megafan drainage patterns. Ninety-four percent of avulsions in the Chaco Plains and Gangetic Plains in the period 1984–2014 occurred within 100 km of the apex (Edmonds et al. 2016), attesting to the importance of avulsion in the apex zone.

Subapexes, in contrast, are centres of radiating drainage lines that are positioned distant from the major apex. Whereas the main megafan apex represents a permanent nodal avulsion point (*sensu* Slingerland and Smith 2004), subapexes can be impermanent in those cases in which they shift position as the fan-forming river moves across the surface of the megafan; these are the random, local avulsion sites *sensu* Slingerland and Smith (2004). Most commonly represented from the alluvial fan literature is the subapex situated at the intersection point where an incised sector of the fan-forming river gives way to an unconfined sector. In the long term this is thought of as a mobile subapex because the formative river avulses to different radial orientations on the megafan cone. An unappreciated type is the permanent type, which is produced by subtle slope components, where the fan-forming river is locally slightly confined, as for example between an obstacle (such as protruding bedrock or an abutting alluvial fan) and the low slope of the megafan cone itself (point c, Fig. 17.4, SOD 17.3a). On the downstream side of the obstacle, flow is once again *unconfined* laterally, thereby forming a subapex, downstream of which a radial drainage pattern develops. This would also be described as a nodal, regional avulsion *sensu* Slingerland and Smith (2004).

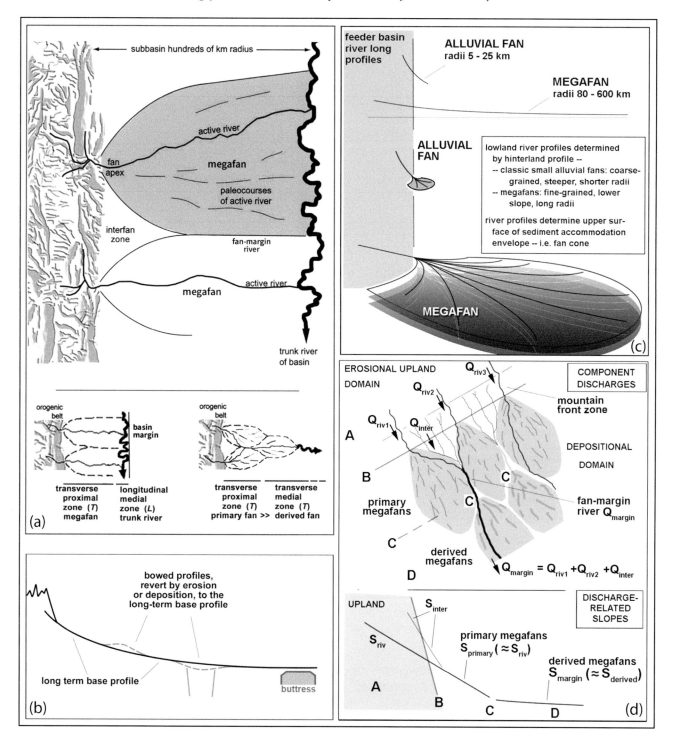

Figure 17.3 **Planform and cross-sectional geometries in single megafan and multi-megafan systems**.
(a) Upper panel: features of megafan landscapes showing upland, and neighbouring sedimentary basin megafans with radial palaeocourses of mountain-fed river (fan-forming river), fan-fed rivers (often occupying palaeocourses), foothills-fed rivers of the interfan zone, and transverse trunk/axial river. Lower panel: nested megafans with T/L configuration and associated triangle/delta-like planform (left); T/T configuration with associated diamond-shaped planforms (right). From Wilkinson et al. (2006).
(b) Long-term parabolic longitudinal profile of megafan-forming river.
(c) River slope and fan slope ('2D–3D model'). Small alluvial fans commonly formed by steep, short streams, compared with megafans formed by larger, lower-declivity rivers with associated much extended horizontal reach.
(d) Primary and derived megafan types. Upper panel: fan-margin river discharge (Q_{margin}) component supplied by foothills-fed (interfan) rivers (Q_{inter}). Episodic additions are supplied by one or both adjacent mountain-fed (fan-building) rivers, Q_{riv1} and Q_{riv2}, at times when these are aligned to the fan-margin depression, as shown here. See text. Lower panel: primary megafan slopes ($S_{primary}$) are associated with upland river slopes; lower fan-margin river slopes (S_{margin}) are apparently determined by the combined discharge of megafan rivers and interfan drainages. These impose very low slopes of the derived megafans.

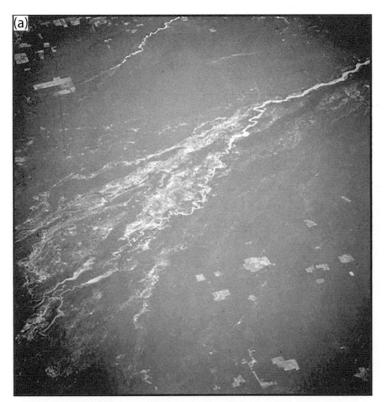
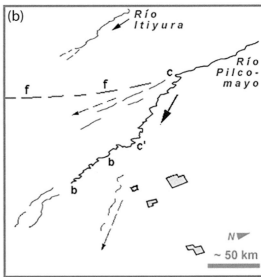

Figure 17.4 **Subapex formed at points of local loss of topographic confinement**. Detail of southern margin of the Pilcomayo megafan, Argentina–Paraguay border. The Pilcomayo River is locally confined by the low slopes of the Pilcomayo megafan itself and that of the smaller neighbouring Itiyuro fan (upper part of image); at point c, the margin of the Itiyuro fan curves away to the south (c–f–f), releasing the Pilcomayo floodplain from the locally confined zone. Several radial courses of the Pilcomayo River are shown: the modern course follows the line c–c'–b–b. Astronaut handheld image STS32–88-93 courtesy of the Earth Science and Remote Sensing Unit, NASA Johnson Space Center. From Wilkinson et al. (2006).

However, their typology accommodates avulsive behaviours on megafans less well because it is designed specifically for geometries seen on floodplains, whereas the distinction between spatially fixed versus mobile avulsion points described above is fundamental for understanding megafan drainage patterns. Both types of subapex give rise to lobes of sediment, the relative ages of which can often be estimated from the cross-cutting relationships of palaeochannels (e.g., Kosi and Tista megafans; Chakraborty et al. 2010).

Change in the direction of stream flow – and hence sediment dispersal – is thought of as almost entirely restricted to the operation of avulsion points. But two unusual styles of avulsion on megafans are now well documented, one on the Okavango megafan in Botswana, and the other on the Pilcomayo megafan on the Chaco Plains, although the associated sediment dispersal system in the longer term is less well understood. On the former, Wilson and Dincer (1976) have described 'avulsion by underflow' which involves diversion of flow from an existing channel as the result of *subsurface* diversion of drainage. This leads to the formation of a new surface channel which, counterintuitively, is physically disconnected from the parent channel (McCarthy 2013; Wilkinson et al., Ch. 4, SOD 4.1.2).

The other phenomenon is termed channel 'self-blockage' – a state of incipient avulsion that migrates upstream without definitive reorientation of the fan-forming river. This phenomenon is documented on the Pilcomayo megafan, where the Pilcomayo River is aggrading its own channel to such a degree that in-channel sedimentation completely blocks the main channel (Meyer 1996). Overbank spillage of the sediment load of this large river (average annual suspended load of 140×10^6 t, one of the largest in the world: Martín-Vide et al. 2014). results in 'a dome-like deposit... extending some 15 km downstream of the point of [blockage], 5 km upstream of it and 10 km... on both sides of the channel' with an annual volume equivalent to 'a ½ m-deep layer of sediment in a square with sides measuring 11.8 km' (Martín-Vide et al. 2014:159). The locus of such deposition has migrated upstream a remarkable 300 km since 1947 (Martín-Vide et al. 2014). The rate of retreat of the river blockage point averages 12 km per year. The net result is serial deposition of vast quantities of sediment as a linear zone ~20 km wide along much of a 300-km stretch of the river.

Hartley et al. (2010a) have demonstrated that all channel 'planforms' are recognisable on megafan surfaces, with braided channels being the most common in the population they examined (>30 km long; Hartley et al. 2010a, 2013; Weissmann et al. 2010, 2011; Davidson et al. 2013; Davidson and Hartley 2014; but see Latrubesse's comments listed in Section 17.6.1.1). There is often a downfan trend from braided to meandering and anastomosing patterns (Ori 1982; Räsänen et al. 1992; DeCelles and Cavazza 1999; Fontana et al. 2008; Ralph and Hesse 2010; Bernal et al. 2013; Davidson et al. 2013). Anabranching patterns have been described also (e.g., Jain and Sinha 2003, 2004, 2005).

The tendency of channel belts to cluster has been partly explained by processes that operate at different scales, as with compensational stacking which tends to force flow pathways and sediment transport toward topographic lows in aggradational settings (e.g., Straub et al. 2009), or more locally by the influence of relict channel belts guiding the courses of successive drainage (Edmonds et al. 2016).

Megafan shape may grow increasingly complex as fan dimension increases. In triangle-shaped fans, drainage orientation remains generally radial down the length of the fan. However, drainage orientation on elongated/diamond-shaped fans (Fig. 17.3a, d) can change distally such that parallel and even convergent patterns emerge. The Loire megafan in France (Gunnell, Ch. 9) and the Rio Grande megafan in Bolivia (May et al., Ch. 7) are examples.

17.3.1.2 Confined Zones and Convergent Drainage Patterns on Megafans

Important for the present discussion is the major distinction proposed by Hartley et al. (2010a, 2015) and Weissmann et al. (2010, 2011, 2015) between styles of sedimentation on unconfined broadly conic surfaces of unincised megafans, as opposed to styles in confined settings such as interfan zones and classical fluvial valleys in degradational landscapes. Confined settings result in phenomena such as amalgamation of point-bar and meander-belt successions (leading to scroll-bar topography near the active channel belt; e.g., Hartley et al. 2018), as well as cutoff avulsions, the significant reworking of floodplain fines by channel-belt migration, and reduced potential for soil maturation due to continuous alluviation and reworking of surfaces within the zones influenced by channel belts. In the formulation of the above-mentioned authors, processes typical of confined settings are either non-existent or much reduced on unconfined DFS surfaces, leading to lesser amalgamation, fewer cut-off avulsions, greater potential for accumulation of floodplain fines, and the probability of mature soil development on those sectors of megafan surfaces where river return times are long. Weissmann et al. (2015) then widen these dichotomous settings to characterise the major morphological distinctions between converging river patterns (valley-confined geomorphic settings of classic dendritic drainage patterns), and the notionally opposite radial drainage patterns in megafan systems (unconfined, non-dendritic, 'distributive' geomorphic settings).

As argued here, however, because these landscapes are so large and complex, various types of *confined* zones in fact occur in megafan landscapes – in which convergent drainage patterns necessarily exist. The following types of negative relief at the sub-megafan scale (partly following Cordini 1947, but excluding small depressions of lengths 10^1–10^2 m) will all display convergent drainage patterns:

(i) zones of fluvial incision that cut radially-oriented, broadly linear trenches (proximal 'top-down' fan-head trench and 'bottom-up' distal incisions which can be narrow or wide; Weissmann et al. 2015:206–207), that can also extend the full length of a megafan;
(ii) local elongate negative relief corresponding to abandoned channels;
(iii) 'floodbasins', large and small, confined between alluvial ridges;
(iv) negative relief between slopes of depositional lobes;
(v) aeolian-scour depressions of varied dimensions;
(vi) negative relief resulting from sediment compaction;
(vii) tectonically induced negative relief.

All of these except (i) can be topographically open or closed depressions, and all except (v) can give rise to locally convergent drainage. Consequently, a mix between confined and unconfined depositional environments will be encountered on megafans, the former relating more commonly to the fan-forming river and the latter to the other types of drainage system on and between megafans.

17.3.1.3 Megafans as Closed Hydrological and Sedimentological Systems

Evidence is accumulating that megafans can often behave as closed sedimentological systems, even though by definition they are topographically open (i.e., individually exorheic, whether or not they are part of wider exorheic or endorheic basins). Indeed, most megafans exist within exhorheic basins. The best known example of a topographically open megafan that is closed both sedimentologically and hydrologically may be the Okavango megafan, from which 2% or less of the discharge that enters at the apex escapes from the toe (McCarthy 2013). Whereas the closed character of the Okavango megafan was thought to be unusual, Latrubesse (2015) and Latrubesse et al. (Ch. 5) have documented the regional-scale example of the Chaco Plains in central South America, in which four of the six rivers that have built very large megafans today fail to reach the trunk Paraná River. Their river end-points are in fact now situated hundreds of kilometres upstream of the respective megafan toes. Latrubesse et al. (Ch. 5) have concluded, therefore, that 'the Chaco plain is ... one of the largest active continental sedimentary sinks of the planet', sequestering 'approximately ~325 Mt/yr of sediment [by] the six rivers generating the largest megafans'. It is often assumed that sedimentation of regional extent requires topographically closed basins (Wilkinson et al., Ch. 16), but megafans provide dramatic evidence that topographically open systems are well able to sequester sediment.

17.3.2 Feeder River, '2D–3D River-Slope' Model, and Megafan Length

Holbrook et al. (2006) adapted the idea of the equilibrium profile of a river, as a longer-term balance between the rate of change in accommodation and the water/sediment discharge flux. An approximately parabolic river profile is controlled respectively by tectonic and climatic influences from the upland zone, regulating the balance between water discharge and sediment supply, and by a 'buttress point' distally, corresponding to a regional or local base level – sea level, lake levels, the altitude of trunk-river courses, or other geologic structures within continental basins (Fig. 17.3b). Buttress movements tend to dominate river aggradational and degradational dynamics in downstream basin sectors (Shanley and McCabe 1994); and complex fluvial responses, after disturbances in the upland buffer or lowland buttress zones, tend to restore a broadly parabolic profile (Holbrook et al. 2006).

Applying the fluvial equilibrium profile as a basic concept for understanding megafan sedimentation, the longitudinal, two-dimensional profile of the feeder river in the upland broadens out into a three-dimensional megafan surface at the point where the river enters the unconfined lowland zone (Fig. 17.3c). This '2D–3D river slope model' simply states that in the longer term (10^3–10^4 yr) the fan-surface gradient is closely related to some long-term average river discharge (equivalent to the > 1000 yr equilibrium/average calculated by Blum et al., 2013) at the mountain front. Of importance here is the observation that steeper upland rivers give rise to steeper fans (i.e., of shorter radius), whereas upland rivers with lower channel gradients give rise to megafans of lower slope and thus potentially longer radii (Fig. 17.3c). This model is based on the general observation (though not universally applicable: SOD 17.3c) that river gradient is inversely related to discharge, compatible with what Hartley et al. (2010a:173, their figs. 3, 6, 7) term a 'clear relationship' that has long been attested (e.g., Drew 1873; Gilbert 1877; Blair and McPherson 1994, 2009; Cooke et al. 2006; Davidson et al. 2013, their fig. 3).

In mountain-front settings, and assuming only transport-limited conditions (i.e., unlimited sediment supply from the catchment over the geological period of interest), the very low hydraulic gradients of large rivers would be associated in the right setting with megafan surfaces that project many hundreds of kilometres into the basin, if the basin is extensive enough. The Pilcomayo River, a major tributary of the Paraná

River in central South America, is the classic example: its low gradient of 0.00036 has enabled the megafan surface to advance across the entire topographic basin, achieving a radius of > 700 km (Weissmann et al. 2015).

Local effects can modify this relationship, as in some documented cases where a significant reduction in gradients of megafan slopes has been reported compared with the feeder river in the upland catchment (e.g., Ganti et al. 2014). However, this is a limited perspective. The general relationship holds; small variations inevitably occur, as in the northern Kalahari research area, where feeder river gradients vary from steeper than the fan gradient (Cassai megafan case), to less than the fan gradient in the Okavango megafan case (Wilkinson et al., Ch. 4). Other local effects are the differences between the fan-surface gradient and that of rivers that flow across the surface, whose discharges and sediment loads change so that rivers adjust their sinuosity accordingly.

Dimensions of the receiving basin, and the competition for space between neighbouring megafans, are two other major controls. The dimension of the receiving basin acts as an independent control of fan length in cases where the basin margin truncates the potential horizontal reach ('horizontal accommodation'; Weissmann et al. 2010:199). The Kosi megafan of northern India is instructive: the megafan has extended as far as the opposite (downstream) margin of the basin, suggesting it would have projected a greater distance had the margin been situated further from the Himalayan mountain front (Figs. 17.1b, 17.3a lower panel).

River discharge volume is another control of fan length, with greater discharge roughly associated with larger fan areas in many basins. The complex issue of the competition for space, which also affects fan length, is dealt with in Section 17.5.1.

17.4 Multi-Fan Scales: Four Basic River Types, Three Discharge Regimes of the Fan-margin River, and Megafan Taxonomy

17.4.1 Four River Types: Fan-Forming, Fan-Fed, Foothills-Fed and Fan-Margin Rivers

Drainage patterns in multi-megafan landscapes show several characteristics that differ so significantly from the well-known dendritic pattern of erosional landscapes that they are set out here in some detail, as identified 25 years ago by Sinha and Friend (1994). Our descriptive terms are the dominant *fan-forming river*, the smaller *fan-fed rivers*, the *foothills-fed rivers*, and the *fan-margin river* ('mountain-fed', 'plains-fed', 'foothills-fed', and 'mixed-fed', respectively, in the original terminology of Sinha and Friend 1994). The fan-forming river is sourced in a major upland drainage basin and supplies the sediment that builds the megafan (Fig. 17.3d). By contrast, the numerous, relatively short fan-fed streams – sourced by water available on the fan surface – from rainfall, seepage from the main channel (where the main channel is near enough), and ground water that may reach the surface – often appear as underfit channels in the abandoned palaeochannels of the fan-forming river. Both river types are connected to the positive topography of the megafan cone, where radial drainage dominates.

In contrast, the following two river types are specifically located in the negative topographic zone between two megafans where convergent drainage is the norm. The interfan zone – also termed 'inter-megafan' area (DeCelles and Cavazza 1999) and 'intercone' area (Geddes 1960) – gives rise to rivers of intermediate discharge (foothills-fed rivers), also termed the *Inter-DFS tributary system* (Weissmann et al. 2015). These streams combine to flow into the marginal depression between the fan cones (Fig. 17.3a, d) to form the single fan-margin river.

It is generally assumed that the dominant rivers in multi-megafan landscapes are the permanent fan-forming (mountain-fed) rivers, but we show in the next section that the fan-margin river is episodically far larger.

17.4.2 The Three Hydrologic Regimes of Fan-Margin Rivers

The fan-margin river (Q_{margin}) is fed permanently by the interfan foothills-fed streams (Q_{inter}) (Fig. 17.3d). Significantly, it is augmented by the discharge of the neighbouring megafan-forming rivers (Q_{riv1} and Q_{riv2}) when one or both avulse into the fan-margin depression in the course of normal fan-forming river behaviour. These distinct components, Q_{inter}, Q_{riv1} and Q_{riv2}, in different combination thus give rise to the following three discharge scenarios in the fan-margin river, each depending on the orientation of the fan-forming rivers.

Scenario 1: in most cases considered in the literature, the interfan streams are the only major contributors to the fan-margin river, so that $Q_{margin} = Q_{inter}$ (Fig. 17.3d) (drainage by small fan-fed rivers off the neighbouring megafan slopes is disregarded in this brief rendering).

Scenario 2: one megafan river avulses into the fan-margin depression, resulting in a significant increase in discharge of the fan-margin river:

$$Q_{margin} = Q_{inter} + Q_{riv1}, \text{ or}$$
$$Q_{margin} = Q_{inter} + Q_{riv2}$$

Scenario 3: when rivers from both adjacent megafans avulse simultaneously into the fan-margin river, discharges from all three sources are combined, such that $Q_{margin} = Q_{inter} + Q_{riv1} + Q_{riv2}$ (Fig. 17.3d).

These scenarios involve relative basin size, and hence varied discharge volumes, that need to be considered in assessing fan-margin hydrology. Assuming conservatively, for example, that Q_{inter}, as a foothills-fed zone, represents an area one-fifth the size of each fan-forming basin, and that the areas of the fan-forming river basins are the same, and that basin area is a proxy for discharge volume, then: $Q_{inter} = 1$, $Q_{riv1} = 5$, and $Q_{riv2} = 5$. In Scenario 1 the discharge of the fan-margin river is $Q_{margin} = 1$; in Scenario 2, $Q_{margin} = 1 + 5$; and in Scenario 3, $Q_{margin} = 1 + 5 + 5$. In Scenario 3, therefore, the fan-margin river experiences discharges far higher than those of the individual fan-forming rivers.

Empirically we know that fan-margin river gradients are usually lower than those of channels on the neighbouring megafans (Fig. 17.3d); this suggests the hypothesis that the highest discharges (Scenario 3) determine the gradient of fan-margin rivers (by reasoning in Section 17.3.2). Another implication is that the fan-margin river, which in many megafan landscapes appears to be one of the smaller tributaries of the trunk river (a foothills-fed river), can carry significantly higher discharges than those of the individual fan-forming rivers (11, 12, Fig. 17.5).

Furthermore, the fan-margin river must generate sediment packages commensurate with the hydrologic regimes of each of the three scenarios, leading to a complex sedimentary history (Fig. 17.6). An associated implication is that fan-margin river mineral assemblages must differ from those in the flanking megafans because the fan-margin sediments are derived from three distinct sources (Q_{inter}, Q_{riv1} and Q_{riv2}).

17.4.3 Megafan Taxonomy: Primary, Derived, Axial, Megafan-Deltas, and Divide Megfans

Primary megafans are the most common type, apexed at the mountain front (Fig.17.3d, upper panel) and with orientations generally transverse to the upland topographic margin. Almost all examples discussed in this volume deal with this type. A common control of primary megafan length is the existence of a structural topographic margin that limits the horizontal extension of megafans, mentioned above with examples from the Chaco and Okavango regions.

Derived megafans are the obverse of the primary type; these develop immediately downstream of the primary megafan set, if sufficient horizontal accommodation exists within the basin to allow such development. The fan-margin river in the groove between two primary megafans then feeds the growth of another megafan downstream, 'derived' thereby from the discharge of the primary megafans and the foothills-fed rivers. Because fan-margin rivers display lower gradients than megafan rivers (Section 17.3.2, Fig. 17.3d), there is a point at which the lower gradient of the fan-margin river intersects the higher slopes of the primary megafans (point C – Fig. 17.3d, upper and lower panels); this determines the apex of a derived fluvial fan, from which, as with the primary type, the base profile of the fan-margin river expands to become the three-dimensional surface of the derived megafan (Fig. 17.3d). An example is the downstream Negro derived megafan in the Pantanal region of Brazil: this is fed by drainage from the primary Negro and Aquidauana megafans upstream (Fig. 17.5). Derived megafans are also arbitrarily defined as having length dimensions > 80 km and widths > 40 km,

The morphology of derived megafans differs from that of primary megafans: coniform morphology is generally lacking, so that alluvial ridges counterintuitively become the dominant landform in terms of vertical dimension, but here they achieve even greater vertical dimension than those on primary megafans

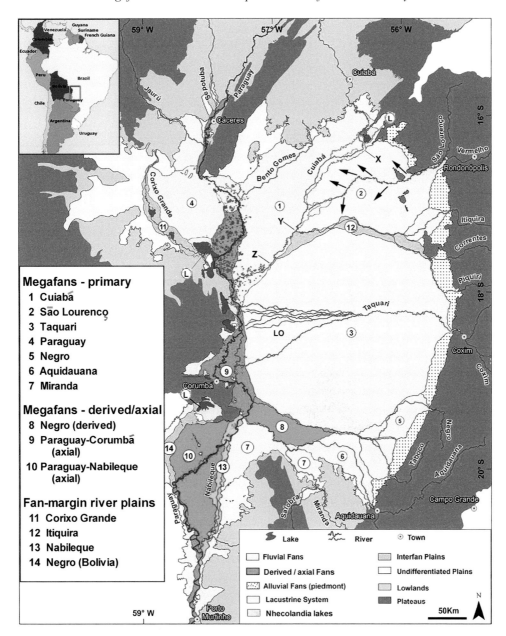

Figure 17.5 Megafans and drainage patterns of part of the Pantanal region, Brazil. Primary megafans: Cuiabá, São Lourenço, Taquari, Paraguay and Aquidauana (1–6). Derived megafan: Negro (7). Axial megafans: Paraguay and Paraguay–Nabileque (9, 10). Fan-margin rivers: Cuiabá between points X and Y, and São Lourenço (between points Y and Z), Corixo Grande, Itiquira, and Nabileque (10, 11, 12). LO: young active lobe of Taquari megafan. L: lakes impounded in depressions formed by megafan cones. Several smaller fans (< 80 km long) are mapped. Short arrows on the São Lourenço fan indicate directions of prior courses visible in imagery and the subapex points from which the prior courses have operated. Similar subapexes can be identified on other megafans. Adapted from Pupim et al. (2017).

because the formative discharges, related to the fan-margin river (*Scenario 3*, Section 17.4.2), can be far higher than those of individual fan-forming rivers (that form the upstream primary megafans). Derived megafans are less common than the primary type and are supplied with sediment from a much wider source area, i.e., from the fan-forming rivers of both upstream primary megafans plus that supplied from the interfan zone (Section 17.4.2). Reasons for distinguishing this type are their morphological difference and the fact they are not anchored at the basin's topographic margin.

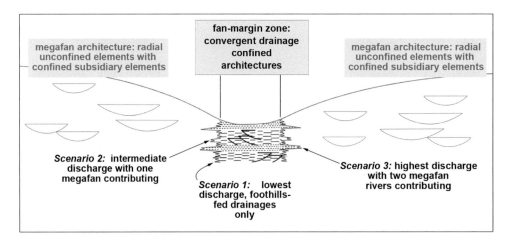

Figure 17.6 **Megafan and fan-margin sedimentary environments**. Diagrammatic section comparing combined megafan successions (combined radial and convergent river-pattern architectures); and one version of a fan-margin sedimentary succession (exclusively convergent-drainage architectures) illustrating the three suggested scenarios imposed on the fan-margin river.

Axial megafans usually form in major valleys where the valley widens because of structural geometries, or because the alluvial fans or fluvial fans of tributaries entering the axial valley are short enough to allow the axial/trunk river sufficient lateral space to develop a megafan (i.e., > 40 km wide). Axial megafans are similar to the derived type in being relatively rare, in not being apexed at an upland margin, and in displaying lower slopes than those of primary megafans because they are related to the lower slopes of trunk rivers (e.g., the Paraguay and Paraguay–Nabileque megafans, Fig. 17.5; the Nile in South Sudan). However, axial megafans are generally smaller than those apexed at an upland margin because they are formed in sectors of a basin which are laterally confined by valley margins.

Divide megafans straddle major watersheds so that river avulsion across the megafan cone results in sediment dispersal from the fan-forming river into different but adjoining major basins. Examples of megafans situated on drainage divides are the Yamuna (Ganga–Indus basin divide), Guëjar (Amazon–Orinoco basin divide) and Parapetí (Amazon–Paraná basin divide). The northern Kalahari drainage systems display four and possibly five such divide-occupying megafans – including the Okavango megafan – in what may be a unique set of drainage patterns on the planet (Wilkinson et al., Ch. 4). Wilkinson (1998, 1999) and Wilkinson et al. (2006, 2010) have investigated the influence of such megafan dynamics on the speciation of aquatic fauna.

Megafan-deltas are now well documented along the shores of the Gulf of Carpentaria (Lane et al., Ch. 12) where the marine effect that operates in the 'backwater' zone is limited to a zone within ~40 km of the coastline. Here, numerous avulsions have been shown to be directly related to the effect of coastal processes, and the avulsion zone has been shown to migrate in sympathy with sea level fluctuations. The Huanghe (Yellow River) displays the same dichotomy of two avulsion zones, one near the apex and another in a coastal zone (Ganti et al. 2014). Similar examples are known from the southeastern coast of India (Gunnell, pers. comm.). In the case of Mesopotamia, historically active megafan-deltas have lost the marine connection (Rzóska 1980, SOD 17.4) – thereby becoming true megafans under solely fluvial influence.

17.4.4 Examples from the Pantanal, Brazil

The drainage geometries of cenarios 2 and 3 (Section 17.4.2) can become complex, as shown by examples in the remarkable cluster of ten primary megafans in the upper Paraguay River basin (Pantanal, SW Brazil: Santos et al., Ch. 6). Here, the modern course of the Cuiabá River, which laid down sediments of the Cuiabá megafan, has avulsed into the fan-margin depression between its own megafan cone and that of the São Lourenço megafan (1 and 2, Fig. 17.5) – and flows in this depression for hundreds of kilometres in the sector marked X–Y in Fig. 17.5. The Cuiabá River thus joins the flow of the interfan drainages, providing an example of a Scenario 2 drainage pattern. Furthermore, the neighbouring fan-forming São Lourenço River joins the Cuiabá River in the same fan-margin depression downstream at point Y (Fig. 17.5), thereby providing an example of Scenario 3 drainage (between points Y and Z, Fig. 17.5), i.e., an example of two megafan-forming rivers combining with the interfan drainage.

At some time in the future, both the Cuiabá and São Lourenço rivers will avulse away from their common fan-margin depression to other orientations, so that the fan-margin river will revert to the status of Scenarios 1 or 2. Several prior avulsion points of the São Lourenço River can be seen in aerial imagery (short black arrows, Fig. 17.5), which give graphic evidence that the river has been connected both to the Cuiabá River to the north, at different points, and to the Piquirí (Itiquira) River to the south–on lobes termed the L1 (Corixo de Bebe River) and L2 (Braço do São Lourenço River) respectively by Assine et al. (2014).

The Pantanal even displays examples of relatively rare derived fans: downstream of point Z (Fig. 17.5), the combined Cuiabá–São Lourenço fan-margin river has given rise to a small triangle-shaped fan (termed active lobe III of the Cuiabá megafan by Pupim et al. 2017). Another example is the abovementioned Negro derived fan. In both cases, the mineralogy of the derived fans can be expected to be demonstrably more diverse than that of the upstream megafans when they are fed during episodic drainage orientations, described as Scenarios 2 and 3.

17.5 Sub-Basin-Scale and Continental-Scale Drainage, Nesting, and Sediment Dispersal Patterns

17.5.1 Sub-Basin-Scale Megafan Drainage and Nesting Patterns

Criteria used in Miall's (1996) nine aggradational, 'basin-fill' landscape patterns are adapted here to describe megafan nesting patterns: drainage orientation with respect to regional geological grain is one criterion (*T:* transverse, *L:* longitudinal), and upstream (megafan) and downstream (trunk river usually) depositional environments (Fig. 17.3a) are another. The global study (Wilkinson and Currit, Ch. 2) reveals

five megafan nesting patterns in basins bounded by either one or two topographic margins (Fig. 17.7). Nesting patterns are also termed tessellation patterns to represent the complex but tight fit that can exist between neighbouring megafans. Weissmann et al. (2011, 2015) use the term coalescing fans apparently in the same way, although the term suggests the merging of the fluvial systems of two megafans, which in the global study appears to be rare because fan-margin rivers (which may shift with time) almost always mark the relatively sharp division between the positive topography of neighbouring megafan cones.

Dual-margin basins (Types 1–4) include continental foreland basins which are the best-known settings for megafan development. As a generalisation, *T/L* configurations are associated with triangular megafan planforms, whereas *T/T* and *L/L* configurations are associated with diamond-shaped planforms (Fig. 17.7).

Type 1 (Chaco/Himalaya type) is the most well-known with a *T/L* configuration (Fig. 17.7). This asymmetric arrangement is extensively developed along the eastern margin of the central and northern Andes, and along the southern margin of the Himalaya Range. Here, megafans have developed along tributaries transverse to a longitudinal axial river (Paraná, Ganges) whose course is forced to an location broadly parallel to the structural margin of the basin opposite the sediment source (Wilkinson et al. 2010, Weissmann et al. 2015). Other examples are the courses of the Orinoco and Mamoré rivers, situated against cratonic scarps (Fig. 17.1a). In these settings, the megafan landscapes of major tributaries consistently occupy the major portion of the basin area, with the axial floodplains confined to a relatively narrow zone.

Types 2 and 4 (Fig. 17.7) correspond to the less common *T/T* drainage configurations. Type 2 is associated notably with Botswana's Okavango megafan, which is oriented transverse to the Okavango Rift – with drainage continuing transversely toward the regional Makgadikgadi basin and megafans occupying most of the depression. The Type 4 (Mesopotamia type) *L/L* configuration is an uncommon pattern with the megafan-forming rivers flowing parallel to the foreland basin axis.

Type 3 displays the *T/L* configuration with three unusual characteristics, as shown in the type area, the Muglad Rift basin of South Sudan (Fig. 3.2b, Wilkinson

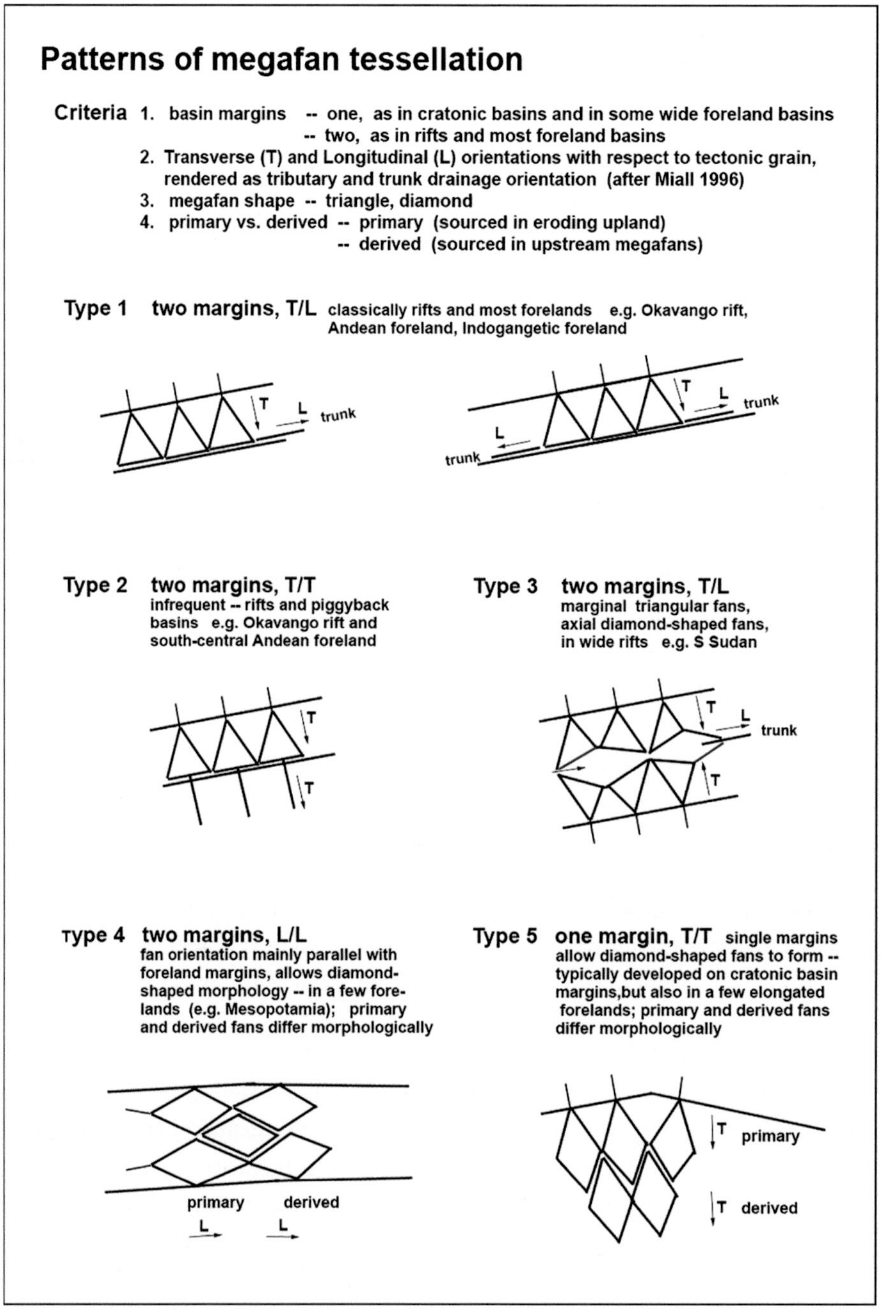

Figure 17.7 **Nesting patterns in double- and single-margin basins**. The most common are Types 1 and 5. Derived megafans are far less common than the primary type.

et al., Ch. 3): megafans of similar size enter the basin symmetrically from both margins; the Bahr el Ghazal trunk drainage consequently flows down the centre of the basin, and this trunk river forms its own axial megafans (undoubtedly due to the lack of a highly asymmetric sediment supply and associated asymmetric tributary pattern of Type 1 basins). Another example is the Balonne megafan in Australia (Fig. 17.8a).

Although the Type 3 configuration is unusual in the megafan population, it is mentioned as common in the large population of smaller fans with radii < 80 km (Weissmann et al. 2011, 2015).

Type 5 (Venezuelan Llanos and Craton types) is the only single-margin configuration, and although not uncommon it is the least well known. Here *T/T* drainage orientations are broadly aligned away from the

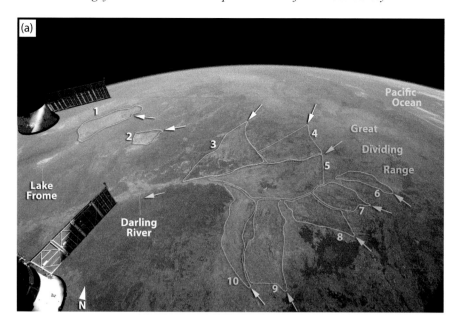

Figure 17.8 **Astronaut images of megafans and** *Large Accretionary Fluvial Systems* **(LAFS) in arid Australia**. (a) Darling River basin and megafan nesting-pattern Type 3 (Fig. 17.7) (also part of the higher-order, subcontinental domain-pattern Type 4, Fig. 17.9). Megafans (outlined) are named after fan-forming rivers: 1– Cooper; 2– Bulloo; 3– Warrego; 4– Maranoa; 5– Balonne; 6– Barwon; 7– Gwydir; 8– Namoi; 9– Castlereagh; 10– Macquarie. Arrows show flow direction of fan-forming rivers. Image centre 27.94 S, 146.68 E. (b) Channel Country. The Cooper Creek, southwestern Queensland, with two wide but confined zones that display one type of *LAFS*. The Cooper is one of several rivers that have generated LAFS. Image centre 26.6 S, 142.11 E. Astronaut images ISS056-E-173717 (a) and ISS041-E-64191 (b) made available courtesy of the Earth Science and Remote Sensing Unit, NASA Johnson Space Center. Colour versions of these figures are available in the SOD for Chapter 17.

upland, typically with diamond-like planforms. The lack of a definite distal basin margin allows for significant nesting variation. In northern Llanos landscapes, derived megafans combine with regional axial drainages in the *T/T* configuration. Similarly, in cratonic basin-swell landscapes of Africa, rivers flowing off swells generate megafan sediment wedges, with drainage usually continuing as transverse axial rivers (Wilkinson et al., Ch. 3). Several megafans in Australia are of this type, as is the Loire megafan of central France (Gunnell, Ch. 9). Type 5 megafans amount to ~40% of those in the global study.

River discharge volume is another control of fan size, with greater discharge roughly associated with larger fan areas in many basins, as for example shown in the Kalahari Basin (Wilkinson et al., Ch. 4) and Chaco Plains (Latrubesse et al., Ch. 5). But this primary relationship is complicated by competition for space. Without competition, even smaller rivers are able to build vast megafan surfaces. Thus, the small, ephemeral Warrego River in southeastern Australia has built a large megafan because it is almost unconfined topographically: 'The huge deltoid [megafan] of the

Warrego River, 1000 miles from the sea, is larger than every true coastal delta except that of the Ganges' (Whitehouse 1944:184). The interplay of discharge and competition for space is well illustrated in a comparison between the Warrego and Kosi megafans: the Warrego River, with an average annual discharge of $4 \times 10^9 \, \mathrm{m}^3/\mathrm{s}$ (New South Wales Dept. of Environment and Conservation 2006) has generated a megafan $> 15,000 \, \mathrm{km}^2$ in area (3, Fig. 17.8a), or $40,000 \, \mathrm{km}^2$ if ancient relict drainage lines are an indication. By contrast, the far larger Kosi River, with a mean annual flow of $52 \times 10^9 \, \mathrm{m}^3$ (Sinha et al., Ch. 11) has built a smaller megafan that measures only ~11,200 km² (Kumar et al. 2014), constricted as it is between neighbouring megafans – which reduce the apical angle of expansion – and the Ganga floodplain. At the other extreme are megafans aligned parallel to the basin axis. These can achieve great lengths, the classic examples being those of the Euphrates in Mesopotamia and the Salamat in southeastern Chad (SW02 and N23 respectively, Wilkinson and Currit, Ch. 2), both $> 500 \, \mathrm{km}$ long. An intermediate geometry, with fans oriented slantwise to the basin axis, has allowed megafans in

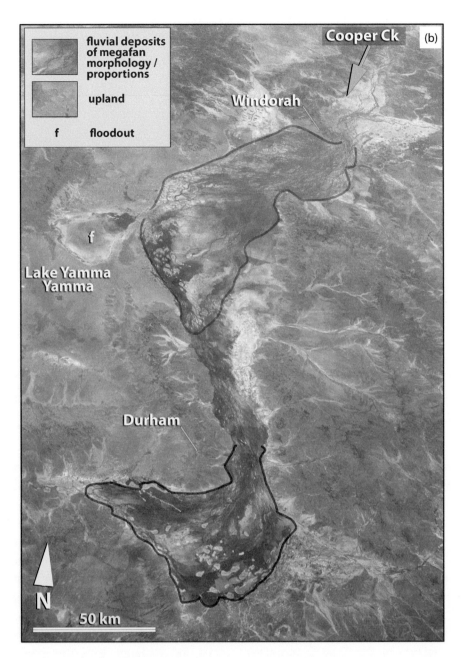

Figure 17.8 (*cont.*)

the Upper Zambezi Basin (S01–S04, Wilkinson and Currit, Ch. 2), to develop much longer axes than the width of the Bulozi Basin. The other control of nesting is the spacing of the exit points of fan-forming rivers from the backing upland or mountain belt (Sinha et al., Ch. 10) – an attribute that ought to be assessed in combination with the altitude of such exit points.

Overall, apart from the intrinsic megageomorphic interest of multi-megafan landscapes, their importance in enabling more realistic paleogeographic reconstructions

of ancient continental landscapes – and therefore the associated depositional environments – cannot be overstated.

17.5.2 Basin and Continental-Scale Drainage/ Sediment Dispersal Models: Schumm, and Megafan Alternatives

At the continental scale, Schumm (1977) identified the now-classic, three-domain model of a continental drainage system as the following sequential zones: *upland erosion → valley-confined floodplain (sediment transport) → depositional coastal delta*. Subsumed in this model were the well-known models of coastal deltas described by Coleman and Wright (1975) and Wright (1985) that implied little or no net alluvial sedimentation on continents. The three-domain model was substantially paralleled by Blum (2008), with Mississippi-like drainages clearly in mind. Bridge and Demicco (2008) used a similar approach in describing continental sediment transport systems, as did Burt and Allison (2010). Megafan zones are poorly represented in all these syntheses of source-to-sink sedimentary systems.

Alternative models seem required to accommodate major alluvial depositional domains located on continents far from any coast. Six continent-scale 'landscape-domain associations' are shown in Fig. 17.9, associations that are known empirically and include single or multi-megafan components.

Three patterns are of particular interest, especially in the matter of palaeogeographic reconstruction. The first is the juxtaposition of megafans or sets of megafans (as a depositional domain), which ought to be expected as a commonly occurring landscape at the foot of upland or mountain belts. The second, as Fig. 17.9 suggests, is the usual status of megafan-forming rivers as tributaries to trunk rivers of major drainage basins (this relationship raises the confusion-laden idea of tributaries that are not dominantly 'tributive', but that also display divergent drainage where they build cones of sediment). The third is the fact that some coastal megafans will act as megafan-deltas when a distal zone is flooded. An alternation between megafan and delta behaviour/process then seems likely for some near-coastal megafans especially during regression-transgression cycles of Pleistocene times (see megafan-deltas of the Gulf of Carpentaria, Section 17.4.3).

More geomorphically varied than Schumm's (1977) schema are Miall's (1996) nine idealised subcontinental-scale 'basin-fill patterns'. The criteria Miall used were orthogonal geometry (*T*: Transverse, *L*: Longitudinal) and three delta types, that is, elaborations that are primarily of the Schumm (1977) type and Galloway's (1975) tripartite delta schema. Small 'fans' of restricted area appear in his figures, but Miall (1996) did not recognise the existence of megafans at that time as normal landscape components (SOD 17.5).

17.6 Discussion

This section addresses definitional issues between various features at the megafan-scale, but also between megafans and features at other levels of the Miall (1996, 2014, 2021) hierarchy (Fig. 17.2). Genetic continuums between these features at different levels

---

Schumm's classic three-domain association without megafans (e.g., Mississippi, Rhine/Meuse basins)
Type 1 *upland [1] → valley-confined floodplain → open/estuarine delta (continental margin or inland)*

Known combinations with megafans as part of the continental drainage system
Type 2 *upland → **megafan(s)** → (axial valley-confined floodplain/dune field) [2]*   (Saharan MFs, Algeria)

Type 3a *upland → **megafan(s)** → waterbody (small)*   (Ili MF, Balkhash basin)
Type 3b *upland → **coastal megafan/delta(s)** (MF alternating as delta)*   (G. Carpentaria MFs, palaeo-North Chad MFs?)
Type 3c *upland → **megafan(s)** → **coastal megafan/delta(s)** (MF alternating as delta)*   (Southern Chad MFs)

Type 4 *upland → **megafan(s)** → axial valley-confined floodplain → coastal/inland delta*   (Paraná, Nile, Zambezi MFs)

Type 5 *upland → axial valley-confined floodplain → **megafan(s)** → delta*   (East Chad, Aral Sea, Mesopotamia MFs)

Notes. [1] upland includes headwater sediment-source basins of all types; [2] Locations where prior landscape has been so modified that distal environments cannot always be specified. MF is used here as abbreviation of megafan.

Figure 17.9 **Regional- to continental-scale landform-domain associations**. Schumm's (1977) classic association, and six models that include megafans (MF).

suggest topics of possible importance in the next phase of megafan studies. The inherent complexity of mega-fan landscapes is emphasised here because such landscapes are orders of magnitude larger than fans even in the mid-size range (30–80 km long: Hartley et al. 2010a; Weissmann et al. 2010, 2011). Such complexity is fundamental to realistic reconstruction of alluvial palaeolandscapes.

17.6.1 Comparison of Megafans and Other Fluvial Depositional Landscapes

Study of larger fluvial systems follows one of the more recent trends in geomorphology, enabled by space-based imagery, that have yielded both regional views and diachronic perspectives (e.g., Gupta 2007; Latrubesse 2008; Ashworth and Lewin 2012). Fluvial sediment bodies that cover areas $> 10^4$ km^2, such as megafans, are now known to be far more significant in continental landscapes; and they probably generate distinctive fluvial architectures with potentially diagnostic features in the rock record.

Megafans are here compared with fluvial fans (fans of intermediate scale ~30–80 km long), valley-confined floodplains, major avulsive fluvial systems (MAFS) and large accretionary fluvial systems (LAFS). More recently the meaning of the term 'alluvial fan' has narrowed to exclude fluvial fans and megafans: Hartley et al. (2010a:168) noted that the present, generally understood view is that small alluvial fans show 'clear differences' with megafans; this view, as they note, has been expressed also by DeCelles and Cavazza (1999), Leier et al. (2005), Gibling (2006), and most recently by Moscariello (2018), and Ventra and Clarke (2018). Several other distinctions between alluvial fans *(sensu stricto)* and megafans are mentioned below (Section 17.6.1.1), one of the most important of which is the increasing complexity that necessarily arises in systems that can reach more than 200,000 km^2 in area.

17.6.1.1 Differences within Large DFS

Hartley et al. (2010a), Weissmann et al. (2010, 2011, 2015), Davidson et al. (2013) and Davidson and Hartley (2014) have made major contributions to the study of larger DFS (> 30 km), leading to significant progress in understanding these sedimentary systems. Further investigation reported here, directed at the largest DFS (megafans > 80 km), now suggests that several attributes that appeared critical to definitions of larger DFS may need modification.

(i) *Convergent vs. divergent/radial river patterns ('tribu-tive' vs. 'distributive'):* convergent drainage (with features representative of the lower hierarchical levels, Fig. 17.2), is common within the broader divergent pattern on megafan cones, and relates to several sub-environments (Section 17.6.2). In contrast, convergent drainage seems far less significant on DFS smaller than megafans.

(ii) *Confined vs. unconfined flow* (SOD 17.3b): related to (i) above, channels on megafans are usually confined within levee-like structures, often for distances of hundreds of kilometres. In contrast, channels on fans at the lower end of the large DFS continuum are often unconfined.

(iii) *Channels:*
  a. Single- *vs.* multiple-channel systems: single- and multiple-channel patterns are known on DFS > 30 km long (Hartley et al. 2010a, Weissmann et al. 2010, 2011; Davidson et al. 2013). However, in the global survey (Wilkinson and Currit, Ch. 2), a single dominant (fan-forming) channel is by far the main characteristic on mega-fans (> 80 km long). Indeed, there is evidence that drainage evolves over time toward a single dominant channel (Fielding et al. 2012).
  b. Radial channel orientations: the requirement that DFS display exclusively radial drainage patterns unnecessarily restricts inclusion of surfaces that are otherwise patently part of continuous surfaces associated with the fluvial sedimentation on a megafan: channels that are radially oriented in proximal zones very often achieve parallel orientations medially and distally, and even convergent orientations on diamond-shaped megafans.
  c. Channel width: many examples of megafans are known in which channel widths do not decline downfan (infiltration prevention and/or inputs from fan-wide water tables are likely major controls in some megafan discharge regimes).

(iv) *Long-profile concavity/convexity:* seldom mentioned is the convex-up profile of some megafans; it seems therefore that concavity ought to be abandoned as a criterion for definition of the largest fans (Wilkinson and Currit, Ch. 2; Wilkinson et al., Ch. 4).

(v) *Coniform morphology:* partial cone morphology is regarded as a major characteristic of DFS, whereas the minimal or non-existent topographic prominence of subapexes on primary megafans and of apexes on derived megafans suggests that this may not be an essential criterion.

(vi) *Complexity:* the variety of subenvironments (Section 17.3.1) is generally missing in fans at the lower end (Latrubesse et al. 2012) of large DFS examined by Hartley et al. (2010a) and Weissmann et al. (2010, 2011, 2015). The distinction between megafans and smaller DFS may be critical in palaeogeographic reconstruction, suggesting that preliminary estimates be made concerning plausible dimensions of subsurface basins since basin width is one of the stronger controls of DFS size. Other aspects are presented in Section 17.6.3.

Latrubesse (2015) has argued that some attributes of larger DFS differ from those proposed by Hartley et al. (2010a, b) and Weissmann et al. (2010, 2011): he noted that neither decreasing channel dimensions nor distributive patterns are 'characteristic' (Latrubesse 2015:19) of larger fans, and that 'many rivers (…) forming the largest megafans keep an almost uniform sandy bed load grain size', failing to display downfan diminution in grain size. He further noted that fluvial style may be little different between DFS and entrenched river settings, and disputed the claim that DFS are not areally larger in aggrading continental basins, arguing that areas of modern confined fluvial belts of the largest rivers, such as the lower Amazon and Mississippi, cover areas greater than or similar to those of unconfined systems.

17.6.1.2 *Megafans* vs. *Valley-Confined Floodplains*

The viability of even the basic megafan criterion of fan-like planform has been questioned (Blair and McPherson 1994; Latrubesse 2015). However, Wilkinson (2015) suggested the following major differences between large fan-like landforms and floodplains, concluding that megafans constitute coherent unitary systems at *SRS* level 9 in the hierarchy of fluvial forms (Fig. 17.2) (megafans are compared with very large floodplains in 17.6.1.3 below). None of the megafan attributes listed can be reasonably expected on valley-confined floodplains as normally understood.

(i) *Channels:* (a) Radial, unconfined drainage orientations are displayed at both the proximal megafan apex and at downfan subapexes, which leads to distinctive sedimentary architectures; (b) closer channel spacing in the (sub)apex zone leads to higher proportions of channel sands, versus high proportions of fine-grained

sediments distally where channel spacing distances are higher (Weissmann et al. 2015), with a dominant downstream trend on megafans but a dominant lateral trend on floodplains; (c) reduced channel dimensions with distance downstream is a common characteristic on many megafans, versus the common increase on floodplains (Weissmann et al. 2015); (d) extensive, unconfined overbank environments are established on the open megafan surface (leading to reduced channel connectivity – see discussion in Weissmann et al. 2015, and SOD 17.3b).

(ii) *Sedimentological trends downfan:* in general, systematic and comparatively rapid grain-size change occurs downfan (compared with floodplains over equivalent distances), with grain-size decreases on many megafans and greater volumes of fine-grained sediment preserved distally (SOD 17.3b).

(iii) *Uniform mineralogical signatures due to lack of tributaries:* the unique mineralogical mix at the entry-point (apex) of each individual fan is translated into relatively consistent compositional signals down the system's length, as opposed to the variable sediment composition consequent upon tributary inputs in most rivers.

(iv) *Spring lines:* these are common on megafans, oriented transverse to slope (Hartley et al. 2013; Weissmann et al. 2015), and associated with changes in channel style and soil type above and below the spring line.

(v) *Dimension of the depositional zone:* for rivers of equivalent discharge, those that generate megafans lay down depositional areas significantly larger than those of valley-confined floodplains (this generalisation appears not to hold for those few very large rivers such as the Amazon basin; e.g., Latrubesse 2015). The difference in affected area is significant because (a) large depositional surfaces result in widely divergent ages for surface sediment emplacement, and hence long channel-return times; which in turn lead to widely varying degrees of soil development (Hartley et al. 2013) compared with more uniform surface ages in floodplains; (b) megafans are probably never completely covered by floods of the main fan-forming river, whereas valley-confined floodplains are often entirely inundated by floods emanating from the mainstem river; (c) sediment body architecture is partly determined by topographic confinement (in the form of valley-confining hillsides) or lack thereof on megafan lobes (SOD 17.3b).

(vi) *Fan-margin zone:* because of its location between megafan cones, the fan-margin depression must experience three identifiably different discharge regimes compared to the more uniform, longer-term hydrological regimes of rivers in valley-confined floodplains, as described in detail in Section 17.4.2.

(vii) *Rheic vs. perirheic zones:* The geomorphically active parts of a river channel are termed the rheic genetic zone of large floodplains (Lewin and Ashworth 2014) with water supplied from a region-wide catchment; this is contrasted with the perirheic genetic zone that is detached (to different degrees) from the main channel geomorphically, and contains hydrologically local water (discharge from tributaries, meteoric water and groundwater). Geomorphic paleoforms can be prominent in the perirheic zone. In proposing the concept of the perirheic zone, Mertes (1997) stressed the significance of the hydrologic convergence of the rheic and perirheic zones as providing boundaries or mixing zones – 'that is, *ecotones*, between resource patches or ecosystems for providing stability and resiliancy to the environment' (Mertes 1997:1751) (emphasis added).

Mertes (1997) and others saw the classic floodplain as the generic setting of the rheic and perirheic zones, that is, a setting confined between the river and higher, neighbouring non-inundated terrestrial systems. However, at the scale of the unitary megafan, 'terrestrial systems' barely operate. The rheic–perirheic dichotomy seems unserviceable applied to megafan landscapes, because the interface of the rheic-perirheic pair along the fan-forming river is areally only a small component of the entire megafan drainage system (SOD 17.6a). In hydrological terms megafans are better thought of as three relatively independent and highly distinctive rheic-perirheic zone pairs: these are the abovementioned spatially restricted fan-forming river pair; the numerous pairs that operate along each fan-fed stream whose discharges are orders of magnitude smaller; and the fan-margin river pair (highly variable discharge in the rheic fan-margin zone and perirheic discharges from neighbouring fan-fed megafan slopes when Scenarios 2 and 3 are not operating – Section 17.4.2).

17.6.1.3 Very Large Axial River Floodplains

Comparisons of megafans with the floodplains of very large rivers present a different focus for the geomorphologist. The following comparison is based on descriptions of the geomorphology and dynamics of such floodplains by Ashworth and Lewin (2012) and Dunne and Aalto (2013). Significant differences between large-river floodplains and megafans are the following: (i) the transverse dimension of the former measures only 'one to several times the width of channel bends on a larger river' (Dunne and Aalto 2013:646), as opposed to megafans for which these

ratios can be orders of magnitude greater; (ii) large-river floodplains are topographically confined; (iii) in a list of the world's twenty largest rivers (Ashworth and Lewin 2012), only five have given rise to a single megafan along their courses, all being relatively small; (iv) conversely, almost all megafan-forming rivers, as tributaries of large rivers, have significantly smaller discharges compared to those associated with the periodic, highly fluctuating inundation levels on major confined floodplains; and (v) sediment supply from tributaries generates irregular, poorly predictable downstream grain-size trends on large fluvial floodplains (Dunne and Aalto 2013), compared to the consistent distal fining registered on fluvial fans.

Surprisingly, as many as ten key attributes of large floodplains listed by Ashworth and Lewin (2012) and Dunne and Aalto (2013) also apply directly to the description of megafans (SOD 17.6c). This observation suggests that similarities and differences between megafan environments formed by inherently smaller rivers (almost uniquely tributaries of axial/trunk rivers) on one hand, and floodplains of rivers orders of magnitude greater on the other, deserve attention. Alternatively, it is possible that megafan characteristics are still poorly enough appreciated that similarities and differences between megafans and very large floodplains need further investigation and clarification.

17.6.1.4 Megafans and Major Avulsive Fluvial Systems (MAFS)

Latrubesse et al. (2012) and Latrubesse (2015) have given a full description of these systems, also termed *fluvial avulsive systems* and *avulsion dominated systems from large rivers* (Latrubesse 2015:23, 21). For brevity, and to avoid confusion with other avulsion-dominated systems such as megafans, these systems are here specifically termed MAFS. Latrubesse (2015:20–21) notes that only a few large MAFS have thus far 'dominated the discussion of avulsion processes in medium to large rivers'. Some features classified as MAFS by Latrubesse (2015) (listed in SOD 17.6d), from which the following summary is drawn, have been classed heretofore as megafans or deltas.

Several differences and similarities between megafans and MAFS are apparent. Areal dimensions of

MAFS typically scale with, or are larger than, megafan areas (e.g., the Ucamara depression of Peruvian Amazonia), and are associated with the largest regional rivers, in distinction to megafan-forming rivers which are typically tributary to axial/trunk/regional rivers (Weissmann et al. 2015). Being formed by axial rivers, MAFS are also usually confined within bedrock valleys. MAFS rivers always display very low gradients, and drainage patterns may be radial or non-radial, similar to those of megafans. But MAFS lack a coniform apex, although avulsion is the major process at the scale of the drainage system, as for megafans. Channel patterns differ from those on megafans, namely anabranching with large semi-permanent or permanent islands, a pattern associated with a non-nodal style of avulsion, many permanent water bodies, and markedly active sedimentation (compared with the sediment bypass of 'conveyor axial systems'; Latrubesse 2015:19).

The sense that MAFS are especially prone to form in rapidly subsiding depressions may explain the frequent development of wetlands associated with some of them. Research on these systems and their distinctions from megafans and deltas is patently needed.

17.6.1.5 Large Accretionary Fluvial Systems (LAFS)

These are a variety of fluvial body specifically to describe certain Australian landscapes, especially one type, the well-known 'Channel Country' in the interior region bordering SW Queensland, the SE Northern Territory, NE South Australia, and NW New South Wales. The Channel Country alluvial plains are identified in new research as one of four types of large accretionary fluvial system ('LAFS'; G. Nanson pers. comm; Fig. 17.8b, SOD 17.6e). This type shows an unusual geomorphic combination of very wide alluvial plains set within an erosional landscape of subcontinental extent (see discussion of erosional *vs.* aggradational landscapes, Section 17.5.2). The Channel Country LAFS appear to be unique in showing characteristics of both megafans and floodplains, i.e.: (i) reaching hundreds of kilometres in length; (ii) attaining up to tens of kilometres in width (without meeting the megafan 40-km width criterion we employed), LAFS active floodplain width compares with the floodplains of rivers with discharges orders of magnitude larger,

(iii) but unlike most megafans, Channel Country LAFS are everywhere confined by bedrock valley slopes; (iv) some display positive conic relief at points along their length (resembling megafans), whereas others do not; (v) they resemble very large floodplains in displaying numerous channels that may all be active during any one flood episode; (vi) like floodplains, they show characteristics of sediment accretion; and (vii) they may be related to environmental conditions specific to arid climates and low regional topography.

17.6.2 Convergent Drainage: An Integral Component of Megafans

The negative relief of several kinds of depressions on megafans is widespread (Section 17.3.1.2), and hence also are the *convergent* drainage patterns and sedimentary architectures associated with confined fluvial settings. Because of the high mobility of rivers on fan cones, depressions such as incised sectors (usually connected to intersection points), inter-lobe depressions and floodbasins must also migrate in the process of megafan cone construction. This means that convergent drainage patterns ought to be widely incorporated within megafan sediment masses. This, in turn, means that both types of architecture, confined/dendritic and unconfined/radial, ought to be represented in the array of megafan architectures. The intersection point is generally thought of as migrating up- and downfan, so that the incised sector is shortened and lengthened. *Lateral* migration, however, probably also occurs in cases where the incised zone is subsequently buried such that the river is once again able to avulse laterally to a new position where a subsequent cut-and-fill cycle may be repeated. In the long life of large megafans such complex surface behaviours are to be expected.

Willett et al. (2018) compared drainage systems of the modern megafan surfaces of the Chaco Plains in central South America. They employed a scaling parameter χ, which 'characterises the integrated steepness of a basin' (p. 528), to distinguish between the drainage systems of the megafan surfaces (high values of χ) with minimal tributary junctions, and the strongly dendritic, younger, actively eroding drainage systems. Importantly for the present argument, subtle surface patterns appear: the maps portray in striking detail that radial and convergent drainages form a tightly integrated mosaic across the surface of each megafan – giving geomorphic detail to the present argument that megafan surfaces include, as a norm, numerous topographically negative zones with convergent drainage. The map also shows that numerous subapexes with avulsion nodes on the megafans give rise to V-shaped, lobe-like zones with specific high-χ aggradational characteristics (SOD 17.6f Figure).

17.6.3 Complexity of Megafan Landscapes

Summarising pioneering work by Wiens (1989) and Levin (1992), Thoms et al. (2007:588) have stated that:

A positive relationship between scale and complexity is an accepted paradigm in the study of natural ecosystems (...). Larger ecosystems can be characterised by an increase in the number of components; an increased number of different types of components; interconnections between components; the presence of both positive and negative feedback loops; and, the number of interactions between different levels of organisation within a system compared with smaller ecosystems.

The following phenomena, some already introduced here, can be seen in terms of the greater inherent complexity of megafans and multi-megafan landscapes, and appear to be unique to these landscapes – compared with most valley-confined floodplains, alluvial fans *(sensu stricto)*, and even the larger fluvial fans with radii up to approximately one-tenth the length of megafans (less than 80 km long).

At the single megafan scale, radial drainage over wide areas – versus valley-confined floodplains – operates at two scales: the main proximal node fixed at the upland–lowland margin, determines broad drainage orientations on the fan cone, whereas subapexes operate at points up to hundreds of km downstream of the apex. Subapexes are patterns of importance in understanding the subsurface distribution of channels (higher frequency proximally and increased mud distribution distally). Within the overriding radial pattern, convergent drainage has been noted as an integral component of the broader divergent drainage setting/environment of megafans, in distinction to drainage on alluvial fans and on larger DFS at the lower end of the size continuum. This combination of fan-wide divergent drainage combined with locally divergent (subapex-related) and convergent patterns also has implications for understanding channel patterns and architectures in the subsurface. Many megafans because of their tectonic setting have the propensity to act as regional sediment sinks for thousands, even tens of thousands of years. They also commonly display large wetlands – proximally in some cases (Okavango) and distally in others (Chaco megafans) due to their very low gradients where small topographic variations act to slow drainage significantly. The propensity for megafans to contain buried units of high porosity and large volume as reservoirs for fluids has been suspected and is becoming rapidly better documented (e.g., Miller et al., Ch. 15, Ventra and Moscariello, Ch. 14).

At the multi-megafan landscape scale, one of the most important drainage features is the tripartite discharge regime of fan-margin rivers (Section 17.4.2), leading to discharges that are episodically far larger than those of fan-forming rivers – fan margin rivers usually appear as minor rivers when in the low-flow mode, far smaller than the permanent and obviously important fan-forming rivers. A series of five nesting models, based on single- and double-margin basins (Section 17.5.1), appears to be unique to megafan landscapes; this suggests that there is a relatively small number of multi-megafan drainage patterns. At the yet wider scale of continental drainage, several models that include megafan components (Section 17.5.2) were contrasted with Schumm's model, which does not include megafan landscapes.

Two relatively rare drainage styles were described in which megafans affect regional drainage patterns. The first, as shown by more than a dozen examples worldwide, is that of active and relict 'divide' megafans whose discharge switches between two major basins. The second is the drainage pattern by which 'derived' megafans evolve downstream from the common primary megafan type, where horizontal accommodation allows (Section 17.4.3).

Megafan complexity is nowhere better illustrated than in the phenomenon of hydrological 'convergence zones' (described in Section 17.6.1.2 (vii) – not to be confused with drainage convergence of dendritic river patterns discussed extensively above). These are important to floodplain ecotone investigations in ecology. Mertes's (1997) rheic-perirheic approach is based on the notion of regional water in a mainstem river and its immediate margins as a *rheic* zone of relatively energetic activity, contrasted with a *perirheic* zone of less active, locally derived water that interfaces with the rheic zone. Adapted in geomorphic studies of large floodplains by Lewin and Ashworth (2014), these authors examined sediment transport with respect to the development of intricate floodplain surface morphologies, and contrasted the geomorphic with the hydrologic systems.

Hydrologic convergence is distinctly more complex on megafans than the rheic–perirheic mixing described by Mertes (1997) for linear floodplains of similar discharge: the figure in SOD 17.6b suggests several other categories of hydrological convergence. A first level of complexity thus arises because not one but three different river types exist in megafan landscapes (fan-forming, fan-margin, and fan-fed types), each of which will display its own rheic-perirheic zones of hydrological convergence per Mertes (1997). These are rendered as linear zones aligned along the course of each river type (*a, b, c*) in the SOD 17.6b Figure.

A second level of complexity arises where the megafan river types interact with one another (rendered *a + b, b + b, b + c*, see SOD 17.6b Figure). Convergences between the fan-fed rivers *(b + b)* are numerous because entire networks of these smaller streams exist on the megafan surface. Less numerous are the major convergences between the fan-forming river where it intersects fan-fed streams (*a + b*), but especially where it intersects with the fan-margin river (e.g., *a + b + c* in the specific drainage conformation of the SOD 17.6b Figure). Yet higher levels of complexity arise where these mixed waters converge with a wetland water body, rendered *b + b + w* (not shown are convergences related to the other two river types that may also intersect with wetlands), or with the trunk river where the most heterogeneous mixing takes place, e.g., *b + c + g + t, b + g + t*, and *a + b + c + t*. Subsurface 'hyporheic water' is shown where it emerges in a spring line zone; this water then potentially mixes with all the aforementioned water types (only one example of which is shown in the figure, namely *b + c + g*).

Change through time adds another level of complexity as most of the convergence zones itemised above shift when the fan-forming river avulses to other radial positions on the megafan cone. The fan-forming river then interacts with a different set of fan-fed streams or seasonal wetlands, and/or shifts to different confluence points with either the other fan-margin river or with the trunk river. Spring lines also shift up- and down-fan as a result of climate changes or local tectonic influence. The complexity of the potential rheic-perirheic convergences thus increases even further when the time dimension is included. The number of convergence zones – rendered in the SOD 17.6b Figure and those mentioned but not rendered – is so large on megafans compared with those of floodplains, that their imporance for aquatic ecology is surely a topic ripe for future enquiry.

17.6.4 Continuums and Disjunctions

At the present state of knowledge, it appears that continuums exist between megafans and other landforms either in terms of morphology or process. The following examples illustrate how confusion has arisen in some cases, but also point to possible avenues for future research. Thus, although the morphological continuum between alluvial fans *(sensu stricto)* and megafans has been demonstrated (Saito 2003; Saito and Oguchi 2005, Hashimoto et al. 2008), the major processual distinctions between these end members are now also quite clear. However, along the transition between the two (DFS radii between ~20 and 80 km) the characteristic processes of both end members should be expected to co-occur; i.e., as a general expectation the mix of debris and sediment-gravity flows and sheetflood processes must give way to floodplain-related processes as river discharge increases. This mix at the intermediate scale presumably produces, sedimentologically and architecturally, what may well be the most complex fan type. Indeed Hartley et al. (2010a) document similarities between channel habits and drainage configuration (typical of alluvial fans *sensu stricto*) appearing on fans with radii as long as 50 km. At minimum therefore, it seems necessary to recognise this overlap of process, considering the great number of fans of intermediate dimension that exist. Such overlap of forms and processes at the intermediate scale may explain why so many characteristics of alluvial fans (*sensu stricto*) have been assigned in the literature to large fans, and *vice versa* – an undoubted source of the confusions noted by Latrubesse (2012, 2015).

Transitional forms and processes likely exist also as valley-confined floodplains become wider and less confined and acquire characteristics of the entirely unconfined megafan setting. At what stage does a floodplain become a megafan, especially since it is megafan-forming rivers that build these areally significant features? Furthermore, considering the controversy that has arisen around large DFS and very large floodplains, examination of the overt differences between the two seems warranted.

In similar vein but at an even larger scale it is apparent that there are obvious major differences between lowland continental landscapes that are today erosional, versus those that are dominantly aggradational (usually megafan-dominated). Examples of these opposites are the almost entirely degradational lowland plains of the modern Mississippi, Congo, and Rhine basins, in which the floodplains of even these large rivers are minor areal entities. At the other extreme are the almost entirely aggradational landscapes of the Chaco and the Gangetic plains. Between these, the landscapes of central Amazonia arguably represent a transitional continental lowland type – suggesting that a continuum exists based on varying proportions of erosional *vs.* aggrading landscapes. This continuum provides the palaeogeographer a flexible and more realistic schema for reconstructing past lowland landscapes at a subcontinental scale.

17.7 Conclusions

17.7.1 Overview and Outlook

The frequency distribution of larger fan-like features/ DFS with radii > 30 km (e.g., Hartley et al. 2010) has influenced perceptions of the megafan phenomenon. Firstly, it suggests a misleading dominance of fans with smaller dimensions, and concomitantly that megafans are relatively unimportant in world landscapes. The global study reported in Wilkinson and Currit (Ch. 2) suggests a different conclusion: with more than 270 megafans (> 80 km long) identified in modern landscapes, and the fact that individual megafans can extend to areas greater than 10^5 km², the result is that when nested, their areal significance

can reach subcontinental scales despite their relatively small numbers (Fig. 17.2). Twenty-two contiguous megafans – stretching from western Amazonia south to the Chaco and then east into the Pantanal basin of southwestern Brazil – cover an area of almost one million square kilometres. The 87 megafans mapped in Africa comprise a similarly large area (> 1.2 M km²), an order of magnitude more extensive than the total area of Africa's major subaerial deltas. Secondly, the notion of the dimensional continuum, although apparent in nature (Saito 2003; Saito and Oguchi 2005; Hashimoto et al. 2008), has masked the disjunction in process at either end, as noted in Section 17.6.4.

Research interest in the largest fan-like features is now sufficient to justify focusing on the megafan subpopulation (> 80–100 km in length) of DFS. Ironically, this focus has benefited from the fact that fluvial processes on megafans – related to the floodplain channel–levee–overbank subenvironments – are some of the best understood in fluvial geomorphology. In contrast, megafans are one of the least well-known depositional environments – as is the fluvial midscale generally, a perspective suggested for example by the proliferating types at the *SRS* level 9 (megafans, MAFS, LAFS) in Fig. 17.2.

Evidence accumulates that individual megafans function morphologically, hydrologically, and sedimentologically, as coherent unitary systems, distinct in significant ways from alluvial fans, floodplains and deltas. Megafans are gaining scientific attention not only for reasons of fundamental research, but also for their importance as landscape elements occupied by major population centres around the world and as centres of rapid agricultural expansion in remote areas such as the Chaco Plains.

The megafan model has recently provided new impetus to the study of such diverse topics as subsurface exploration for geo-energy resources and biological aquatic evolution (Wilkinson et al. 2006, 2010). An attempt was made in this chapter to explore the variety of hydro-ecological zones of enhanced biological activity that must accompany megafan river dynamics. As more attention is paid to well-described successions in the rock record, it seems that the DFS model is also increasingly invoked as an alternative palaeogeographic setting, one that is unconfined by valleys at a scale of hundreds of kilometres, as shown for example in numerous reinterpretations by Davidson et al. (2013), Weissmann et al. (2013), and Ventra and Moscariello (Ch. 14). Blair and McPherson (1994) dismissed larger fans as uninteresting on the grounds of component facies being analogous to those of valley-confined floodplains. Fielding et al. (2012) did not dismiss larger fans, but argued that they are less representative in the subsurface than claimed by Hartley et al. (2010a, b) and Weissmann et al. (2010, 2011). It seems more than possible that the lack of mention of DFS-style settings in the literature – given by Fielding et al. (2012) and Miall (2014) as evidence for their probable lack of importance in the subsurface – may point rather to a lack of recognition than to the insignificance of these features (Hartley et al. 2013).

Megafans create very low-relief surfaces that either have not been recognised as fluvial in geomorphic surveys or have been confused with low-angle, erosional pediment forms of subregional extent. Remote sensing of the Earth's surface, combined with the capability of generating three-dimensional topographic maps, has opened a way for recognising these vast aggradational landscapes. Imagery from space provided the first indications that megafans are a topographic norm rather than an inconsequential oddity. The low number of active megafans in modern Europe and North America, the birthplaces of modern geology, undoubtedly contributed to the relative lack of awareness of the megafan feature.

The infancy of megafan research is evident from papers on fan-shaped geomorphic bodies that have appeared in the last twenty years – especially as represented by research performed on different continents (Part II of this volume) – and from the ensuing proliferation of special meanings and terminological confusion. Megafan and multi-megafan landscapes, in particular, seem to be *terra incognita* in Earth science texts and the wider technical literature. Studies of fans in general have yielded the unsettling opinion that the usually simple distinction between allogenic and autogenic processes/controls becomes blurred (Ventra and Nichols 2014). This chapter, and several others in this volume, emphasise the relatively novel autogenic dynamics at megafan and sub-megafan scale as a necessary basis from which to understand the

behaviour and geometries of drainage patterns, and sub-megafan-scale depositional environments in particular. Longer-term dynamics have come into focus with studies of allogenic aspects such as tectonic forcing and source-to-basin controls of megafan morphology and subtype. The grand topic of what is finally preserved in the rock record is generally beyond the scope of this volume, but the array of new data and syntheses presented here (especially Ch. 14 and Ch. 15) are expected to promote interpretation of buried fluvial sediments, on both Earth and other planetary bodies where rivers have operated.

17.7.2 Terminology

The issue of confusion in definitions underlay the proposal to redefine all fan-like forms as DFS (Hartley et al., 2010a, b, and Weissmann et al. 2010, 2011). Misunderstanding arises as one term may be used with different meanings (e.g., *alluvial fan*) or as meanings change over time (e.g., *terminal fan*). When well-known features thought of as megafans are classified anew as MAFS (major avulsive fluvial system – Section 17.6.1.4), the cross-currents of definition can also lead to confusion.

For fan-like features, scale-based distinctions in terminology seem to be needed at the present stage – as in *small alluvial fan*, or *alluvial fan (sensu stricto)* – or by stating dimensions explicitly, as in *DFS < 20 km long*. The terms *confined* and *unconfined* are used in this review primarily to denote valley-confined *vs.* unconfined settings. They are also commonly used, however, to distinguish within-channel from overbank processes, so that the distinction between these scales of analysis needs to be very clearly stated. Despite careful definition by some writers (e.g., Hartley et al. 2010a, b; Weissmann et al. 2010, 2011, 2015), the terms *distributive*, and sometimes its opposite, *tributive*, are easily confused with the established terms *distributary* and *tributary*, terms of significantly different and almost unrelated connotation in the lexicon of most users. Unambiguous adjectives such as *radial* or *divergent* for *distributive* could also be employed. *Convergent* in the sense of drainage by a system of tributaries is unambiguous.

A partial glossary of commonly used terms in present megafan and DFS writing is given in Table 17.1.

Table 17.1 *Fan-like features: existing definitions and suggested usage. Underlined features are cross-referenced within the Table*

Definitions as of 2022	Suggested usage (see text)
Alluvial fan – strict definition (Blair and McPherson 1994) of a coarse-grained coniform sediment mass less than 20 km long. More inclusive definitions (e.g., Miall 1996) subsume all fan-like sediment bodies from small to large, i.e., from small alluvial fans to the largest *megafans*. See also *DFS*	*small alluvial fan* (as lithologically distinct from other fan forms); or *alluvial fan* with statement of length (usually less than 10–20 km)
Avulsive system – see *MAFS, Major Avulsive Fluvial System* below. Not defined in EG [1]	*major avulsive (fluvial) system* with statement specifically excluding *megafans*
Bajada – 'compound or coalesced' relatively small alluvial fans that form a piedmont surface between a mountain mass and a basin floor (EG [1]). Occasionally now used for coalesced megafans	*bajada*; retain traditional use, i.e. restricted to smaller alluvial fans
Derived megafan – fed by discharge from upstream (primary) *megafans*; see *primary megafan* (Section 17.4.3)	*derived megafan*
Distributive, distributary, tribative, tributary – *distributive* is proposed by HW [2] as a synonym for radial drainage to avoid confusion with the usual delta-related meaning of distributary (channels smaller than the feeder channel and usually simultaneously functioning). *Tribative* is denoted by the same authors to mean a system of tributaries flowing into ever larger rivers, as seen in upland erosional landscapes. *Tributary* is sometimes used, confusingly, in the same sense	retain *distributary* and *tributary* for their presently precise and accepted meanings. Use *radial* in preference to *distributive* as less open to confusion
Distributive fluvial system, 'DFS' – term coined by HW encompassing all fanlike forms from small *alluvial fans* as small as a few km long to *megafans* hundreds of km long, but excluding deltas. Definition overlaps with the "inclusive" definition of alluvial fans (e.g., Stanistreet and McCarthy 1993)	distinguish between *small DFS, intermediate DFS, megafan*
Fan delta – an *alluvial fan* (of unstated length) that enters a water body, i.e., with proximal subaerial and distal marine/lacustrine depositional environments. The term *megafan delta* follows logically from the distinction suggested here; some of the first usages appear in Lane et al. (Lane et al., Ch. 12) are examples from the global study	*fan delta (_ km long)* with length designation; *megafan delta, megafan-delta*
Floodplain – a zone of fluvial deposits held within some confining topography (valley side slopes, flanking fans/megafans), dominated by a single river that can receive tributaries	*floodplain; flood plain*
Fluvial fan – another term for larger fans, to distinguish them from small fans where non-fluvial processes also operate. Also defined more specifically as a fan measuring between ~30 and ~80 km in length (HW [2]). See *DFS*	*fluvial fan (30–100 km long)*, i.e., with general length designation
Inland delta – term loosely used for larger fan-like fluvial deposits, especially for the Okavango megafan, Inland Delta of the Niger and the Sudd, all of which qualify in terms of criteria used by this author (Wilkinson and Currit, Ch. 2) as *megafans*	abandon term; use appropriately specific term such as *megafan, fluvial fan-delta, megafan delta, delta*

Definitions as of 2022	Suggested usage (see text)
Large accretionary fluvial system (LAFS) (Section 17.6.1.5) – 'large accretionary system' to describe four types of Australian areally expansive zones of fluvial sediment accumulation (Nanson and Nanson in prep.) at *SRS* levels 9–10 (Fig. 17.2). Three LAFS are equivalent to *megafans* as described in this chapter; the fourth type describes the valley-confined 'Channel Country' floodplains of the eastern half of arid Australia, that display some attributes that resemble megafans; this type is distinct from but equivalent to, or larger in areal terms, than a *megafan*	*large accretionary fluvial system (LAFS); Channel Country floodplain*
Major avulsive fluvial system (MAFS) (Section 17.6.1.4) – 'avulsive system', used by Latrubesse (2015) for areally expansive zones of fluvial sediment accumulation at *SRS* levels 9–10 (Fig. 17.2), distinct from and equivalent to, or larger in areal terms, than a *megafan*. This feature should not be confused with smaller fluvial avulsive systems that are a basic component of floodplain, delta and megafan evolution	*major avulsive fluvial system (MAFS)*, with some indication of length/area [in preference to 'avulsive system']
Megafan – distinguished by HW[2] from the *fluvial fan* and *alluvial fan*. Suggested as the preferred term by longer usage for larger fans, usually > 80, or > 100 km long, whose processes are specifically not the mass-flow type (alluvial fan complex of processes). *Primary* and *derived* types are distinguished (Section 17.4.3). See *primary megafan* and *derived megafan*	*megafan* (preferably with length designation)
Primary megafan – fed directly from the source highland; see *megafan* and *derived megafan* (Section 17.4.3)	
Terminal fan – used for low-angle but small sandy fan-like deposits at the failing end of a relatively small drainage line; originally described from the western Gangetic Plain but known elsewhere; disfavored by North and Warwick (2007) as being applied uncritically to the rock record and to features such as *megafans*. Not defined in EG[1]	*terminal fan* with indication of size and slope, to allow distinction from *alluvial fans* (significantly steeper in the strict definition), and from the larger *fluvial fans/megafans*
Tributary – see *distributive, distributary, tributary*	
Wet fan – synonym for *fluvial fan* and *megafan;* falling out of use	*fluvial fan; megafan*

Notes: [1] EG: Encyclopedia of Geomorphology (Goudie 2004). [2] HW: Hartley et al. (2010a, b). Weissmann et al. (2010, 2011).

Apart from the central term *alluvial fan*, the glossary includes *fluvial fan, wet fan, megafan, inland delta, distributive fluvial system* (*DFS:* Hartley et al. 2010a, b, and Weissmann et al. 2010, 2011), *terminal fan, fan delta* and *avulsive fluvial system*, and new terms such as *major avulsive fluvial systems (MAFS)* and *large accretionary fluvial systems (LAFS)*, among others. Features associated with definitional arguments such as *delta* and *floodplain* are included. Terms such as *alluvial fan-delta* (with dimension stated) and *megafan-delta* seem needed.

Acknowledgements

MJW thanks chapter authors from whom he has gained much by their skepticism and enthusiastic discussion during preparation of this volume. Several authors and my co-editor Yanni Gunnell have my heartiest thanks for critical editing. I thank especially Andrea Moscariello, Dario Ventra, Gary Weissmann, Louis Scuderi, and Gerald Nanson for discussions over the years.

References

Ashworth, P. J. and Lewin, J. (2012). How do big rivers come to be different? *Earth-Science Reviews*, 114, 8–107.

Ashworth, P. J. and Lewin, J. (2014). The negative relief of large river floodplains. *Earth-Science Reviews*, 129, 1–29.

Assine, M. L. (2005). River avulsions on the Taquari megafan, Pantanal wetland, Brazil. *Geomorphology*, 70, 357–371.

Assine, M. L., Corradini, F. A., Pupim, F. N., and McGlue, M. M. (2014). Channel arrangements and depositional styles in the São Lourenço fluvial megafan, Brazilian Pantanal wetland. *Sedimentary Geology*, 301, 172–184.

Bernal, C., Christophoul, F., Darrozes, J., et al. (2013). Crevassing and capture by floodplain drains as a cause of partial avulsion and anastomosis (lower Rio Pastaza, Peru). *Journal of South American Earth Sciences*, 44, 63–74.

Blair, T. C. and McPherson, J. G. (1994). Alluvial fans and their natural distinction from rivers based on morphology, hydraulic processes, sedimentary processes, and facies assemblages. *Journal of Sedimentary Research*, A64, 450–489.

Blair, T. C. and McPherson, J. G. (2009). Process and Forms of Alluvial Fans, In A. J. Parsons and A. D. Abrahams, eds., *Geomorphology of Desert Environments*. Springer, Dordrecht, 2nd edn, p. 354–402.

Blum, M., Martin, J., Milliken, K., and Garvin, M. (2013). Paleovalley systems: Insights from Quaternary analogs and experiments. *Earth-Science Reviews*, 116, 128–169.

Bridge, J. and Demicco, R. (2008). *Earth Surface Processes, Landforms and Sediment Deposits*. Cambridge University Press, Cambridge, 815 pp.

Burt, T. P. and Allison, R. J. (2010). *Sediment Cascades: An Integrated Approach*. Wiley, Chichester, 471 pp.

Chakraborty, T., Kar, R., Ghosh, P., and Basu, S. (2010). Kosi megafan: Historical records, geomorphology and the recent avulsion of the Kosi River. *Quaternary International*, 227, 143–160.

Coleman J. M. and Wright L. D. (1975). Modern river deltas: variability of processes and sand bodies. In M. L. Broussard, ed., *Deltas: Models for Exploration*. Houston Geological Society, 99–149.

Cooke, R. U., Warren, A., and Goudie, A. S. (2006). *Desert Geomorphology*. University College London Press, London, 2nd edn, 526 pp.

Cordini, R. (1947). Los Ríos Pilcomayo en la Región del Patiño. *Anales I, Dirección de Minas y Geología* (Buenos Aires), 82 pp.

Davidson, S. K., Hartley, A. J., Weissmann, G. S., Nichols, G. J., and Scuderi, L. A. (2013). Geomorphic elements on modern distributive fluvial systems. *Geomorphology*, 180–181, 82–95.

Davidson, S. K. and Hartley, A. J. (2014). A quantitative approach to linking drainage area and distributive-fluvial-system area in modern and ancient endorheic basins. *Journal of Sedimentary Research*, 84, 1005–1020.

DeCelles, P. G., Gray, M. B., Ridgway, K. D., et al. (1991). Controls on synorogenic alluvial-fan architecture, Beartooth Conglomerate (Paleocene), Wyoming and Montana. *Sedimentology*, 38, 569–590.

DeCelles, P. G. and Cavazza, W. (1999). A comparison of fluvial megafans in the Cordilleran (Upper Cretaceous) and modern Himalayan foreland systems. *Geological Society of America Bulletin*, 111, 1315–1334.

Drew, F. (1873). Alluvial and lacustrine deposits and glacial records of the Upper-Indus Basin. *Quarterly Journal of the Geological Society of London*, 29, 441–471.

Dunne, T. and Aalto, R. E. (2013). Large river floodplains. In J. Shroder (Editor in Chief), E. Wohl, ed., *Treatise on Geomorphology, vol. 9, Fluvial Geomorphology*, 645–678. Academic Press, San Diego, CA.

Edmonds, D. A., Hajek, E., Downton, N., and Bryk, A. B. (2016). Avulsion flow-path selection on rivers in foreland basins. *Geology*, 44, 695–698.

Fielding, C. R., Ashworth, P. J., Best, J. L., Prokocki, E. W., and Sambrook Smith, G. H. (2012). Tributary, distributary and other fluvial patterns: what really represents the norm in the continental rock record? *Sedimentary Geology*, 261–262, 15–32.

Fontana, A., Mozzi, P., and Bondesan, A. (2008). Alluvial megafans in the Venetian-Friulian Plain (north-eastern Italy): evidence of sedimentary and erosive phases during Late Pleistocene and Holocene. *Quaternary International*, 189, 71–90.

Ganti, V., Chu, Z., Lamb, M. P., Nittrouer, J. A., and Parker, G. (2014). Testing morphodynamic controls on the location and frequency of river avulsions on fans versus deltas: Huanghe (Yellow River), China. *Geophysical Research Letters*, 41, 7882–7890.

Geddes, A. (1960). The alluvial morphology of the Indo-Gangetic plains. *Transactions of the Institute of British Geographers*, 28, 253–276.

Gibling, M. R. (2006). Width and thickness of fluvial channel bodies and valley fills in the geological Record: a literature compilation and classification. *Journal of Sedimentary Research*, 76, 731–770.

Goudie, A. S. (ed.) (2004). *Encyclopedia of Geomorphology*. Routledge, London, 2 vols.

Gilbert, G. K. (1877). *Report on the Geology of the Henry Mountains*. US Geological Survey, Washington, DC.

Gupta, A. (2007). *Large Rivers: Geomorphology and Management*. Wiley, Chichester, 720 pp.

Hartley, A. J., Weissmann, G. S., Nichols, G. J., and Warwick, G. L. (2010a). Large distributive fluvial systems: characteristics, distribution, and controls on development. *Journal of Sedimentary Research*, 80, 167–183.

Hartley, A. J., Weissmann, G. S., Nichols, G. J., and Scuderi, L. A. (2010b). Fluvial form in modern continental sedimentary basins: distributive fluvial systems: reply. *Geology*, 38, e231.

Hartley, A. J., Weissmann, G. S., Bhattacharayya, P., et al. (2013). Soil development on modern distributive fluvial systems: preliminary observations with implications for interpretation of paleosols in the rock record. In S. Driese, ed., *New Frontiers in Paleopedology and Terrestrial Paleoclimatology*, SEPM Special Publication, 104, 149–158.

Hartley, A. J., Owen, A. E., Swan, A., et al. (2015). Recognition and importance of amalgamated sandy meander belts in the continental rock record. *Geology*, 43, 679–682.

Hartley, A. J., Owen, A., Weissmann, G. S., and Scuderi, L. (2018). Modern and ancient amalgamated sandy meander-belt deposits: recognition and controls on development. In M. Ghinassi, L. Colombera, N. P. Mountney, and A. J. H. Reesink, eds., *Fluvial Meanders and Their Sedimentary Products in the Rock Record*. International Association of Sedimentologists, Special Publication, 48, 349–384.

Hashimoto, A., Oguchi, T., Hayakawa, Y., et al. (2008). GIS analysis of depositional slope change at alluvial-fan toes in Japan and the American Southwest. *Geomorphology*, 100, 120–130.

Holbrook, J., Scott, R. W., and Oboh-Ikuenobe, F. E. (2006). Base-level buffers and buttresses: a model for upstream and downstream control on fluvial geometry and architecture within sequences. *Journal of Sedimentary Research*, 76, 162–174.

Iriondo, M. H. (1984). The Quaternary of northeastern Argentina. *Quaternary of South America*, 2, 51–78.

Iriondo, M. H. (1987). Geomorfolgía y Cuaternario de la Provincia Santa Fé (Argentina). *D'Orbignyana*, 4, 54 pp.

Iriondo, M. (1993). Geomorphology and late Quaternary of the Chaco (South America). *Geomorphology*, 7, 289–303.

Jain, V. and Sinha, R. (2003). River systems in the Gangetic Plains and their comparison with the Siwaliks: a review. *Current Science*, 84, 1024–1033.

Jain, V. and Sinha, R. (2004). Fluvial dynamics of an anabranching river system in Himalayan foreland basin, Baghmati River, north Bihar plains, India. *Geomorphology*, 60, 147–170.

Jain, V. and Sinha, R. (2005). Response of active tectonics on the alluvial Baghmati River, Himalayan foreland basin, eastern India. *Geomorphology*, 70, 339–356.

Klausen, T. G., Ryseth, A. E., Helland-Hansen, W., Gawthorpe, R., and Laursen, I. (2014). Spatial and temporal changes in geometries of fluvial channel bodies from the Triassic Snadd Formation of offshore Norway. *Journal of Sedimentary Research*, 84, 567–585.

Kumar, R., Jain, V., Prasad Babu G., and Sinha, R. (2014). Connectivity structure of the Kosi megafan and role of rail-road transport network. *Geomorphology*, 227, 73–86.

Latrubesse, E. M. (2008). Patterns of anabranching channels: The ultimate end-member adjustment of mega rivers. *Geomorphology*, 101, 130–145.

Latrubesse, E. M. (2015). Large rivers, megafans and other Quaternary avulsive fluvial systems: A potential "who's who" in the geological record. *Earth-Science Reviews*, 146, 1–30.

Latrubesse, E., Stevaux, J. C., Cremon, S., et al. (2012). Late Quaternary megafans, fans and fluvio–aeolian interactions in the Bolivian Chaco, Tropical South America. *Palaeogeography, Palaeoclimatology, Palaeoecology*, 356–357, 75–88.

Leier, A. L., DeCelles, P. G., and Pelletier, J. D. (2005). Mountains, monsoons, and megafans. *Geology*, 33, 289–292.

Lewin, J. and Ashworth, P. J. (2014). The negative relief of large river floodplains. *Earth-Science Reviews*, 129, 1–23.

Levin, S. A. (1992). The problem of pattern and scale in ecology. *Ecology*, 73, 1943–1967.

Martín-Vide, J. P., Amarilla, M., and Zarate, F. J. (2014). Collapse of the Pilcomayo River. *Geomorphology*, 205, 155–163.

McCarthy, T. S. (2013). The Okavango Delta and its place in the geomorphological evolution of Southern Africa. *South African Journal of Geology*, 116, 1–54.

Mertes, L. A. K. (1997). Documentation and significance of the perirheic zone on inundated floodplains. *Water Resources Research*, 33, 1749–1762.

Meyer, L. A. (1996). El Río Pilcomayo: un caso de estudio, In Atti del Corso: Sviluppo e Gestione dei Bacini Idrografici (Course text: Watershed development and management). *Cuaderno ILLA, Serie Cooperación No. 6 (Istituto Italo-Latino Americano)*, Rome, 303–314.

Miall, A. D. (1996). *The Geology of Fluvial Deposits*. Springer, New York, 582 pp.

Miall, A. D. (2014). *Fluvial Depositional Systems*. Springer, New York, 316 pp.

Miall, A. D. (2015). Updating uniformitarianism: stratigraphy as just a set of 'frozen accidents'. In D. G. Smith, R. J. Bailey, P. M. Burgess, and A. J. Fraser, eds., *Strata and Time: Probing the Gaps in our Understanding*. Geological Society of London, Special Publication, 404, 11–36.

Miall, A. D., Holbrook, J. M., and Bhattacharya, J. P. (2021). The stratigraphic machine. *Journal of Sedimentary Research*, 91, 595–610.

Moscariello, A. (2018). Alluvial fans and fluvial fans at the margins of continental sedimentary basins: geomorphic and sedimentological distinction for geoenergy exploration and development. In D. Ventra and L. E. Clarke, eds., *Geology and Geomorphology of Alluvial and Fluvial Fans: Terrestrial and Planetary Perspectives*. Geological Society of London, Special Publication, 440, 215–243.

New South Wales Dept. of Environment and Conservation (2006). Warrego River. <http://wiserivers.nationalparks.nsw.gov.au/Multimedia/vBlob.jsp?id=668> accessed June 2020.

Ori, G. G. (1982). Braided to meandering channel patterns in humid-region alluvial fan deposits, River Reno, Po Plain (northern Italy). *Sedimentary Geology*, 31, 231–248.

Pupim, F. N., Assine, M. L., and Sawakuchi, A. O. (2017). Late Quaternary Cuiabá megafan, Brazilian Pantanal: Channel patterns and paleoenvironmental changes. *Quaternary International*, 438, 108–125.

Ralph, T. J. and Hesse, P. P. (2010). Downstream hydrogeomorphic changes along the Macquarie River, southeastern Australia, leading to channel breakdown and floodplain wetlands. *Geomorphology*, 118, 48–64.

Räsänen, M., Neller, R., Salo, J., and Jungner, H. (1992). Recent and ancient fluvial deposition systems in the Amazonian foreland basin, Peru. *Geological Magazine*, 129, 293–306.

Saito, K. (2003). Model of alluvial fan development based on channel pattern and gravel size. *Report of Research Project, Grant-in-Aid for Scientific Research*, 138 (in Japanese).

Saito, K. and Oguchi, T. (2005). Slope of alluvial fans in humid regions of Japan, Taiwan and the Philippines. *Geomorphology*, 70, 147–162.

Schumm, S. A. (1977). *The Fluvial System*. Wiley Interscience, New York, 338 pp.

Shanley, K. W. and McCabe, P. J. (1994). Perspectives on the sequence stratigraphy of continental strata. *American Association of Petroleum Geologists Bulletin*, 78, 544–568.

Sinha, R. and Friend, P. F. (1994). River systems and their sediment flux, Indo-Gangetic plains, Northern Bihar, India. *Sedimentology*, 41, 825–845.

Slingerland, R. and Smith, N. D. (2004). River avulsions and their deposits. *Annual Review of Earth and Planetary Sciences*, 32, 257–285.

Stanistreet, I. G. and McCarthy, T. S. (1993). The Okavango fan and the classification of subaerial fan systems. *Sedimentary Geology*, 85, 115–133.

Straub, K. M., Paola, C., Mohrig, D., and Wolinsky, G. T. (2009). Compensational stacking of channelized sedimentary deposits. *Journal of Sedimentary Research*, 79, 673–688.

Thoms, M. C., Rayburg, S. C., and Neave, M. R. (2007). The physical diversity and assessment of a large river system: The Murray-Darling Basin, Australia. In A. Gupta, ed., *Large Rivers: Geomorphology and Management*. Wiley, Chichester, 861–890.

Ventra, D. and Clarke, L. E. (eds.), 2018. Geology and geomorphology of alluvial and fluvial fans: Terrestrial and planetary perspectives. In D. Ventra and L. E. Clarke, eds., *Geology and Geomorphology of Alluvial and Fluvial Fans: Terrestrial and Planetary Perspectives*. Geological Society of London, Special Publication, 440, 1–21.

Ventra, D. and Nichols, G. J. (2014). Autogenic dynamics of alluvial fans in endorheic basins: Outcrop examples and stratigraphic significance. *Sedimentology*, 61, 767–791.

Weissmann, G. S., Hartley, A. J., Nichols, G. J., et al. (2010). Fluvial form in modern continental sedimentary basins: distributive fluvial systems. *Geology*, 38, 39–42.

Weissmann, G. S., Hartley, A. J., Nichols, G. J., et al. (2011). Alluvial facies distributions in continental sedimentary basins—distributive fluvial systems. In S. K. Davidson, S. Leleu, and C. P. North, eds., *From River to Rock Record: The Preservation of Fluvial Sediments and their Subsequent Interpretation*. SEPM Special Publication, 97, 327–355.

Weissmann, G. S., Hartley, A. J., Scuderi, L. A., et al. (2013). Prograding distributive fluvial systems—Geomorphic models and ancient examples. In S. G. Driese, and L. C. Nordt, eds., *New Frontiers in Paleopedology and Terrestrial Paleoclimatology*. SEPM Special Publication, 104, 131–147.

Weissmann, G. S., Hartley, A. J., Scuderi, L. A., et al. (2015). Fluvial geomorphic elements in modern sedimentary basins and their potential preservation in the rock record: a review. *Geomorphology*, 250, 187–219.

Whitehouse, F. W. (1944).The natural drainage of some very flat monsoonal lands. *Australian Geographer*, 4, 183–196.

Wiens, J. A. (1989). Spatial scaling in ecology. *Ecology*, 3, 385–397.

Wilkinson, M. J. (1996). Large subaerial distributary systems: global distribution and implications. *Internal Report, Earth Sciences & Image Analysis Laboratory, NASA-Johnson Space Center, Houston, Texas*, 33 pp.

Wilkinson, M. J. (1998). River behavior on large fluvial distributary systems and putative dynamics for fish speciation in South America. *Internal Report, Earth Sciences & Image Analysis Laboratory, NASA-Johnson Space Center, Houston, Texas*, 48 pp.

Wilkinson, M. J. (1999). Fluvial dynamics in the Andean foreland: Megafans and mechanisms for biodiversity in aquatic biota. *Geological Society of America, Fall Meeting Abstracts, 25–28 October 1999, Denver, Colorado*.

Wilkinson, M. J. (2001). Where large fans form: interim report of a global survey. *Fluvial Sedimentology 2001, 7th International Conference on Fluvial Sedimentology, Program and Abstracts, University of Nebraska-Lincoln, Lincoln, Nebraska (USA), 6–10 August 2001, p. 282. (University of Nebraska-Lincoln, Institute of Agriculture and Natural Resources, Conservation and Survey Division, Open-file Report 60)*. www.unl.edu/geosciences/ICFS/ICFS.html *29 November 2003*.

Wilkinson, M. J. (2002). Modern river and sedimentation patterns in Africa – subbasin-scale models derived from astronaut photography, Poster, *Petroleum Exploration Society of Great Britain and Houston Geological Society, First Annual International Symposium, 17–18 September 2002*.

Wilkinson, M. J. (2003). A geomorphic classification of sedimentary subbasin types in the south American foreland—tectonic and drainage pattern controls. *Alluvial Fans 2003, Conference Abstracts, Sorbas,*

Almería, Spain, 8–13 June 2003. <http://alluvialfans .net/Abstracts.htm>, *29 November 2003.*

Wilkinson, M. J. (2005). Large fluvial fans and exploration for hydrocarbons. *NASA Tech Briefs* 29, 64 [NASA Tech Briefs Online, No. MSC-23424 www.nasatech .com/Briefs/ps.html 30 March 2004].

Wilkinson, M. J. (2006). 'Method for Identifying Sedimentary Bodies from Images and Its Application to Mineral Exploration' – US Patent Office, #6,851,606, issue date 10 January 2006.

Wilkinson, M. J. (2015). Large fluvial fans: Aspects of the attribute array. *AAPG Annual Convention and Exhibition (American Association of Petroleum Geologists), Conference Abstracts, Theme 4: Large Fluvial Fans, Denver, Colorado, 31 May–3 June 2015.* http://ace.aapg.org/2015

Wilkinson, M. J. and Cameron, N. R. (2002). Global geomorphic survey of large modern fans: distribution and exploration implications, *American Association of Petroleum Geologists, Program and Abstracts, Annual Meeting, Houston, Texas, 11–15 March 2002.*

Wilkinson, M. J., Cameron, N. R., and Burke, K. (2002). Global geomorphic survey of large modern subaerial fans. *Houston Geological Society Bulletin*, 44, 11–13.

Wilkinson, M. J., Marshall, L. G., and Lundberg, J. G. (2006). River behavior on megafans and potential influences on diversification and distribution of aquatic organisms. *Journal of South American Earth Sciences*, 21, 151–172.

Wilkinson, M. J., Marshall, L.G., Lundberg, J. G., and Kreslavsky, M. H. (2010). Megafan environments in northern South America and their impact on Amazon Neogene aquatic ecosystems. In C. Hoorn, and F. P. Wesselingh, eds., *Amazonia, Landscape and Species Evolution: A Look into the Past.* Blackwell, London, 162–184.

Willett, S. D., McCoy, S. W., and Beeson, H. W. (2018). Transience of North American High Plains landscape and its impact on surface water. *Nature*, 561, 528–532.

Wilson, B. H. and Dincer, T. (1976). An Introduction to the hydrology and hydrography of the Okavango Delta. *Symposium on the Okavango Delta and Its Future Utilization, Proceedings*, National Museum, Gaborone, Botswana, 33–48.

Wright, L. D. (1985). River Deltas. In R. A. Davis, ed., *Coastal Sedimentary Environments.* Springer, New York, 2nd edn, 1–76.

18

Some Future Megafan Research Directions

M. JUSTIN WILKINSON and YANNI GUNNELL

Abstract

New findings around fluvial megafans have accrued from the world survey presented in this book, and challenge some commonly accepted generalisations. Among a list of unexpected results are that (i) megafans constitute a landform and sedimentary body of regional significance on Earth, subsidiarily on Mars, despite their relatively small number; (ii) any topographic step, high or low, can provide the anchor point for a megafan apex; (iii) most megafans are associated with tributary drainages, seldom with axial drainages; (iv) megafans form in all climates. Megafan sizes, shapes, nesting patterns, drainage configurations, tectonic settings, and sediment dispersal styles are summarised, classified, and compared to other large fluvial sediment bodies. Finer mosaics of landscape elements and landforms belonging to the rheic zone (belts of fluvial incision, narrow or wide, that cut into fan surfaces) and perirheic zone (extensive land surface beyond the reach of the fan-forming river) are reviewed from modern analogues, and their implications for identifying megafans and other distributary fluvial systems in the rock record are examined. Vocabulary defining megafans and their environments has been sharpened as a result, with some avenues for further investigation laid out in this closing chapter.

18.1 Introduction

This summary is a selection of perspectives culled from the chapters of this book with topics that may warrant further research. Megafans are now recognised as normal components of aggrading lowlands at subcontinental scales. Megafans constitute the largest of fan-like fluvial forms termed DFS (distributive fluvial systems) by Hartley et al. (2010) and Weissmann et al. (2010). Two perspectives arise that seem to us particularly important. The first is the major geomorphic division that exists, or has existed, in continental interiors between *erosional upland domains* typically with dendritic drainage, and *aggradational lowland domains* with divergent drainage patterns. The latter domain, it is worth repeating, can reach subcontinental dimensions. The main component of lowland aggrading landscapes is nested fans, including very large fans (megafans). The second perspective is that megafan-forming rivers can shift laterally over distances of hundreds of kilometres in brief geological time spans (10^2–10^3 years). Although this statement appears to be a truism, river courses are still generally thought of as static features of the landscape, fixed within the immovable limits of valley margins, as they appear in classical erosional landscapes. But lateral channel migration at the larger scale produces drainage patterns on megafan cones that are divergent, radiating from an apex, rather than convergent, a fact that dominates the entire complex configuration of drainage in such landscapes. Several implications of these often unappreciated patterns are summarised below.

The lack of recognition of the significance of megafans may relate to the fact that the landscapes of Europe and North America, the homes of modern geology, are devoid of major non-coastal aggradational geomorphic domains – despite the fact that this dichotomy absolutely dominates major parts of South America, Asia, Australia, and lesser but still significant parts of Africa. Geddes (1960:259) stated that large alluvial plains (multi-megafan landscapes) 'have tended to be almost completely ignored in geomorphology, instead of providing a central theme. . .'. Our sense is rather that even six decades later the significance of most remains unrecognised. Their intermediate scale between smaller features such as river reaches and alluvial fans (*sensu stricto*) on one hand, and entire sedimentary basins on the other, adds to their importance – at what is the junction between the autogenic and allogenic sets of controls that operate to emplace fluvial sedimentary bodies (Ventra and Clarke 2018).

The novelty of lowland aggradational landscapes – at least in the kind of detail presented in this volume – is evident in models used to reconstruct the palaeogeographies upon which interpretations of the rock record rely. A striking example from this volume lies in the long-standing attempts at understanding the origin of the extensive body of rocks at Sinus Meridiani on Mars, in the lowland immediately adjacent to the southern highlands of that planet. This suite of layered rocks is sufficiently unique that the decision was taken to land the rover *Opportunity* there for direct exploration, but several major analyses of the origins these rocks have proved inconclusive. One reason is that interpretations were based generally on models based in *erosional* geomorphic domains (dendritic/convergent fluvial geometries). Such models have diverted interpretations from those based on what could be argued is the more feasible *aggradational* fluvial domain that ought arguably to apply to a major set of layered rocks immediately downstream of undoubted valley networks (Wilkinson, Ch. 16). On Earth and Mars, the erosional-domain model has been a deficient analogue for understanding the vast aggradational plains of multi-megafan landscapes (Sections 18.2 and 18.5.2).

Half the studies in this volume focus on one or a few clustered megafans. One chapter (Gunnell, Ch. 9) provides greater time depth than the others because it documents the entire Neogene history and stratigraphy of the Loire megafan. The others are based on regional and global studies, and in one case (Ventra and Moscariello, Ch. 14) the chapter is general in the sense of encapsulating a wide literature review in the related area of interpretation of the rock record. All the studies are aimed at the same fan-like feature, which is larger by orders of magnitude in areal terms than the small, coarser-textured alluvial fan typical of mountain piedmont locations. Megafans fit best at the mid-level (level 9) in Miall's (1996, 2014) geometrically-scaled *SRS* hierarchy of fluvial depositional forms, whereas smaller alluvial fans (*sensu stricto*) are distinct in terms of process, grain size, and lesser complexity (*SRS* levels 7 or 8) (Wilkinson, Ch. 17).

The criteria used for megafan recognition and mapping have centred on two fundamentally different purposes that had been employed thus far, namely (i) the definition of megafans as lithological units, or (ii) as morphological bodies, a distinction exemplified in two global studies (Wilkinson and Currit, Ch. 2, and Hartley et al. 2010, respectively). Definition of the distal margin of megafans and concavity of slope are two criteria that differ between these studies. Proximal coniform morphology is not questioned as a common feature, but the recently recognised major avulsive fluvial systems ('MAFS' – see Section 17.6.1, Wilkinson, Ch. 17) do not display coniform apex topography. This raises the hydraulic question of why some large fan-like features display coniform proximal zones and others do not.

Studies of megafans and megafan landscapes have accelerated sufficiently that several assumptions concerning megafan origins and attributes have been laid to rest (Table 18.1). The common assumption that megafans form at the foot of major mountain chains – based on the setting of the Kosi and Gangetic Plains megafans (Fig. 18.1) – fails to recognise dozens of megafans apexed at low, apparently insignificant topographic breaks of slope. Examples are the very large Paraná megafan (Fig. 18.2) and the well-known Okavango megafan (Fig. 4.1b, Wilkinson et al., Ch. 4). Another assumption has been the frequently invoked climatic control of megafan development. These prior perspectives, among others, may have derived from the following facts: (i) active fans are well displayed in piedmont zones at the foot of the Himalaya and Andes mountains; (ii) wetland vegetation on active megafans is more easily identified

Table 18.1 *Commonly held assumptions concerning megafans* vs. *results from global studies: a comparative overview*

Commonly accepted generalisation	Unexpected results from global studies
Megafans are associated with major mountain range fronts (Fig. 18.1)	Any topographic step, high or low, can provide the anchor point for megafan apexes (Fig. 18.2)
Megafans are associated with the largest rivers	Most megafans are associated with tributary drainages, seldom with major trunk/axial drainages
Wet climates are a necessary condition for development of 'wet fans' (often synonymous with large fans in the literature)	Megafans form in all climates
Upland feeder basin area is strongly related to megafan size (as for small alluvial fans)	Lowland basin dimension can be a more significant control of megafan dimension (seldom the case for small alluvial fans)
Megafans occupy the tail of the fan size/frequency distribution, and are therefore insignificant	Nested megafans constitute a landform and sedimentary body of regional significance on the planet, despite their relatively small number
Radial/diverging drainage criterion 1 – *single megafan cones*: produce radial drainage patterns	1 – true of proximal sectors; often not true of medial and distal patterns (drainage can be parallel and contributory)
2 – *multi-fan landscapes*: dominated by radial drainage patterns	2 – smaller fan-fed drainages on megafan surfaces can display contributary/dendritic drainage in depressions (e.g., between lobes and alluvial ridges)
Long profiles are concave	Long profiles can be convex and convexo-concave
Planforms are triangular	Planforms are often diamond-shaped
Sheetfloods operate on megafans	Generally untrue despite the well-known Okavango megafan displaying a seasonal sheet-like flood

remotely in zones with drier savanna climates; (iii) relict fans without active lobes (a large minority) are difficult to detect; and (iv) remotely-sensed topographic and other data are now much more readily available for detecting their distribution.

The Okavango and Kosi, two of the best known megafans, are not typical of the wider population, a fact that may have promoted some of the misconceptions we have encountered. The first characteristic is their dimension, which at 150 km long lies at the smaller end of the megafan size continuum (maximum length recorded: 720 km). The reputation of these two as very large fans was undoubtedly derived from comparisons with the ubiquitous small alluvial fan. The Okavango and Kosi both display triangular planforms while many megafans, especially those in less topographically restricted basins, are diamond-shaped. The Okavango and Kosi are also less complex morphologically than larger fans. Furthermore, the river behaviours on both are atypical. The Okavango River discharge flows as a sheetflood across the upper

megafan surface during the annual high-flow phase and displays distributaries in the true sense (Fig. 18.3). Neither attribute, however, is typical of drainage on most megafans. The Okavango megafan is also a closed sedimentary system, which, although the relevant research has not been conducted, seems to be a less common state for most megafans.

In the case of the Kosi River, engineering works in the form of a damming barrage and two marginal but discontinuous embankments now control avulsive stream behaviour. Prior to the completion of the engineering works, the Kosi River was a 'fan-margin river' once it occupied its present course on the extreme western margin of the megafan cone. This implies that discharges from exterior sources – the major foothills-fed rivers to the west, including the Baghmati, would have undoubtedly increased its discharge. One effect of fan-margin status, as theory suggests (Section 17.4.1, Wilkinson, Ch. 17), is that its gradient is lower than that of courses in a mid-fan orientation, a difference in slope that has been commented upon in

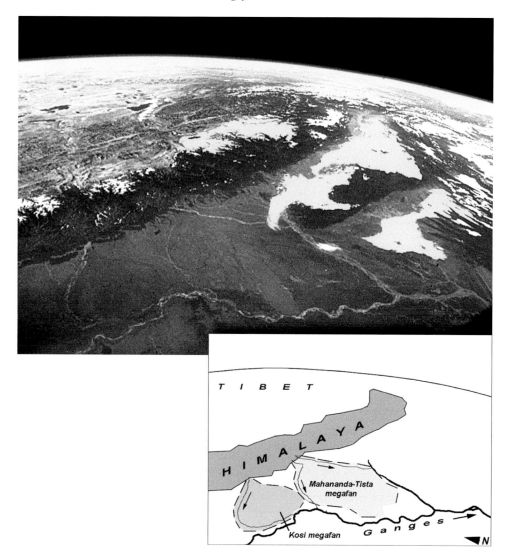

Figure 18.1 **Astronaut image of Ganges Valley megafans**. The Kosi and Tista megafans are fed from Himalayan mountain catchments. The fan-forming rivers are tributaries of the regional trunk river, the Ganga (foreground). Astronaut handheld image STS27–39-27, courtesy of the Earth Science and Remote Sensing Unit, NASA Johnson Space Center. A colour version of this figure is available in the SOD for Chapter 18.

previous work but not in the context of fan-margin *vs.* fan-forming river discharges.

18.2 Landscape Elements and Drainage Configurations

The immensity of megafan landscapes becomes more apparent to the human eye when seen from a light aircraft. Even from the vantage of hundreds of metres above the ground, the horizon of the vast flat surface is not broken by even the smallest irregularity, no matter in which direction one looks. Even flying for hours the moving circle of the horizon remains strikingly flat and

absolutely smooth. This broad view nevertheless masks the characteristic components that are found on most megafans. Megafan and multi-megafan landscapes are sufficiently unique that the basic features and drainage styles are summarised in the following brief compendium.

18.2.1 Landscape Elements

Aggradational features in terms of decreasing dimension are the following.

(i) The partial cone of the megafan is the largest feature, even though the cone may only achieve vertical

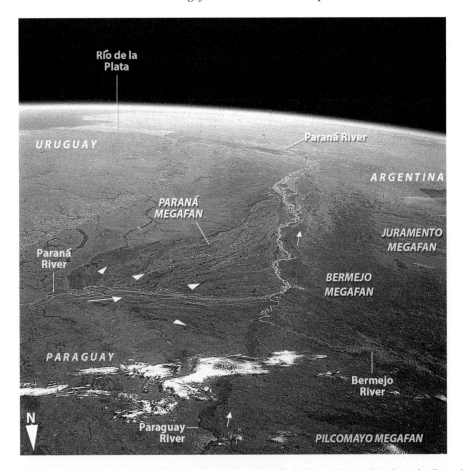

Figure 18.2 **Astronaut image of the Paraná megafan**. This oblique handheld image shows the triangular Paraná megafan (left centre of the image). Present course of the Paraná River appears as a tan line; the river flows west, then angles south in the middle of the image (flow is towards top of the image). The fan apex is located where the river crosses the low margin of the Brazilian craton (left margin of the image); long radius is 320 km. Short arrows show the radial drainage pattern in the apex zone. Lower slopes of three Chaco megafans appear along the right margin (from bottom to top: Pilcomayo, Bermejo, Juramento). Astronaut handheld image STS51D-46-22, courtesy of the Earth Science and Remote Sensing Unit, NASA Johnson Space Center. A colour version of this figure is available in the SOD for Chapter 18.

dimensions of a few metres in transverse profile despite widths and lengths of tens or hundreds of kilometres. Such convexity is one of the most consistent diagnostic features. The narrow apex zone of the cone necessarily concentrates the area of channel-dominated environments at the expense of overbank/floodbasin environments. The same concentration operates at subapexes (below), whether these be obstacle-related or shoreline-induced. Zones of channel-body concentration are of specific interest in interpretation of the rock record.

(ii) The next largest feature of the active fan is the sediment lobe, which also often shows convexity of the transverse profile, although of very small vertical dimensions. Lobes are probably the main architectural element that builds megafans. Active megafan surfaces show lobes of sediment related to phases of sedimentation emanating from the main apex (nodal avulsions) and from downfan subapexes ('regional avulsions',

sensu Slingerland and Smith 2004). Lobes are often identifiable in detailed DEMs, with relative lobe age ascertainable from drainage network patterns.

(iii) At the next smaller scale, alluvial ridges, up to tens of kilometres wide, appear to be responsible for the construction of sediment lobes, but their vertical dimensions are often *greater* than those of the lobe of which they form a part. The fan-forming river within its levees is often located along the summit of the alluvial ridge in what is seen as an inherently unstable condition prone to avulsion. Lobes and alluvial ridges of different ages have been identified, but the dating of such features is still in its infancy and very important for further understanding of aggradational landscapes. Alluvial ridges appear as clusters of more elevated topography, amounting to the areas of similar scale as lobes.

(iv) At a usually smaller scale, crevasse splays are so ubiquitous in some imagery that the idea has been

Figure 18.3 **Astronaut image of the Okavango megafan in flood**. Oblique image shows the Okavango megafan (150 km from apex to toe) in sunglint, looking northwest, on 6 June 2014. Distributary drainage and the generalised flood are atypical of the megafan population. Flow is south and east (from top to bottom of the image): the Okavango and Cuito rivers (top left), also in sunglint, lead water from Angola. Linear fault lines demarcate megafan toe. The Boteti River drains vestigial water from the toe to the light-toned salt flats of the Makgadikgadi basin (foreground, lower right). Astronaut handheld image ISS040-E-8209, courtesy of the Earth Science and Remote Sensing Unit, NASA Johnson Space Center. A colour version of this figure is available in the SOD for Chapter 18.

expressed that megafan surfaces may be primarily composed of these sand-rich bodies.

The zone where the fan-forming river is active can include sectors of fluvial incision, most commonly thought of as trenchlike zones that terminate on the fan at an intersection point, cut into fan surfaces, frequently occurring in the proximal zone, but seen in other locations. Where these types of incision operate for long periods, perhaps permanently, the fan-forming river is unable to avulse to any point flanking the incised zone. But wide and shallow, rather than trenchlike, depressions are also known (e.g., Okavango and Kosi megafans). Such depressions presumably *can be refilled* during aggradational phases, at which stage avulsion by the fan-forming river on the fan surface once again becomes possible.

If the river incises repeatedly, terraces develop upon which soils of different ages develop. The Gandak megafan on the Gangetic Plains shows a sequence of terraces cutting the length of the megafan (Mohindra et al. 1992). The Kosi megafan is unusual in displaying a small mesa-form remnant of a higher alluvial surface, suggesting the existence of an extensive prior surface. In later phases of erosion, widespread incision is instigated along all major streams (e.g., the Loire megafan; Gunnell, Ch. 9).

We find that the extreme mobility of the fan-forming river is frequently underestimated, compared with rivers in erosional geomorphic domains where rivers occupy geographically fixed valleys. The periodicity of the 'return time' of these shifts in the course of the fan-forming river is thus of great interest. Thousands of years may separate the avulsions that led to construction of the major alluvial ridges on the megafan cone. Since the mid Holocene, for example, May et al. (Ch. 7) were able to give a sense of the length of time taken to rework the entire surface of the Río Grande megafan in Bolivia, as six to seven thousand years. During this period eight alluvial ridges (with single or multiple arms) have migrated laterally across the megafan. This implies that the low-energy 'plains-fed' streams and pedogenesis are the main morphogenetic processes between periods when the fan-forming river returns to a specific location. Research on 'return time' periodicity suggests extreme variability from tens of thousands of years, implied, for example, by Latrubesse et al. (Ch. 5) for the Chaco megafans, to a few thousands for the Río Grande megafan (May et al., Ch. 7), to a few hundreds for the Kosi megafan (Chakraborty et al. 2010; Sinha et al., Ch. 11).

The zone beyond the reach of the fan-forming river is a critical component of megafan landscapes for two main reasons. The first is that it comprises the great majority of the surface of most megafans. For example, 97% of the Pilcomayo megafan surface today lies in this zone (having increased from 87% in the early twentieth century as a result of the 300-km retreat of

the river end point in the last 100 years: see Section 18.2.3). Geomorphic activity here is much less energetic because relatively small fan-fed streams dominate this zone, often occupying larger palaeochannels. Soil development and aeolian dynamics are relatively important. The fan-fed zone is significant also because it includes megafan surfaces that are partly or entirely relict (i.e., beyond the reach of avulsions of the fan-forming river). Relict status characterises fully one third of the global population of megafans.

Depressions of all kinds, open and closed, large and small, abound in megafan landscapes as normal components of these landscapes – in sharp distinction from smaller fans where complex surfaces are almost non-existent.

18.2.2 Drainage Types and Configurations

The elegantly simple textbook stream classification based on stream order from headwater tip to trunk river, and the acute junction angles of the dendritic drainage systems, seems an insufficient system for river classification in megafan landscapes. As described by Wilkinson (Ch. 17), the component river types in megafan landscapes require a different and often unfamiliar terminology (original terms from Sinha and Friend 1994, are shown in parentheses):

(i) the largest rivers that build megafan sediment cones are termed fan-forming; these rivers drain major drainage basins in eroding highlands and mountain chains (mountain-fed rivers);
(ii) small fan-fed rivers rise are the numerous small rivers that rise on megafans (plains-fed);
(iii) intermediate-sized rivers rise in the foothills facing the plains (foothills-fed);
(iv) the seldom mentioned fan-margin river type downstream of foothills-fed streams where it gathers the foothills-fed streams into a single channel (mixed-fed).

The fan-forming river determines the radial drainage at the cone scale which, in detail, often gives way in distal sectors of the cone to parallel patterns (where megafans are laterally confined by neighbouring megafans). The radial drainage pattern of an alluvial fan or a megafan is a simple and well-known pattern. But associated drainage configurations are radically different and unfamiliar in comparison to the dendritic pattern of erosional 'valleyed' landscapes. Most important is the simple fact mentioned above that radial drainage on a cone necessitates extreme repositioning of the fan-forming river by even hundreds of kilometres laterally, as it switches to a different radially aligned course. This dynamic often appears to be unappreciated, perhaps because of the scale of such course reorientations, a scale implied by avulsions at the megafan apex. Another unfamiliar morphology is that of megafan rivers flowing along alluvial ridges, which are topographically the *highest* sectors of a megafan, quite unlike rivers flowing in topographically low valleys.

Apart from the radial drainage pattern at the cone apex, other nodes of radial drainage are known; these 'subapexes' form at three kinds of locations on megafans: most typically at the 'intersection point' (where the proximal incised zone that typifies many megafans, comes to an end); at those locations where drainage on the megafan diverges downstream of any zone that is locally confined by an obstacle (such as a protruding bedrock hill); and at numerous nodes along both the main river and plains/fan-fed rivers where avulsions take place. It has been noted that the styles of avulsion known from floodplains (e.g., Slingerland and Smith 2004) do not fully describe the fixed and mobile styles on megafans (Section 17.3.1.1, Wilkinson, Ch. 17).

One of the significant aspects of apexes and subapexes is their influence on channel distribution, because channels are necessarily heavily concentrated immediately downstream of any apex. Such patterns hold specific importance for interpretation of the rock record. Furthermore, subapexes are the source points of lobes of sediment that build megafans. Further investigation on the topic of subapexes is thus indicated as lobes are a key building block of megafan cones.

Some fan-forming rivers are known to deposit their entire load on the megafan because they are unable to reach the trunk river of the basin – effectively establishing internal/endorheic drainage regimes, despite being situated on the topographically open slope of the megafan surface (internal drainage is commonly assumed to imply closed topographic depressions lacking an outlet to the sea). Systems such as these can persist as sediment sinks for tens of thousands of years (e.g., most of the Chaco Plains megafans, where they form one of the largest continental sediment sinks

on the planet; Latrubesse et al., Ch. 5). The Okavango megafan of northwest Botswana is another (Wilkinson et al., Ch. 4).

In multi-megafan landscapes the discharge of fan-margin rivers alternates between three modes, i.e., it is permanently foothills-fed but its discharge rises significantly on those occasions when one mountain-fed (fan-forming) river avulses into it, and rises yet more when both fan-forming rivers avulse into it. Examples are given in Wilkinson, Ch. 17, Section 17.4.2). Other than the trunk river of a major basin, the fan-margin river is thus at times the largest river in multi-megafan landscapes; but also at times far smaller than the fan-forming rivers when these latter switch away from discharging into the fan margin zone. Such extremes of discharge variability – as a non-meteorological fluctuation in a component of the drainage network – is quite foreign to dendritic drainages.

Furthermore, tributaries as normally understood do not operate because they cannot reach fan-forming rivers that are located on a topographic high, namely the megafan cone. The drainage pattern is even more complex in multi-megafan landscapes that include fan-margin rivers in interfan zones. At those times when fan-forming rivers avulse into this river the fan-forming rivers themselves act, confusingly, as tributaries. The fan-margin river then loses these very significant 'tributaries' when the fan-forming rivers avulse away to other orientations on their respective cones.

At the scale of the smaller individual fan-fed streams, drainage patterns vary, including parallel, divergent, and convergent. These typically form the smaller, convergent drainage systems on megafans. They usually appear as underfit streams within the long wavelength palaeo-meanders of the fan-forming river with meanders orders of magnitude smaller than those of the fan-forming river. Their drainage systems operate for the long periods between the return episodes of the fan-forming rivers, and as such deserve scientific attention: fan-fed streams are the unique drainage type of the largest areal component of the megafan surface, increasingly occupied by the expanding human populations that cultivate them.

A related and probably more common phenomenon is that the very low slopes of megafans promote the development of wetlands, whether seasonal or permanent, that can reach immense sizes. The total flooded area of the six major Chaco megafans during the wet season amounts to more than 300,000 km², a remarkable 48% of their combined surface area (Latrubesse et al., Ch. 5). The Pantanal megafans are a smaller but better-known system of wetlands. Megafans as the setting for many wetlands worldwide is a topic that deserves further scientific attention, particularly from the angle of biodiversity conservation and ecosystem services, in contexts such as the large-scale megafan reclamation for ranching and monoculture in Paraguay (https://earthobservatory.nasa.gov/images/92078/deforestation-in-paraguay), Argentina, Bolivia and parts of Africa.

18.2.3 Avulsions and Other Sediment Dispersal Styles

Various chapters have documented the detail of major avulsions that have affected fan-forming rivers, a topic that has required remote-sensing-based comparisons and is important because both the periodicity and speed of avulsions remain poorly understood. Avulsions on the Río Grande are elucidated by May et al. (Ch. 7) for the vast area of the Río Grande megafan (eastern Bolivia) from sequential satellite imagery and detailed sedimentological and dating work. Latrubesse et al. (Ch. 5) mention several historical records of avulsions of the past four centuries for the large fan-forming rivers of the Chaco Plains. A set of marine-induced avulsions is fully documented along the distal margin of the youngest lobe of the lower Mitchell megafan in northeastern Australia (Lane et al., Ch. 12). The study of avulsions is at least as important for understanding the evolution of megafan morphology and sediment lobes as it is for understanding the evolution of valley-confined rivers and their sediments.

Two unusual styles of avulsion and sediment dispersal on megafans suggest another avenue for future research. One is *avulsion by underflow* on the Okavango megafan, and the other is 'self-blockage' of the episodically endorheic Pilcomayo megafan in the Chaco by its own sediment (Section 17.3.1.1, Wilkinson, Ch. 17). Although the processes are well documented on these two megafans, the style of sediment dispersal in the former case is unclear since flow disappears underground. In the latter case what has been termed 'incipient avulsion' results in the full load of the river being laid down along the immediate margin of the river. The zone of deposition has migrated *upstream* over an extended distance (300-km) as the river end-point has migrated away from the trunk river over several decades.

Multi-megafan landscapes often transgress the boundaries between major continental drainage basins. In these cases, it is common for one megafan to be situated on the drainage divide between the basins ('divide megafans') – such that the fan-forming river flows at times into one basin and at other times avulses into the other basin. For example, Bolivia's Parapetí River has alternated between draining into the Paraná and the Amazon basins. These rivers may have particular importance for the dispersal of aquatic life.

18.3 Megafan Size and Shape

Latrubesse et al. (Ch. 5) and Santos et al. (Ch. 6) illustrate that the expected relationships hold to some degree, in the Chaco and Pantanal megafans, between feeder basin area and fan area, between fan slope and fan area, and between fan slope and the circularity index of the megafans. The strongest relationship for these megafans is the negative relationship between fan slope and feeder-basin area. This research highlights two aspects apparent thus far in morphometric studies of megafans: the first is the array of independent variables that appear to act in shaping megafan morphology, including feeder-basin area, geology, hypsometry, and climate. These variables include less obvious effects such as megafan planform interactions in which megafans are nested adjacent to one another.

The second aspect is the consistently low number of megafans in any one basin that make up the sample being studied – e.g., six in the Chaco Plain, six in the Pantanal basin, four in the Gangetic Plains, and ten in the Kalahari Basin. Although several relationships usually can be subjectively judged, low sample sizes limit the value of statistical comparison, which in turn emphasises the value of global inventories such those of Hartley et al. (2010) and Wilkinson and Currit (Ch. 2). The global population of nearly 300 megafans allows for future efforts in broadening the basis for morphological understanding.

Nevertheless, one major factor consistently stands out as a control of megafan size compared with studies of smaller fans: because megafans can reach lengths of hundreds of kilometres and areas of tens to hundreds of thousands of square kilometres, they often impinge on the topographic margins of the host basins. *Basin dimension* can thus act as a dominant control of mega-fan dimension, where this is seldom a control on smaller fans that are orders of magnitude smaller than their host basins (see Wilkinson, Ch. 17). Thus, where the positive relationship between river discharge (using basin area as a proxy) and fan size is expected, the opposite is seen in the narrow Okavango Rift basin (90–150 km): here the Zambezi megafan is the smallest of the ten studied in the northern Kalahari Basin due to constriction by basin margins, despite being built by one of Africa's largest rivers (Wilkinson et al., Ch. 4).

Other, apparently more complex, controls also oper-ate and require future attention. Megafan size and shape are strongly modified where there is competition between neighbouring megafans for space. The con-trols are different sediment yields from neighbouring basins, spacing between feeder-river exit points, exit-point altitude, and fan slopes (smaller rivers producing steeper fans), all of which appear to play a part in the ultimate nesting configuration which includes fan area. Alignment of the megafan axis with respect to the basin axis also plays a part. For example, the axis of the Kosi megafan is orthogonal to the Himalaya and Ganga trunk river, and displays a roughly triangular to oval planform. In contrast, the length of the nearby Gandak megafan is twice that of the Kosi, partly because of its alignment oblique to the Himalayan mountain front.

Another probably important control is that of stream power. If they are close enough to their neighbours, smaller, steeper rivers appear to build fans that com-pete against their bigger, lower-gradient neighbours by maintaining their conical form at the expense of the larger cone whose margin is modified by the smaller cone. The reason smaller fans successfully compete for space may lie in their steeper slopes which endow them with locally greater stream power. A prime example is that of the Río Parapetí, whose sediment yield is far lower than that of its neighbours (Pilcomayo and Grande) but whose megafan area is

slightly larger than that of the Grande (Latrubesse et al., Ch. 5). The controls are undoubtedly complex. Applying reasoning from megafan controls identified in the Gangetic Plains by Sinha et al. (Ch. 10), it is suggested here that reasons for the size of the Parapetí megafan probably include (i) its relatively high sedi-ment yield, (ii) an exit point that lies at an altitude significantly higher than those of its neighbours (~150 m higher than the Pilcomayo apex and ~275 m higher than the Grande), and (iii) an exit point at a comparatively great distance from its neighbours' exit points (140 and 125 km respectively) – all of which improve this small river's competition for space. It seems, however, that considerations of stream power, as pioneered by Latrubesse et al. (Ch. 5) for the largest Chaco rivers, need to be included in understanding the size of megafans in a space-sharing context.

Four of the ten megafans in the Kalahari population (Wilkinson et al., Ch. 4) show overtly *convex-up* lon-gitudinal profiles, and three others are straight or slightly convexo-concave. One reason may be the relationship that obtains in more desertic climates, that channel profiles steepen with the downstream reduc-tion of flow (Cooke et al. 2006) – i.e., the opposite of the commonly accepted concave-up profile. This indi-cates that the generalisation that megafans show concave-up long profiles is by no means universal.

The influence specifically of rock type in feeder basins is documented in the Pantanal megafans, where sediment calibre derived from sedimentary source rocks has apparently contributed to the development of larger fans with more complex bar-form develop-ment, compared with megafans fed from catchments dominated by Precambrian basement rocks (Santos et al., Ch. 6). The controls appear to be more complex in the Andes Mountains, where not only source rock differences but also variation in climate, catchment area, and degree of relief have all affected sediment yield, which in turn has influenced megafan areal dimensions (Latrubesse et al., Ch. 5).

The storage of coarse sediment in mountain basins has materially affected the supply of gravel to the Gangetic Plains megafans, the coarser sediment fraction in piggyback basins of the Terai belt having been sequestered upstream of the Gandak megafan. However, this upstream filter does not exist in the case of the Kosi River (Sinha et al., Ch. 11). The results of the lack of such sequestration on the Kosi channel are thought to be a more continuous supply of sediment, a higher proportion of coarse sediment, with increased bed aggradation, and a more avulsive style of river behaviour on the megafan.

18.4 Megafans and Other Fluvial Sediment Bodies

A task for future research is investigation of similarities and differences between the various alluvial landforms at scales both smaller than, the same as, and larger than megafans (the mid-level fluvial forms of Miall's 1996, 2014, hierarchy: Wilkinson, Ch. 17).

Differences between megafans and other depositional environments, such as smaller DFS and valley-confined floodplains, have been examined in Ch. 17). At their simplest, these are distinctions between dominant sets of processes, such as sheet floods and sediment gravity flows, etc., in the case of relatively small alluvial fans (*sensu stricto*) (Blair and McPherson 1994); versus the classic associations of channel–levee–overbank depositional environments on megafans. Differences also relate to the greater variety of morphological components identified on megafan surfaces.

Differences between megafans and confined floodplains are more subtle because megafans comprise the same alluvial-belt components as floodplains. As elucidated by Weissmann et al. (2015), differences relate at least partly to controls exerted on the avulsive dynamics by valley-wall and terrace constriction, or lack thereof.

There also seem to be identifiable differences between megafans and other fluvial bodies of similar scale such as megafan-deltas, major avulsive fluvial systems ('MAFS'), bajadas (i.e., alluvial aprons), and large accretionary fluvial systems ('LAFS') (Wilkinson, Ch. 17). The settings in which these develop also need to be better understood. Features apparently most closely related to megafans are megafan-deltas, the coalesced floodplain, MAFS, and LAFS.

(i) Mitchell megafan-delta, Queensland, Australia. Lane et al. (Ch. 12) give a detailed account of the controls that operate on the modern 'megafan-delta' of the Mitchell River, whose distal zone lies at the interface between continental fluvial and fluvial-deltaic systems

on the Gulf of Carpentaria in northeast Australia. This is a zone of demonstrably increased frequency of avulsions today that must have migrated seaward with coastline regressions.

(ii) Coalesced floodplains. The upper strata of the western Gangetic Plain (WGP) in northwest India comprise the 900-km-long 'interfluve' between major incised valleys of the Yamuna and Ganga rivers, dated as between 90 and 27 ka (Gibling et al. 2005). This interfluve, almost 100 km wide, has been termed by Sinha et al. (Ch. 10) a 'coalescing floodplain' formed by three closely spaced rivers, the Ganga, Ramganga, and Yamuna, rather than as a megafan. Sinha et al. (Ch. 10) observe that the exit-points from the Himalaya of the two very large Ganga and Yamuna rivers are located only 75 km apart, which from the perspective of the global study constitutes an unusual geometry (Wilkinson and Currit, Ch. 2). Only two other megafan landscapes appear to be similar in terms of rivers of roughly the same size, though significantly smaller than the Yamuna and Ganga. These are found in the Indus basin of northern Pakistan and the northwest Llanos of Venezuela. In the former case, the Ravi and Beas river exit-points from the Himalaya lie less than 15 km apart, and both appear to have contributed to the development of a single piedmont surface of remarkably low relief that extends more than 250 km from the exit points, roughly along the international boundary with India. Both rivers have incised wide valleys, producing a landscape similar to that of the Yamuna–Ganga interfluve. These valleys incise a prior surface of floodplain sediments (the Older Alluvium) likely derived from the largest rivers (Gibling et al. 2005). In the Llanos case, the Guanare, Sabaneta, and Santo Domingo river exit-points from the northern Andes are also comparatively closely spaced (50 km and 25 km). These rivers have also generated what appears to be a single piedmont feature of coalesced floodplain surfaces reaching megafan dimensions (extending more than 100 km from the mountain front). This surface extends fully 150 km along the mountain front. Such megafan-like features appear to be formed by combined fluvial discharges of rivers large enough to generate megafans, but in the unusual setting of short distances between the points at which the formative rivers exit the mountain mass. Further research on these variants of megafan-related landscapes is needed because they contradict the usual pattern of megafan cones being sharply demarcated from each other by a fan-margin river.

(iii) MAFS and LAFS – other midscale fluvial features. These two newly described features, like megafans, fall within Miall's (1996, 2014) fluvial midscale (*SRS* levels 9, 10, Fig. 17.2). One is the 'major avulsive fluvial system' (MAFS, Latrubesse 2008, 2015; Latrubesse et al. 2010), erected to encompass fan-like features sometimes larger than large megafans but that do not display coniform morphology. The other is the 'large accretionary fluvial system' (LAFS, G. Nanson, pers. comm. to MJW) which encompasses, *inter alia*, the extremely long and wide coniform sediment bodies, confined within bedrock valleys, that characterise eastern Australia's arid Channel Country. The attributes of these features are outlined briefly in Wilkinson (Ch. 17). Both need more research to flesh out their characteristics.

The comparison of megafans with very large floodplains shows that several similarities between them exist despite the fact that megafans are, first, generated by rivers of significantly smaller discharge; and second, lack the relatively more frequent and variable inundation by flood waters (Wilkinson, Ch. 17). The rising interest in very large confined floodplains should thus benefit from comparison with megafan landscapes.

Investigation of these features is driven by intrinsic geomorphic interest, but also by the need for integrated landscape models to promote improved palaeogeographic reconstructions that include and integrate a more complete set of known geomorphic landscapes.

18.5 Regional Setting

18.5.1 Tectonic Setting

Hartley et al. (2010) and Weissmann et al. (2011) found large DFS (i.e., > 30 km long) to be associated with basins of every tectonic style. Megafans in the studies in this volume lie in two major tectonic environments. The very large Chaco megafans and the smaller central European megafans of the Alpine and Carpathian piedmonts lie in the best-known setting, that of foreland basins of various types. In contrast, cratonic settings show only a generalised connection between elevated landscapes and juxtaposed megafans in those relatively rare locations where megafans occur. Africa's surface morphology since the Miocene has been dominated by basin-and-swell structures generally a few to several hundred kilometres in length (Burke and Gunnell 2008; Burke and Wilkinson 2016). Swells are as a megafan-forming environment are significant because almost all of Africa's

eighty-seven megafans, amounting to a third of the global megafan total, show a direct association with the flanks of medium-sized swells. Interestingly, no identifiable megafans lie on the flanks of the largest swell, the East African, counter to the expectation based on mountain-front megafan clusters such as those adjacent to the Andean and Himalayan orogens. The reason probably involves swell-related uplift that is sufficient in the East African case to impose an *erosional* geomorphic regime on a wide zone of surrounding terrain – a regime inimical to megafan development – but not large enough to provide the accommodation of an adjacent, foreland-style basin.

Gunnell (Ch. 9) has shown that the same direct relationship exists between the Loire megafan of central France and the Massif Central, which is likewise a geologically young swell structure. The tectonic settings of the megafan sub-population of large DFS may not follow the patterns identified for all fans > 30 km long, as witnessed by the fact that the great longitudinal valleys of the East African Rift, amounting to thousands of kilometres in length, host numerous smaller fans but not one megafan. The tectonic setting of megafans thus appears to warrant further research.

18.5.2 Drainage and Nesting Patterns that Include Megafans: Basin and Continental Scales

Megafan-forming rivers tend to be aligned transverse to the source upland or mountain front and feed a trunk river that is frequently aligned parallel to the mountain front. Megafan-filled basins are usually strongly asymmetric because the mountain-belt margin of a basin supplies more sediment than the opposite (lower) margin. As megafan sediment wedges are constructed, they displace the trunk river to the opposite margin that is poorly supplied with sediment. The floodplains of trunk rivers such as the Ganga and Paraguay–Paraná, are consequently reduced to small percentages of the basin area in megafan-dominated basins. In contrast, areas genetically connected to the megafan drainages in these two basins are much larger (sampled areas are 84% and 87% of basin areas respectively). Importantly, the largest areas undergoing active deposition, for example in foreland basins, are related specifically to *tributary megafan drainages* rather than to the trunk drainage of the basin. This pattern is counterintuitive because it is the opposite of dendritic (i.e., erosional) landscapes where progressively larger valleys host progressively larger floodplain sediment bodies. The connection between the megafan-in-foreland distribution and what is ultimately preserved in the rock record is a complex issue beyond the scope of this discussion.

At the basin scale, the configuration of megafans and trunk rivers can be simplified into five dominant types (based on transverse, *T*, and longitudinal, *L*, orientations, after Miall 1996), the main variables being: (i) single and double-margined basins, and (ii) fan orientation with respect to basin margins (Fig. 17.7, Wilkinson, Ch. 17). Four of the configurations display fans orthogonal to the upland margin, i.e., variants of the transverse/longitudinal (*T tributary river (megafan)/L axial river*) configuration. Less common is the *L/L* (*L tributary river (megafan)/L axial river*) configuration that typifies megafan orientation in most cratonic basins.

At the wider subcontinental scale, fully six depositional 'domain' patterns are known from the global survey (Fig. 17.9, Wilkinson, Ch. 17) in which megafans form a major domain – in variable geographic arrangement with valley-confined floodplains of trunk rivers, and large coastal deltas. All these patterns exist in addition to the well-known three-zone model of Schumm (upland sediment supply zone–valley-confined floodplain transport zone–coastal delta depositional zone).

18.6 World Survey

A repeated refrain is heard that the flatness of megafans in some way has prevented their recognition (Cameron pers. comm. to MJW; Miller pers. comm. to MJW) as widespread landforms. In stark contrast is Drew's (1873) comment that small alluvial fans are 'among the most conspicuous forms of superficial deposits' (p. 445), a view that helps explain the large corpus of scientific literature on alluvial fans (*sensu stricto*).

A few megafans have been well known for decades. The work of identifying the entire global population, however, is recent and has been based almost entirely on remotely sensed data. Criteria for recognising megafans have been the identification of

very low gradient, smooth surfaces, many tens to hundreds of kilometres in length, displaying partial cone morphology, sometimes with recognisable radial drainage, with a river exiting an upland near the apex as a source of sediment (Hartley et al. 2010, Wilkinson and Currit, Ch. 2). Data sources for the world survey reported by Wilkinson and Currit (Ch. 2) were astronaut images and detailed topographic maps in the early days of data collection, and latterly aerial imagery and contoured topographic data. Topographic roughness data (see examples in Wilkinson et al., Ch. 3; Wilkinson et al., Ch. 4), originally developed to characterise the surface of Mars, provided significant interpretive detail.

Before megafans were recognised as ubiquitous, extensive, and very low-gradient surfaces (especially in remoter parts of the world), they tended to be interpreted as pediment surfaces. Other techniques have been employed, especially airborne radiometric data, in the identification of megafan sediment bodies. For example, Kapteinis et al. (Ch. 13) combined these techniques with airborne radiometric and lidar imagery, allowing three megafans to be recognised for the first time, and in their full extent. In cases where ground truthing has been possible, the interpretation of megafans is secure, as in most studies in this volume. Exceptions are those identifications in chapters based solely on remotely sensed data, i.e., some on the global map (Wilkinson and Currit, Ch. 2), in the Africa survey (Wilkinson et al., Ch. 3), and some of the features in the Upper Zambezi basin (Wilkinson et al., Ch. 5). We note that the criteria used for identification in these latter studies are repeatedly vindicated by field studies, and thus appear to be robust.

Almost three hundred megafans have been identified in our world survey (Wilkinson and Currit, Ch. 2), doubling the number reported by Hartley et al. (2010). One of the main reasons for the increase appears to be (i) that relict megafans of reduced prominence in remotely sensed imagery have been included; and (ii) that we used a less restrictive definition of the distal margin of megafan features (our criteria were aimed at identification of entire megafan sediment bodies rather than employing the morphological criterion of radial drainage, which restricts selected areas on some megafans to proximal-medial sectors).

The best-known pattern is the clustered distribution in basins along the Andean and Himalayan mountain fronts, with others in similar tectonic settings accounting for 40% of the total. The majority of megafans in Asia are related to major mountain ranges, and are almost exclusively oriented and nested transverse to structural grain. A major exception is Mesopotamia, where the largest megafans are aligned parallel with the foreland depression. The high megafan count in basin-and-swell Africa was unexpected, where the number is as high as that of the foreland basin setting (38%, $n = 84$). Megafans in Africa are highly dispersed, many erosionally degraded, with only three clusters that number more than four megafans. This may explain why many were not recognised until recently.

North America displays only one megafan (although numerous fans of smaller dimensions), due mainly to regional uplift having imposed a continent-wide lowland 'erosional geomorphic domain' (see Section 17.5, Wilkinson, Ch. 17). For the same reason, active megafans in Europe are also few in number.

18.7 Rheic and Perirheic Hydrological Convergence Zones

Mertes (1997) identified a rheic zone for fluvial activity in and near the mainstem river of a floodplain, and a perirheic zone for the less active and more locally derived water distant from the main river (floodplain lake/wetland water, inundation water, ground water, rainfall, and tributary streams). Lewin and Ashworth (2014) applied the same dichotomous system to the geomorphology of large floodplains. The megafan drainage template provides a starkly different set of perspectives, especially with respect to hydrological analysis of floodplain waters and the 'hydrological convergence' zones where rheic water meets perirheic water, zones identified by Mertes (1997) as ecologically critical. The three river types in megafan landscapes – the much-mentioned fan-forming, fan-fed and fan-margin rivers – suggest a far greater number and a more complex set of convergence zones in such landscapes. First, each of these three types displays its own specific set of rheic–perirheic interactions – with numerous convergence zones scattered widely across the megafan surface in the case of fan-fed streams. At a wider scale, convergence zones exist wherever each of the three river types intersect the other types – for example where the fan-forming river meets the fan-margin river, or where fan-fed rivers meet extensive wetlands. Furthermore, the three river types and the intervening wetland and lakes all intersect the trunk river of the basin, producing yet another series of convergence points. A diachronic dimension further obtrudes when the fan-forming river shifts with time to different radial positions on the fan surface. Interactions with fan-fed drainages in one place then stop functioning but start where the fan-forming river intersects fan-fed streams distant from its prior course (Section 17.6.3, Wilkinson, Ch. 17). These interactions combined with seasonal hydrological variability, as stressed by Mertes (1997), all offer further avenues of investigation.

18.8 Megafans in the Rock Record

18.8.1 Impact of Large DFS on Rock-Record Thinking

Ventra and Moscariello (Ch. 14) evaluate megafans from the perspective of the explorationist. They state that 'fluvial-fan studies are opening a renewed approach to river geomorphology and sedimentology at the scale of entire drainage networks rather than... single channel patterns, and have increasing relevance for analyses of continental basins where clastic fills might be represented by fan successions in much larger proportion than previously thought'. They provide a synopsis of characteristic architectural patterns within megafan deposits and consider that 'recurrent stratigraphic patterns within fluvial-fan successions offer good opportunities for subsurface exploration and reservoir development to be framed in a more predictable stratigraphic context than is commonly the case...'. They comment on the opinion that large fluvial fans and their components may be fundamental building blocks of alluvial facies tracts on scales of tens to hundreds of kilometres. In their view, the recognition of the widespread existence of modern landscapes based on large DFS (distributive fluvial systems; Hartley et al. 2010, Weissmann et al. 2010) has reinvigorated interpretation of the rock record, which they illustrate with a rich literature review.

18.8.2 Megafans and Confined vs. Unconfined Fluvial Environments

Large DFS (> 30 km long – Hartley et al. 2010) are often presented as being dominated by architectures that reflect radially-organised, unconfined, coniform (positive relief) depositional environments. It is there-fore counter-intuitive that the largest features (> 80 km long) display a variety of topographically confined negative relief features, from small to large, which are themselves all prone to fluvial deposition just as are floodplains. 'Chi' maps produced by Willett et al. (2018) have revealed the strongly integrated character of subtle zones of positive and negative relief on the Chaco megafans, a pattern that is difficult to appre-ciate in contoured data (SOD Fig. 17.5.2, Wilkinson, Ch. 17). The drainage system can be interpreted as a combination of radial *and* convergent, and comprises triangular zones of divergent drainage pointing up-fan, several of which accord with known apexes and sub-apexes; interspersed are zones of slightly convergent drainage. This pattern is unexpected in the sense that it reveals first, the coexistence of both types on megafan surfaces, but also that these are tightly integrated spatially, and that the pattern persists across several major megafans – appearing as a normal component of the morphology of unincised megafan surfaces.

Apart from the inherent geomorphic interest in this aspect of megafan surface complexity, the confined/unconfined distinction has peculiar salience now because Hartley et al. (2010) and Weissmann et al. (2011) have used it to distinguish models of sediment-ary architectures in valley-confined settings from those developed in the unconfined setting of large DFS. Confined fluvial systems are seen in this view as repre-senting broadly upland erosional domains, whereas unconfined (fan-related) fluvial systems are seen as representing aggradational lowland domains. Although these writers have noted the frequently occur-ring confined environments of incised sectors above the intersection point on large DFS, Wilkinson (Ch. 17) argues that the variety of negative relief features on megafans implies that megafan architectures encoun-tered in the rock record are likely to be a complex interfingering of unconfined and confined types.

Distinguishing dendritic river patterns from radial patterns in subsurface data is likely to be difficult without (i) well-developed theory about complexity of the mid-scale features we call fluvial megafans, and (ii) without significantly dense well spacing. Because the scale of the depositional system can be difficult to assess from limited subsurface data, sub-megafan-scale depressions can host sediments that might be misinterpreted from palaeoflow data as regionally convergent drainage patterns set in degrada-tional alluvial landscapes. Despite being fairly intensely documented, the Loire megafan (Gunnell, Ch. 9), for example, shows that subsurface architec-tural reconstructions are still relatively crude. Interpretation of ancient stratigraphic records based on observations of modern systems is made more complex due to partial preservation of resulting sedi-ment bodies on longer timescales, and also because of modifications due to burial and diagenesis. Similarly, stratigraphic architectures in fan-margin/interfan rivers will be dominated by confined alluvial settings. At the wider multi-megafan scale, this leads to an interfinger-ing of fan-margin sedimentary units with units of the neighbouring divergent-drainage megafans (and their subsidiary, non-radial depositional environments) to produce a complex set of architectures. This is yet another set of emergent relationships that are neither a part of alluvial fan (*sensu stricto*) landscapes nor of valley-confined floodplains.

18.9 Conclusion: Megafans as a Global Geomorphic Norm with Emergent Properties

Although the existence of fans of larger size appears to have been accepted after the papers describing 'large DFS' by Hartley et al. (2010) and Weissmann et al. (2011), many implications of the megafan subpopula-tion of DFS treated in this volume have as yet not been explored. Reasons may be the lack of recognition of the widespread distribution of megafans globally, and possibly also the novelty of major landscapes in which rivers occupy the highest topographic points in the landscape, and are laterally extremely mobile over tracts of terrain hundreds of kilometres wide. Nor do river types in megafan landscapes correspond to the elegantly simple stream-order systems of dendritic drainage. An unfamiliar terminology for describing components of major landscapes, with confusing cross-currents of meaning, adds difficulty. The megafan landscape contrasts in several important ways with the well-known configurations of the erosional domain – with its valley-confined rivers, interfluves and dendritic drainage.

Megafans as a significant geomorphic norm (in those continental lowlands undergoing wide fluvial sedimentation) are illustrated simply by the number of now-known megafans ($n = 272$; Wilkinson and Currit, Ch. 2). Their significance is evident from areal totals: megafans in Africa alone reach > 1.2 M km^2, broadly equivalent to the areal total of 23 deltas of some of the largest rivers on the planet (Wilkinson and Currit, Ch. 2).

The complexity of morphology and of the geomorphic system operation at megafan scales, especially compared to smaller alluvial fans, is evidence of several emergent characteristics that endow megafans with properties not found on alluvial fans (*sensu stricto*), nor on smaller fluvial fans, nor on valley-confined floodplains. For example, the relatively large lateral dimensions of megafan cones increase the recurrence time of the formative river returning to any one of its prior courses – which in turn allows for more advanced soil maturation and longer operation of the numerous, small fan-fed rivers. Neither of these effects is seen on valley-confined floodplains formed by rivers of similar discharge. At a different scale, even very large floods, such as that experienced on the Kosi in 2008, fail to cover the entire surface of a megafan. In contrast, decadal and hundred-year floods regularly inundate valley-confined floodplains in wetter parts of the world. Subapexes located far from the main megafan apex are phenomena that give rise to entire zones of radial drainage, tens or hundreds of kilometres long, that are not seen on valley-confined floodplains.

The complexity of multi-megafan landscapes is particularly well illustrated not only by the abovementioned five megafan nesting patterns, but also by the six domain associations at the continental scale, of which they form a major part (Wilkinson, Ch. 17). Neither alluvial fans (*sensu stricto*), nor deltas, nor major valley-confined floodplains present these patterns. It has been suggested that river behaviours on megafans may have been instrumental in promoting the speciation of fish populations (Wilkinson et al. 2006, 2010).

There seems little doubt that the implications of megafans and multi-megafan landscapes will extend beyond geomorphology to interest practitioners in other fields. We give a sampling of these wider interests in Part III of this volume, in which applications to subsurface exploration for water (Miller et al., Ch. 15) and hydrocarbons (Ventra and Moscariello, Ch. 14) and to recognition of aggradational features on other planets (e.g., Radebaugh et al. 2018; Wilkinson et al., Ch. 16) are presented. With the dense human populations presently supported on megafans, especially for example in India, and with the rapid opening up for agriculture of the Chaco megafans in central South America, human exploitation in all its complexity becomes a topic of special interest.

Megafans and multi-megafan landscapes thus appear more complex than the characterisation formulated twenty-five years ago, according to which megafans 'should simply be classified as rivers with locally expanding reaches' (Blair and McPherson 1994:481); and that they are indeed distinctly more complex than small alluvial fans (*sensu stricto*), in which processes are quite different from those that operate on megafans (Wilkinson, Ch. 17).

References

Blair, T. C. and McPherson, J. G. (1994). Alluvial fans and their natural distinction from rivers based on morphology, hydraulic processes, sedimentary processes, and facies assemblages. *Journal of Sedimentary Research*, A64, 450–489.

Burke, K. and Gunnell, Y. (2008). *The African Erosion Surface: A Continental-scale Synthesis of Geomorphology, Tectonics, and Environmental Change over the Past 180 Million Years*. Geological Society of America Memoir, 201, 66 pp.

Burke K. and Wilkinson M. J. (2016). Landscape evolution in Africa during the Cenozoic and Quaternary—the legacy and limitations of Lester C. King. *Canadian Journal of Earth Sciences*, 53, 1089–1102.

Chakraborty, T., Kar, R., Ghosh, P., and Basu, S. (2010). Kosi megafan: Historical records, geomorphology and the recent avulsion of the Kosi River. *Quaternary International*, 227, 143–160.

Cooke, R. U., Warren, A., and Goudie, A. S. (2006). *Desert Geomorphology*. University College London Press, London, 2nd edn, 526 pp.

Drew F. (1873). Alluvial and lacustrine deposits and glacial records of the Upper-Indus Basin. *Quarterly Journal of the Geological Society, London*, 29, 441–471.

Geddes, A. (1960). The alluvial morphology of the Indo-Gangetic plains. *Transactions Institute of British Geographers*, 28, 253–276.

Gibling M. R., Tandon S. K., Sinha R., and Jain M. (2005). Discontinuity-bounded alluvial sequences of the southern Gangetic plains, India: aggradation and degradation in response to monsoonal strength. *Journal of Sedimentary Research*, 75, 369–385.

Hartley, A. J., Weissmann, G. S., Nichols, G. J., and Warwick, G. L. (2010). Large distributive fluvial systems: characteristics, distribution, and controls on development. *Journal of Sedimentary Research*, 80, 167–183.

Latrubesse, E. (2008). Patterns of anabranching channels: the ultimate end-member adjustments of mega-rivers. *Geomorphology*, 101, 130–145.

Latrubesse, E. M., Cozzuol, M., da Silva-Caminha, S. A. F., et al. (2010). The Late Miocene palaeogeography of the Amazon Basin and the evolution of the Amazon River system. *Earth-Science Reviews*, 99, 99–124.

Latrubesse, E. M. (2015). Large rivers, megafans and other Quaternary avulsive fluvial systems: A potential 'who's who' in the geological record. *Earth-Science Reviews*, 146, 1–30.

Lewin, J. and Ashworth, P. J. (2014). The negative relief large river floodplains. *Earth-Science Reviews*, 129, 1–23.

Mertes, L.A.K. (1997). Documentation and significance of the perirheic zone on inundated floodplains. *Water Resources Research*, 33, 1749–1762.

Miall, A. D., 1996). *The Geology of Fluvial Deposits*. Springer, New York, 582 pp.

Miall, A. D., 2014). *Fluvial Depositional Systems*. Springer, New York, 316 pp.

Mohindra, R., Parkash, B., and Prasad, J. (1992). Historical geomorphology, and pedology of the Gandak Megafan, Middle Gangetic Plains, India. *Earth Surface Processes and Landforms*, 17, 643–662.

Sinha, R. and Friend, P. F. (1994). River systems and their sediment flux, Indo-Gangetic plains, Northern Bihar, India. *Sedimentology*, 41, 825–845.

Slingerland, R. and Smith, N. D. (2004). River avulsions and their deposits. *Annual Review of Earth and Planetary Sciences*, 32, 257–285.

Ventra, D. and Clarke, L. E., eds. (2018). *Geology and Geomorphology of Alluvial and Fluvial Fans: Terrestrial and Planetary Perspectives*. Geological Society of London, Special Publication, 440, 1–21.

Weissmann, G. S., Hartley, A. J., Nichols, G. J., et al. (2010). Fluvial form in modern continental sedimentary basins: distributive fluvial systems. *Geology*, 38, 39–42.

Weissmann, G. S., Hartley, A. J., Nichols, G. J., et al. (2011). Alluvial facies distributions in continental sedimentary basins – distributive fluvial systems. In S. K. Davidson, S., Leleu, and C. P. North, eds., *From River to Rock Record: The Preservation of Fluvial Sediments and their Subsequent Interpretation*. SEPM Special Publication 97, 327–355.

Weissmann, G. S., Hartley, A. J., Scuderi, L. A., et al. (2015). Fluvial geomorphic elements in modern sedimentary basins and their potential preservation in the rock record: a review. *Geomorphology*, 250, 187–219.

Wilkinson, M. J., Marshall, L. G., and Lundberg, J. G. (2006). River behavior on megafans and potential influences on diversification and distribution of aquatic organisms. *Journal of South American Earth Sciences*, 21, 151–172.

Wilkinson, M. J., Marshall, L. G., Lundberg, J. G., and Kreslavsky, M. H. (2010). Megafan environments in northern South America and their impact on Amazon Neogene aquatic ecosystems. In C. Hoorn and F. P. Wesselingh, eds., *Amazonia, Landscape and Species Evolution: A Look into the Past*. Blackwell, London, 162–184.

Index

Printed in the United States
by Baker & Taylor Publisher Services